QC
793.3
C56T55
1982
c.1

Lecture Notes in Physics

Edited by H. Araki, Kyoto, J. Ehlers, München, K. Hepp, Zürich
R. Kippenhahn, München, H. A. Weidenmüller, Heidelberg
and J. Zittartz, Köln

DISCARDED MATERIAL
FROM UVU LIBRARY

171

Time-Dependent Hartree-Fock and Beyond

Proceedings of the International Symposium
Held in Bad Honnef, Germany, June 7–11, 1982

AF768230

Edited by K. Goeke and P.-G. Reinhard

Springer-Verlag
Berlin Heidelberg New York 1982

Editors
Klaus Goeke
Institut für Kernphysik
KFA Jülich, D-5170 Jülich
and
Physik Department
Universität Bonn, D-5300 Bonn

Paul-Gerhard Reinhard
Institut für Kernphysik
Universität Mainz, D-6500 Mainz

ISBN 3-540-11950-7 Springer-Verlag Berlin Heidelberg New York
ISBN 0-387-11950-7 Springer-Verlag New York Heidelberg Berlin

This work is subject to copyright. All rights are reserved, whether the whole or part of the material is concerned, specifically those of translation, reprinting, re-use of illustrations, broadcasting, reproduction by photocopying machine or similar means, and storage in data banks. Under § 54 of the German Copyright Law where copies are made for other than private use, a fee is payable to "Verwertungsgesellschaft Wort", Munich.

© by Springer-Verlag Berlin Heidelberg 1982
Printed in Germany

Printing and binding: Beltz Offsetdruck, Hemsbach/Bergstr.
2153/3140-543210

PREFACE

The present proceedings contain the talks of the International Symposium on "Time Dependent Hartree-Fock and Beyond" which was held in Bad Honnef (Bonn), June 7-11, 1982. The conference was organized by the Institut für Kernphysik, Kernforschungsanlage Jülich, and the Institut für Kernphysik, Universität Mainz.

The Symposium was concerned with the microscopic description of large amplitude nuclear collective motions. In a series of lectures and seminars, discussions were held on the present status of time dependent Hartree-Fock and the various ways to go beyond it. These included Monte Carlo methods, the formulation of reaction theories, incorporation of collision terms, evaluation of correlations and dispersion, use of path integral formalisms, extraction of quantized collective Hamiltonians, and related subjects.

On behalf of the organizing committee we gratefully acknowledge the generous financial support of the Stiftung Volkswagenwerk and the Kernforschungsanlage Jülich.

The organization of the conference would not have been possible without the tireless effort of many members of both host institutions and the staff of the Physikzentrum in Bad Honnef. Particular thanks are due to Ms M. Heese, KFA, and Dr. J. Debrus, Physikzentrum, for their enthusiasm and efficient handling of all the tasks necessary to keep the conference running smoothly.

Jülich/Mainz, August 20, 1982 K. Goeke P.-G. Reinhard

TABLE OF CONTENTS

ORGANIZING COMMITTEE

K. Goeke, Chairman	Jülich
P.-G. Reinhard	Mainz
J. Speth	Jülich

INTERNATIONAL ADVISORY BOARD

D.M. Brink	Oxford
M. Irvine	Manchester
A.K. Kerman	Boston
D.J. Rowe	Toronto
M. Vénéroni	Orsay
H.A. Weidenmüller	Heidelberg

THE TIME-DEPENDENT HARTREE-FOCK DESCRIPTION OF HEAVY-ION COLLISIONS: A PROGRESS REPORT

S. J. Krieger
University of California, Lawrence Livermore National Laboratory
Livermore, California 94550 U.S.A.

1. INTRODUCTION

Although the time-dependent Hartree-Fock (TDHF) approximation was formulated by Dirac[1)] over 50 years ago, it is only in the last six years that TDHF calculations have actually been effected. The initial application of the TDHF approximation, by Bonche, Koonin, and Negele[2)], to reactions between semi-infinite slabs of nuclear matter produced a rich collision phenomenology which included compound "nucleus" formation, resonances, and highly inelastic reactions. In this seminal calculation, the dynamics were conspicuously dominated by the propagation and reflection of single-particle wave functions in the mean HF potential. Indeed, much of the work in the past several years has been devoted to the question as to what extent the single-particle dynamics and collision phenomenology would persist in more realistic calculations, and it is only recently that emphasis has shifted to comparison of TDHF results with experiment. Prior to discussing these results of the greater part of a decade of research, however, it is useful to review the motivations for applying the TDHF approximation to nuclei.

The ultimate goal of any microscopic theory of nuclear motion is to provide a unified description of the reaction dynamics for such diverse physical phenomena as giant multipole resonances, fission, fusion, and compound nucleus formation. TDHF is indeed a candidate for such a microscopic theory, since it requires as input only an effective nuclear interaction, and the specification of initial conditions. No assumptions need be made as to the relevant collective or intrinsic coordinates. Also, TDHF offers an additional bonus. Because TDHF is a quantal independent particle approximation, it admits to a semiclassical interpretation, thus offering possible insight into the underlying physics, which would not be obtainable in calculations using more complicated wave functions. This bonus, however, is not without cost. Given the physical picture of independent nucleons moving in their self-generated, time-dependent one-body potential, it is clear that the approximation can only be valid at energies below that in which the dominant interaction is that of nucleon-nucleon collisions. At first sight this limitation on the domain of applicability of the theory might appear to severely restrict its usefulness. For unlike the static HF approximation which has been so successful over the past two decades in calculating the properties of the ground states and low-lying excited states of nuclei throughout the periodic table, the

---

*Work performed under the auspices of the U. S. Department of Energy by the Lawrence Livermore National Laboratory under contract No. W-7405-ENG-48.

TDHF approximation is used in situations in which the same strongly interacting fermion systems are far from equilibrium. Fortunately the regime in which the approximation may be applied need not be so severely limited. For although the mean free path for a nucleon at high energies, $E/A \gg \epsilon_F$, is indeed only a small fraction of the nuclear diameter, at energies of but a few MeV per nucleon above the Coulomb barrier the suppression of the number of accessible final states due to the Pauli principle insures that at such low energies, $E/A \ll \epsilon_F$, the mean free path is several times greater than the nuclear size.[3)] To iterate, even though the applications of interest deal with strongly interacting systems, far from equilibrium, the Pauli blocking is nevertheless effective enough to render the independent-particle TDHF theory a reasonable first approximation to the solution of such problems.

Having motivated the use of TDHF in studying the processes observed in heavy ion collisions, it is perhaps useful to comment on some of the other methods, such as resonating groups,[4)] generator coordinates,[5)] the random phase approximation (RPA),[6)] adiabatic time-dependent Hartree-Fock,[7)] and fluid dynamics[8)] which have previously been employed to describe these processes. As we shall see, none of these methods is without its drawbacks. Thus, computational complexities restrict the application of the resonating group method to all but the lightest nuclei, while in the case of generator coordinates it is necessary, in order to make calculations feasible, to select, a priori, the allowed collective degrees of freedom. The RPA, being a linearized version of TDHF, is applicable only to small amplitude disturbances, and ATDHF is limited to the description of low-lying excitations in the case of large amplitude motion,[9)] and may even be further limited.[10)] Finally the application of fluid dynamics to nuclei depends upon approximations whose validity can be determined only by testing them by a more complete microscopic theory.[11)]

Information concerning the question as to whether the TDHF approximation indeed yields the <u>correct</u> mean field theory is now being generated by a number of works,[12),13)] treating the time-dependent mean field problem. These works,[12),13)] which seek both to offer a rigorous justification of the TDHF method, as well as to overcome some of the serious conceptual and practical problems[14)] inherent in the method, are making it increasingly plausible that TDHF <u>is</u> the correct mean field theory for the calculation of one-body observables.

Such conformation is indeed important. For while the TDHF equations can be derived from a time-dependent variational principle,[15)] there is an important conceptual difference in its application as compared to the principle which gives rise to the static HF equations. In the static case it is known that one obtains the best wave function in the sense that the solution of the HF equations yields the determinant in which the expectation value of the energy is closest to the exact energy. While the above does not insure that the wave functions necessarily will yield accurate values of other observables, it nevertheless is a stronger statement than one can make in the dynamic situation. Perhaps the best that can be done in the latter case is to show that if, at any time, the exact wave function is a Slater determinant, then the TDHF wave function yields the Slater determinant which maximizes the overlap with the exact wave function in the interval about that time.[16)]

Given the theoretical difficulties in establishing the validity of the TDHF approximation, it is perhaps most effective at this time to assess the results of the TDHF calculations by comparing them, insofar as is possible, with experimental results. In this task we shall be limited by the fact that the TDHF approximation does not yield an inclusive description of nuclear reactions, but rather an exclusive description of nuclear collisions.[13)] Thus the semi-classical nature of the approximation which offers such a simple picture of certain gross properties, at the same time effectively prohibits the acquisition of detailed channel information. In spite of this we shall still succeed in showing rather good agreement between theory and experiment for the particular reactions which result in fusion or in deep inelastic collisions. For a more complete (and more balanced) review of TDHF results the reader is referred to four recent review articles.[17)]

The structure of this review is as follows. In Section 2 the TDHF equations are derived and briefly discussed. The effective interaction employed in the calculations is described in Section 3, and some technical aspects of the calculations are discussed in Section 4. Fusion results are presented in Section 5 along with a brief discussion of deep inelastic collisions. Finally, the results are summarized in Section 6.

2. THE TDHF EQUATIONS

The TDHF equations may be derived from a variational principle[15)] which bears a formal resemblance to the stationary action principle which gives rise to the time-dependent many-body Schrödinger equation. To be specific, the TDHF equations follow from the requirement

$$\delta \int dt\, \Psi^{\dagger}\, (i\hbar \frac{\partial}{\partial t} - H)\, \Psi = 0 \qquad (2.1)$$

where H is the full many-body Hamiltonian, but Ψ, the many-body wave function, is restricted at all times to the subspace of Slater determinants. In the coordinate representation, with this restriction on Ψ (and with spin and isospin labels suppressed), the linear Schrödinger equation for the A particle system is replaced by A coupled, non-linear differential equations for the A occupied orbitals $\psi_i(\mathbf{r},t)$

$$i\hbar \frac{\partial}{\partial t} \psi_i\, (\mathbf{r},t) = \frac{\delta\langle\Psi|H|\Psi\rangle}{\delta\psi_i^*\, (\mathbf{r},t)} \qquad i = 1,2,\ldots A \qquad (2.2)$$

Equations (2.2) conserve in time a number of important quantities including the expectation value of the energy, H, the total momentum, the total angular momentum, and the scalar products of the single-particle wave functions, $\langle\psi_i|\psi_j\rangle = \delta_{ij}$. Insight into the underlying physics governed by equations (2.2) can be obtained by considering the situation in which the Hamiltonian, H, consists of the kinetic energy plus a local central two-body potential, V(r). In this case the energy functional

$$\langle\Psi|H|\Psi\rangle = \int d^3r\, \frac{\hbar^2}{2m}\, \tau(\mathbf{r},t) + \frac{1}{2}\int d^3r\, d^3r'\, \rho(\mathbf{r},t)\, V(\mathbf{r}-\mathbf{r}')\, \rho(\mathbf{r}',t)$$

$$- \frac{1}{2}\int d^3r\, d^3r'\, |\rho(\mathbf{r},\mathbf{r}',t)|^2\, V(\mathbf{r}-\mathbf{r}')\ , \qquad (2.3)$$

where $\rho(\mathbf{r},\mathbf{r}',t) = \sum_{i=1}^{A} \psi_i(\mathbf{r},t)\, \psi_i^*(\mathbf{r}',t)$ is the one-body density matrix, whose diagonal elements yield the particle density $\rho(\mathbf{r},t) \equiv \rho(\mathbf{r},\mathbf{r},t)$, and $\tau(\mathbf{r},t) = \nabla_{\mathbf{r}} \cdot \nabla_{\mathbf{r}'}\, \rho(\mathbf{r},\mathbf{r}',t)|_{\mathbf{r}=\mathbf{r}'}$ is the kinetic energy density. Variation of equation (2.3) as called for by equations (2.2) then results in the TDHF equations

$$i\hbar \frac{\partial}{\partial t}\, \psi_i(\mathbf{r},t) = -\frac{\hbar^2}{2m}\nabla^2\psi_i(\mathbf{r},t) + \int d^3r'\, V(\mathbf{r}-\mathbf{r}')\, \rho(\mathbf{r}',t)\, \psi_i(\mathbf{r},t)$$

$$- \int d^3r'\, V(\mathbf{r}-\mathbf{r}')\, \rho(\mathbf{r},\mathbf{r}',t)\, \psi_i(\mathbf{r}',t)\ . \qquad (2.4)$$

Neglecting for the moment the third term on the right of equation (2.4), it may be noted that each single-particle wave function simply evolves through a Schrödinger equation in which the potential is given by the mean field generated by the instantaneous nuclear density. The third term then gives the exchange term required by the Pauli principle.

3. THE EFFECTIVE INTERACTION

Although calculations have been effected using a realistic phenomenological K-matrix model[18)] for head-on collisions of $^{12}C + {}^{12}C$, these calculations are technically complicated, and have neither been extended to collisions at finite impact parameters nor to collisions of heavier ions. In this review we shall therefore discuss calculations which have been effected using effective interactions of the Skyrme type.[19)] Skyrme's interaction[19)] consists of a two-body term $V^{(2)}$ including a spin-orbit potential, and a three-body term $V^{(3)}$. In the coordinate representation, $V^{(2)}$ and $V^{(3)}$ are expressed as

$$V^{(2)}(\mathbf{r}_1,\mathbf{r}_2) = t_o(1+x_oP_\sigma)\ \delta(\mathbf{r}_1-\mathbf{r}_2) + \frac{1}{2} t_1[\delta(\mathbf{r}_1-\mathbf{r}_2)\ k^2 + k'^2\ \delta(\mathbf{r}_1-\mathbf{r}_2)]$$

$$+ t_2\,\mathbf{k}' \cdot \delta(\mathbf{r}_1-\mathbf{r}_2)\mathbf{k} + iV_{s.o.}(\sigma_1+\sigma_2) \cdot \mathbf{k}' \times \delta(\mathbf{r}_1-\mathbf{r}_2)\ \mathbf{k}\ , \qquad (3.1)$$

$$V^{(3)}(\mathbf{r}_1,\mathbf{r}_2,\mathbf{r}_3) = t_3\ \delta(\mathbf{r}_1-\mathbf{r}_2)\ \delta(\mathbf{r}_2-\mathbf{r}_3)\ . \qquad (3.2)$$

In equation (3.1) P_σ is the spin exchange operator and σ represents the vector of the Pauli spin matrices; $\mathbf{k}$ is the operator $\frac{1}{2i}(\nabla_1-\nabla_2)$ acting on the right and $\mathbf{k}'$ is $-\frac{1}{2i}(\nabla_1-\nabla_2)$ acting on the left. The quantities x_o, t_o, t_1, t_2, t_3 and the spin-orbit coupling strength $V_{s.o.}$ are parameters of the specific Skyrme interaction. A particular virtue of the zero-range nature of the interaction is that the expectation value of the many-body Hamiltonian can simply be written as an integral of an energy density $H(\mathbf{r})$,

$$\langle\Psi|H|\Psi\rangle = \int d^3r\ H(\mathbf{r}) \qquad (3.3)$$

which itself depends only upon various special densities which can be constructed from the single-particle density matrix. Full details are given in reference 9. However, in the special case of a spin-saturated and charge conjugated nucleus, the energy density reduces to

$$H(\mathbf{r}) = \frac{\hbar^2}{2m}\tau + \frac{3}{8}t_o\,\rho^2 + \frac{1}{16}(3t_1+5t_2)\ (\rho\tau-\mathbf{j}^2) - \frac{1}{64}(9t_1-5t_2)\ \rho\,\nabla^2\rho$$

$$+ \frac{1}{16}t_3\,\rho^3 \qquad (3.4)$$

In equation (3.4) $\mathbf{j}$ is the momentum density

$$\mathbf{j}(\mathbf{r},t) = \frac{1}{2i}[(\nabla-\nabla')\ \rho(\mathbf{r},\mathbf{r}',t)]_{\mathbf{r}=\mathbf{r}'}\ . \qquad (3.5)$$

In this case the TDHF equations are extremely simple. The Hamiltonian, $h(\mathbf{r})$, governing the time evolution of the single-particle wave functions then takes the form[9)]

$$h(\mathbf{r}) = -\nabla \cdot \left[\frac{\hbar^2}{2m^*(\mathbf{r})}\nabla\right] + U(\mathbf{r}) + \frac{1}{2i}[\nabla \cdot \mathbf{I}(\mathbf{r}) + \mathbf{I}(\mathbf{r}) \cdot \nabla]\ . \qquad (3.6)$$

In equation 3.6, $m^*(\mathbf{r})$ is an effective mass whose inverse is linearly proportional to the density, $U(\mathbf{r})$ is a potential linearly proportional to the density and kinetic energy density, and $I(\mathbf{r})$ is linearly proportional to the momentum density. Thus, by using the Skyrme interaction, the generally non-local HF Hamiltonian is reduced to a second order differential operator, and the numerical solution of the TDHF equations is thereby not only significantly eased, but made considerably faster.

The use of the Skyrme interactions in TDHF calculations is motivated not only by the resulting computational facility outlined above, but also by the fact that the interactions have been utilized in static HF calculations to give an excellent description of nuclear ground state properties for a broad range of nuclei throughout the periodic table.[20)] Finally, while the Skyrme interaction has been introduced strictly as a phenomenological effective interaction, its form can be justified theoretically by the density matrix expansion[21)] and density-dependent HF approach.[22)]

For completeness we should note that in some of the calculations to be discussed, the surface energy term proportional to $\nabla^2\rho$ in equation 3.4 has been replaced for numerical reasons by a double convolution of the density with a short-range Yukawa interaction. In such cases the parameters t_0 and x_0 are renormalized as the double convolution of the density also contributes a term proportional to ρ^2.[23)] Though we did not so indicate, the direct Coulomb contribution to the energy has been included in the calculations. Inclusion of the Coulomb exchange term, in the local Slater approximation, has been shown to have no effect on the dynamics of the relatively light $^{40}Ca + {}^{40}Ca$ system.[24)] For reasons of computational efficiency, none of the calculations to be reviewed have utilized the spin-orbit term in the interaction. However, recent improvements in the algorithm for calculating the orbital angular momentum suggest the possibility of including this term in future calculations.[25)]

4. METHOD OF CALCULATION

A. Description of Models

Although technological advances have made possible the solution of the TDHF equations in full three-dimensional geometry, such calculations are extremely time-consuming, and it is therefore desirable, where possible, to reduce the number of non-trivial spatial dimensions from three to two. For this reason, as well as for the reason that almost all calculations effected to date have utilized a two-dimensional geometry, we shall review the various two-dimensional models and comment upon the regime in which they are expected to yield good approximations to the exact three-dimensional geometry. We shall restrict our discussion to coordinate space models since models which introduce[26] deformed oscillator bases are limited to light systems because of the large number of basis states required to achieve convergence.

In the clutching model[27] the assumption is made that the colliding nuclei retain axial symmetry about the (rotating) line joining their respective centers. In this model the single-particle wave functions are assumed to be of the form

$$\psi_i(\mathbf{r}) = \psi_i(r,z)\, e^{i|\mu_i|\phi} \quad . \tag{4.1}$$

As in the case of the ordinary Schrödinger equation with axial symmetry, the dependence upon ϕ is eliminated upon substitution of the wave function into the equations of motion, and the desired goal of reducing the problem to one involving two non-trivial spatial dimensions, in this case r and z, has been achieved. In order to acount for the energy of rotation of the body fixed axis, the energy functional is supplemented through the addition of the classical rotational energy

$$H \to H + \frac{L^2}{2I(\rho)} \, , \tag{4.2}$$

where L is the conserved total angular momentum along the axis of rotation which is perpendicular to the reaction plane, and the moment of inertia $I(\rho)$ is a prescribed function of the density. Initially, when the colliding ions are well separated, the moment of inertia is assumed to be that of two point masses. However, when the minimum density along the symmetry axis between the two ions exceeds a value arbitrarily taken to be one-half the saturation density of nuclear matter, the ions are assumed to have clutched and the moment of inertia is then taken to be that of a rigid body. Other prescriptions for the moment of inertia have been studied[28] as well. In this model the symmetry axis is assumed to rotate in space with the classical angular frequency

$$\omega = \frac{L}{I(\rho)} \tag{4.3}$$

and Coriolis forces are completely omitted.

Some of the arbitrariness in the clutching model is removed by multiplying the wave functions, equations 4.1, by the function $e^{i\chi_i(r,z,\phi)}$.[28),29)] With the introduction of the additional phase χ, the model is completely self-consistent with all quantities derivable from the variational principle. Thus no assumption as to the nature of the rotational energy (as in equation 4.2) need be made, and the moment of inertia arises naturally, again without requiring assumptions as to its form.

Unfortunately if the χ_i for the single-particle wave functions are allowed to differ, the resulting equations are then almost as complicated to solve as the full three-dimensional equations. On the other hand, when the same χ, taken to be

$$\chi(r,z,\phi) = g(r,z)\cos\phi \ , \tag{4.4}$$

is used for each wave function, the χ term in the velocity field corresponding to these wave functions is irrotational and therefore cannot describe the physically expected rigid body motion.

The separable approximation[30),31)] is motivated by the fact that in collisions described by the TDHF approximation very little momentum is transferred in the z-direction perpendicular to the reaction plane. Guided by this observation one separates each single-particle wave function into a time-dependent function ϕ_i of the coordinates in the reaction plane and a time independent function χ_i

$$\psi_i(\mathbf{r},t) = \phi_i(x,y,t)\ \chi_i(z) \ . \tag{4.5}$$

The functions χ_i are chosen once and for all at time t = 0 as the appropriate harmonic oscillator wave functions. In the case of mass asymmetric systms, the appropriate oscillator parameter for the wave functions may be chosen, for example, to be that which minimizes the combined energy of the two non-interacting ions. The two non-trivial coordinates in this model are clearly the cartesian coordinates in the reaction plane. In contrast to the clutching models which impose axial symmetry in the rotating frame, the separable approximation makes no such restriction and non-axial configurations are allowed.

B. Accuracy of Models

For closed shell nuclei the axially symmetric calculations are in exact agreement with three-dimensional calculations for head-on collisions. (In open shell nuclei, small deviations can occur due to the different filling prescriptions[32)] employed in the distinct geometries.) Thus the results of calculations employing the axially-symmetric codes will necessarily agree with the results using the three-dimensional codes for small values of the angular momentum, in the vicinity of the L = 0 head-on collisions, as well as for grazing collisions in which the ions interact only weakly.

In order to assess the accuracy of the axially symmetric approximation for other values of the angular momentum, it is instructive to examine Figure 4.1,

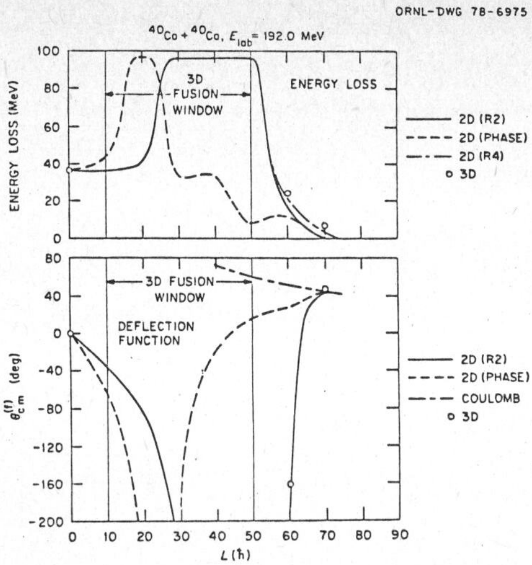

Figure 4.1 Energy loss and deflection function. The full curve is for the clutching model, the dashed for the phase model and the open circles for three dimensional results.[28)]

which displays the energy loss and deflection functions curves for ^{40}Ca + ^{40}Ca, calculated[28)] using various axially symmetric models, as well as the three-dimensional code, at the laboratory energy of 192 MeV. At this energy, the three-dimensional code predicts that the colliding ions will fuse when the total angular momentum, L, is in the region $10\hbar \lesssim L \lesssim 50\hbar$. The results of calculations using the clutching model are in reasonably good agreement, predicting fusion in the region $30\hbar \lesssim L \lesssim 50\hbar$. However the phase theory predicts fusion only in a very narrow range of L values centered about $L \sim 20\hbar$. Although it is by no means obvious from this limited comparison, this figure does illustrate a behavior which has been generally observed. Namely the phase theory best reproduces the three-dimensional results at the lower end of the fusion region, while the clutching model more accurately reproduces the results at the upper part of the region. From this behavior it can be inferred that the moment of inertia is essentially irrotational for low values of L, but becomes predominantly rigid body in nature for high values of the angular momentum.

At higher energies and especially in collisions of highly mass asymmetric systems, non-axial deformations become increasingly more important, and the axial approximations break down. For example at a laboratory energy of 278 MeV, three-dimensional calculations of ^{40}Ca + ^{40}Ca result in fusion for $25\hbar \lesssim L \lesssim 70\hbar$, while the clutching model exhibits but a narrow orbiting region and no fusion whatsoever. The conclusion is that the clutching model can usefully be employed for head-on collisions, but in the case of collisions at finite impact parameter must be restricted to calculations at energies not much higher than the interaction barrier.

The separable approximation has successfully reproduced the results of the three-dimensional calculations much more closely than the axial symmetric models over a wide range of energies.[31)] This accuracy is illustrated in Figure 4.2 in

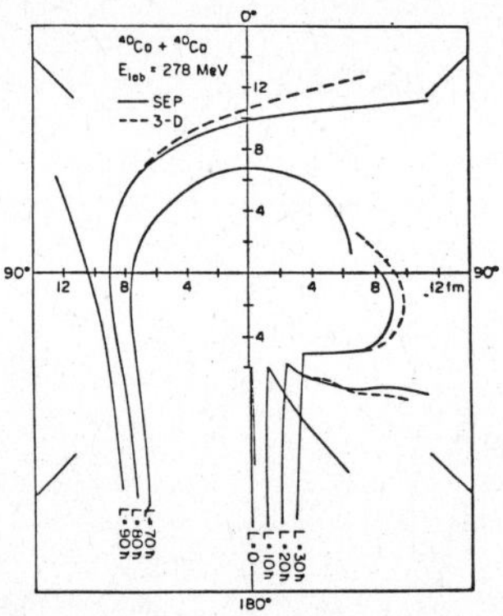

Figure 4.2 Comparison of trajectories calculated in the separable approximation (full curves) with three dimensional results (dashed).[31)]

which trajectories are compared for several values of the angular momentum for $^{40}Ca + {}^{40}Ca$ at a laboratory energy of 278 MeV.[31)] Except for the minor differences observed at $L = 20\hbar$ and $L = 30\hbar$, and the difference at $L = 80\hbar$, where the test is perhaps unreasonably strict as the scattering angle changes extremely rapidly in this orbiting region, the results of the separable approximation are indistinguishable from the three-dimensional results. Similar agreement is obtained in comparing the deflection function and energy loss curves at this energy. Moreover, whereas the clutching model gives no fusion at this energy, the separable approximation results in fusion for the same values of angular momentum as the three-dimensional calculation. In Figure 4.3 the fragment separation coordinate, defined as in

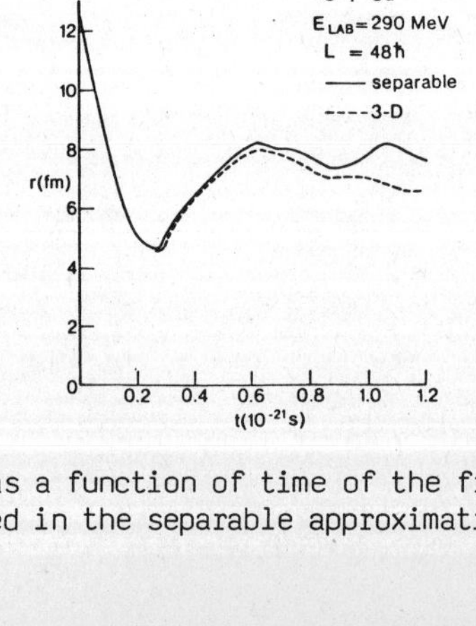

Figure 4.3 Comparison as a function of time of the fragment separation coordinate as calculated in the separable approximation and three dimensions.[45)]

reference 27, is plotted as a function of time in a collision between ^{16}O and ^{40}Ca ions at a laboratory energy of 290 MeV with $L = 48\hbar$. The separable approximation yields results in excellent agreement with the results of the fully three-dimensional calculation, until $t \approx 9 \times 10^{-22}$s, by which time the system has fused.

Although the separable approximation has been shown to yield excellent results over a wide range of energies, further tests of the approximation are still required. For some recent calculations for the highly mass asymmetric system $^{16}O + {}^{93}Nb$ indicate discrepancies between the separable and axially symmetric models for the case of head-on collisions.[33)]

5. RESULTS

Fusion cross-sections cannot be extracted with total unambiguity from either the experimental data or from the results of TDHF calculations. Experimentally, the fusion cross section is defined as the cross-section for observing a composite system with mass close to that of the total mass of the system. For light systems at low bombarding energies the compound nucleus so formed decays almost entirely by light particle and γ-ray emission, so that the fusion cross section is in fact equal to the evaporation residue cross section.[34] At the other end of the spectrum, for heavy systems, $A > 200$, the dominant decay mode of the compound system is fission, and the fusion cross section can be equated to the cross section for the observation of fusion-fission events. The latter, at least in the case of highly asymmetric collisions, can be distinguished from deep inelastic events by the symmetric mass distribution, as the compound system exists for a sufficiently long time to allow mass equilibration.[35]

In the TDHF calculations it is practical only to follow the composite system through a few rotations, or, in the case of axially symmetric calculations, through a few oscillations of its rms radius. The assumption is then implicitly made that by this time the system no longer retains a memory of the entrance channel, and it is therefore meaningful to compare the cross section for such events with the evaporation residue cross section for light systems, and the fusion-fission cross section for heavy systems. Note too that if this argument is accepted, then the question as to the accuracy with which TDHF can describe the long-time behavior of the system is no longer pertinent to the calculation of fusion cross sections.

The TDHF fusion cross section is calculated at a given energy by determining the minimum and maximum values of angular momentum, $\ell_< \hbar$ and $\ell_> \hbar$ respectively, for which the system fuses and then using the sharp cut-off approximation[36]

$$\sigma_f = \frac{\pi}{k^2} \sum_{\ell_<}^{\ell_>} (2\ell + 1)$$

$$= \frac{\pi\hbar^2}{2\mu E_{cm}} \left[(\ell_> + 1)^2 - (\ell_< + 1)^2\right] . \qquad (5.1)$$

In the above, k is the relative wave-vector of the colliding ions, and E_{cm} the center of mass energy. Note that if the TDHF approximation yields fusion for the same values of the angular momentum as can occur experimentally, then equation 5.1 overestimates the fusion cross section as it assumes that all of the reaction strength lies in the fusion channel.

We begin by discussing fusion in the light systems, where by far the greatest number of calculations have been effected. In Figure 5.1 the fusion cross section calculated[37] in the axially symmetric approximation is compared with the results of experiment for the reactions $^{12}C + {}^{16}O$ [38] and $^{12}C + {}^{18}O$.[39] The calculations are in good agreement with experiment except for the most energetic point of each calculation, and for the fact that the fluctuations in the experimental cross section of the $^{12}C + ^{16}O$ system are not reproduced. It could not be determined in this calculation whether the latter defect is due to the determinantal nature of the TDHF wave function or to the filling approximation which was employed, and which masks the shell structure presumably responsible for the fluctuations.

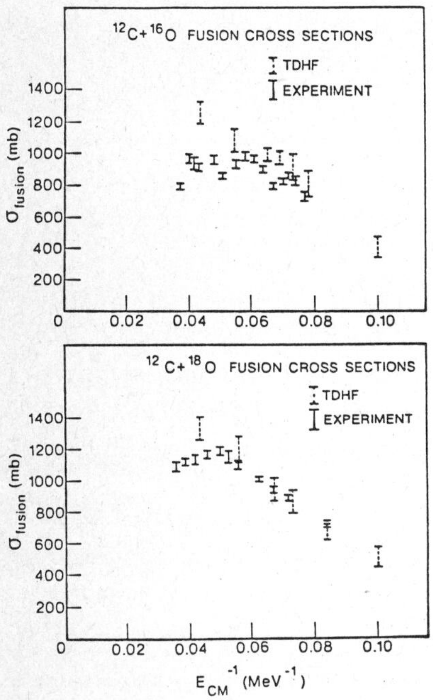

Fig. 5.1 Comparison of fusion excitation functions with experimental results.[37]

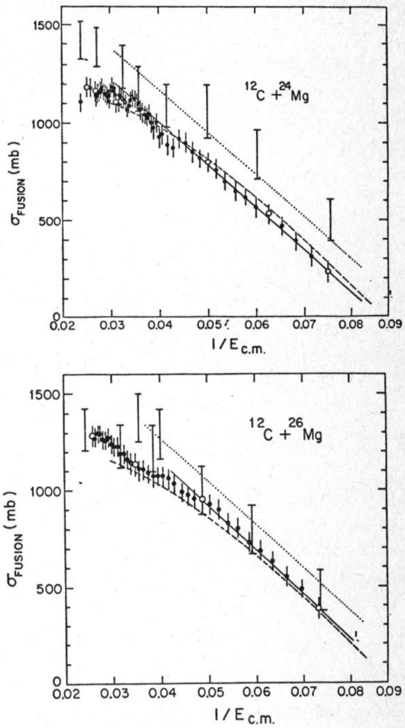

Fig. 5.2 Comparison of fusion excitation functions with experimental results.[40]

In order to clarify the above point a comparison has been made of the calculated and experimental cross sections for the systems $^{12}C + ^{24}Mg$ and $^{12}C + ^{26}Mg$.[40] In these calculations the open-shell nuclei are treated by introducing a residual pairing interaction within the constant G model.[41] Even so, the fluctuations in the experimental cross section in the alpha-like system $^{12}C + ^{24}Mg$ are not reproduced by the calculations. Referring to Figure 5.2, it may be additionally observed that the calculated cross sections consistently exceed the experimental values. As the geometry utilized in the calculation, that of an oblate carbon nucleus incident upon a prolate magnesium nucleus, tends to yield a result lower than that expected from a correct averaging over relative orientation, this difference is presumably real, and worthy of further study. The fact that the calculations do not reproduce the fluctuations should not be regarded as surprising. For the resonant structure may be traced to the alpha evaporation channels, and the TDHF calculations which, at best, reproduce the average behavior of the system, are not expected to fit structure due to specific exit channels.

The dependence of the cross section on the parameters of the Skyrme interaction is illustrated in Figure 5.3 for the reaction $^{16}O + ^{24}Mg$. While the TDHF

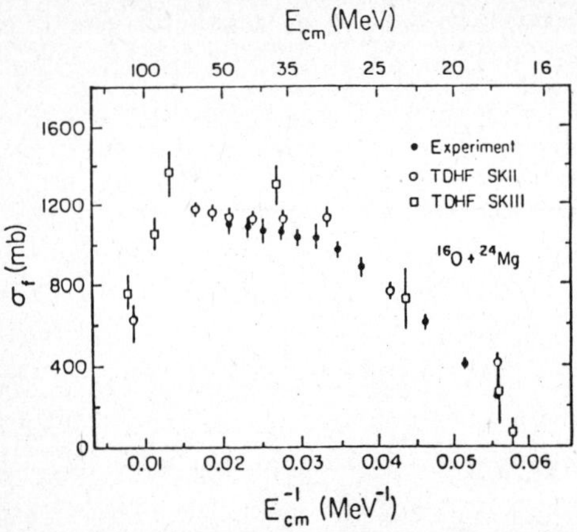

Fig. 5.3 Comparison of fusion excitation functions calculated using the interactions Skyrme II and Skyrme III with experiment.[44]

results obtained[42] using the interaction Skyrme II are in good agreement with experiment,[43] the results with Skyrme III exceed the experimental results by some 15-20% at intermediate energies.[44] The underlying reason for this enhancement has not been understood, and is of some importance, as the interaction Skyrme III, with its effective mass ratio $m^*/m = 0.76$ in nuclear matter, is the generally favored interaction.

The precipitous drop in the calculated cross section at the high energy end of the spectrum is due to the opening of an angular momentum window for fusion. Above the energy $E_{cm} \simeq 70$ MeV low angular momentum collisions result in inelastic scattering, rather than fusion. As illustrated in Figure 5.4, the minimum angular

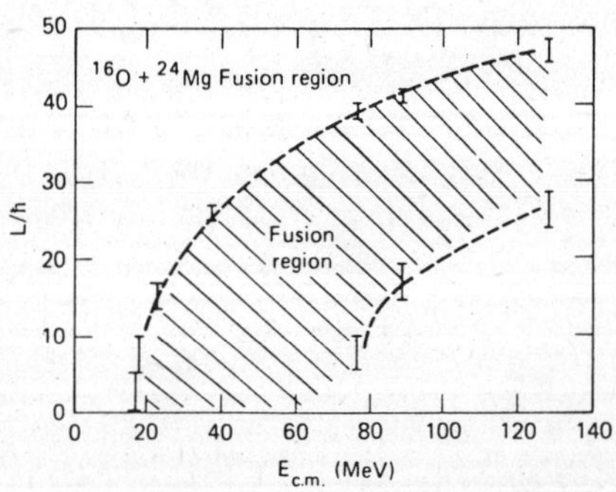

Fig. 5.4 Fusion region in the angular momentum-energy plane as calculated using the interaction Skyrme III.[44]

momentum for which fusion occurs is an increasing function of energy above the threshold of 70 MeV. This effect which has previously been observed in TDHF calculations for the systems $^{16}O + {}^{40}Ca$ and $^{28}Si + {}^{28}Si$,[45] $^{86}Kr + {}^{139}La$ and $^{84}Kr + {}^{209}Bi$[46] as well as in the present system[42),44)] remains an experimentally unverified prediction of the theory. Although experiments have been undertaken in order to observe the additional deep inelastic scattering cross section due to the non-fusion events they have not truly been conclusive. For example, the most recent experiment reported[47] an experimental upper limit of 6 mb deep inelastic for the system $^{16}O + {}^{16}O$ at E_{lab} = 68 MeV, whereas TDHF calculations as cited by Davies et al.[17] give a value of 132 mb. While at first sight the difference between experiment and theory might seem sufficient to negate the theory, it must be noted that the energy 68 MeV is very close to the low energy threshold for the effect, which itself is a very sensitive function of the symmetries enforced and interactions used in the calculations.[24),46)]

The effect that the imposition of symmetries can have is illustrated in Figure 5.5 where the $^{40}Ca + {}^{40}Ca$ fusion cross section is plotted as a function of

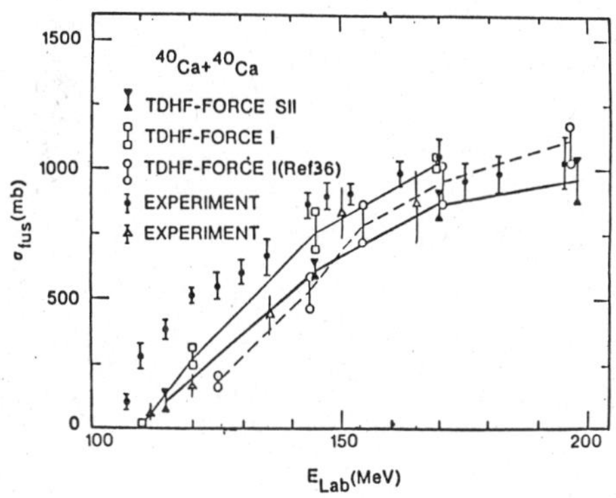

Fig. 5.5 Fusion cross section for $^{40}Ca + {}^{40}Ca$.[24] See text.

energy for several calculations. To be compared are the axially symmetric calculations[24] labeled by the open squares in which isospin symmetry is <u>not</u> enforced, and the three dimensional calculations, labeled by the open circles, in which isospin symmetry <u>is</u> enforced. It may be observed that the relaxation of the isospin symmetry results in an increase of $\approx$200 mb at the lower energies studied. Indeed, this figure represents a lower estimate, as the axial approximation is known[28] to underestimate the fusion cross section. In general, the relaxation of any symmetry has resulted in an increase in the calculated fusion cross section, and it will be of interest to examine the effect of removing the spin degeneracy through the introduction of the spin-orbit interaction as calculated with the improved algorithm.[25] The difference in cross section as calculated with the interactions Skyrme I and Skyrme II is presumably due to the fact that the rms radius and 10% - 90% surface thickness of ^{40}Ca are 1% and 10% less respectively calculated with Skyrme II as compared with Skyrme I.

Of particular interest is the comparison of theory with experiment. The experimental results depicted by the black circles[47] were completed prior to the TDHF calculations, and are in clear disagreement with the TDHF results at the lower energies. However the experiment recently has been repeated by two groups[48],[49] whose results are in agreement with one another but not with the results of reference 47. The results of reference 49, indicated by the open triangles, are in excellent agreement with the theoretical results.

As the dominant process in the collision of heavy systems is deep inelastic scattering, with fusion contributing perhaps 1/10 of the total reaction cross section, it is an open question as to whether or not TDHF calculations can yield valid information concerning fusion in heavy systems. While much more work must be done in order to answer this question, the results of calculations effected to date are certainly not discouraging. We exemplify the fusion behavior found in such systems with a discussion of head-on collisions of ^{86}Kr + ^{139}La.

The first striking feature observed is the existence of an energy threshold for fusion. Unlike the situation in light systems, in which fusion occurs at an energy immediately above the Coulomb barrier, in heavy systems it is found[46] that there exists an energy region above the barrier below which fusion does not occur. In this region very long collision times are observed and the system develops a pronounced neck reminiscent of a molecular configuration. At higher energies fusion is observed over a broad range $\approx$200 MeV. Although the fusion cross sections calculated[46] using the interaction Skyrme II are in considerable disagreement with the available experimental results the defect may well be in the interaction rather than the calculation. Indeed, calculation with the interaction Skyrme III shows that there exists a 150 MeV difference in the fusion threshold as compared to that obtained using Skyrme II. As shown in Table 5.1, reproduced from

Table 5.1

E_{lab} (MeV)	σ_f(expt.) (mb)	σ_f(axi. sym) Skyrme II	σ_f(sep.) Skyrme III
505	170±50	no fusion	201±55
610	225±50	no fusion	
710	170±50	141±20	89±15

Davies et al,[17] the Skyrme III results are in reasonable agreement with experiment.

To illustrate the deep inelastic processes, as calculated in TDHF, we compare the calculated cm energies and scattering angles with the experimental Wilczynski plot as given in Davies et al.[17] for the $^{86}Kr + ^{139}La$ system. As may be seen in Figure 5.6, the experimental energy losses and scattering angles are fairly well produced in the grazing, and quasi-elastic regions, although the low angular momentum points lie approximately 50 MeV above the experimental ridge. At 710 MeV, the calculations are in similar agreement. However at 505 MeV, the calculated points have too little dissipation and are not sufficiently forward in angle.

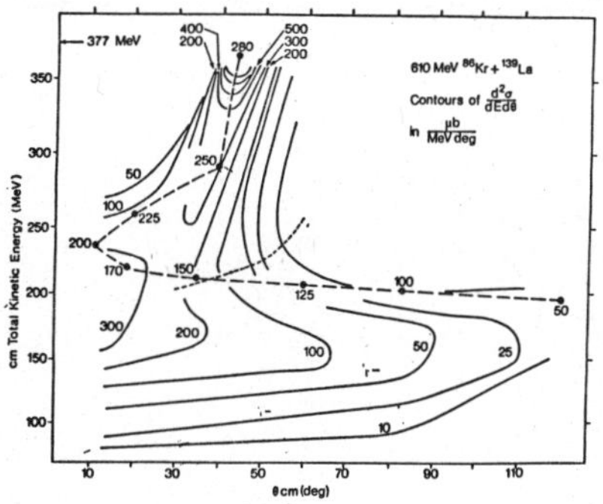

Fig. 5.6 Comparison of calculated points labeled by the initial L with the experimental Wilczynski plot.[17]

Other observables which can be compared with experiment include the charge and mass distribution widths. Although the TDHF widths are always much smaller than the experimental widths, such comparisons represent overly stringent tests of the mean-field theory, since they require the evaluation of two-body matrix elements.

6. SUMMARY

The domains of validity of the various approximations to full three-dimensional TDHF calculations are fairly well understood at this point. Further work needs to be done to determine the appropriate effective interaction to employ, and it is important to investigate the effects of relaxing the spin symmetry. Nevertheless significant progress has been made in effecting realistic calculations, and comparison with experiment, especially for fusion in light systems, where most of the data exists, is extremely encouraging. Finally, it is important that better calculations, as well as experiments at higher energy be effected in order to settle the question as to the existence or non-existence of an angular momentum window for fusion.

REFERENCES

1. P.A.M. Dirac, Proc. Camb. Phil. Soc. 26, 376 (1930).
2. P. Bonche, S. E. Koonin, and J. W. Negele, Phys. Rev. C13, 1226 (1976).
3. G. Wegmann, Physics Lett. 50B, 327 (1974); S. E. Koonin, Prog. in Part. and Nucl. Phys. 4, 283 (1980).
4. J. A. Wheeler, Phys. Rev. 52, 1083, 1107 (1937).
5. D. L. Hill and J. A. Wheeler, Phys. Rev. 89, 1102 (1953).
6. A. de-Shalit and H. Feshbach, Theoretical Nuclear Physics, (Wiley, New York, 1974) Vol. 1, p. 541.
7. S. T. Belyaev, Nucl. Phys. 64, 17 (1965); F. M. H. Villars, in Dynamic Structure of Nuclear States, Proceedings of the 1971 Mont Tremblant International Summer School, edited by D. J. Rowe (Univ. of Toronto Press, Toronto, 1972); M. Baranger and M. Veneroni, Ann. Phys. (N.Y.) 114, 123 (1978).
8. K. T. R. Davies, A. J. Sierk, and J. R. Nix, Phys. Rev. C6, 2385 (1976).
9. Y. M. Engel, D. M. Brink, K. Goeke, S. J. Krieger, and D. Vautherin, Nucl. Phys. A249, 215 (1975).
10. S. J. Krieger and K. Goeke, Nucl. Phys. A234, 269 (1974).
11. S. E. Koonin, Ph.D. thesis, Massachusetts Institute of Technology, 1975 (unpublished); C. Y. Wong, J. A. Maruhn, and T. A. Welton, Nucl. Phys. A256, 469 (1975).
12. S. Levit, Phys. Rev. C21, 1594 (1980); S. Levit, J. W. Negele, and Z. Paltiel, Phys. Rev. C21, 1603 (1980); Y. Alhassid, B. Müller, and S. E. Koonin, Phys. Rev. C23, 487 (1981).
13. Y. Alhassid and S. E. Koonin, Phys. Rev. C23, 1590 (1981).
14. J. W. Negele, in Proceedings of the Meeting on Heavy Ion Collisions, Fall Creek Falls State Park, Pikeville, Tennessee (1977); P. C. Lichtner and J. J. Griffin, Phys. Rev. Lett. 37, 1521 (1976).
15. A. Kerman and S. E. Koonin, Ann. Phys. 100, 332 (1976).
16. G. E. Brown, Unified Theory of Nuclear Models (North-Holland, Amsterdam, 1964).
17. J. W. Negele in Theoretical Methods in Medium Energy and Heavy-Ion Physics, K. McVoy and W. A. Friedman, ed. (Plenum Press, New York, 1978); S. E. Koonin, Progress in Part. and Nucl. Phys. 4, 283 (1980); J. W. Negele (to be published in Reviews of Modern Physics); K. T. R. Davies, K. R. S. Devi, S. E. Koonin and M. R. Strayer, Cal-Tech preprint MAP-23 (1982).
18. R. Y. Cusson, H. P. Trivedi, H. W. Meldner R. Wright and M. S. Weiss, Phys. Rev. C14, 1615 (1976).
19. T. H. R. Skyrme, Phil. Mag. 1, 1043 (1956); Nucl. Phys. 9, 615 (1959).
20. P. Quentin and H. Flocard, Ann. Rev. of Nuc. and Part. Science 28, 523 (1978).
21. J. W. Negele and D. Vautherin, Phys. Rev. C5, 1472 (1973).

22. J. W. Negele, Phys. Rev. C1, 1260 (1970).

23. P. Hoodbhoy and J. W. Negele, Nucl. Phys. A288, 23 (1977).

24. S. J. Krieger and K. T. R. Davies, Phys. Rev. C18, 2567 (1978).

25. P. H. Heenen, private communication.

26. J. A. Maruhn and R. Y. Cusson, Nucl. Phys. A270, 471 (1976); R. Y. Cusson and J. A. Maruhn, Phys. Letts. B62, 134 (1976).

27. S. E. Koonin, K. T. R. Davies, V. Maruhn-Rezwani, H. Feldmeier, S. J. Krieger, and J. W. Negele, Phys. Rev. C15, 1359 (1977).

28. K. T. R. Davies, H. T. Feldmeier, H. Flocard, and M. S. Weiss, Phys. Rev. C18, 2631 (1978).

29. H. T. Feldmeier, ORNL Report ORNL/TN6053 (1977).

30. K. R. Sandhya Devi and M. R. Strayer, Phys. Letts. B77, 135 (1978).

31. S. E. Koonin, B. Flanders, H. Flocard, and M. S. Weiss, Phys. Letts., B77, 13, (1978).

32. In the filling approximation, a spherical nucleus is constructed by assigning equal, time-independent fractional occupation probabilities, n_i to the orbitals of the last shell. Although the density $\rho = \Sigma_j \; n_j \psi_j \psi_j^*$ then no longer satisfies the HF condition $\rho^2 = \rho$, this technique has been adopted in order to obtain spherically symmetric solutions for open-shell nuclei.

33. K. R. Sandhya Devi, M. R. Strayer, K. T. R. Davies, S. E. Koonin, and A. K. Dhar (to be published in Phys. Rev. C).

34. E. Tomasi, D. Ardouin, J. Barreto, V. Bernard, B. Cauvin, C. Magnago, C. Mazur, C. Ngô, E. Piasecki, and M. Ribrag, Nucl. Phys. A373, 341 (1982).

35. M. Lefort, Rep. Prog. Phys. 39, 129 (1976).

36. P. Bonche, B. Grammaticos, and S. E. Koonin, Phys. Rev. C17, 1700 (1978).

37. S. J. Krieger and K. T. R. Davies, Phys. Rev. C20, 167 (1979).

38. P. Sperr, S. Vigdor, Y. Eisen, W. Henning, D. G. Kovar, T. R. Ophel, and B. Zeidman, Phys. Rev. Lett. 36, 405 (1976).

39. P. Sperr, T. H. Braid, Y. Eisen, D. G. Kovar, F. W. Prosser, Jr., J. P. Schiffer, S. L. Tabor, and S. Vigdor, Phys. Rev. Lett. 37, 321 (1976).

40. K. Daneshvar, D. G. Kovar, S. J. Krieger, and K. T. R. Davies, Phys. Rev. C25, 1342 (1982).

41. J. Blocki and H. Flocard, Nucl. Phys. A273, 45 (1976).

42. K. R. S. Devi, A. K. Dhar, and M. R. Strayer, Phys. Rev. C23, 2062 (1981).

43. S. Tabor, D. Geesaman, W. Henning, D. Kovar, K. Rehn, and F. Prosser, Phys. Rev. C17, 2136 (1978).

44. S. J. Krieger and M. S. Weiss, Phys. Rev. C24, 928 (1981).

45. P. Bonche, K. T. R. Davies, B. Flanders, H. Flocard, B. Grammaticos, S. E. Koonin, S. J. Krieger, and M. S. Weiss, Phys. Rev. C20, 641 (1979).

46. K. T. R. Davies, K. R. Sandhya Devi, and M. R. Strayer, Phys. Rev. C24, 2576 (1981).

47. A. Lazzarini, H. Doubre, K. T. Lesko, V. Metag, A. Seamster, R. Vandenbosch, and W. Merryfield, Phys. Rev. C24, 309 (1981).

48. J. Barreto, G. Auger, H. Doubre, M. Langevin, and E. Plagnol, Annual Report IPN Orsay (1980) to be published.

49. E. Tomasi, D. Ardouin, J. Barreto, V. Bernard, B. Cauvin, C. Magnago, C. Mazur, C. Ngô, E. Piasecki, and M. Ribrag, Nucl. Phys. A373, 341 (1982).

TDHF CALCULATION OF NUCLEAR MOLECULES FROM C12 + C12 TO U238 + U238

R. Y. Cusson[+], H. Stöcker,
G. S. I. Darmstadt, W. Germany

M. R. Strayer, S. Umar, D. A. Bromley
Yale University, New Haven, Conn., U. S. A.

J. A. Maruhn, and W. Greiner
Institut für Theoretische Physik der Universität Frankfurt
D-6000 Frankfurt am Main, West Germany

Abstract

A review of recent applications of the TDHF method to the problem of determining the dynamics of nuclear molecules is made. Two types of nuclear molecular motion are considered; diatomic and polyatomic. Polyatomic motion involes alpha clusters and diatomic motion is defined by requiring a long-lived well defined neck. Some interpretation of the available data is made along these types of dynamics.

1. *Introduction*

The Time Dependent Hartree Fock(TDHF) method[1-7] has shown itself to be rich in nuclear phenomena and continues to stimulate our understanding of nuclear dynamics, in spite of the various conceptual and practical difficulties involved in its use. The first results which came out of TDHF showed that one could make a reasonable account of the fusion cross-section in light systems and the Wilczynski plots of medium ion reactions. Later, higher energy calculations showed that incomplete fusion and preequilibrium nucleon emission[5] were also predicted by the model. In this paper we discuss the latest findings connected with molecular effects in nuclear dynamics.

We will be interested in two types of nuclear molecules. The first type is a diatomic molecule where the nuclear composite system has a well-defined neck and a total rms radius substantially greater than that of the normal compound nucleus. Such molecular shapes may depend on angular momentum and their lifetime could vary greatly as a function of the total J. It is also to be expected that potential pockets connected with the details of the shell correction and spin-orbit potentials will affect not only their stability, but also the prospects for exciting them in nuclear heavy-ion reactions. The second type will be called a polyatomic molecule because it has so far been predicted to

---

[+]Invited speaker; permanent address, Physics Dept., Duke U., Durham, N. C. 27706, U. S. A.

appear as elaborate motions of several alpha clusters in the composite system. This second type had not been predicted so often as the first type in the past, but in view of the strong role played by alpha-like clusters of 2 protons and 2 neutrons in nuclei(quartets), its occurence is, in the end, not so surprising. Polyatomic motions involving bigger clusters such as whole O16 or Ca40 fragments have been postulated[8)] but will not be discussed here.

Three different computer codes were used to obtain the results presented here. They use somewhat different interactions and representations for the wavefunction, so that effects seen in several codes have a smaller chance of being an artifact of a particular force and reppresentation. The first code[2)] is a momentum space 3-dimensional code using a simplified Skyrme plus Yukawa and Coulomb interaction. This code represents the nucleons as quartets with an effective charge $e(Z_1 + Z_2)/(A_1 + A_2)$ and is in some sense biased toward alpha clusters. It does the time evolution using an extremely accurate 6-th order predictor-corrector and uses the momentum space oportunity to smooth the wavefunctions by cutting off momentum components $\geq 2.\mathrm{fm}^{-1}$. This can be viewed as a gross approximation to the time smoothing effects[9)] of the residual interaction, which is otherwise neglected in this work. The second code used is similar to the first one. However, it operates entirely in coordinate space using a 5-point formula for the second derivative on the coordinate grid, and it propagates the states in time using an exponential expansion and an effective Hamiltonian accurate to second order in time. Perhaps because of the high (spurious?) momenta propagated by this code, the polyatomic motions were the least stable with it. The third code[3)] works in a rotating frame with axial symmetry and uses a centrifugal potential based on a rigid clutching model for the moment of inertia. Its chief virtue is to allow the use of the full Skyrme interaction without spin-orbit force, and the separation of neutrons and protons. Both SkyrmeII and SkIII interactions were used for the polyatomic motions with similar results.

The TDHF method can only compute the semiclassical trajectories of the collective moments of the composite system as a function of time. The presence of residual interactions induces collisions which, after $t \gtrsim 3\times10^{-21}$ sec. may produce fluctuations and correlations which affect the mean values of the collective trajectories. These collision terms are not included so that the long time behaviour of our results could be adversely affected. Another source of uncertainties stems from the fact that any trajectory calculation involves a potential surface and a set of mass parameters. It is well known that mass parameters are affected by level crossings. Our TDHF work assumes orbit occupation numbers which are time independent, whereas the collision would dictate that levels coming down should see their occupation increased, and so on. This probably affect the fission and fusion behaviour at long time into the reaction. In many TDHF reactions the incident energy is several hundred Mev above the composite system potential energy surface. Yet the single particle friction can quickly absorb this into the thermal doorway states. These states form a small subset of all the compound states at this high excitation energy. On this subset, it appears that the influence of shell effects is still substantial. As long as the average single particle excitation energy per nucleon in the thermal doorway state is less than the shell energy (≃ 4-8 MeV), the details of the ground state potential energy surface are felt, so that shell correction effects may still influence the dynamics even though the excitation energy is very high. Our work has incorrect shell correction energies for heavy systems because we do not have a spin-orbit term. Thus we can not expect to reproduce reliably those potential pockets which can trap the molecular states[3] and produce long-lived molecules.

Keeping in mind the above pittfalls we can still pursue our study of nuclear molecules provided we exercise care in stating those conclusions which are most sensitive to the errors of the method. We have selected 5 reaction types where TDHF and experiment help us obtain information about nuclear molecules. These are summarised in Table 1 and include i) polyatomic molecules with alpha clusters, ii) medium weight high J_{tot} and E_{tot} diatomic molecules, iii) barrier top potential trapped molecules, iv) asymmetric molecules, and finally v) the di-Uranium system.

2. *Polyatomic Molecules*

The resonances in C12 + C12 scattering[10] are shown in Fig. 1 and a fit using Iachello's U(4) symmetry model is shown[11]. Apart from the missing odd parity levels the agreement with experiment is remarkable and suggests that one should consider intrinsic radial vibrational motion with a dissociation threshold near 7 MeV. The required Morse potential[12] is illustrated in Fig. 2 together with the more conventional C12-C12 real part of the optical potential[13] for two C12 nuclei in their ground state. In spite of the Morse potential's ability to represent the data so well, Fig. 2 shows some of its unphysical properties. To begin with, it has a minimum value where the top of the Coulomb berrier should be. Then, the potential is still non-Coulomb at separations of order 10.fm. It is difficult to se how two C12 nuclei with radii of order 3fm should have a nuclear interaction at such large separation. Finally, it remains puzzling that the potential should be repulsive and non-absorbtive at small radius.

We have made a sudden approximation constrained static Hartree-Fock calculation of the potential between a spherical C12 and a deformed one. These two states of C12 are illustrated in Fig. 3, which shows the H.F. energy vs deformation[13] in C12, and the ground and 4p-4h excited states shapes. The 4p-4h state, (configuration B) shows 3 alpha-like clusters in line and could provide an attractive nuclear interaction at large separations. The sudden potential[14] between A and B, with the long axis of B pointing toward A, and using the simplified Skyrme interaction[1] is shown in Fig. 4. In contrast with Fig. 2, the minimum now occurs at the correct separation of ~7fm. However, the radial width of the potential is smaller than that of the Morse potential and it is somewhat too deep (the Coulomb term is not shown in Fig. 4). In order to fit the observed $\hbar\omega_{vib}$ with this potential we would need a mass parameter larger than the reduced mass of C12+C12. This is puzzling and so is the fact that one can not really expect a sudden potential to exibit the long-lived resonances shown in Fig. 1. An attempt to resolve this puzzle was made by initialising the TDHF code with the 9fm configuration of Fig. 4. This, if was expected, would give an indication of the time constant for the relative motion of these 2 clusters.

The TDHF motion which resulted from this is shown in Fig. 5. The top left hand frame is the initial configuration and the later times are shown immediatly below and then to the right. The first obervation is that the repulsive part of the sudden potential of Fig. 4 is absent. We have instead, a successive tunnelling of the 3 alphas across the spherical C12 until the initial configuration is reversed. This motion appears quasiperiodic in the TDHF code with a period of about $2.3x10^{-21}$sec. corresponding to a harmonic vibrational excitation energy of $\hbar\omega_{vib} \simeq 1.8$MeV. This may be compared with the coefficient a = 1.44MeV in Fig. 1. This is reasonably close in view of the inevitable anharmonicity of the motion. A schematic representation of the potential energy surface of $C12+C12^{*}(7.6MeV) \leftrightarrow Mg24^{*}$ is shown in Fig. 6, as a function of the mass octupole fragmentation coordinate Q_3 and the mass quadrupole coordinate Q_2. Apart from the expected Mg24 ground state valley, there should be a deep quasimolecular valley which is narrow in the Q_2 direction and long and symmetric about zero in the Q_3 direction. We have indicated schematically with arrows the various observed decay channels of the states in

Table 1: Summary of Reactions Considered.

Reactions	Energy Range	L values	Feature of Interest
C14+α	<<1MeV/A	0	polyatomic vibrations
C12+C12	above Coul. barr.		large dipole transitions
Ar36+α			suspension of s.p. viscosity
Pb208+α			quasi periodic motion
Sr88+Sr88	2-15MeV/A above barr.	$\sim\frac{1}{2}L_{gr}$	very high L-values
Kr86+Er166			diatomic molecules
Kr86+Pb208	top of Coul. barr.	0	long-lived diatomic res.
			potential pocket trapping
Pb208+Ni58	1-3MeV/A above barr.	$0 \le L \le L_{gr}$	symmetric breakup of
		~100h	long-lived diatomic molecule
U238+U238	barr. top(~5.9MeV/A)	0<L<150h	deformed U+U entrance channel
	to 6.5MeV/A		→ fast 3-body breakup of
			diUranium composite(6.4MeV/A)

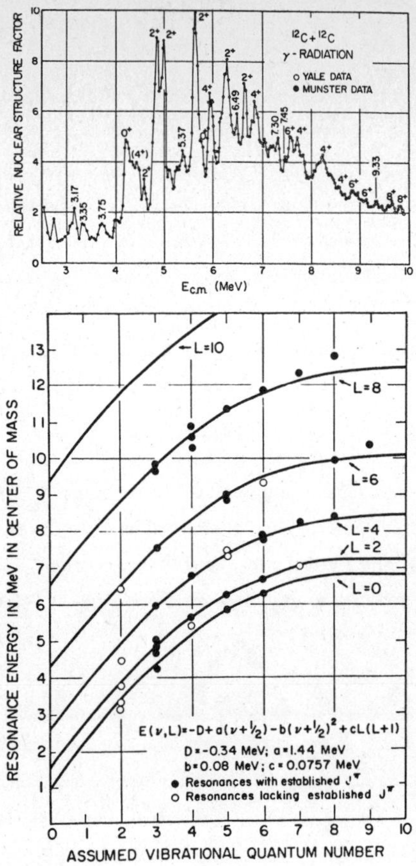

Figure 1. Experimental resonances in C12 + C12 scattering[10].

this valley. Constrained Hartree-Fock calculations of this potential are in progress. It would interesting to also obtain the mass parameter for the octupole motion.

Several types of alpha-cluster motions have been predicted[15] in the light nuclei; this leads us to chose our next example from the reaction C14+α. This reaction is of special interest because the extra 2 neutrons could allow for charge dipole oscillations. A TDHF calculation for this system was performed and showed time-dependent configurations with one alpha-like cluster oscillating across the C14 nucleus. In particular, it was remarkable that this moving cluster did not charge equilibrate while traversing the C14 but retained its 2p-2n character; thus the charge dipole moment shows strong undampened oscillations vs time. These are shown in Fig. 7, where one observes a complex quasiperiodic motion with little damping. The Fourier analysis of this result is in progress; the autocorrelation function and the power spectum should provide additional insight on the nature of the TDHF motion. The time dependent charge quadrupole moment is shown in Fig. 8, the octupole moment is in Fig. 9, and the hexadecaple one is shown in

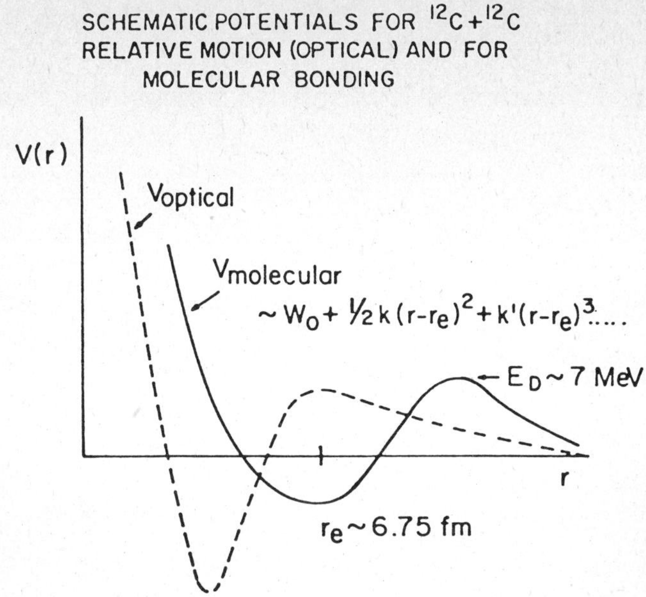

Figure 2. Potentials for C12 + C12 system.

Fig. 10. Whereas the odd multipoles oscillate about zero, the even ones do so about a large average value, so that their % change is much smaller. The crosscorrelation function between the various multipoles is beeing generated and should help determine if the higher multipole oscillations are well correlated with the dipole ones. Neglecting for the moment the energy dependence of the dipole transitions implied by the time dependence, we simply extract the RMS dipole moment from Fig. 7. This is then used to estimate the dipole transition strength inside the O18 rotational band[15] starting at 3.63 MeV(see Fig. 11). This agrees with the experimental values within a factor of 2. We note that there is some odd-even staggering of the energies shown in the inset. This could be

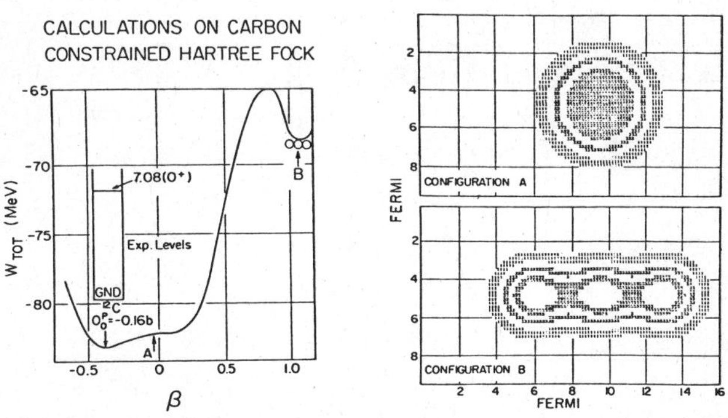

Figure 3. Constrained HF calculations for C12.

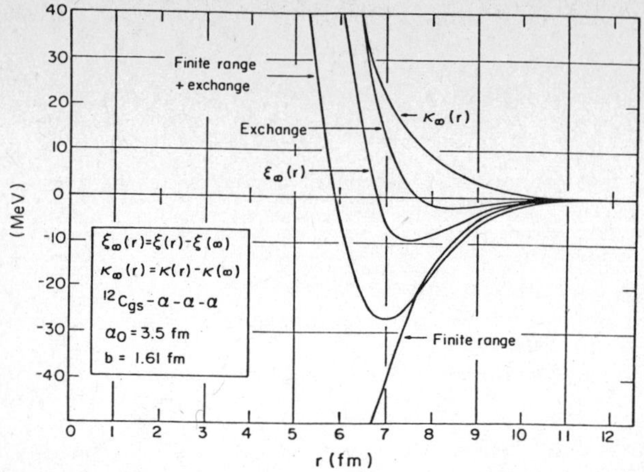

Figure 4. Sudden potential for C12(def.) + C12(gnd).

accounted for by the fact that the even L states are generated by an intrinsic state having no nodes in its vibrational wavefunction, while the odd L one must have one node and one quantum of vibrational excitation. The period observed in the C14+α reaction corresponds to a harmonic excitation energy $h\omega_{vib}$ ≃ 3Mev. In a recent[15)] U(4) fit of this and the associated rotational band, vibrational energies of this magnitude were required. The moment of inertia implied by the rotational band of Fig. 6 is also consistent with

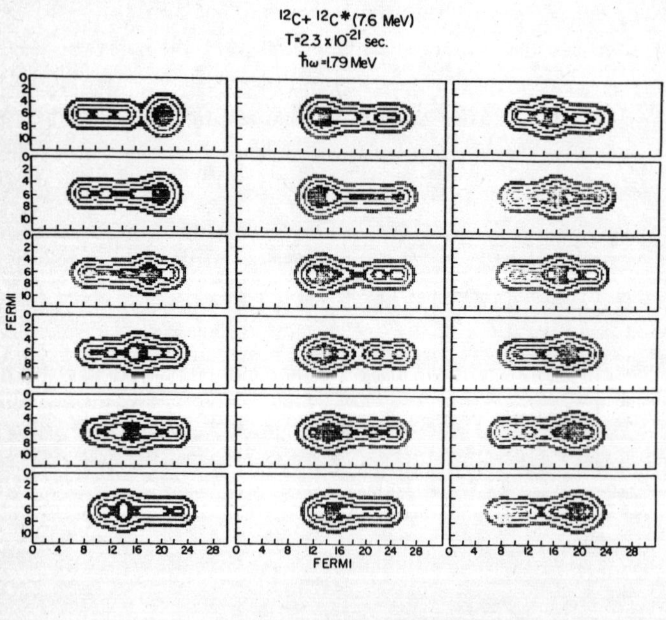

Figure 5. TDHF development of the molecular C12 + C12 system.

the mass quadrupole moment observed in the TDHF calculation. Similarly, the quadrupole transition strengths are consistent with the charge quadrupole moment of Fig. 3.

The motion of alpha-like clusters is also observed with targets of Ar36 and Pb208. Again, no charge equilibration and no damping of the oscillations seem to be taking place. In the case of Ar36, one can clearly see several alpha-like clusters appear in the target during the motion. For a Pb208 target, the transit time of the alpha-like cluster across the nucleus is about 5×10^{-21} sec. Such a picture suggests estimates for the barrier attack frequency in alpha decays. Preliminary calculations[8)] imply enhanced and correlated alpha-decay properties for a number of rotational states in Po212.

3. *High Angular Momentum Diatomic Molecules.*

The TDHF method predicts high L-value capture into fast rotating and vibrating composite systems. Figure 12 shows this effect[6)] in the reaction[16)] Kr86 + Er166 at 12.2 MeV/A. A capture region from $100 \le L_{in} \le 270\hbar$ is observed. The time dependence of the density is illustrated in Fig. 13 for a non-capture event at $L_{in} = 335\hbar$. The capture events show a similar behaviour at intermediate times: two well-defined clusters, separated by a well-formed neck, rotate about each other. In addition, strong surface waves can usually be seen going around each cluster. The long time behaviour is somewhat uncertain in TDHF, but Fig. 14 illustrates that, even after 5×10^{-21}sec., the system shows no clear indication of a coming separation(for $L_{in} = 215\hbar$). The flux contained in the capture window is not necessary[7)] to account for the experimental deep inelastic cross-section. This is illustrated in Fig. 15, where the Wilczynski plots at 6 and 12 MeV are compared with experiment. The motions obtained in the capture region all lie well above the liquid drop limiting values[18)] for stable rotations, but the decay mode

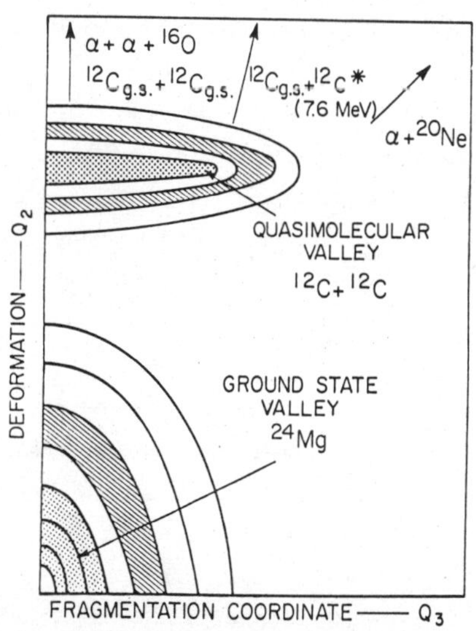

Figure 6. Quasimolecular valley of C12 + C12.

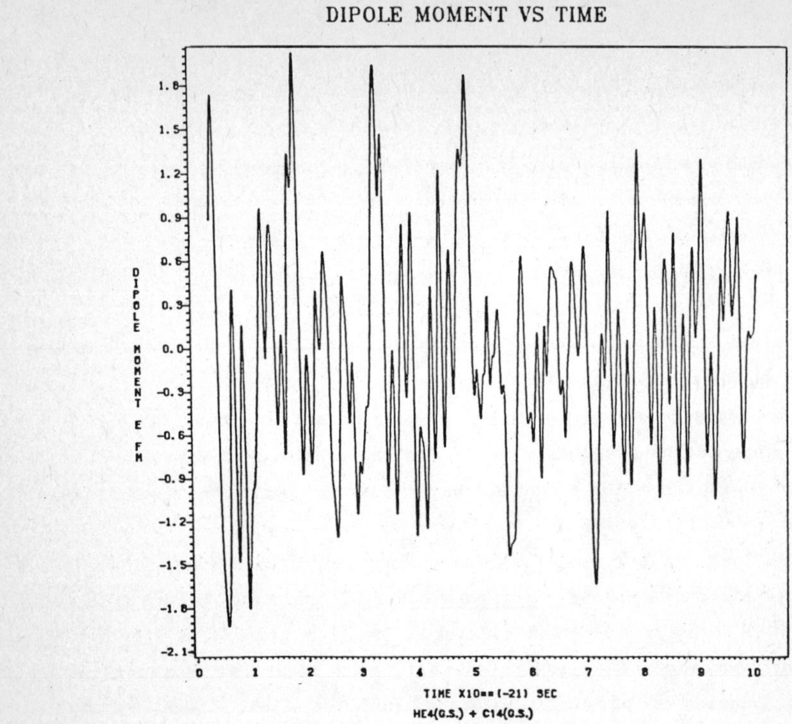

Figure 7. Electric dipole moment of O18 vs time.

need not be two-body, and the lifetime is not known. Although we have observed fast 3-body breakup only once in the U + U system with TDHF, experiments suggest[16,17] that this breakup mode may alrady be be present in the Kr + Er reaction at 12.2 Mev/A. An even stronger suggestion of 3-body breakup is obtained experimentally[17] in the reaction Mo92(14.8MeV/A) + Mo92. Our TDHF calculations for this reaction shows a small capture region. Indeed, a similar calculation of the Sr88 + Sr88 reaction at 20MeV/A also shows a capture region. Of course the TDHF capture process at high L and moderate bombarding energy is subject to substantial fluctuations. They could be due to residual influence of shell effects or to a complicated systematic of hydrodynamic instabilities. So far however, the Mo + Mo 3-body breakup has not been seen in TDHF.

4. *Barrier Top Resonances.*

Heavy ion reactions near the top of the Coulomb barrier and with $L_{in} = 0$ nearly always exibit the phenomenon of sudden neck formation, in TDHF. Depending on the systematic of Coulomb energies, the liquid drop model energy surface for the resulting necked molecular shape can be used to predict the subsequent behaviour of the system. This smooth part of the total energy surface usually predicts fission for heavy systems and fusion for light ones. This picture can be modified by shell effects however. The TDHF shell effects, although incorrect, can thus increase the lifetime of some barrier top states by a considerable amount. This is illustrated[3] in Fig. 16 for the Kr + La reaction at L = 0. It is believed that there occurs a potential pocket at the fully relaxed energy which traps the system for a long time. A similar effect for Kr + Pb208 has been observed. Such a mechanism could lead to cold fusion in heavy systems, although it is not expected to yield completely stable nuclear molecules.

5. *Symmetric Break up of Nuclear Molecules.*

An interesting class of nuclear molecules shows a stable diatomic necked shape as a result of a balance between the Coulomb force and the centrifugal barrier[6]. This occurs in the Pb + Ni region at 1-3MeV/A above the Coulomb barrier and for $L \leq L_{gr} \sim 100h$. Figure 17 illustrates the possible reaction types under those circumstances. The first horizontal sequence represents ordinary deep inelastic scattering. The second one is fusion-fission where the breakup takes place with a mass drift of some 25 units towards symmetry[19]. The third panel shows actual capture into a long-lived nuclear molecule, while the fourth one shows actual fusion. The different behaviour of these panels are further illustrated in Fig. 18, showing the total RMS radius vs time. All these systems are above the liquid drop limit and should therefore decay. However, the time scale for that is well beyond the capabilities of TDHF even if the calculations could be believed at such long times. Presuming that the dominant decay modes are symmetric

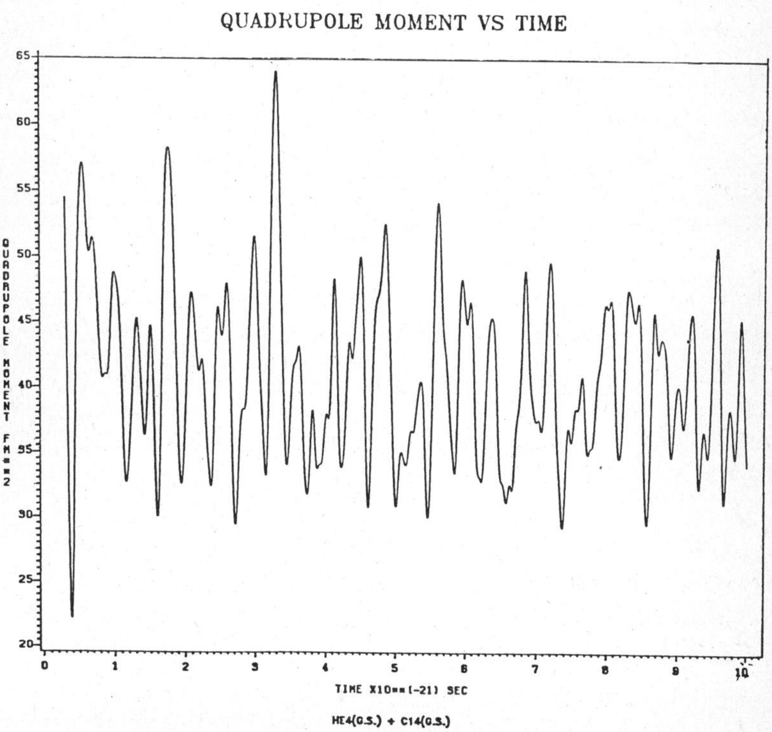

Figure 8. Electric quadrupole moment of O18 vs time.

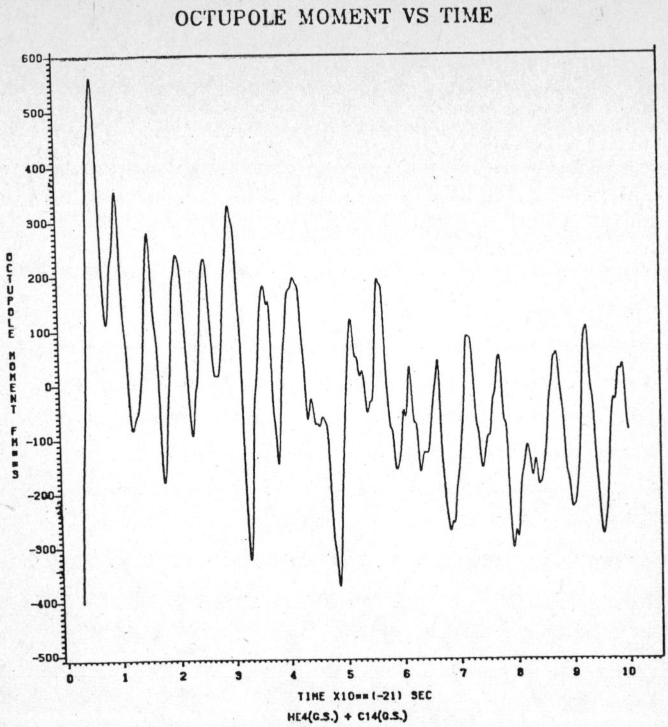

Figure 9. Electric octupole moment of O18 vs time.

fragmentations, we can compare our capture cross- sections with the experimental symmetric fragmentation one. This is done in Fig. 19. The error bars on the computed Pb208 + Ca48 and the U238 + Ca48 cross-sections come from having few impact parameters at each bombarding energy. The dashed line shows the experimental symmetric fragmentation cross-section[19] for Pb208 + Ca48. Here the TDHF result substantially overestimates the cross-section. This appears to be due to a rather unrealistic shell energy for the effective charge quartet model force used for this calculation.

6. *The Di-Uranium Molecule?*

Following recent positron data[20] on the U + U collision, it is interesting to consider again the suggestion[21] of a di-Uranium nuclear molecule. At the barrier top experimental energy of 5.9MeV/A, we have verified that two spherical Uranium nuclei do not even go through a neck formation phase, in TDHF, because of the very large Coulomb repulsion. However, there is a probability that Coulomb excitation can align the deformed Uraniums and even vibrationally excite them into the highly deformed second minimum. Under such circumstances, a neck does form at 5.9MeV/A and it is conceiveable that a fortuitously large amplitude shell correction could lead to a long-lived state of the type seen in Fig. 16. We have looked without success for such a resonance at various initial energies from 5.9 to 6.9 Mev/A, with L=0, at several deformations, using the axial 2-D code. This of course does not preclude its possible presence if one uses a realistic force with spin-orbit interactions. One novel feature observed in this reaction is the 3-body breakup of the di-Uranium near 6.4MeV/A. The outer fragments each have A ~ 125 and the remaining one has A ~ 226, corresponding to nearly symmetric breakup of each U fragment. These breakups have not so far been detected experimentally for U + U but they have probably been seen for lighter systems such as Kr + Er and Mo + Mo as discussed earlier.

Acknowledgements

We are grateful for the oportunity to utilise the Floating Point Array Processing facilities at the GSI Computation Center where most of the calculations were performed.

References

1. P. Bonche, S. E. Koonin, and J. W. Negele, Phys. Rev. *C13* (1976) 1226.
2. R. Y. Cusson, J. A. Maruhn, and W. A. Meldner, Phys. Rev. *C18* (1978) 2589.
3. K. T. R. Davies,K. R. Sandhia Devi, and M. R. Strayer, Phys. Rev. *C20* (1978) 1372; Phys. Rev. *C24* (1981) 2576.
4. M. Flocard, S. E. Koonin, and M. S. Weiss, Phys. Rev. *C17* (1978) 1682.
5. H. Stöcker, R. Y. Cusson, J. A. Maruhn, and W. Greiner, Phys. Lett. *101B* (1981) 379.
6. H. Stöcker, R. Y. Cusson, H. J. Lustig, A. Gobbi, J. Hahn, J. A. Maruhn, and W. Greiner, Zeitschrift f. Physik *A* (1982) (in press).
7. R. Y. Cusson, J. A. Maruhn, H. Stöcker, Nuclear Physics *A* (1982) (in press).
8. M. Sandalescu (1981) (private communication).
9. R. Balian, and M. Veneroni, Annals of Physics,(N.Y.) *135* (1981) 270.
10. K. A. Erb, and D. A. Bromley, Phys. Rev. *C23* (1981) 2781.
11. F. Iachello, Phys. Rev. *C23* (1981) 2778.
12. K. A. Erb, private communication (1981).
13. R. Y. Cusson, R. Hilko, and D. Kolb, Nucl. Phys. *A270* (1976) 437.
14. P. G. Zint, and U. Mosel, Phys. Lett. *56B* (1975) 563.
15. M. Gai, and D. A. Bromley, (1982), (to be published).
16. A. Olmi, et. al. Phys. Rev. Lett. *44* (1980) 383.
17. A. Gobbi, (1981) (private communication).
18. S. Cohen, F. Plasil, and F. Swiatrcki, Ann. Phys.(N.Y) *82* (1974) 557.
19. A. Olmi et. al., 15th Bormio Winter Meeting on Nuclear Physics, Bormio, Italy (1978) p.724; and to be published.
20. H. Backe, et. al. Phys. Rev. Lett. *40* (1980) 1443.
21. K. Smith, H. Peltz, B. Mueller, and W. Greiner, Phys. Rev. Lett. *32* (1974) 554.

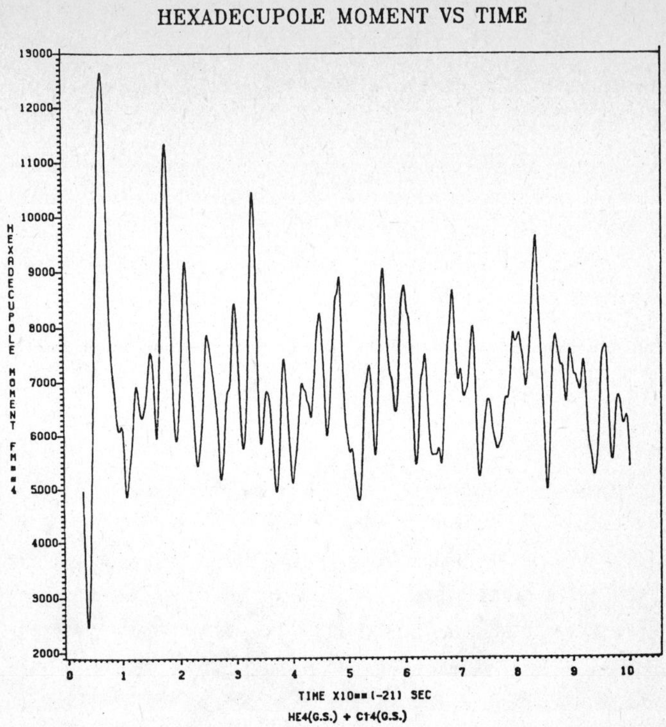

Figure 10. Electric hexadecapole moment of O18 vs time.

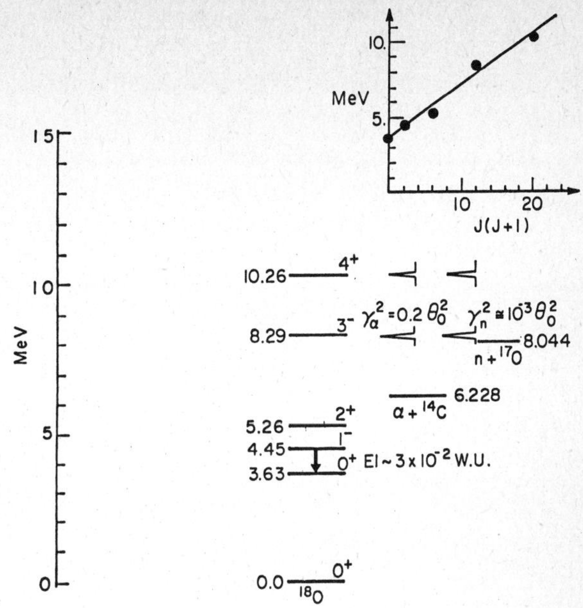

Figure 11. Levels in O18.

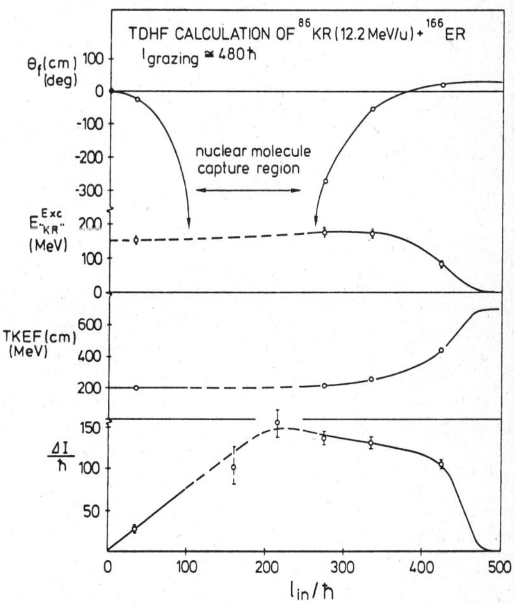

Figure 12. Kr + Er quantities vs L_{in} in TDHF.

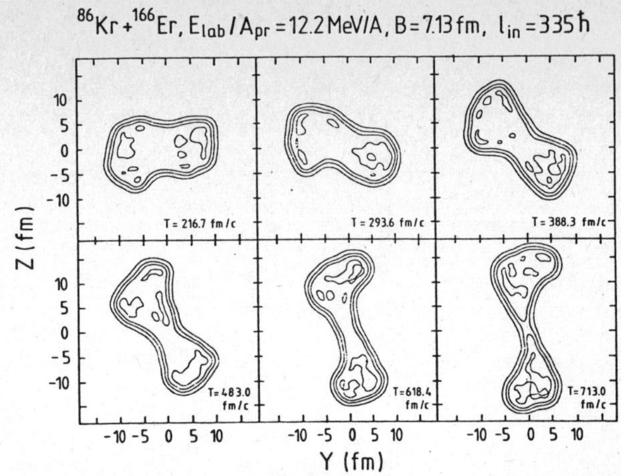

Figure 13. Snapshot of the reaction vs time.

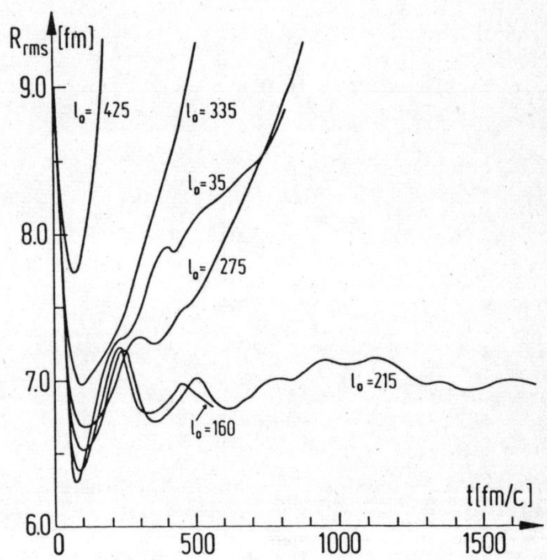

Figure 14. RMS radius of the Kr + Er reaction vs time.

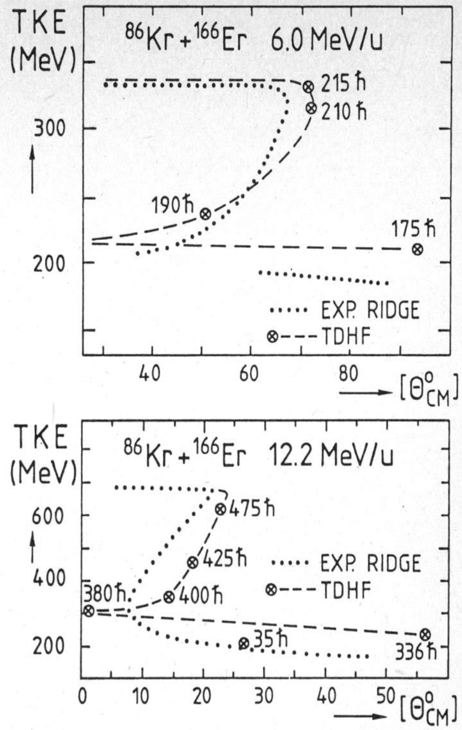

Figure 15. Wilczynski plots ridges compared to experiment[16).]

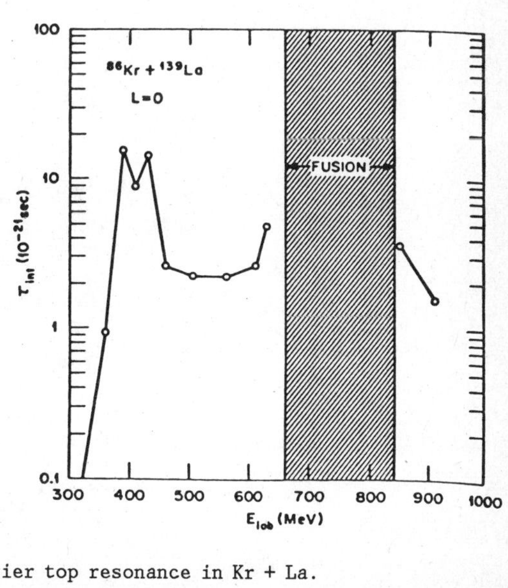

Figure 16. Barrier top resonance in Kr + La.

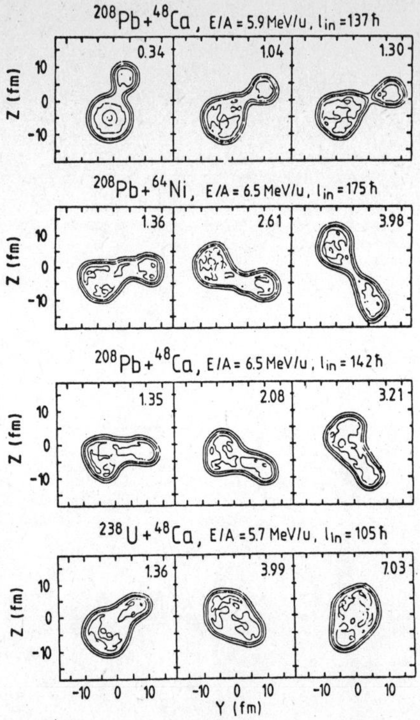

Figure 17. Four types of heavy ion reactions.

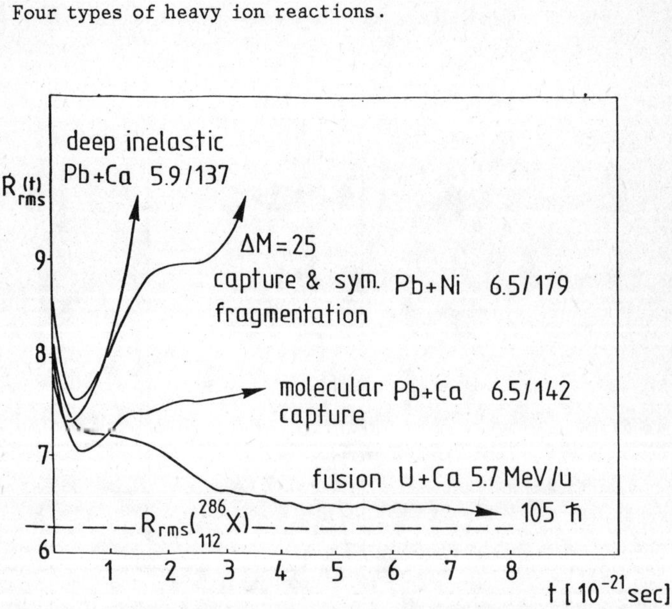

Figure 18. Rms radii vs time.

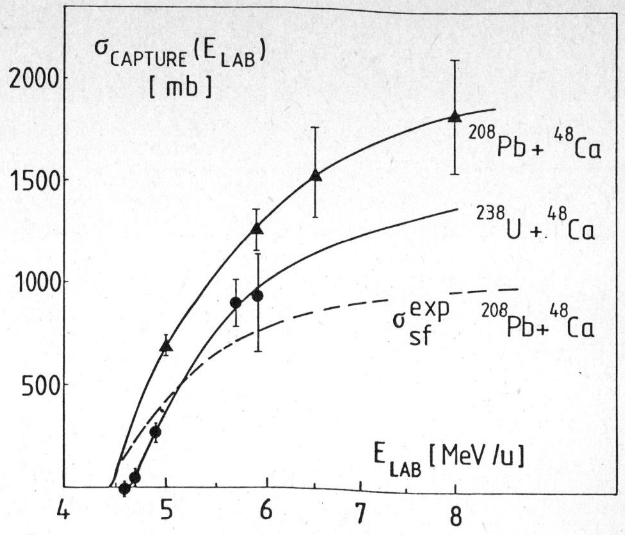

Figure 19. Experimental symm. fragm. cross-section and TDHF values.

Science and Art in Heavy Ion Collisions

M. S. Weiss
University of California, Lawrence Livermore National Laboratory
Livermore, California 94550

One of the more intriguing phenomena discovered in heavy-ion physics is the seeming appearance of high energy structure in the excitation spectra of inelastically scattered heavy ions. For reasons to be illustrated below, these may well be a phenomena unique to heavy ions and their explanation perhaps unique to TDHF.

The calculations which I will subsequently describe were motivated by the experiments of Nimet Frascaria[1] and her group at Orsay. Figure 1 shows the energy spectra of specific fragments from the collision of 400 MeV laboratory energy ^{40}Ca incident upon ^{40}Ca. Figure 1a displays the spectra for ^{36}Ar. The apparent independence of the excitation spectra over a range of scattering angles lend credence to the observed structure being "giant resonance like" in their origin. This interpretation is supported in Fig. 1b where the spectra of

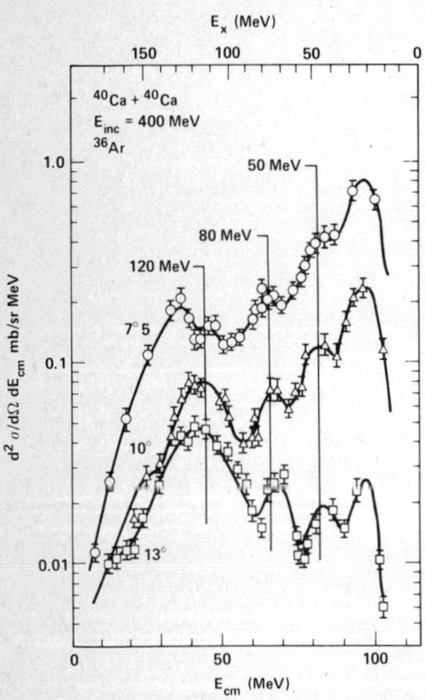

Fig. 1a

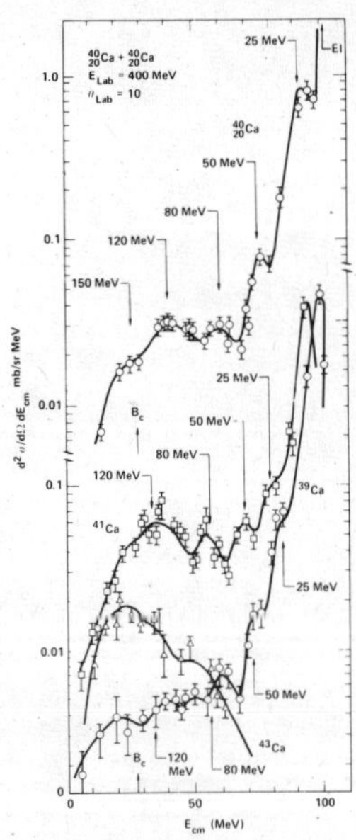

Fig. 1b

several species at fixed angle are displayed. A laboratory scattering angle of 10° corresponds (for this system) to a peripheral collision. The experiment was also performed for a more central scattering angle (and correspondingly more inelastic collision) and the excitation spectra remained unaltered in energy but the "peaks" were reduced in size by an order of magnitude.

Besides being beaten on by the experimenters, Flocard and I were motivated to explore this phenomena by the work of Van Gai[2] who argued that there could not be giant resonances above 40 MeV. His position was based upon the finiteness of the shell model potential with the consequence that there could not be any collective strength at high energy. His argument is very convincing because it is so simple.

Our approach was to model the experimental situation in TDHF.[3] After choosing the initial conditions to duplicate the experimental incident energy and final scattering angle, the calculation was terminated with the fragments well separated and therefore interacting only through a central Coulomb potential. The calculation was then re-started in a frame fixed on the center of mass of one of the fragments which was then allowed to evolve in time while the cartesian moments of the (ordinary) density were computed and stored. The buzz words to describe this calculation are symmetric fully three dimensional TDHF with BKN Force in coordinate space with 4N symmetry and the details and references are available.[4] The calculation was repeated for a more central collision and Table 1 shows the results for both collisions. Figure 2 displays the density

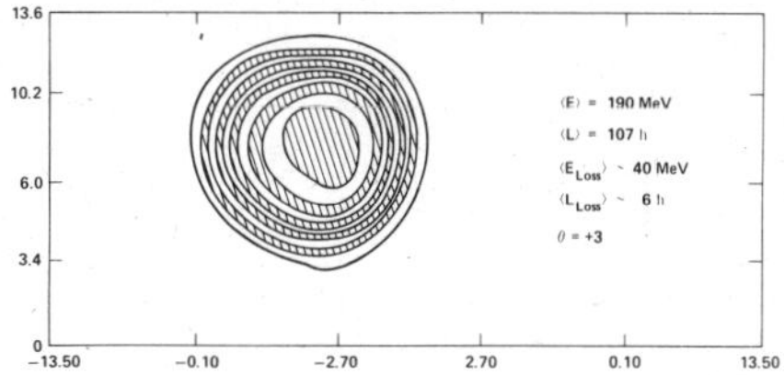

Fig. 2

contours projected onto the reaction plane for an isolated fragment. Table 1 shows the initial angular momentum transfer and energy loss for both fragments, and the centre of mass scattering angle for the two cases run. $L_{initial} = 107h$ is the more peripheral collision and corresponds to the experiment.

Our procedure was to Fourier transform the Cartesian moments of the density and look at the spectrum. Figure 3 displays the rms radius of the system for the $2 \cdot 10^{-21}$ sec it was followed. That such a system might have a rich Fourier spectrum is confirmed in Fig. 4 which displays the Fourier transform of $\int d^3r$

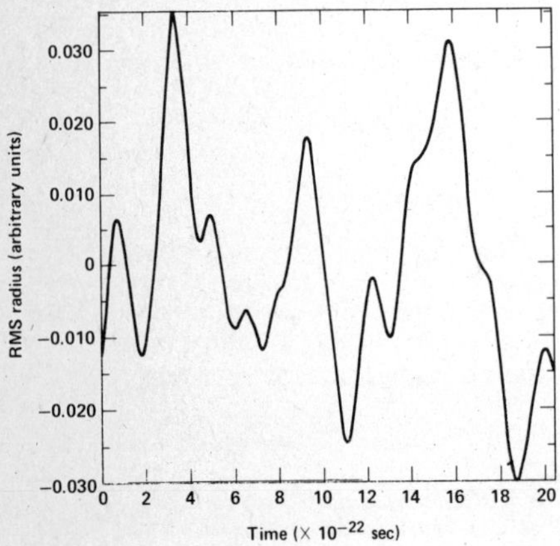

Figure 3

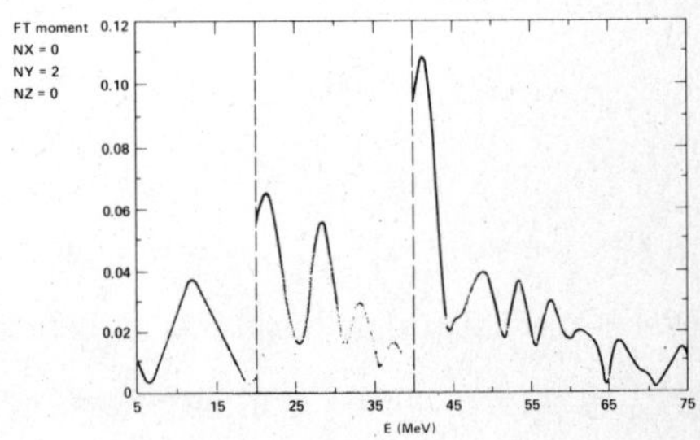

Figure 4

$y^2 \rho(r,t)$. Note the change in scale by a factor of 10 at 20 MeV and again at 40 MeV. While the observed spectra are too broad to isolate which are harmonics and which are primitive vibrations, the fact is there are excitations at high energy. Figure 5 shows spectra for a variety of moments which again exhibit high

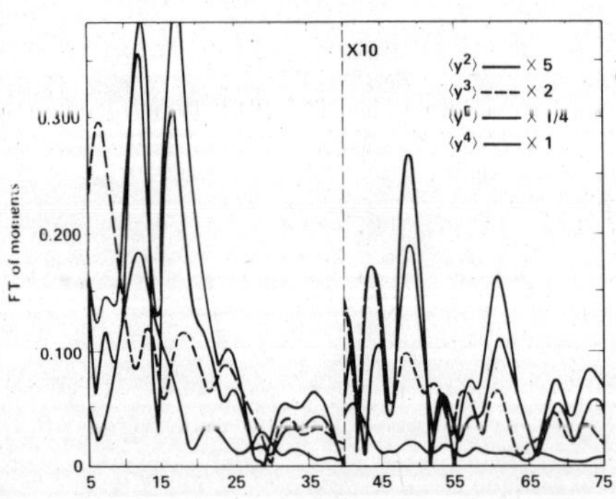

Figure 5

energy (more properly, high frequency) motion. While we have not shown that the observed structure is the one seen in our calculation, it is supportive that the TDHF calculation exhibits high energy peaks and when the calculation is repeated at a more central collision ($L = 97\hbar$), the Fourier spectrum remains but the amplitude of the high energy peaks goes down on order of magnitude, mimiking experiment.

Before proceeding I would like to remind you of the viewpoint of an American sage:

"There is something fascinating about science. One gets such wholesale returns of conjecture out of such a trifling investment of fact.

Mark Twain in "Old Times on the Mississippi"

Because up till now I have retold what I know and I will now spend a few minutes on plans and speculations.

Either Frascaria is wrong or we are dealing with a very interesting and new phenomena. Narrow resonances (if that is not a tautology) at 50 MeV and above would be intensely interesting even if they were "simple" giant resonances. If one accepts Van Gai's argument that they are not, one must conclude they must be a different and perhaps new type of state. Let us attempt to delineate the possibilities:

(a) They represent the appearance of some new symmetry (but I can't produce a candidate).

(b) They represent giant resonance built upon giant resonances. That is, the heavy ion collision having excited the target or projectile, to say, the dipole state

$$|\Psi_D\rangle = D|0\rangle$$

Then proceeds to build, say, a quadrupole state upon it

$$|\Psi_{Qd}\rangle = Q|\Psi_D\rangle$$

etc. These are not "true" multi-boson states because of blocking and changes in the size of the mean field, but approximately, the energy of the observed state would be the sum of its constitutents

$$E_{OBS} = \sum_{i=1}^{N} E_{GRi}$$

and the probability of formation

$$P_{OBS} = \sum_{i=1}^{N} P_{GRi} \,,$$

the product of probabilities for excitation of the individual resonances. This picture and the data are sufficiently imprecise that the measurements are not inconsistent with this idea. In this model, these states would be improbably seen in other than heavy ion reactions because it necessitates the sequential application of one body operators.

(c) The observed states are "true" multiple particle-hole states,

$$|\Psi_{obs}\rangle = \sum_{ijk\ell} \alpha_{ijk\ell}\, a_i^+ a_j^+ a_k a_\ell |0\rangle$$

$$+ \sum_{ijk\ell mn} \beta_{ijk\ell mn}\, a_i^+ \ldots\ldots$$

which can be excited only by "true" many body operators, not products of single particle operators. These would truly be unique to heavy ion physics because only a heavy ion will behave as a many body operator.

(d) Lastly, the states are large amplitude semi-classical motion.

Clearly spectulations (b), (c) and (d) are not independent in the sense that linear combinations of the (a) can make the (c) and (d), etc. but in the sense of which is the leading of principal term the descriptions are quite disparate.

Within the context of Mark Twain's admonition, there is one preliminary datum[5] which can be used either to support hypothesis (b) and (c) or discredit the entire picture. Figure 6 shows the proton spectrum in coincidence with the

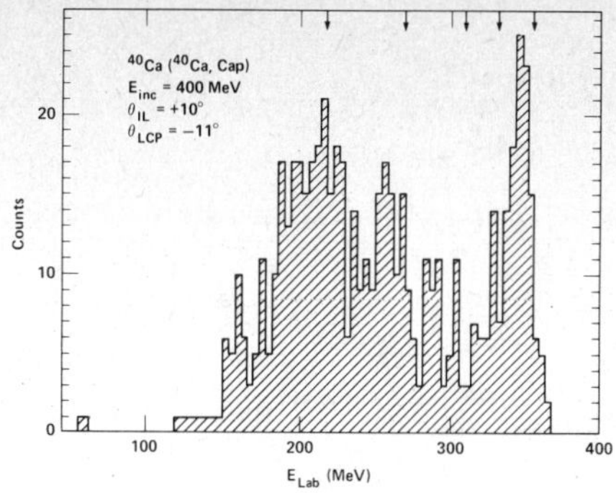

Figure 6

peaks in the spectra of the heavy ion excitation. The detected protons have an energy of approximately 10 MeV corresponding to prompt emmision from a low-lying particle-hole pair. This would support the idea of a giant resonance decay. However, the observed energy also corresponds to the beam velocity which is not obviously in any of the pictures we have described. Clearly one must await the results at other incident energies.

TDHF provides us with a unique tool to attempt to discriminate among these choices or perhaps even rule them out.

By repeating the previous calculation in which the density was evolved and its Cartesian moments calculated, we could instead calculate

$$I(r,L,M,z) = \int_{|\vec{r}|=r} d\Omega \, Y_{LM}(\hat{\Omega}) \, \rho(\vec{r})$$

and Fourier analyze I. This will tell us where in r and what in (L,M) the Fourier spectra observed in our previous (Cartesian) calculation originated. An alternative is to project the TDHF wave function over a giant resonance and Fourier analyze the resultant form factor.

A somewhat more ambitious program that will delineate among some of the options would involve replacing the TDHF collision calculation by the same trajectory for the Hartree-Fock potential of the projectile.

This would approximate the collisions by an external potential acting on the target (see Fig. 7). Gogny[6] and collaborators has examined the RPA spectra as

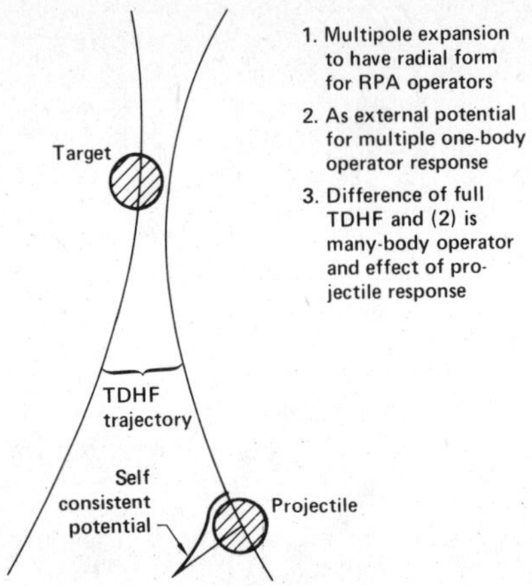

Figure 7

a function of momentum transfer for a calculation in which the usual radial moments of the electric multipole operators are replaced by $j_L(qr)$. While his procedures are controversial, he does find high energy spectra for $q \sim 1$ fm. While his choice of $j_L(qr)$ as the heavy ion operator is probably idiosyncratic, the conclusion is that a sufficiently complicated radial operator will produce a high energy spectra. Van Gai's[2] calculation is only for normal electric multipoles. Hence calculating the "correct" potential form factors in this way would test hypothesis (6) and extract the appropriate input for a more realistic RPA calculation. The fixed external potential of the projectile would act as a sequential single body operator. The comparison between the results of this external potential and the full TDHF response of the two interacting heavy ions would delineate the many body operator component, testing hypothesis (c). One can expect a variety of these ideas to be explored by various linear combinations of people at Orsay and Livermore certainly in time for the next TDHF meeting.

Now switching topics to a more artistic application of TDHF I would like to show you a color movie of the scattering of ^{86}Kr + ^{139}La at E_{lab} = 505 MeV. The calculation used the Skyrme III force in the frozen[7] and filling[8] approximation in three dimensions by an amalgam of workers.[9] Table II summarizes the results for the 13 impact parameters studied and Fig. 8 shows the requisite

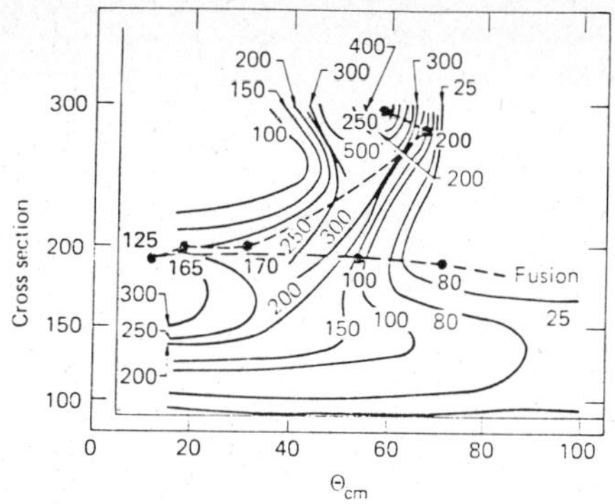

Figure 8

juxtaposition of calculation and experimental[10] Wilshinski plot. The calculated fusion cross section and deeply inelastic cross section agree with experiment.[10] The movie was prepared by the LLNL Computer Graphics Group's Nelson Max.

Acknowledgment

This work was performed under the auspices of the U. S. Department of Energy by the Lawrence Livermore National Laboratory under contract No. W-7405-ENG-48.

TABLE I			
L/h	$\frac{\Delta L}{h}$	ΔE MeV	θ_{cM} deg
97	10	60	-6
107	6	40	+3

TABLE II				
L	E_{final}	θ_f	t interacting 10^{-22}sec.	Z_{LF}
0	198	180	44.8	34.3
5	fused		>44.5	
10	fused		>44.5	
25	fused		>44.5	
50	fused		>54.5	
60	fused		>44.5	
80	199	70	34.2	34.3
100	200	52	32.0	33.9
125	201	-7	38.7	34.4
165	207	14	21.6	34.6
170	210	28	18.4	34.6
200	284	66	3.2	36.0
250	294	57	0	36.0
∞	311			

References

1. N. Frascaria et al., Z. Physik A294 (1980) 167. J. C. Roynette, et al., Z. Physik A299 (1981) 73.

2. Nguyen Van Gai, Phys. Lett. 105B (1981) 11.

3. H. Flocard and M. S. Weiss, Phys. Lett. 105B (1981) 14.

4. H. Flocard, S. E. Koonin and M. S. Weiss, Phys. Rev. C17 (1978) 1682.

5. N. Frascaria, private communication.

6. J. Dechaye, D. Gogny, G. Grammaticos and L. Sips, preprint and private communication.

7. S. Koonin, B. Flanders, H. Flocard, M. S. Weiss, Phys. Lett. 77B (1978) 13.

8. P. Bonche et al., Phys. Rev. C20 (1979) 641.

9. P. Bonche, B. Flanders, S. Koonin and M. S. Weiss (in preparation).

10. R. Vandenbosch et al., Phys. Rev. C17 (1978) 1672.

THE TDHF EVOLUTION OF COLLECTIVE MOTION IN HEAVY NUCLEI

J. Nemeth

Institute of Theoretical Physics, Eötvös University, Budapest, Hungary

and

S. Faidi and J.M. Irvine

Department of Theoretical Physics, The University, Manchester, M13 9PL, U.K.

INTRODUCTION

It is experimentally observed that excited states of compound nuclei in the rare earth and actinide regions produced in heavy ion reactions are more likely to decay by fission than states at a similar excitation energy produced in light ion reactions. The most probable explanation for this phenomena is that a much greater degree of collectivity of excitation is engendered by the heavy ions.

We have begun a programme to study the evolution of collective motion in heavy nuclei and in particular how this may be transferred to the fission mode within the framework of time-dependent Hartree-Fock theory. Some preliminary results are presented for ^{240}Pu and a 'model' rare earth nucleus ^{168}Gd.

THE TDHF CODE

In order to address the problem of the TDHF evolution of collective motion in heavy nuclei we have written two computer codes. Both contain the restriction of axial symmetry. The first is a static H.F. code with the facility to impose various constraints on the H.F. state via Lagrange multipliers, e.g. we could constrain various moments of the one-body density or a component of the angular momentum perpendicular to the symmetry axis. The second code uses the output from the static code as initial data for a TDHF calculation. The trigger for the time evolution can either be the relaxation of a static constraint or the superposition of a collective velocity field on the static configuration. For example, we could spin the H.F. state and watch its time evolution in the rotating frame, we could start it vibrating in various modes or we could initiate the time evolution with a combination of rotation and vibration.

The basic structure of the two codes is similar with the static code employing the imaginary time ansatz. The numerical techniques we have used are familiar from the works of Davies, Koonin and Feldmeier on axially symmetric rotating frame calculations and employs the Peaceman-Rachford method of approximating unitary operators. We shall, therefore, restrict ourselves to a brief review of the basic TDHF equations.

We approximate the true many-body wavefunction by a Slater determinant at each instant of time

$$\Psi(\underline{x}_1,\underline{x}_2 \ldots \underline{x}_A;t) = (A!)^{-\frac{1}{2}}\det \psi_i(\underline{x}_j,t) \quad (1)$$

where the single-particle states ψ_i are solutions of the TDHF equation

$$i\hbar\, \dot{\psi}_i(\underline{x},t) = \int d\underline{x}'\, h_{HF}(\underline{x},\underline{x}';t)\psi_i(\underline{x}',t) \quad (2)$$

As with most other TDHF calculations we have restricted ourselves to Skyrme interactions plus a direct Yukawa and Coulomb contribution thus reducing h_{HF} to a local operator

$$h^{\alpha}_{HF} = -\frac{\hbar^2}{2m}\nabla^2 + U^{\alpha}_s + U^{\alpha}_y + U^{\alpha}_c \quad (3)$$

where we have used

$$U^{\alpha}_s(x,t) = \tfrac{1}{2}t_o[(1+x_o/2)\rho(x,t) - (\tfrac{1}{2}+x_o)\rho_{\alpha}(x,t)]$$

$$+ \tfrac{1}{4}t_3[\rho^2(x,t) - \rho^2_{\alpha}(x,t)]$$

$$U^{\alpha}_y(\underline{x},t) = V_u\int d\underline{x}'\exp[-|\underline{x}-\underline{x}'|/a][|\underline{x}-\underline{x}'|]^{-1}a\rho(\underline{x},t)$$

$$+ (V_L - V_u)\int d\underline{x}'\exp[-|\underline{x}-\underline{x}'|/a][|\underline{x}-\underline{x}'|]^{-1}a\rho_{\alpha}(\underline{x}',t)$$

$$U^{\alpha}_c(\underline{x},t) = \delta_{\alpha,p}\, e^2\int d\underline{x}'[|\underline{x}-\underline{x}'|]^{-1}\rho_p(\underline{x}',t)\ .$$

Here $\rho(\underline{x},t)$ is the total one-body density

$$\rho(\underline{x},t) = \sum_{\alpha=n,p}\rho_{\alpha}(\underline{x},t) = \sum_{i=1}^{A}|\psi_i(\underline{x},t)|^2\ , \quad (4)$$

where the label α denotes neutrons n or protons p. The force parameters employed were

t_o = -497.66 MeVfm^3	t_3 = 17288 MeVfm^6	x_o = 0.373
a = 2.175 fm	V_L = -55.25 Mev	V_u = -277.5 MeV

With the restriction to axial symmetry and the use of cylindrical polar co-ordinates (r,z,θ) we may write

$$\psi_i(\underline{x},t) = \phi_i(r,z,t)\exp(in_i\theta) \quad (5)$$

where the component $n_i\hbar$ of angular momentum along the symmetry axis is conserved and the states n_i and $-n_i$ are degenerate. Thus in the absence of spin orbit forces the single-particle level $|n_i|$ contains four nucleons $n_i \neq 0$ and two nucleons for $n_i = 0$.

If the single particle wavefunctions (5) are each multiplied by a phase factor $\exp[i\,\chi(\underline{x},t)]$ this corresponds to a collective velocity field

$$V_{col}(x,t) = \hbar\,\nabla\chi(\underline{x},t) \tag{6}$$

We have initiated TDHF evolution by vibrations

$$\chi_{vib} = \alpha\,(z - z_o)^2 \tag{7}$$

and rotations

$$\chi_{rot} = g(r,z,t)\cos\theta \quad . \tag{8}$$

We have considered only symmetric vibrations so that z_o is the centre of mass of the nucleus. In the case of rotations the TDHF equations (2) may be written in the body fixed axially symmetric frame as

$$\begin{aligned} i\hbar\mathring{\phi}_i &= -\hbar^2/_{2m}\left[\frac{\partial^2}{\partial r^2} + \frac{1}{r}\frac{\partial}{\partial r} + \partial^2/_{\partial z^2} - n_i^2/_{r^2}\right]\phi_i \\ &\quad + U_{HF}\,\phi_i + \hbar^2/_{2m}\cdot\tfrac{1}{2}[(\partial g/_{\partial r})^2 + (\partial g/_{\partial z})^2 + (g/_r)^2]\phi_i \\ &\quad - \tfrac{1}{2}\hbar w\{z[(\partial g/_{\partial r}) + g/_r] - r\,\partial g/_{\partial z}\}\phi_i \end{aligned} \tag{9}$$

where the phase function g is the solution of

$$\begin{aligned} &\partial\rho/_{\partial r}\;\partial\tilde{g}/_{\partial r} + \partial\rho/_{\partial z}\;\partial\tilde{g}/_{\partial z} + \rho\{\partial^2\tilde{g}/_{\partial r^2} + \frac{1}{r}\,\partial\tilde{g}/_{\partial r} \\ &+ \frac{\partial^2\tilde{g}}{\partial z^2} - \frac{1}{r^2}\tilde{g}\} = \frac{m}{\hbar}\left\{z\frac{\partial\rho}{\partial r} - r\frac{\partial\rho}{\partial z}\right\} \end{aligned} \tag{10}$$

subject to the boundary conditions

$$\lim_{r\to\infty}\partial\tilde{g}/_{\partial r} = {}^{mz}/_{\hbar}\,, \qquad \lim_{|z|\to\infty}\frac{\partial\tilde{g}}{\partial z} = -{}^{mr}/_{\hbar} \tag{11}$$

where $\tilde{g} = g/w$ and the rotational velocity w is determined by the conserved angular momentum about the y-axis

$$L_y = \hbar\int\rho\{z\left[\frac{\partial g}{\partial r} + g/_r\right] - r\,\partial g/_{\partial z}\}\pi r\,dr\,dz \tag{12}$$

An effective moment of inertia can then be defined by

$$\mathcal{J} = \hbar\int\rho\,\{z[\frac{\partial\tilde{g}}{\partial t} + \tilde{g}/_r] - r\frac{\partial\tilde{g}}{\partial z}\}\,\pi r\,dr\,dz \tag{13}$$

so that the usual collective rotational energy is given by

$$E_{rot} = L_y^2/2\mathcal{J} \tag{14}$$

RESULTS

We first present the results of some preliminary studies on the nucleus ^{240}Pu. The valence neutron shell then consists of the N = 6 states while valence protons occupy states in the N = 5 shell. We are then faced with the choice of a number of static H.F. configurations depending on the occupancies of the valence orbits. We have chosen to consider the three configurations indicated in figure 1: The state S1 is the most stable H.F. configuration of valence protons in N = 5 levels and valence neutrons in N = 6 levels. This H.F. ground state lies within 1% of the experimental binding energy of ^{240}Pu indicating that our choice of force parameters is not unreasonable. The static H.F. configuration is then put in a rotating frame and the state of 35 units of angular momentum is generated. This lies at the exceptionally high excitation energy of 288 MeV. Next we searched on static H.F. configurations to find the state S2 which supports the H.F. Yrast state of 35ħ. S2 lies at an excitation energy of 30 MeV with respect to S1 but the 35ħ state which it supports lies at only 125 MeV excitation. Indeed the 55h state supported by S2 at an excitation energy of 256 MeV is still more bound than the 35ħ state based upon S1. Finally, we have constructed the static H.F. state S3 which maximises the elongation along the symmetry axis, this lies at 52 MeV with respect to S1 and supports a 35ħ level at 148 MeV excitation. We have used our TDHF code to follow the time evolution of the 35ħ states based upon S1, S2 and S3 and the 55ħ state based upon S2.

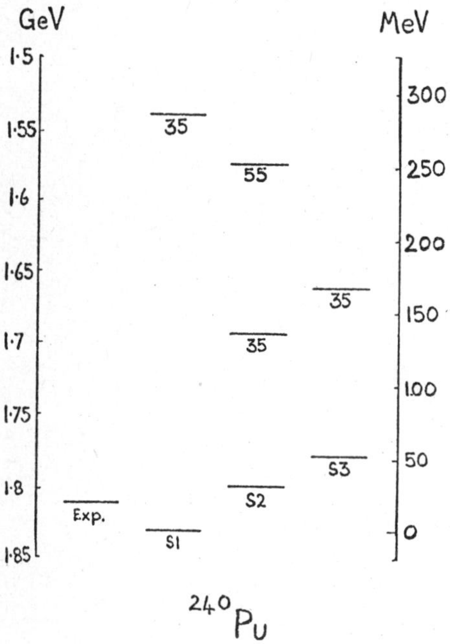

Figure 1

In figure 2 we show the time evolution of the one body density of the 35ħ state based upon S1 projected onto the z-axis, i.e. we plot $\int 2\pi r\rho(r,z,t)dr$. The time steps are in units of 10^{-21}s and the calculation is terminated at 0.5×10^{-21}s due to the appearance at the edge of our mesh, which is 58fm long, of detectable nuclear densities. This is traced to the emission of a few unbound nucleons from the nucleus. In figure 3 we plot the time evolution of $\langle z^2 \rangle$ and of the collective velocity of the right hand half of the one-body density in the z-direction V_c and

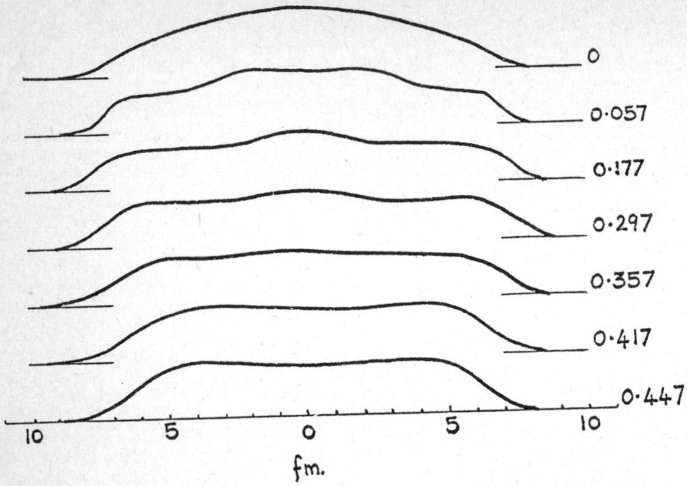

Figure 2

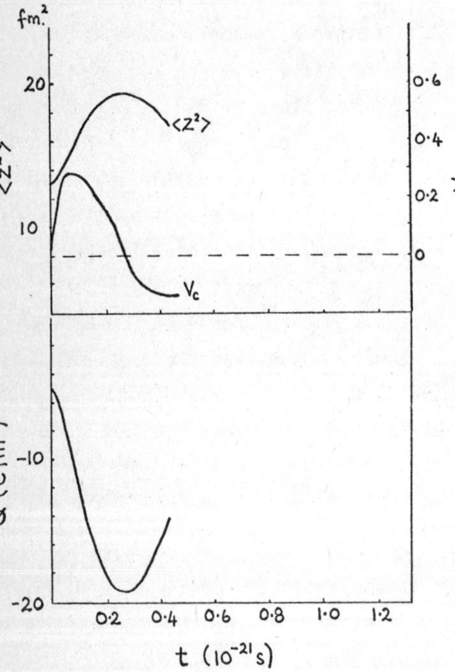

Figure 3

the quadrupole moment Q. We see that initially V_c increases rapidly as the nucleus elongates. V_c reaches a maximum and starts to decrease, it reaches zero when $\langle z^2 \rangle$ and $|Q|$ have reached their maximum values. As V_c becomes negative and the nucleus starts to retract $\langle z^2 \rangle$ and $|Q|$ begin to decrease.

In figure 4 we show the time evolution of the one-body density for the 35ℏ H.F. Yrast state. The structure of this state is completely different from that of the ground state. There is a large compact core to the nucleus with relatively low density highly extended arms which give it a large moment of inertia and hence a low rotational excitation energy. This configuration is so stable that the one-body density shows little change with time. Even at an angular momentum of 55ℏ (figure 5) the same stability is exhibited although the arms are slightly more extended. In figure 6 we show the time evolution of $\langle z^2 \rangle$, V_c and Q for both the 35ℏ (solid lines) and 55ℏ (dashed lines) states. For the 35ℏ state there is a hint of an oscillation in $\langle z^2 \rangle$ and Q, but following the initial stretching the amplitude of the second cycle is very small.

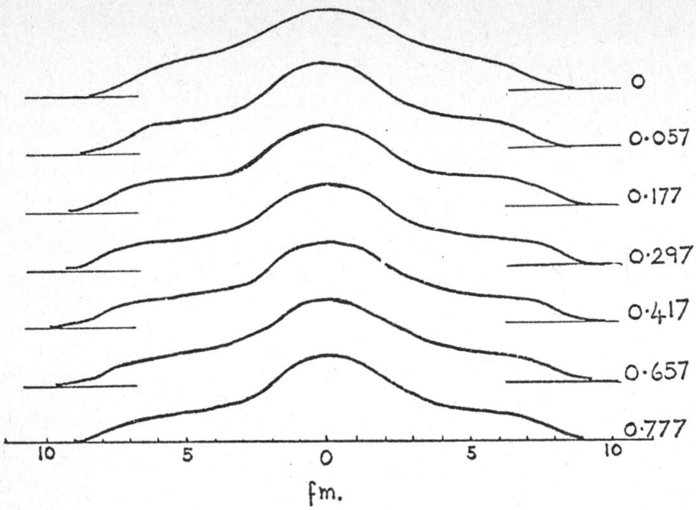

Figure 4

For the 55ℏ state the initial stretching is greater and the amplitude of the second cycle is even smaller. The evolution of the 55ℏ state is terminated at 0.6×10^{-21}s as unbound nucleons are again shaken off and reach the edge of the mesh.

Finally, we consider the 35ℏ state based upon S3, figures 7 and 8. Again we have a different intrinsic structure with a larger quadrupole moment and more of the bulk of the nucleus pushed away from the centre. This system exhibits a well defined highly damped oscillatory behaviour. It appears stable against particle emission but its behaviour is different from the other stable 35ℏ state, i.e. the Yrast level in that for the Yrast level there is little single particle level crossing as the nucleus elongates whereas the 35ℏ state based upon S3 has, after a few oscillations,

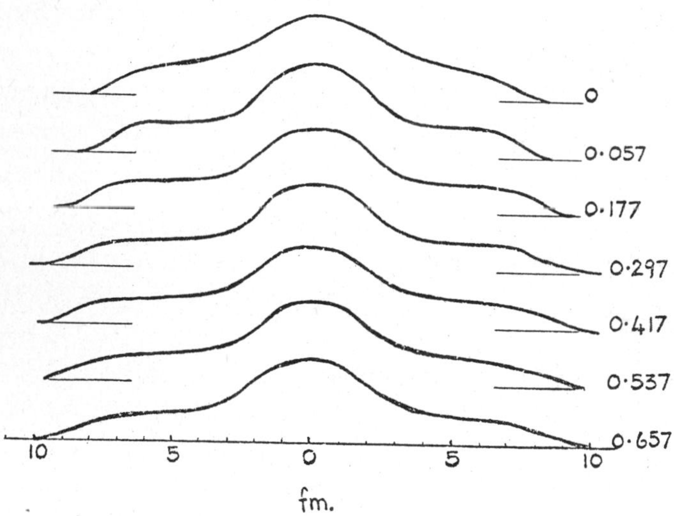

Figure 5

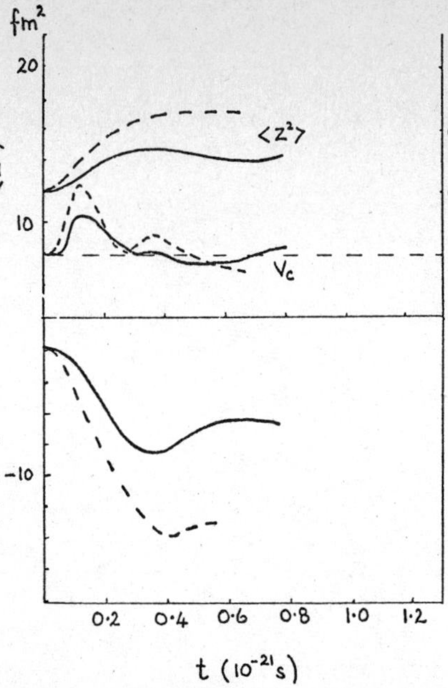

Figure 6

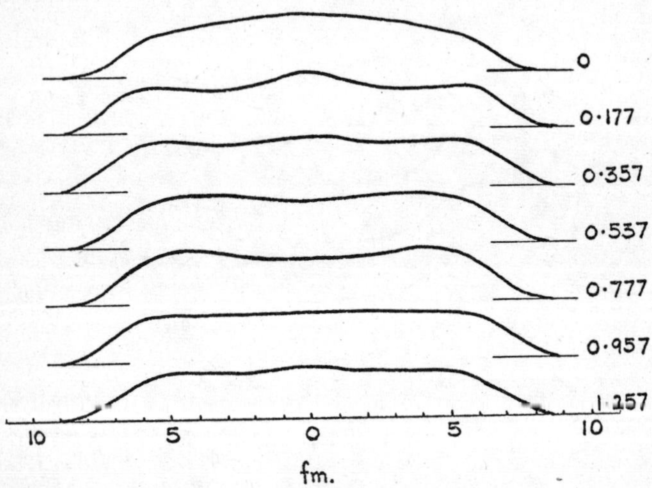

Figure 7

experienced considerable heating in the sense of much crossing of occupied and unoccupied single particle levels.

We next consider a 'model' rare earth nucleus ^{168}Gd in which we assume complete neutron-proton degeneracy and distribute the charge equally amongst all the nucleons. In these studies we initiated the TDHF evolution not only with rotations but also vibrations. Using eqtn.(7) the parameter α is chosen to give a certain amount of initial collective kinetic energy

$$E_{vib} \quad = \quad 4\alpha^2 \langle (z - z_0)^2 \rangle \hbar^2 A/2m \tag{15}$$

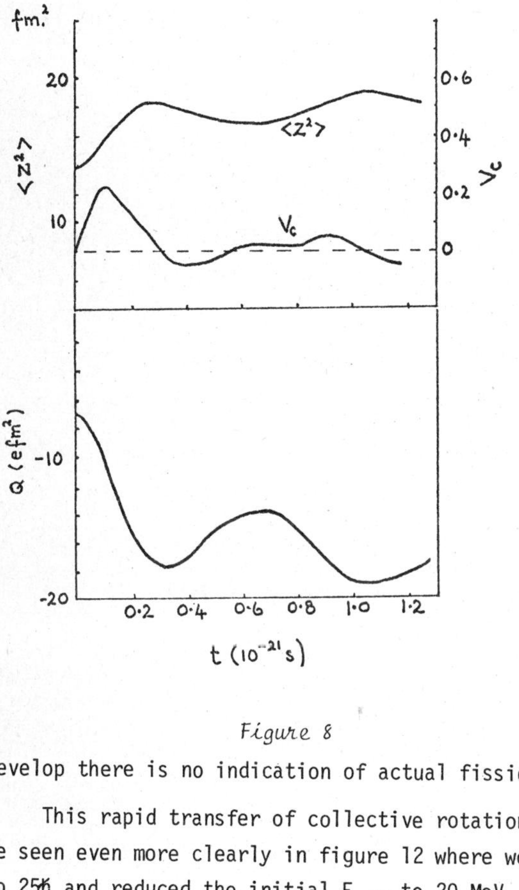

Figure 8

In figure 9 we plot the time evolution of the collective vibrational energy starting with an initial excitation of 162 MeV and in the absence any rotation. As with the rotational excitations we see a rapid damping of the collective motion. In figure 10 we plot the collective energy evolution of an initial configuration of 15ℏ angular momentum (collective rotational energy 98.4 MeV) plus a vibrational excitation of 111.6 MeV. We see that the rotational mode damps first transferring its energy initially into the vibrational mode but that this then is also rapidly damped out. In figure 11 we plot the projected density profiles and while a pronounced dip in the central density does develop there is no indication of actual fission.

This rapid transfer of collective rotational energy into the vibration mode can be seen even more clearly in figure 12 where we have cranked up the angular momentum to 25ℏ and reduced the initial E_{vib} to 20 MeV.

Finally we consider the role of the Coulomb force by doubling the effective charge on all the nucleons and adjusting the Skyrme parameters to give the same total binding energy as before. The rotational energy of the 15ℏ state is raised from 98.4 MeV to 140 MeV and the density profiles are reproduced in figure 13. Again we see the development of a distinctive neck but still no fission.

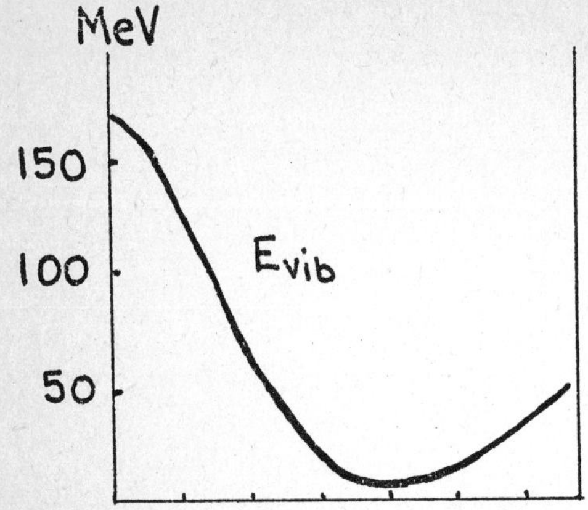

Figure 9

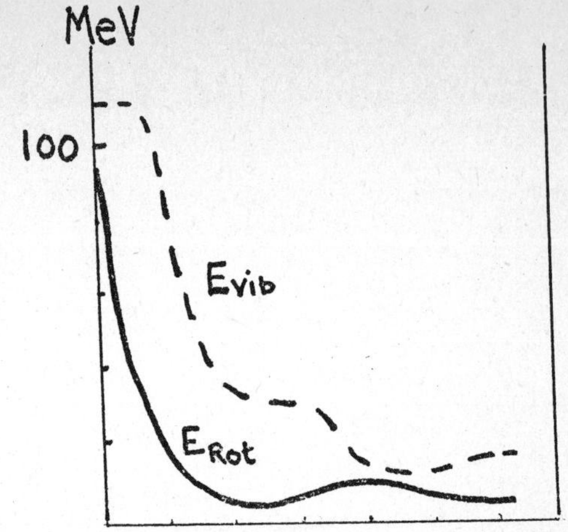

Figure 10

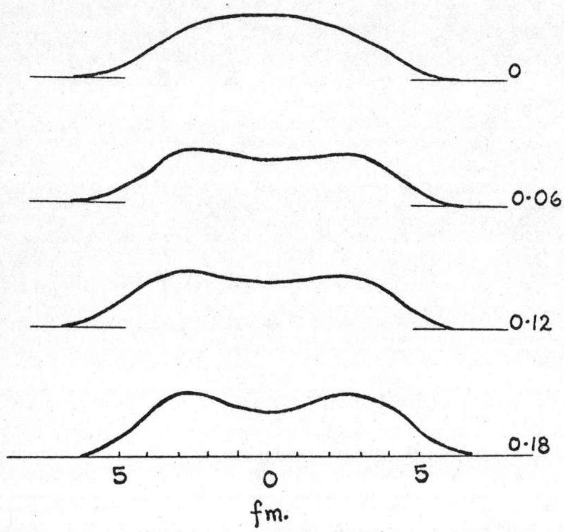

Figure 11

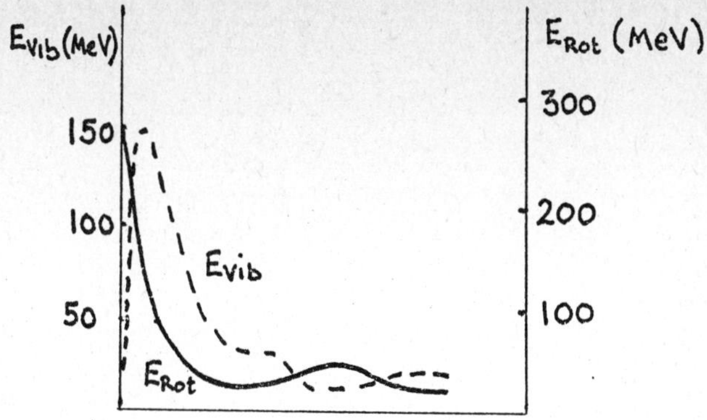

Figure 12

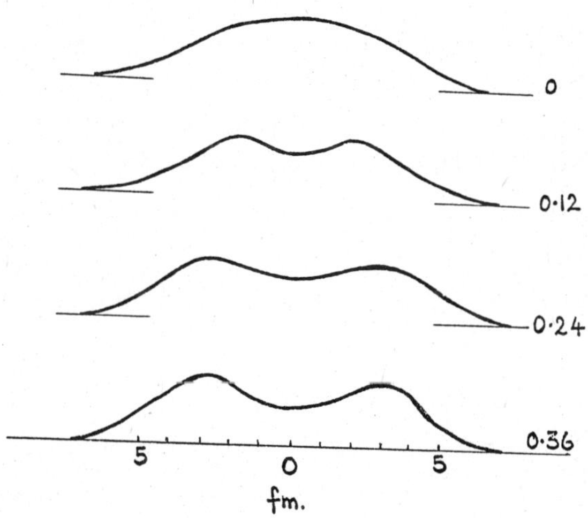

Figure 13

DISCUSSION

We have seen that within a TDHF description of collective nuclear motion any initial collective rotational energy is rapidly converted into stretching motion which can give rise to vibrations but these are always severely damped so that fission does not occur for modest initial excitations. If initially energy is put directly into the vibrational mode this is also severely damped. It is of course possible to initiate the evolution with sufficient collective excitation energy to guarantee fission but in our TDHF calculations at such excitation energies individual nucleons rapidly become unbound and the calculations terminate when these reach the edge of the mesh.

We are continuing these studies in two directions: We have seen that the vibrational mode most likely to lead to fission is heavily damped and the nucleus 'heats up' in the sense of there being many single particle level crossings. This one-body dissipation by the mean field can be avoided by going beyond TDHF to include collision terms which avoid the level crossings and should hence keep the nucleus 'cool' and its excitation energies in collective modes. Secondly, all the results quoted here are for collective excitations of H.F. configurations close to the ground state. We now intend to carry out a series of constrained static H.F. calculations for a range of deformations and cranked angular momenta in order to plot out the shape of the fission barrier. This will allow us to initiate the TDHF evolution close to the saddle point and hence hopefully to study the route to scission.

We are grateful to Dr. I.S. Grant for helpful discussions on the experimental background to heavy ion induced fission. One of us (J.N.) would like to express her appreciation to Dr. Nörenberg for his hospitality at G.S.I.

STUDY OF THE ^{16}O + ^{16}O REACTION IN THE TIME-DEPENDENT HARTREE - FOCK - BOGOLYUBOV APPROXIMATION

S. Drożdż, J.Okołowicz and M.Ploszajczak
Institute of Nuclear Physics, PL - 31-342 Kraków, Poland

Microscopic description of the dynamics of a large amplitude collective motion arising in the heavy-ion (HI) collision or in the fission process has been extensively investigated during recent years. It is generally believed that the time-dependent mean field approach provides a natural and accurate method for treating the dynamics of many-nucleon system. Extensive calculations performed in this framework achieved considerable successes in describing inclusive properties of the HI reaction [1,2]. Probably the largest achievement of the time-dependent Hartree - Fock (TDHF) calculations with the density dependent Skyrme interaction was the reproduction of experimental fusion excitation functions for different reactions. This feature suggests that the one-body dissipation mechanism contained in TDHF approach is dominating and, consequently, the TDHF theory with appropriately defined effective force provides a good basis for systematic refinements of the description of nuclear dynamics by including the two-body collision term [3] and (or) the residual nucleon-nucleon interaction. It is indeed an undesired feature of TDHF evolution that the nuclear system even moving infinitely slowly cannot end up in its ground state due to the absence of residual nucleon-nucleon interaction which can cause transitions between single particle (s.p.) levels at the frequent crossings in the average potential well evolving in time. This residual interaction causes a two-body nuclear dissipation and leads for example to the frequent pair scattering between different conjugate s.p. orbits [4,5]. It is not clear to what extend one needs the two-body dissipation to describe various inclusive reaction properties. However, one expects that a low angular momentum cutoff which limits the complete - fusion cross section at high energy and is predicted by TDHF calculations [1] can depend sensitively on details of the dissipation mechanism. In particular, the precise value of the kinetic energy of ions above which the nonfusion events at low-L appears can be an accurate measure of the nuclear dissipation. In case of ^{16}O + ^{16}O reaction the TDHF calculations using the Skyrme force supplemented by contributions from the direct parts of Yukawa and Coulomb interactions predict the existence of a dynamical low-L limit to fusion for $\varepsilon_{CM}^{(f)}=E_{CM}/A \gtrsim 0.84$MeV [2]. Slightly larger value of $\varepsilon_{CM}^{(f)}$ ($\varepsilon_{CM}^{(f)} \gtrsim 1$ MeV) can be obtained if the density dependence of an effective mass is taken into account [8]. However, numerous experimental investigations attempting to isolate and detect nonfusion events do not confirm the existence of a fusion L-window in the predicted range of kinetic energies and put the lowest limit for $\varepsilon_{CM}^{(f)}$ equal 1.06 MeV or higher [6]. Also the studies of the shapes of angular distributions for the reaction $^{12}C(^{16}O, \ )^{24}Mg$ (in the ground state) do not show a clear evidence for the low-L limit to fusion until $\varepsilon_{CM}^{(f)} \simeq$ 1.53 MeV [7]. Thus, it seems one has to go beyond a simple one-body dissipation mechanism contained in TDHF approximation to describe the nuclear dissipation accurately. A straightforward possibility is provided by the time-dependent Hartree - Fock - Bogolyubov (TDHFB) approximation which allows to investigate this part of a two-body dissipation which is contained in the average quasi-particle field. To solve TDHFB eqs. in its full glory is a formidable task even in the static case. Thus, one has to impose several restrictions upon properties of the residual pair field Here, we approximate the complicated nucleon-nucleon residual interaction by the state-independent monopole pairing interaction. This simple form of the pairing force gives a reasonable account of various static nuclear properties. There is however a considerable uncertainty in choosing values of the strength parameter G for various configurations of the dinucleus. Since the average value of G is inversely proportional to the number of particles then it can change by factor ~2 going from the initial state of two separated ^{16}O ions to the dinucleus ^{32}S close to the ground state. Corresponding variations of the average value of the gap parameter Δ are smaller since $\Delta \sim A^{-1/2}$ [9]. Moreover, even if all pairing matrix elements $G_{\nu\nu'}$ for various s.p. configurations remain unchanged during the collision, then the appropriately defined average pairing matrix element $G(t) = \sum_{\nu\nu'} G_{\nu\nu'} U_\nu(t) V_\nu(t) U_{\nu'}(t) V_{\nu'}(t) / (\sum_\nu U_\nu(t) V_\nu(t))^2$ depends on time due to the explicit time-dependence of occupation amplitudes U_ν, V_ν. Thus, we prefer to solve TDHFB eqs. for the monopole pairing field with a constant value of the deformation Δ in the gauge space. In principle, the pair field deformation Δ may depend on the instantenuous properties of the mean field which are expressed in terms of various multipole deformations $\beta_{\mu\nu}(t)$. This desired dependence is not given by the time-dependent variational principle which governs the evolution of the system but assumes no renormalization of the effective force during the evolution. Thus, one would need in addition an appropriate self-consistency condition to recover from the time-dependent average field the field producing components of the residual interaction. The simplest candidate for such a condition, which was shown also to be extremly accurate in the static calculations, is the canservation of the average density of particles in the nucleus [10]. Here, however, we neglect the apparent dependence $\Delta(\beta_{\mu\nu}(t))$ and solve TDHFB eqs. for a constant value of Δ [11].

TDHFB eqs. can be derived by varying the action integral:

$$\delta \int_{t_1}^{t_2} \langle \Phi | H - i\hbar \frac{\partial}{\partial t} | \Phi \rangle dt = 0 \tag{1}$$

where H is a nuclear many-body Hamiltonian here given by the sum of a Skyrme Hamiltonian and a monopole pairing field $-\sum_{\tau} \Delta_{\tau} \sum_{\nu > 0} (c_{\nu}^{+} c_{\tilde{\nu}}^{+} + c_{\tilde{\nu}} c_{\nu})$. For a central collision the most general Bogoliubov trial wave function $|\Phi\rangle = \prod_{k>0} (U_k + V_k a_k^+ a_{\tilde{k}}^+) |0\rangle$, where a_k^+, $a_{\tilde{k}}^+$ are creation operators for conjugate canonical s.p. states k, $\tilde{k}$. In this canonical representation the density matrix $\hat{\rho}_{kk'}$ and the pairing tensor $\hat{\varkappa}_{kk'}$ take the form:

$$\rho_k \equiv \hat{\rho}_{kk} = \rho_{\tilde{k}\tilde{k}} = |V_k|^2$$

$$\varkappa_k \equiv \hat{\varkappa}_{k\tilde{k}} = U_{\tilde{k}} V_k = - \hat{\varkappa}_{\tilde{k}k} \qquad (V_{\tilde{k}} = -V_k \, ; U_{\tilde{k}} = U_k) \tag{2}$$

where amplitudes U_k, V_k satisfy the relation $|V_k|^2 + U_k^2 = 1$. Therefore, as independent variational quantities in eq.(1) one can choose the occupation amplitudes V_k (V_k^*) of conjugate canonical states $|k\rangle, |\tilde{k}\rangle$ and the amplitudes of canonical s.p. wave functions $\psi_k(\vec{r}) = \langle \vec{r} | a_k^+ | 0 \rangle$, $\psi_{\tilde{k}}(\vec{r}) = \langle \vec{r} | a_k^+ | 0 \rangle$ ($\psi_k^*(\vec{r})$, $\psi_{\tilde{k}}^*(\vec{r})$). The resulting equations become [4,5]:

$$i\hbar \dot{\psi}_l = (\hat{\mathcal{U}}(\rho) - \varepsilon_l) \psi_l$$

$$i\hbar \dot{\rho}_l = \Delta (\varkappa_l - \varkappa_l^*) \tag{3}$$

$$i\hbar \dot{\varkappa}_l = 2 \varkappa_l (\varepsilon_l - \lambda) + \Delta (2\rho_l - 1)$$

where $\hat{U}(\rho)$ is the one-body Hamiltonian governing the changes of s.p. states ψ_l and ε_l is a s.p. energy. More about properties of those eqs. can be found in refs. 4,5.

TDHFB calculations presented here are performed for a constant gap parameter (Δ = 3 MeV) which for ^{16}O corresponds ot an average pairing gap $\tilde{\Delta} = 12/A^{1/2}$ MeV [9]. Parameters of the Skyrme force Hamiltonian are same as used in ref. 8 (Skyrme II parameters) where largest one-body dissipation in ^{16}O - ^{16}O dinuclear system was found. S.p. wave functions $\psi_k(\vec{r},t)$ (eq.(3)) are expanded in a static, two-center basis [12] which is obtained as a product of one-dimensional basis $\varphi_n^{(z_0)} = \sqrt{G_{z_0}(z)}\, P_n^{(z_0)}(z)$ in the z-direction with the set of eigenfunctions of the cylindric oscillator potential in x - y plane. $P_n^{(z_0)}(z)$ form a set of orthogonal polynomials for the two-center weight function:

$$G_{z_o}(z) = \left[(0.5 + z_o^2)/(1 + z_o^2)\right]\left\{ \exp\left[-(z-z_o)^2\right] + \exp\left[-(z+z_o)^2\right]\right\}^{1/2} \quad (4)$$

,where $2z_o$ is the separation distance between centers of the basis. In this way the equation of motion for ψ_k (ψ_k^*) is reduced to a system of differential equations for the expansion coefficients. Those eqs. can be solved using the Crank - Nicholson method as described in ref.13. It was found by Flocard [12] that in the static constrained HF calculations for $^{16}O + {}^{16}O$ the total energy and the quadrupole moment of the dinucleus converge fast with increasing shell number for each separation distance between oxygen ions and reach their asymptotic values already for small N. In our calculations values of both the separation distance between centers of the basis $2z_o$ and the deformation $q \equiv \omega_\perp/\omega_z$ of the basis are fixed and have to be chosen to yield an optimal description of the dy-

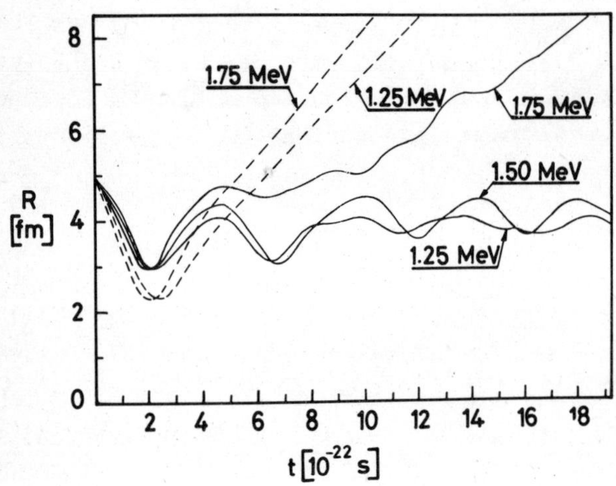

Fig.1 The fragment separation coordinate (2R) is shown as a function of time for the $^{16}O + {}^{16}O$ reaction at E_{CM}/A = 1.25, 1.50 and 1.75 MeV. The solid lines depict the TDHFB trajectory whereas the dashed lines show R(t) for the TDHF evolution.

namics of the $^{16}O + {}^{16}O$ collision as well as static properties of ^{16}O. For $2z_o$ =5 fm, q=4 and N=13 we achieve an excellent agreement with results of TDHF calculations of ref. 8 where the evolution of $^{16}O - {}^{16}O$ system was studied on the discrete spatial mesh (compare the curve drawn with a dashed line in fig.1 for E_{CM}/A = 1.25 MeV (E_{LAB} = 80 MeV) with the corresponding curve in fig.2 of ref.8). Thus, we use this basis for

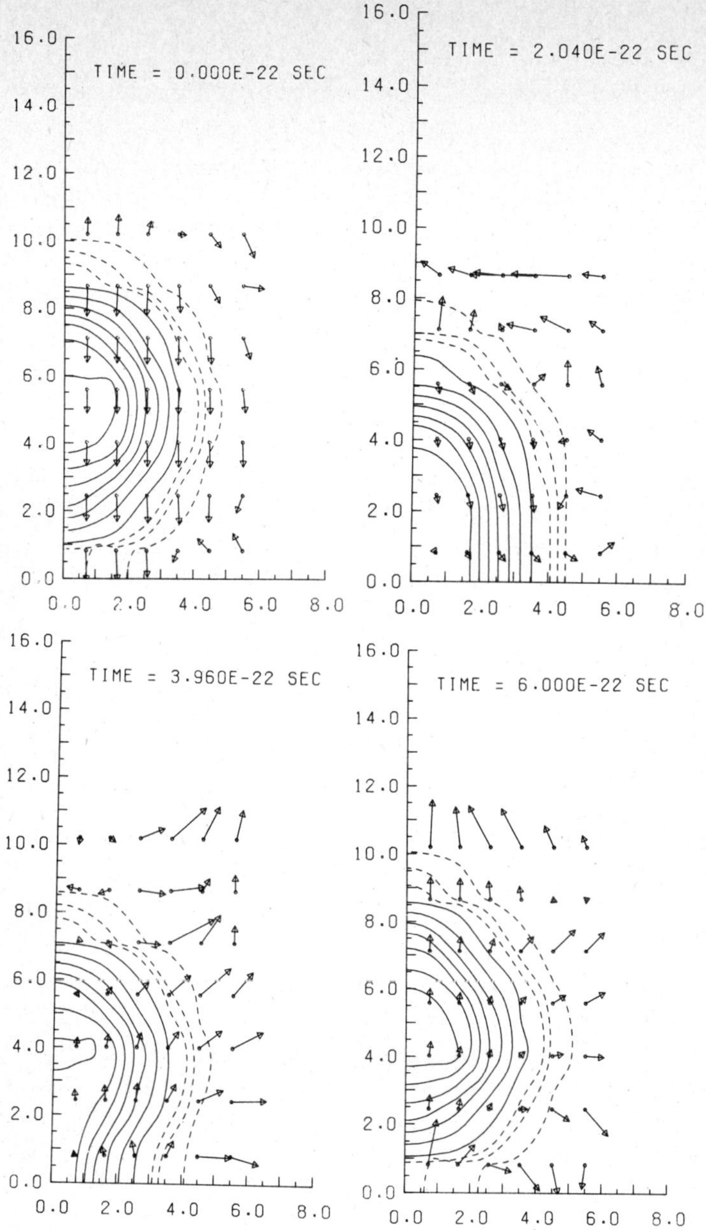

Fig.2 The TDHF density contours $\varrho(z,r)$ and the local velocities of the nuclear matter $\vec{v}(z,r) \equiv \vec{j}(z,r)/\varrho(z,r)$ ($\hbar/m=1$) for the symmetric half of the dinucleus $^{16}O - {}^{16}O$. The initial kinetic energy of ions is $E_{CM}/A=1.25$ MeV. The equidensity lines are drawn in step of $\Delta\varrho=0.02$ nucl/fm^3 starting with the minimal value 0.02 nucl/fm^3. The dashed lines denote equidensity contours plotted in step of $\Delta\varrho=0.004$ nucl/fm^3 beginning with 0.002 nucl/fm^3. The arrows show velocities

of various fragments of a nucleus. Direction of an arrow points to the direction in which the fragment is moving with velocity $|\vec{v}|$. Length of the arrow depict value of the velocity $3|\vec{v}|$ in units $\hbar/m = 1$. Thus, the vector of a length 1 fm represents $|\vec{v}| = 1/3\ fm^{-1}$.

solving TDHFB as well as TDHF equations.

Variations of the fragment separation coordinate $2R = 2A^{-1}\int d^3r|z|\rho(\vec{r})$ in time is displayed in fig.1 for various kinetic energies per particle in the center of mass. The solid lines denote the TDHFB evolutions whereas the dashed lines present results for the TDHF evolution. The central collision of ^{16}O-ions at $E_{CM}/A = 1$ MeV described in the TDHF limit leads to the fusion (see fig.2 of ref.8) whereas at $E_{CM}/A = 1.25$ MeV the dinuclear system splits into fragments. Additional dissipation induced by pairing correlations is sufficiently large to fuse the two ^{16}O-ions up to $E_{CM}/A \lesssim 1.75$ MeV. For $E_{CM}/A \gtrsim 1.75$ MeV the two ions fly apart after a few giant shape oscillations ($t = 2\text{-}14.5\cdot 10^{-22}$s). It is instructive to compare various features of the collective motion described by TDHFB approximation at $E_{CM}/A = 1.25$, 1.50 and 1.75 MeV. At $E_{CM}/A = 1.25$ MeV and for $t = 2\text{-}6.5\cdot 10^{-22}$s the separation coordinate R exhibits half of the oscillation period. For $t > 6.5\cdot 10^{-22}$s the collective shape vibrations are strongly damped. One sees only the small irregular variations of R around $R_0 \simeq$ 3.9 fm with amplitude $\sim$0.15 fm. An increase of the kinetic energy from 1.25 to 1.50 MeV per nucleon in the center of mass leads to the appearance of slowly damped oscillations of R with an amplitude $\sim$0.4 fm. These oscillations are centered at around $R_0 = 4.1$ fm and R_0 increases slowly in time. This characteristic increase is strongest for the nonfusion TDHFB evolution at $E_{CM}/A = 1.75$ MeV. A more detailed picture of TDHF and TDHFB evolutions can be seen in figs.2 and 3 respectively. Fig.2 shows the TDHF density contours $\rho(z,r)$ as well as the velocities $\vec{v}(z,r)=\vec{j}(z,r)/\rho(z,r)$ ($\hbar/m = 1$) of different parts of a dinucleus for various instants of time. At t = 0.0 the ^{16}O-fragments are connected with a thin neck having density $\rho \lesssim 0.006$ nucl/fm^3. After forming a dinucleus ^{32}S ($t=2.04\cdot 10^{-22}$s) the system splits again. (Due to the initial condition the collision process is symmetric with respect to z=0 and, therefore, it is sufficient to plot a quarter of the dinuclear system.) At $t = 3.96\cdot 10^{-22}$s one sees first the formation of a neck ($\rho \lesssim 0.12$ nucl/fm^3) and then at $t = 6\cdot 10^{-22}$s the two ^{16}O-like fragments. In the early stage of a fission process one can notice the appearance of fast fragments at low densities which try to escape in the direction perpendicular to the symmetry axis of a dinucleus. At the later stage of this process ($t \gtrsim 6\cdot 10^{-22}$s) more such fragments are seen along the z-axis. The TDHFB evolution (fig.3) at $E_{CM}/A = 1.25$ MeV leads to fusion (see fig.1). It is interesting to notice

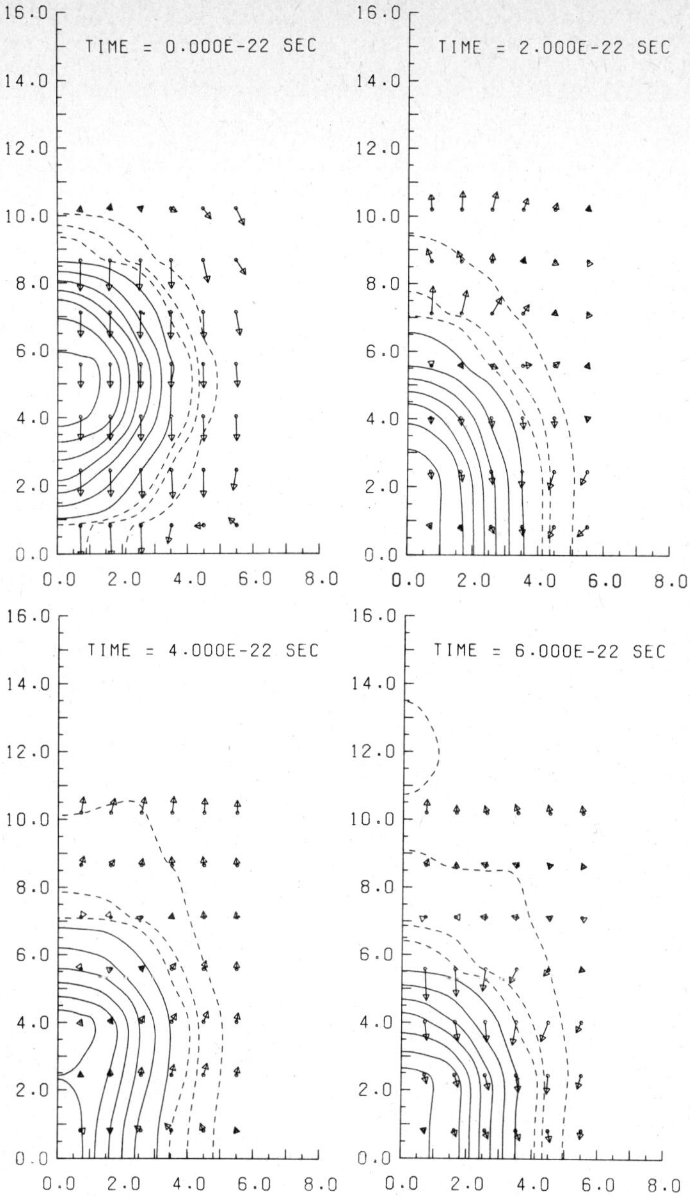

Fig.3 The TDHFB (Δ = 3 MeV) density contours $\rho(z,r)$ and the local velocities of the nuclear matter $\vec{v}(z,r) \equiv \vec{j}(z,r) / \rho(z,r)$ ($\hbar/m=1$) for the symmetric half of the dinucleus ^{16}O - ^{16}O. The initial kinetic energy of ions is $E_{CM}/A=1.25$ MeV. For more details see the caption to fig.2.

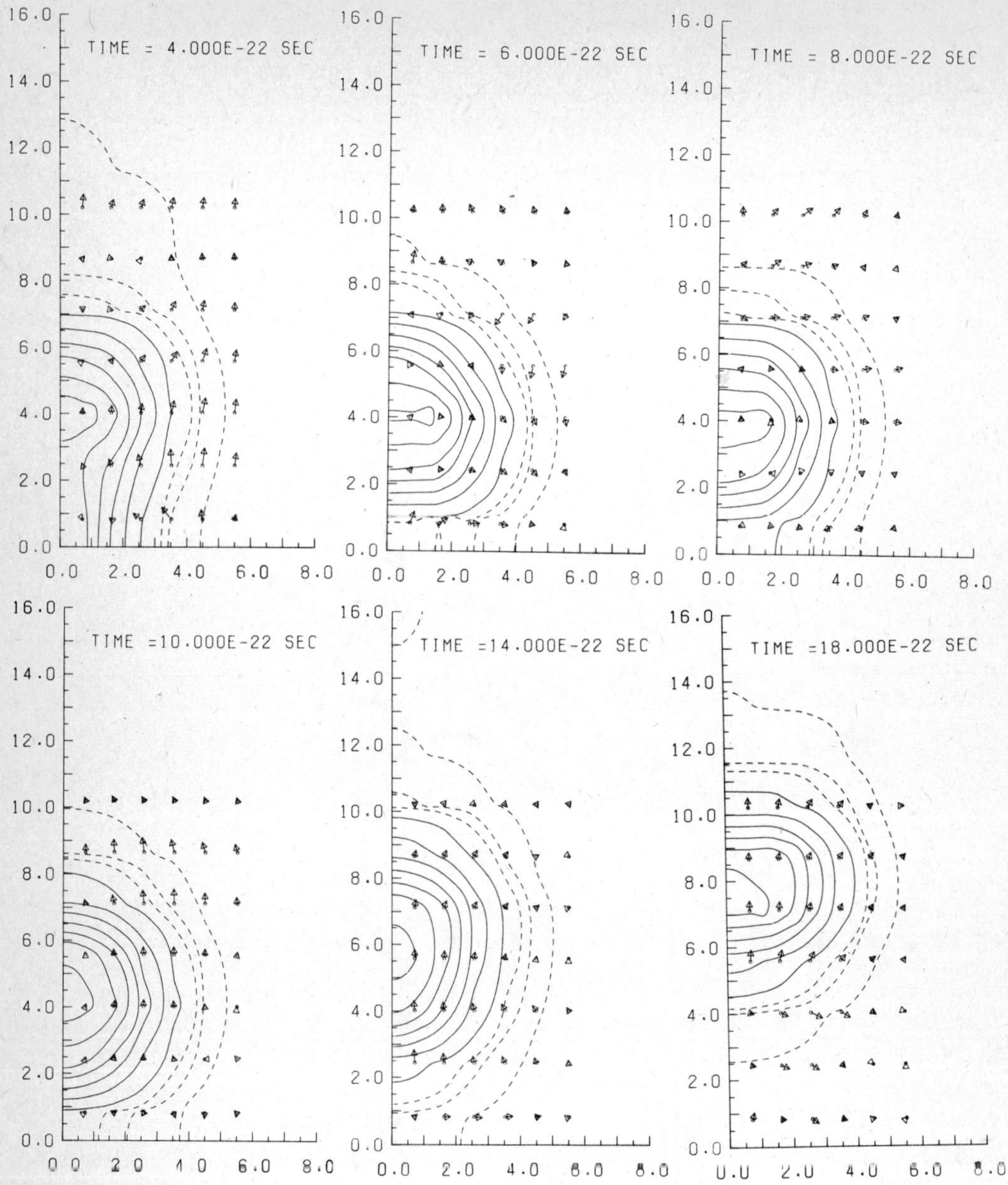

Fig.4 The TDHFB (Δ = 3 MeV) density contours $\rho(z,r)$ and the local velocities of the nuclear matter $\vec{v}(z,r)\equiv\vec{j}(z,r)/\rho(z,r)$ ($\hbar/m=1$) for the symmetric half of the dinucleus $^{16}O - {}^{16}O$. The initial kinetic energy of ions is E_{CM}/A=1.75 MeV. For more details see the caption to fig.2.

both a complete absence of promptly emitted particles (PEP's) [14] which are preemptive to the fast fission [15] and a very large spatial extension of the fused system particularly at low nucleon densities. This latter property results from the inclusion of residual pairing correlations and leads to the large increase of nuclear multipole moments $Q_{\lambda\mu} \sim \int r^2 Y_{\lambda\mu} \rho(r) d\tau$ of the system [11]. The TDHFB evolution at $E_{CM}/A = 1.75$ MeV does not lead to fusion but instead the dinucleus splits gradually into fragments (see the corresponding trajectory R(t) in fig.1) exhibiting the large amplitude shape oscillations. Fig.4 allows to analyse the fission process in details. At $t = 4\cdot 10^{-22}$s the dinuclear system elongates in the z- direction ($\rho \lesssim 0.12$ nucl/fm^3) and its total quadrupole moment becomes even larger than at t = 0.0 when the separation distance of the ^{16}O- ions equals 10fm.

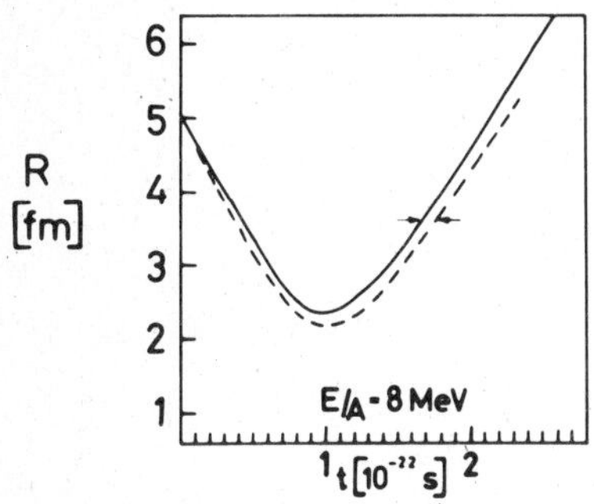

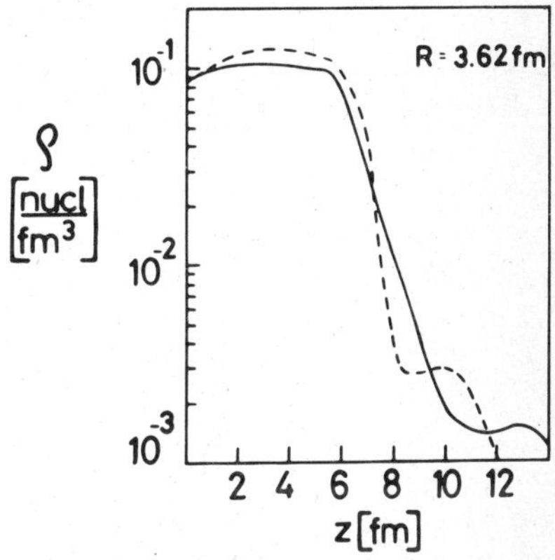

Fig.5 Fragment separation coordinate R is pltted as a function of time for the ^{16}O + ^{16}O collision at $E_{CM}/A = 8$ MeV. The solid line denotes results for TDHFB evolution at Δ=3 MeV whereas the dashed line corresponds to the TDHF evolution. Arrows denote the value of a separation coordinate of oxygen ions for which the density profiles $\rho(z;x=y=0)$ are shown for TDHF and TDHFB evolutions respectively.

This is partially due to the unusually large spreading of a nucleon density along the symmetry axis of the system. In the time period $t = 4\text{-}14\cdot 10^{-22}$s the dinucleus exhibits shape oscillations which at t = 6, 10 and $14\cdot 10^{-22}$s lead to the formation of a molecular configuration with two ^{16}O-ions ($\rho \lesssim 0.14$ nucl/fm^3) sticking to each other and connected with a thin neck ($\rho \lesssim 0.01$ nucl/fm^3).

The role of short range correlations contained in the mean quasi-particle field is expected to decrease while increasing the excitation energy of the system. Fig.5 shows the fragment separation coordinate R as a function of time as well as the nucleon density at various coordinates

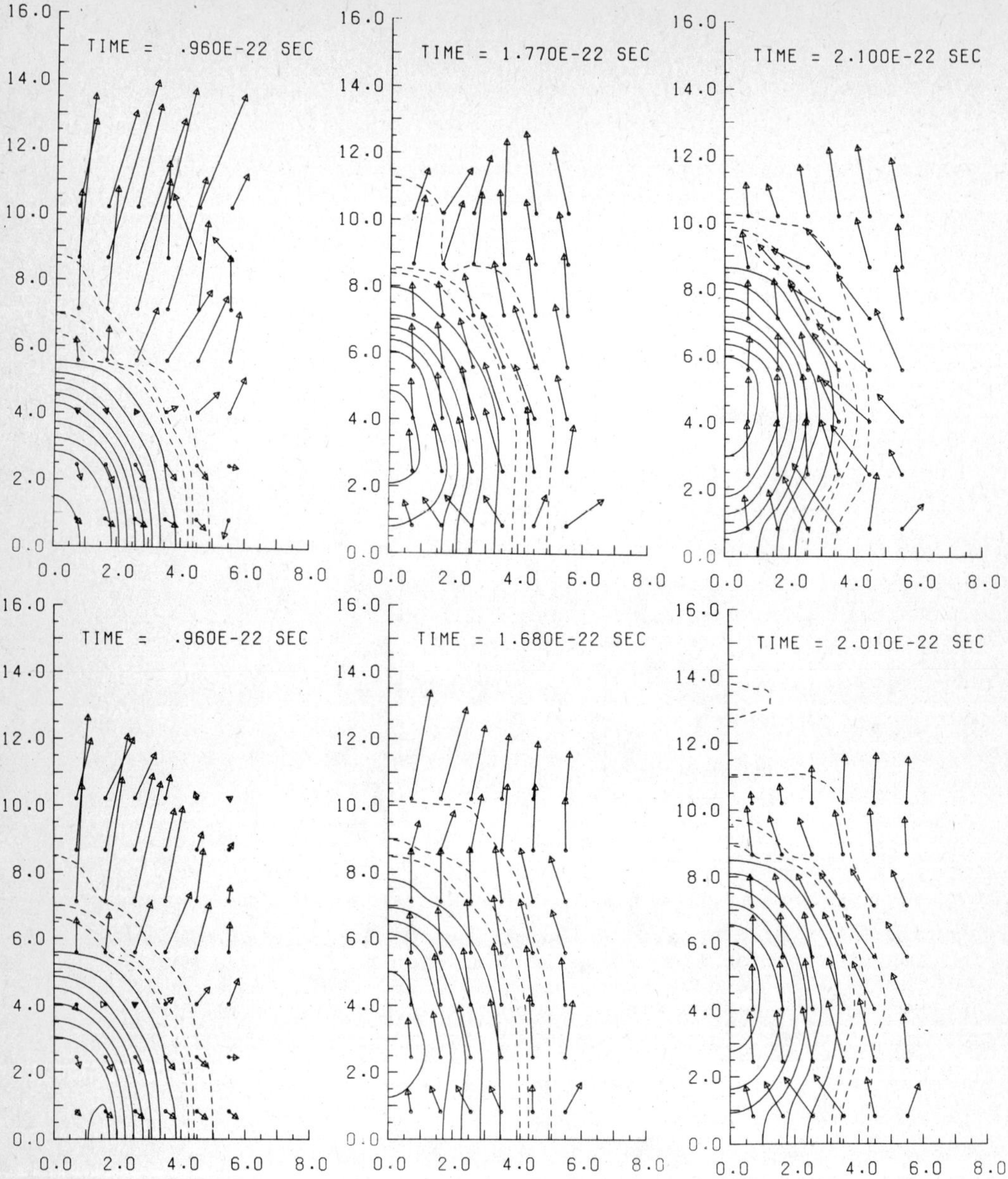

Fig.6 Density contours $\varrho(z,r)$ and local velocities of the nuclear matter $\vec{v}(z,r) \equiv \vec{j}(z,r)/\varrho(z,r)$ ($\hbar/m = 1$) for a symmetric half of the dinucleus $^{16}O - ^{16}O$. The initial kinetic energy of ions is $E_{CM}/A = 8$ MeV. Plots in the upper part of a figure present results for the TDHF evolution at t = 0.96, 1.77 and $2.1 \cdot 10^{-22}$ s whereas the TDHFB evolution is shown in the lower part of this figure. Functions $\varrho(z,r)$ and $\vec{v}(z,r)$ are plotted for instants of time which are different

in case of TDHF and TDHFB evolutions. However, in this way those distributions can be compared at approximately the same fragment separation coordinate.

z (x=y=0.0) for the separation distance between centers of fragments equal 2R = 7.24 fm. Results plotted with the dashed line correspond to the TDHF evolution whereas the solid line exhibits the TDHFB evolution. Comparing the curves R(t) at E_{CM}/A = 8 MeV with those at E_{CM}/A = 1.75 MeV (fig. 1) one concludes that apparently both evolutions become more similar at high energy. This observation is also confirmed by comparing the amount of dissipated kinetic energy of the relative motion. Whereas at E_{CM}/A = 1.75 MeV the initial energy is dissipated in 87% and 98% for TDHF and TDHFB approximations respectively, at E_{CM}/A = 8 MeV the dissipation amounts to ~50% of the initial kinetic energy for both approximations. On the right hand side of fig.5 values of the nuclear density along the z- axis (x=y=0.0) are plotted for the paired and unpaired systems respectively. The picture presents half of the density profile $\rho(z)$ which is symmetric with respect to the reflection in the x - y plane at z = 0. TDHFB and TDHF nucleon densities in fig.5 are shown here for the same fragment separation coordinate 2R. This allows to compare properties of both systems at the similar stage of the reaction. For the unpaired system the nucleon density in the middle of each fragment is approximately 20% larger than in the paired system. On the contrary, the spreading of the nucleon density in the z- direction and consequently, a value of the quadrupole moment is larger for the paired system. The density contours $\rho(z,r)$ and the velocity fields $\vec{v}(z,r)$ for TDHF and TDHFB evolutions at E_{CM}/A = 8 MeV are compared in fig.6 for instants of time which differ in both evolutions but nevertheless, they correspond to the same fragment separation distance (see fig. 5). The dynamics of the TDHF and TDHFB collisions at E_{CM}/A = 8 MeV show remarkable similarities both in the nucleon density distributions $\rho(z,r)$ and in velocity fields $\vec{v}(z,r)$. The only significant discrepancies can be seen in the region of low nucleon densities where PEP's appear but those differences as well as the existence of nucleonic jets depend rather sensitively on the properties of a s.p. basis and particularly on its size. It has been shown in ref. 15 that an increase of the number of basis vectors leads to the decrease of density in the quasi-free peak and in some cases it removes this peak completely. It seems that if at all the formation mechanism of PEP is present in central TDHF collision then the number of such nucleons divided by the total number of nucleons is less than $\Delta A/A = 2.5\cdot10^{-4}$. The inclusion of pair residual interaction between nucleons does not modify this estimate significantly. Apparently, the coupling of relative motion with the Fermi motion in HI collision is extremly weak at least for central collisions of symmetric ions. Thus, it is questionable whether the TDHF or TDHFB theory can be used to explain energetic nucleons observed in the single nucleon spectra [16].

In conclusion, the two-body dissipation mechanism which is contained in the time-dependent average quasi-particle field seems to be significant for explaining the energy loss in HI collisions at small impact parameters. For the ^{16}O - ^{16}O system and for a reasonable value of the Δ-parameter in the pairing field the two-body dissipation mechanism amounts to more than 40% of the total energy dissipated in the central collision. Consequently, the limiting kinetic energy $\varepsilon_{CM}^{(f)}$ for nonfusion at low L is shifted up significantly ($\varepsilon_{CM}^{(f)} \gtrsim 1.7$ MeV) as compared with values obtained in the TDHF approximation [8]. It is expected that also in the peripheral collisions the two-body dissipation mechanism plays an important role increasing a value of the maximum fusion angular momentum $L_>$ and, consequently, the fusion cross section $\sigma_{fus} \sim (L_>+1)^2 - (L_<+1)^2$. This should happen mainly at low bombarding energies when the collision time is long enough to permit frequent scattering of nucleons from s.p. levels of each fragment separately to the s.p. levels of a dinucleus. It is plausible that the absence of dynamical pair correlations in TDHF approximation is responsible for systematically the too low values of calculated σ_{fus} as compared with experimental data [2] for low bombarding energies.

Acknowledgements

Authors wish to thank Dr. J. Błocki for many helpful and stimulating discussions.

References

1. P. Bonche, S.E. Koonin and J. Negele, Phys. Rev. C13 (1976) 1226
S.E. Koonin, K.T.R- Davies, V. Maruhn - Rezwani, H. Feldmeier, S.J. Krieger and J.W. Negele, Phys. Rev. C15 (1977) 1359
H. Flocard, S.E. Koonin and M. Weiss, Phys. Rev. C17 (1978) 1682
2. P. Bonche, B. Grammaticos and S.E. Koonin, Phys. Rev. C17 (1978) 1700
3. P. Grange, J. Richert, G. Wolschin and H. Weidenmüller, Nucl. Phys. A356 (1981) 260
H.S. Köhler, Nucl. Phys. A343 (1980) 315
4. S.E. Koonin and J.R. Nix, Phys. Rev. C13 (1976) 209
5. J. Błocki and H. Flocard, Nucl. Phys. A273 (1976) 45
6. A. Lazzarini, H. Doubre, K.T. Lesko, V. Metag, A. Seamster, R. Vandenbosch and W. Merryfield, Phys. Rev. C24 (1981) 309
S. Kox, A.J. Cole and R. Ost, Phys. Rev. Lett. 44 (1980) 1204
7. A. Szanto de Toledo, T.M. Cormier, M. Herman, B. Lin, P.M. Stwertka, M.M. Coimbra and N. Carlin Filho, Phys. Rev. Lett. 47 (1981) 1881

8. A.K. Dhar and B.S. Nilsson, Nucl. Phys. A315 (1979) 445
9. A. Bohr and B.R. Mottelson, Nuclear Structure, Vol. I, p. 170, ed. W.A. Benjamin, Inc. (1969), New York, Amsterdam
10. Z. Bochnacki, Phys. Lett. 31B (1970) 187
E.B. Balbutsev and Z. Bochnacki, Acta Phys. Pol. B5 (1974) 397
Z. Bochnacki, E.B.Balbutsev and M. Ploszajczak, Acta Phys. Pol. B7 (1976) 649
E.B. Balbutsev and M.Ploszajczak, Acta Phys. Pol. B6 (1975) 761
11. S. Drożdż, J. Okołowicz and M. Ploszajczak, to be published
12. H. Flocard, Phys. Lett. 49B (1974) 129
13. J. Błocki and H. Flocard, Phys. Lett. 85B (1979) 163
14. I.P. Bondorf, I.N. De, G. Fai, O.A.T. Karvinen, B. Jakobsson and I. Randrup, Nucl. Phys. A333 (1980) 285
I.P. Bondorf, I.N. De, G. Fai, A.D.T. Karvinen and B. Jakobsson, Phys. Lett. 84B (1979) 163
15. S. Drożdż, J. Okołowicz and M. Ploszajczak, submitted to Physics Letters B
16. A. Gavron, R.L. Ferguson, F.E. Obenshain, F. Plasil, G.R. Young, G.A. Petitt, K.A. Geoffroy, D.G. Sarantites and C.F. Maguire, Phys. Rev. Lett. 46 (1981) 8
H.C. Britt and A.R.Quinton, Phys. Rev. 124 (1961) 877
J. Gallin, B. Gatty, D. Guerreau, C. Rousset, U.C. Schlotthauer-Voos and X. Tarrago, Phys. Rev. C9 (1974) 1126
L. Westerberg, D.G. Sarantites, D.C. Hensley, R.A. Dayras, M.L. Halbert and I.H. Baker, Phys. Rev. C18 (1978) 796

MATTER, MOMENTUM AND ENERGY FLOW IN HEAVY-ION COLLISIONS

M. Prakash[+]
Department of Physics, State University of New York at
Stony Brook, Stony Brook, New York 11794, USA

S. Shlomo[++]
Department of Physics and Cyclotron Institute
Texas A&M University, College Station, Texas 77843, USA

B. S. Nilsson and J. P. Bondorf
The Niels Bohr Institute, University of Copenhagen,
DK-2100 Copenhagen Ø, Denmark
and
F. E. Serr[+++]
Center for Theoretical Physics, Laboratory for Nuclear Science,
and Department of Physics, Massachusetts Institute of Technology,
Cambridge, Massachusetts 02139, USA

Abstract

We study the flow of matter, momentum and energy in low energy heavy-ion collisions. This is done by using the Wigner phase space distribution function to calculate quantum mechanical analogs of the classical distributions of the observables. We apply the Wigner transformation to time dependent Hartree-Fock calculations of $^{16}O + ^{16}O$ and $^{40}Ca + ^{40}Ca$ reactions. Both static and dynamic features of the distribution functions are demonstrated.

1. Introduction

There now exist many studies of both static and dynamic properties of nuclei using the Wigner distribution function (WDF)[1-8]. Our aim here is to obtain new insights into the dynamics of heavy-ion collisions through the WDF. Rather than evolving the WDF itself, we construct it from wave functions resulting from a dynamical theory. For this we use the time dependent Hartree-Fock (TDHF) approximation[9]. We summarise results of calculations for head-on collisions of $^{16}O + ^{16}O$ and $^{40}Ca + ^{40}Ca$ reactions. TDHF studies of these reactions have been done before using other methods[9].

2. The Wigner Distribution Function

The Wigner phase space distribution function $f(\vec{r},\vec{k},t)$, is given by the shifted Fourier transform of the one-body density matrix $\rho(\vec{r},\vec{r}',t)$;

$$f(\vec{r},\vec{k},t) = (1/2\pi)^3 \int d\vec{s}\, e^{-i\vec{k}\cdot\vec{s}} \rho(\vec{r}+\vec{s}/2, \vec{r}-\vec{s}/2, t) \qquad (2.1)$$

In the present work, we consider $\rho(\vec{r},\vec{r}',t)$ obtained in the TDHF approximation. In this case,

$$\rho(\vec{r},\vec{r}',t) = \sum_j n_j\, \phi_j^*(\vec{r},t)\, \phi_j(\vec{r}',t) \qquad (2.2)$$

where n_j are the occupation numbers, and $\phi_j(\vec{r},t)$ are the single particle wave functions calculated from the TDHF mean field. Since we deal with head-on collisions of heavy-ions, we make use of the axial symmetry and write the single particle wave functions in the form

$$\phi_j(\vec{r},t) = \phi_j(\vec{r}_\perp, z, t)\, e^{i m_j \Phi} \qquad (2.3)$$

where $\vec{r}_\perp$ is the position vector perpendicular to z, Φ is the azimuthal angle, and m_j is the angular momentum projection on z. Physical observables are obtained by evaluating integrals of the form $\langle O \rangle = \int f(\vec{r},\vec{k})\, O(\vec{r},\vec{k})\, d\vec{r}\, d\vec{k}$. Therefore, of special interest are projections $\tilde{f}$, in which some of the arguments are integrated out. We consider a few such projections in the following.

(i) $$\tilde{f}(z, k_z, t) = \int d^2\vec{r}_\perp\, d^2\vec{k}_\perp\, f(\vec{r}_\perp, z, \vec{k}_\perp, k_z, t) \qquad (2.4)$$

(ii) $$\tilde{f}(z, \vec{k}_\perp, k_z, t) = \int d^2\vec{r}_\perp\, f(\vec{r}_\perp, z, \vec{k}_\perp, k_z, t) \qquad (2.5)$$

These projected distribution functions are predominantly positive. We will also consider the WDF along the symmetry axis of the system, viz.,

$$f(\vec{r}_\perp = 0, z, \vec{k}_\perp, k_z, t) \qquad (2.7)$$

The Static WDF

A knowledge of the static WDF is very useful for understanding the dynamic WDF. The static HF wave functions $\phi_j(r)$ in eq. (2.2) were calculated using an effective interaction consisting of the direct Coulomb interaction and modified Skyrme force II, with no spin orbit interactions. The details of the numerical calculation of the WDF are described in ref. 10. The HF results can be compared with those for Fermi gas (FG) and Harmonic oscillator (HO) models for which analytical results can be derived. For the infinite Fermi gas, we have $f^{FG}(\vec{r},\vec{k}) = 4/(2\pi)^3\, \theta(k_F^2 - k^2)$ where k_F is the Fermi momentum. For finite nulei, the WDF oscillates around f^{FG}, due to the presence of shell and surface effects[3,12]. This is illustrated clearly by the exact expression for the WDF for nucleons in closed shells of a spherically symmetric HO potential[8], viz.

$$f^{HO}(\vec{r},\vec{k}) = f^{HO}(\varepsilon) = (4/\pi^3)\sum_K n_K(-1)^K e^{-\varepsilon} L_K^2(2\varepsilon) \qquad (2.8)$$

where $L_K^{\alpha}(x)$ is the associated Laguerre polynomial and $\varepsilon = k^2/\nu + \nu r^2$, with $\nu = m\omega/\hbar$. The factor 4 takes into account spin-isospin degeneracy. The expressions of eq. (2.8) are

$$\begin{aligned} f^{HO}(\varepsilon) &= (8/\pi^3)\, e^{-\varepsilon}(\varepsilon-1) && \text{for } A=16 \\ f^{HO}(\varepsilon) &= (8/\pi^3)\, e^{-\varepsilon}(\varepsilon-1)(\varepsilon-2) && \text{for } A=40 \end{aligned} \qquad (2.9)$$

We see that for finite nuclei, the WDF acquires negative values in specific regions of the phase space, and unlike f^{FG}, extends beyond the Fermi momentum, k_F. For open shell nuclei in the HO model, the WDF is more complicated than eq. (2.8) in that it depends on the angle between $\vec{r}$ and $\vec{k}$[8]. The WDF for any numerical single particle wave function can be calculated by expanding it in a HO basis and using eq. (28) of ref. 8.

The solid lines of fig. 1 show the projected WDF $\tilde{f}(z,k_z)$, for the HF wave functions, at the center and near the surface. The corresponding results shown for the Fermi gas model are obtained by integrating $f^{FG}(\vec{r},\vec{k})$ to give $\tilde{f}^{FG}(z,k_z)=(1/2\pi)(R^2-z^2)(k_F^2-k_z^2)$. We have determined the equivalent sharp surface radius, R, from the relation $R^2=(5/3)\langle r^2\rangle_{HF}$, and $k_F=(9\pi A/8)^{1/3}/R$, from the particle number normalization. We also show in fig. 1 the projected WDF for the harmonic oscillator, $\tilde{f}^{HO}(z,k_z)$, which is obtained from eq. (2.9). The size parameters $\nu = m\omega/\hbar$ were determined by using the root mean square radii of the HF calculations. We see in fig. 1 that the FG model roughly matches the magnitude of $\tilde{f}^{HF}(z,k_z)$ at $z = 0$

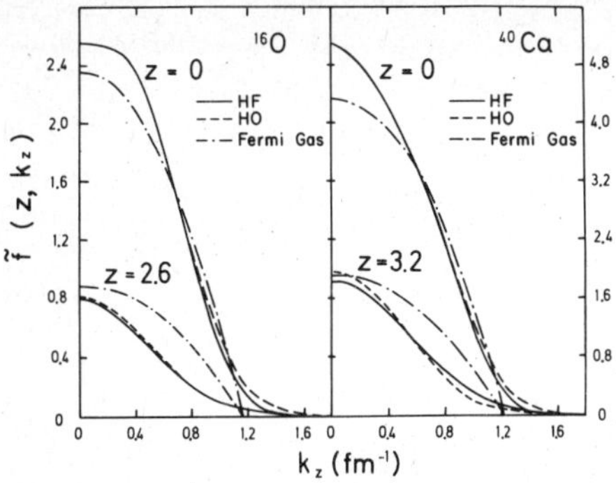

Fig. 1. Projected Wigner distribution functions (eq.(2.4)) for static nuclei at the center and near the nuclear radius.

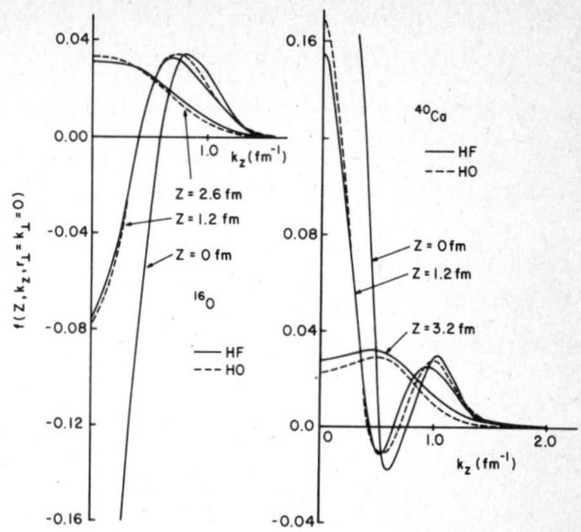

Fig. 2. Cross-sectional cuts for the WDF (eq.(2.7)) for static nuclei at different distances.

and for small k_z, but cannot give the high momentum tail of the distribution. Understandably, the agreement between the two models is worst in the surface of the nucleus. The HO results more accurately reproduce the HF calculations. There is an excess of high momentum components in $\tilde{f}^{HO}(z,k_z)$ at z = 0, and of lower k_z in the nuclear surface.

We notice that $f^{HO}(\vec{r},\vec{k})$ given in eq. (2.9) has negative values for certain values of $\vec{r}$ and $\vec{k}$. This feature also appears in the HF calculations. In fig. 2, we plot $f(z,k_z,\vec{r}_\perp = \vec{k}_\perp = 0)$, obtained from eq. (2.7) for different values of z, along with the corresponding HO results. The HO model reproduces the HF results very well. In the surface, however, the HF calculations predict more high momentum components than the HO model. In both models, the WDF is positive in the surface.

The Dynamic WDF

Using TDHF wave functions, we have calculated the WDFs given in eqs. (2.4), (2.5) and (2.7). The effective Hamiltonian in the TDHF calculations was the same as that used in the static HF calculations.

The contours of the projected WDF, $\tilde{f}(z,k_z,t)$, in fig. 3 show the space momentum correlation during the course of the $^{40}Ca+^{40}Ca$ reaction at E/A = 5.0 MeV. At t = 0, the shift along the k_z axis is equal to k_{rel} = 0.94 fm^{-1}. For 0.20×10^{-21}sec $<t<$ 0.4×10^{-21}sec, the distribution function is that of a dinuclear complex. During this time there is a spilling over of nucleons from one ion to the other. This can most easily be seen by following the maximum $\tilde{f}(z,k_z,t)$ contour as a function of time. Beyond t = 0.4 x 10^{-21} sec, rapid changes of the WDF in the neck region are observed. For $t>0.6\times10^{-21}$ sec, the centers of the two distributions are well separated along the z axis. The shift along the k_z axis is now smaller than at t = 0, which shows that energy loss in relative motion has taken place.

In fig. (4) we show the cross-sectional cuts, $f(z,k_z,\vec{r}_\perp=\vec{k}_\perp=0)$ at three representative values of z, for the ^{16}O + ^{16}O reaction at E/A = 1.25 MeV and the ^{40}Ca + ^{40}Ca reaction at E/A = 5.0 MeV, respectively. Note that, for the symmetric A reactions we consider, $f(-\vec{r},\vec{k}) = f(\vec{r},-\vec{k})$. The time in both figures has been chosen such that the two nuclei overlap considerably with each other. The figures show that the dynamic WDF also has large negative values. As in the static case, the WDF is positive in the surface of the dinuclear complex.

In both figs. 3 and 4, we see that the WDF vanishes for $\vec{r}=\vec{k}=0$. This follows from the idempotent condition for $f(\vec{r},\vec{k})$ and the prevailing symmetry. Given that the density matrix comes from a Slater determinant, we have[9a)]

$$f(\vec{r},\vec{k}) = (2\pi)^3 e^{-i/2\left(\nabla_r^{(1)}\cdot\nabla_k^{(2)} - \nabla_r^{(2)}\cdot\nabla_k^{(1)}\right)} f^{(1)}(\vec{r},\vec{k})\, f^{(2)}(\vec{r},\vec{k}) \qquad (2.10)$$

where the superscripts (1) and (2) indicate which WDF's the derivatives act upon. Taking the real and imaginary parts (remember f is real):

$$\begin{aligned} f(\vec{r},\vec{k}) &= (2\pi)^3 \cos\left[\tfrac{1}{2}\left(\nabla_r^{(1)}\cdot\nabla_k^{(2)} - \nabla_r^{(2)}\cdot\nabla_k^{(1)}\right)\right] f^{(1)}(\vec{r},\vec{k})\, f^{(2)}(\vec{r},\vec{k}) \\ 0 &= (2\pi)^3 \sin\left[\tfrac{1}{2}\left(\nabla_r^{(1)}\cdot\nabla_k^{(2)} - \nabla_r^{(2)}\cdot\nabla_k^{(1)}\right)\right] f^{(1)}(\vec{r},\vec{k})\, f^{(2)}(\vec{r},\vec{k}) \end{aligned} \qquad (2.11)$$

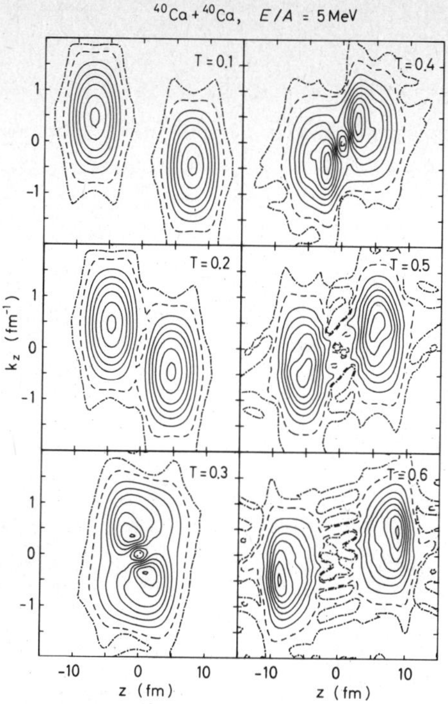

Fig. 3.

The projected WDF $\tilde{f}(z,k_z,t)$ (eq. (2.5)). Time is in units of 10^{-21} sec. The dash-dot and dashed contours correspond to f=0.02 and 0.2 respectively. For t=0.3, 0.4 and 0.6, the seven solid contours are associated with f=0.8(0.8)5.6 starting from the outermost one. For other times, we show six solid contours with f=0.8(0.8)4.8. Note that for all times, $\tilde{f}(z=k_z=0)=0$.

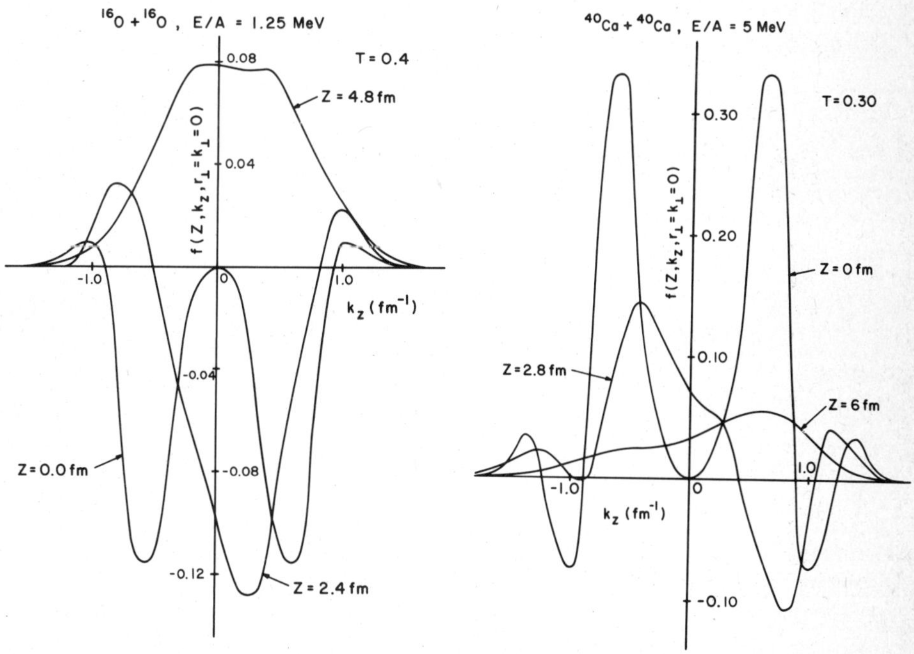

Fig. 4. Cross-sectional cuts of the WDF during the dynamics. Time is in units of 10^{-21} sec.

While the second condition is trivially satisfied, the first condition is nontrivial, since it contains terms like $(\nabla_r f^{(1)} \nabla_k f^{(2)})^2 - (\nabla_r f^{(1)} \nabla_r f^{(2)})(\nabla_k f^{(1)} \nabla_k f^{(2)})$ which in general do not vanish. Now for collisions between identical nuclei, in the c.m. system (not necessarily head on) the symmetry $f(\vec{r},-\vec{k}) = f(-\vec{r},\vec{k})$ implies that $[\nabla_r f]_{\vec{k}=\vec{r}=0} = [\nabla_k f]_{\vec{r}=\vec{k}=0} = 0$. At the origin, then, the Slater determinant condition becomes $f(0,0) = (2\pi)^3 [f(0,0)]^2$ independent of time. This requires $f(0,0) = 0$ or $1/(2\pi)^3$. Since $f(0,0,t=0) = 0$, it will stay that way.

3. Momentum and Energy Flow

In this section, we investigate the flow of momentum during the reaction. In fig. 5 we show the quantity

$$\tilde{f}(k_z, k_\perp, t) = \int_{z\geq 0} \tilde{f}(z, k_z, \vec{k}_\perp, t)\, dz \tag{3.1}$$

for the $^{16}O+^{16}O$ reaction at E/A = 1.25 MeV and the $^{40}Ca+^{40}Ca$ reaction at E/A = 5 MeV. Eq. (3.1) gives the momentum distribution in the positive half of z space only. Therefore, at the lowest value of t we have a smooth distribution corresponding to the ground state momentum distribution of one nucleus centered around $k_z=(1/2) k_{rel}$. The plateau in the region $|\vec{k}-(1/2)\vec{k}_{rel}| \sim 0.6\ fm^{-1}$ in fig. 5(b) is due to shell structure. During the interaction time, momentum components associated with nucleons transferred from the other nucleus, as well as components associated with collective excitations are present in the $z>0$ space. These are clearly seen by the peaks and shoulders of the curves in fig. 5. The distribution for the largest t still differs considerably from a totally thermalized distribution. The absence of two-body collisions in the TDHF approximation naturally leads to very slow thermalization of the separated fragments.

The flow of momentum and energy in the perpendicular direction might have important consequences on the dynamics. To quantify this aspect of heavy-ion collision dynamics, we investigate the behaviour of the parallel and perpendicular components of the total kinetic energy during the reaction. In terms of the WDF, the total kinetic energy T is defined by

$$T = \int \tau_W(\vec{r})\, d\vec{r} = \int (\hbar^2 k^2/2m)\, f(\vec{r},\vec{k},t)\, d\vec{r}\, d\vec{k} \tag{3.2}$$

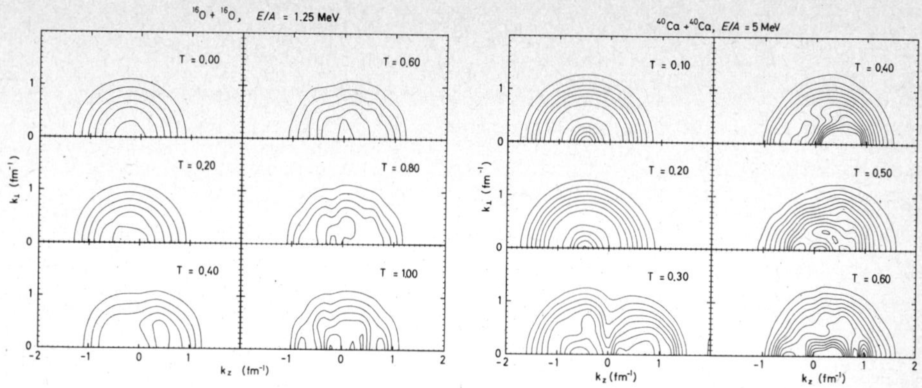

Fig. 5. The momentum distribution in the z>0 space (eq. (3.1), as a function of the parallel and perpendicular components of momentum. Time is in units of 10^{-21} sec. Starting from the outermost one, the contours are associated with $\tilde{f}$ = 0.4(0.4)2.4 in fig.(a). Contours in fig.(b) are associated with $\tilde{f}$ = 1.0(1.0)12.0.

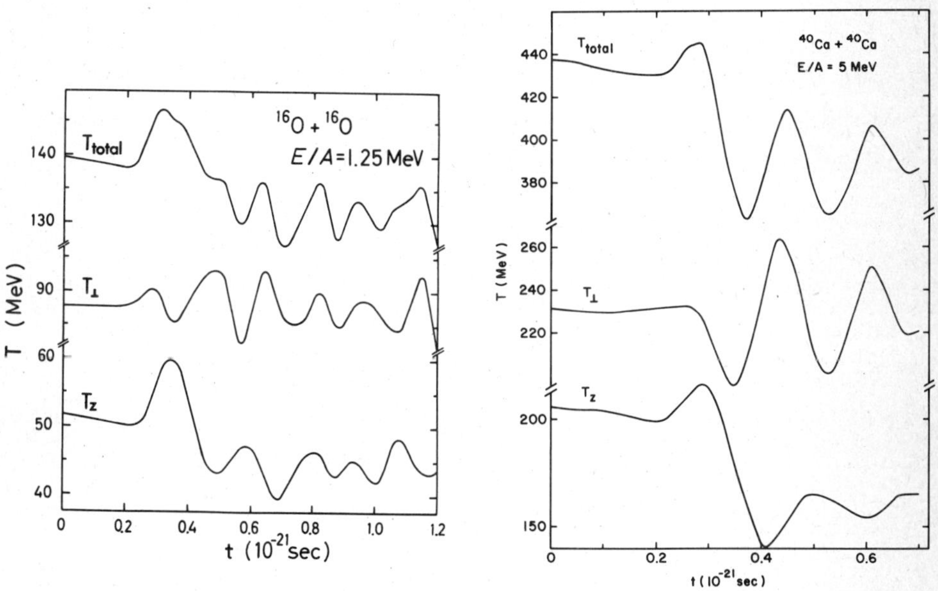

Fig. 6. The kinetic energy as a function of time. T_z and $T_\perp$ are the components of kinetic energy parallel and perpendicular to the reaction axis z ($T = T_z + T_\perp$).

We separate the parallel and perpendicular components of T, denoted T_z and $T_\perp$, by replacing k^2 by k_z^2 or $k_\perp^2$ in eq. (3.2). In fig. 6, we show T_z, $T_\perp$ and T during the $^{16}O + {}^{16}O$ reaction at $E/A = 1.25$ MeV, and $^{40}Ca + {}^{40}Ca$ reaction at $E/A = 5.0$ MeV, respectively. At $t = 0$, $T_z = (1/2)T_\perp + T_{rel} - V_c$, where T_{rel} is the initial kinetic energy with which the ions are boosted towards each other, and V_c is the Coulomb interaction energy at the initial separation. As the two nuclei overlap, T_z, $T_\perp$ and T begin to oscillate. The change in $T_\perp$ during the interaction time is comparable to that for T_z. In the exit channel, T_z and $T_\perp$ oscillate out of phase. These results imply that there is a considerable redistribution of energy between the parallel and perpendicular components of kinetic energy during the reaction. In so far as the outcome of the TDHF dynamics is dependent on what happens during the interaction time, allowing dynamics to take plae only in a plane and freezing the wave functions in the third direction could lead to significantly different results.

4. Matter flow

The exchange of nucleons between the colliding ions is studied by using the concept of a one-way current defined in terms of the WDF[11]. A semi-classical interpretation is given to the one-way current, which allows for a comparison with static approaches. In terms of the WDF, the current density $\vec{j}(\vec{r},t)$ is given by

$$\vec{j}(\vec{r},t) = \int (\hbar\vec{k}/m)\, f(\vec{r},\vec{k},t)\, d\vec{k} \tag{4.1}$$

The one-way current density in the direction of a unit vector $\hat{z}$ is defined as

$$j_+(\vec{r},t) = \int_{k_z \geq 0} d\vec{k}\; (\hbar k_z/m)\, f(\vec{r},\vec{k},t) \tag{4.2}$$

From eq. (4.2), one has that $j = j_+ + j_-$. In the Fermi gas model, $j_+ = \hbar k_F^4/8\pi^2 m)$, which is the bulk flux in nuclear matter. Closed expressions have also been derived for j_+ in the HO model[10,12]. Integrating eq. (4.2) over the perpendicular components of space the one-way current $\Phi_+(z,t)$ across a plane at z is given by

$$\Phi_+(z,t) = \int_0^\infty (\hbar k_z/m)\, \tilde{f}(z,k_z,t)\, dk_z \tag{4.3}$$

In the dynamic situation, our interest is in the flow of nucleons between the two ions. For a reaction between nuclei of identical masses this is given by the one-way current of particles

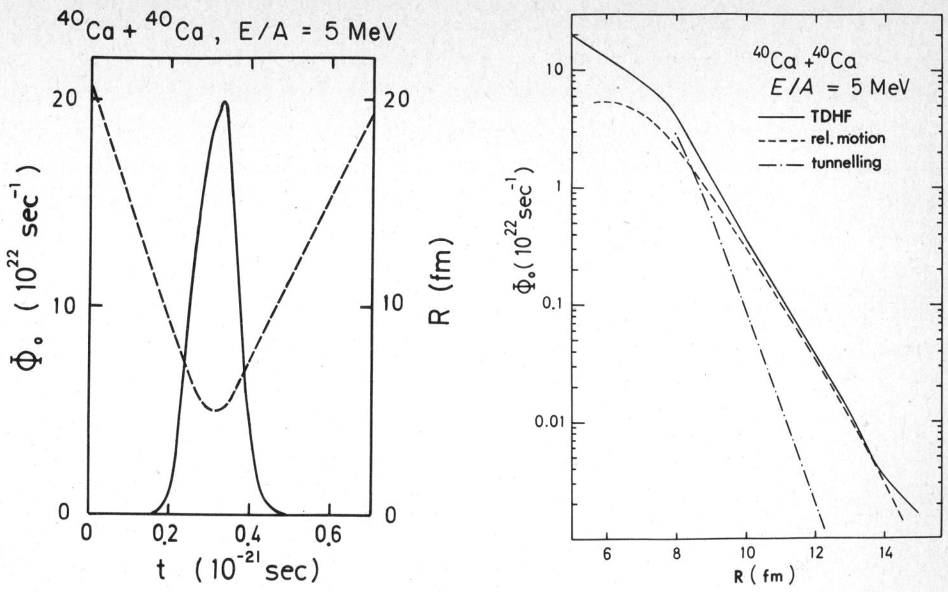

Fig. 7. (a). The solid curve shows the TDHF one-way current across the z = 0 plane dividing the two colliding nuclei as a function of time. The dashed curve shows the fragment separation R (scale on right).
(b). Currents as a function of the fragment separation R. The models are explained in the text.

$\Phi_0 = \Phi_+(z=0,t)$ across the plane z = 0 dividing the two ions. Since Φ_0^{TDHF} is found to be practically unchanged when averaged over a length $\sim k_F^{-1}$, we expect it to give a good description of the one-way current in a semi-classical sense. In fig. 7, we show Φ_0^{TDHF} (solid line) vs.t, and the fragment mass separation coordinate R(t), for the ^{40}Ca + ^{40}Ca reaction at E/A = 5.0 MeV.

We can understand the TDHF one-way current in terms of some simpler models. For example, for large separations we expect the flux to come primarily from the tails of wave functions being carried along by the relative motion of the two self-consistent potential wells. This flux due to relative motion must be extracted before comparing the TDHF results with static models[13,14] for the tunnelling through and over the potential barrier. Also, static models can only give the one-way current as a function of the distance between the ions. We therefore compute Φ_0^{TDHF} vs.R during the approach phase of the reaction, shown by the solid curve of fig. 7, for the ^{40}Ca reaction at E/A = 5.0 MeV. We concentrate on the approach phase because the simple models we now consider for the contributions from relative motion and nucleon tunnelling do not explicitly include deformation of the ions. Deformation effects should be smaller during the approach phase than during the exit phase.

First, we study the contributions from relative motion alone. Motivated by the expected large-distance behaviour, we move the frozen initial HF density profile of one of the ions past the z = 0 plane, with a time dependent average relative velocity determined from the TDHF calculations. The dashed curve in fig. 7 shows $\Phi_0^{rel.motion}$ for the ^{40}Ca reaction at E/A = 5.0 MeV. For distances much larger than the contact distance of the spherical initial half-potential radii R_C = 8 fm, the contribution from relative motion within the frozen density approximation accounts for most of the one-way current Φ_0^{TDHF}. However, when the densities of the two ions begin to overlap considerably, $\Phi_0^{rel.motion}$ falls below Φ_0^{TDHF}. At the minimum values of R, dR/dt, and hence $\Phi_0^{rel.motion}$, will fall to zero.

In the second step, we study the one-way current due to nucleon tunnelling through and over the potential barrier, ignoring relative motion. We use a model based on the approach of Ko et al.[14)]. It is important that our static calcculations mimic the details of the forces in the TDHF calculations as closely as possible. We have therefore used the potential along the z-axis resulting from the TDHF calculations to determine the transmission probability. The results for the ^{40}Ca reaction are shown by the dash-dotted curve in fig. 7. Near the contact distance R_C, $\Phi_0^{tunnelling}$ is comparable to $\Phi_0^{rel.motion}$. However, for $R > R_C + 2$ fm, the one-way current due to tunnelling is extremely small.

Comparing the combination of the one-way currents due to relative motion and tunnelling to the TDHF results, we find that, at distances large compared to the contact distance R_C, Φ_0^{TDHF} is almost entirely due to the relative motion. Near R_C, the sum of the contributions from relative motion and tunnelling is close to the calculated TDHF one-way current. The comparison is not as meaningful for very compact configurations because some of our approximations are no longer valid. Better agreement should be obtained by refining the models for the two contributions.

Integrated One-Way Currents

We denote by N_{ex} the time integrated one-way current across the z = 0 plane and list it in table 1. Roughly speaking, N_{ex} may classically be taken to represent the total number of particles exchanged from one side to the other during the course of the reaction. In order to understand the qualitative behaviour of N_{ex}^{TDHF}, we consider two nuclei in free motion boosted towards each other by a momentum $(1/2)k_{rel}$. When there is no interaction and no antisymmetrization between the two nuclei we have,

$$N_{ex}^{\text{Free motion}} = A + (4/k_{rel}) \int_{(1/2)k_{rel}}^{\infty} (k_z - (1/2)k_{rel}) \hat{f}(k_z)\, dk_z \tag{4.10}$$

where A is the number of particles in one of the nuclei. With the projected momentum distributions from the static HF configurations as inputs, the calculated results are shown in table 1. $N_{ex}^{\text{Free motion}}$ saturates at A for large k_{rel}. This kinetic effect may also be seen in the TDHF results. However, the simple free motion model overestimates N_{ex}. Considerable reduction may be expected from the Pauli exclusion principle. Estimates from the free Fermi gas model which-are based on the plane wave Born approximation (PWBA) predict too large reduction factors for the low enrgy reactions. Results for finite nuclei calculated using the HO model and the PWBA are not very different from the FG predictions. Table 1 summarises these findings. More realistic estimates for finite nuclei which are based on the DWBA approximation[15)] are expected to yield better agreement with the TDHF results.

Table 1.

Reaction	$(E/A)_{CM}$ (MeV)	N_{ex}^{TDHF}	$N_{ex}^{\text{Free motion}}$	N_{ex} Pauli blocking F.G. model	N_{ex} Pauli blocking H.O. model
$^{16}O+^{16}O$	1.25	15.1	35.81	9.74	8.95
	5.00	10.5	18.85	10.5	13.76
	10.0	10.2	16.68	12.73	15.68
$^{40}Ca+^{40}Ca$	5.00	15.9	47.83	26.64	38.5

5. Outlook

We have seen that with the help of the WDF, it is possible to study several aspects of the heavy-ion collision dynamics. Its application is by no means restricted to the TDHF approximation we have chosen to illustrate the course of dynamics. While in principle it provides no more information than obtained from wave functions, the WDF sometimes provides a convenient way of calculating quantities which have a classical or semi-classical analog, emphasizing at the same time the essential differences in the quantum world[7]. Other applications of such calculations would be to determine the momentum region of cluster formation and prompt particle emission in heavy-ion collisions.

+ Supported by the U.S. Department of Energy under contract DE-ACO2-76ER with the State University of New York.

++ Supported by the National Science Foundation under contract PHY-8109019.

+++ Supported by the U.S. Department of Energy under contract DE-ACO2-76ER03069, and the National Science Foundation under contract PHY-7801164.

References

1. E. P. Wigner, Phys. Rev. 40 (1932) 749;
2. G. F. Bertsch, Nucl. Phys. A249 (1975) 253; Many Body Dynamics of Heavy-Ion Collisions, in: Les Housches Lecture Notes, ed. R. Balian et al., (North-Holland, Amsterdam, 1978), pp. 178-262.
3. N. L. Balazs and G. G. Zipfel, Jr., Ann. Phys. 77 (1973) 139; N. L. Balazs and H. C. Pauli, Z.Phys.A281(1977)395;A277(1976)265.
4. P. Ring and P. Schuck, The Many-Body Problem (Springer-Verlag, New York, 1980) Chap. 13, and references therein.

4a. V. M. Kolomietz and H. H. K. Tang, Physica Scripta 24 (1981) 915.

4b. D. M. Brink and M. D. Toro, Nucl. Phys. A372 (1980) 151.

4c. C. Y. Wong, Phys. Rev. C25 (1982)1460.

5. J Richert et al., Phys.Lett.87B(1979);Nucl.Phys.A356(1981)260.
6. H. S. Köhler and H. Flocard, Nucl. Phys. A323 (1979) 189; H. S. Köhler, Nucl. Phys. A343 (1980) 315.
7. H. H. K. Tang, et al., Phys. Lett. 101B (1981) 10.
8. S. Shlomo and M. Prakash, Nucl. Phys. A357 (1981) 157.

8a. J. Hüfner and M. C. Nemes, Phys. Rev. C23 (1981)2538.

9. S. E. Koonin, in Progress in Particle and Nuclear Physics, Vol.4, edited by D. Wilkinson (Pergamon, Oxford, 1980) pp. 283.

9a. S. E. Koonin, Ph.D. Thesis, 1975, Massachusetts Institute of Technology (unpublished).

10. M. Prakash et al., Nucl. Phys. A (in print).
11. M. Prakash, Phys. Rev. Lett. 47 (1981) 898.
12. M. Prakash et al., Nucl. Phys. A370 (1981) 898.
13. J. Randrup, Nucl. Phys.A307(1978)319, and Ann.Phys.112(1978)356.
14. C. M. Ko, Phys. Lett. 77B (1978) 174.
15. S. Shlomo and P. J. Siemens, to be published; G. Bertsch and O. Scholten, MSU preprint 1981.

TIME-DEPENDENT GROUND-STATE CORRELATIONS IN HEAVY ION SCATTERING

P.-G. Reinhard
Institut für Kernphysik
Universität Mainz
D-6500 Main, W.-Germany
and
K. Goeke
Institut für Kernphysik
Kernforschungsanlage Jülich
D-5170 Jülich, W.-Germany
and
R.Y. Cusson
Department of Physics
Duke University
Durham, North-Carolina, U.S.A.

Abstract

Using a time-dependent generator-coordinate method, we derive a theory for time-dependent collective ground-state correlations which account for some quantum fluctuations about a TDHF trajectory. This theory is particularly suited for evaluating spreading widths of collective one-body operators. As an application we study head-on collision of heavy ions in a one-dimensional model. As one of the prominent results we find a substantial enhancement of the spreading width of the internal excitation energy due to the correlations.

1. Introduction

In the last years time-dependent mean field approaches (TDHF) have been widely used for the microscopic description of large amplitude collective motion, in particular for heavy ion collisions. They have provided useful insight into the time evolution of density distributions, currents, multipole moments etc. It has turned out that most one-body obervables are described fairly well by TDHF, e.g., the average mass drift in a heavy ion collision or the ridge in a Wilshinsky plot. That is not surprising since TDHF has been built to optimize the time evolution of the one-body density matrix, and hence, is expected to yield the proper time evolution of averages of one-body operators. In general, however, TDHF will fail to reproduce properties which are sensitive to two-body or higher density matrices, as e.g., speading widths of one-body observables or particular S-matrix elements.

There are several attempts to cure TDHF for this deficiency. It is obvious that one needs to include some correlations into the wave functions, or in other words, to allow for quantum fluctuations about the (classical) TDHF path. One can try to approximate those by quantum statistics[1] or to describe all of them by means of Monte-Carlo integration techniques[2]. However, we expect that there will be some correlations of particular importance, namely those which fluctuate about the TDHF path in the direction of the possible collective excitation, we will call these collective correlations[3].

Thus, we are going to set up a theory for describing the time-evolution of collective ground state correlations by a coherently correlated state. The formalism is derived in section 2 within the framework of a time-dependent generator-coordinate method (TDGCM). In section 3 we introduce shortly the one-dimensional dynamics, which are used to study the TDGCM and discuss its possible collective modes. Finally, in section 4 we present the results of the numerical evaluation of TDGCM in one-dimensional dynamics.

2. Time-dependent generator-coordinate method (TDGCM)

In generalization of the GCM for stationary states we now consider a time-dependent correlated state as a superposition

$$|\psi(t)\rangle = \int dq \; |\phi_q(t)\rangle \; f(q,t) \tag{1}$$

where the collective superposition function f and the basis states $|\phi_q\rangle$ depend explicitly on time. The deformation basis $\{|\phi_q(t)\rangle\}$ consists out of Slater-states which are collectively deformed compared to the uncorrelated TDHF trajectory $|\phi_o(t)\rangle$; each one of the $|\phi_q(t)\rangle$ is supposed to move in its own average field $U_q(t)$ (yet to be determined).

We now want to determine the equations-of-motion for the $|\phi_q(t)\rangle$ and f(q,t) by means of the time-dependent variational principle

$$\langle\delta\psi|H - i\partial_t|\psi\rangle \; . \tag{2}$$

In this general form the emerging variational equations are untractable since complicated integral-kernels $\langle\phi_q(t)|H - i\partial_t|\phi_{q'}(t)\rangle$ arise. And even worse, the path variation $\langle\delta\phi_q|$ turns out to be impractical, if no further restrictions on $|\phi_q(t)\rangle$ are made, since in general, the ansatz (1) still embraces the full Hilbert-space. (Think for example of some peculiar basis $|\phi_q\rangle$ winding successively through all possible many-particle-many-hole states).

An obvious requirement for a tractable $|\phi_q\rangle$ is that it is analytically in q. Furthermore, since $|\phi_q\rangle$ ought to represent a collective deformation, where a small displacement of many particles adds up coherently to a substantial reordering of the matter, the overlap $\langle\phi_q|\hat{O}_p|\phi_{q'}\rangle$ will fall off rapidly with $(q - q')^n$. Therefore, it is reasonable to assume Gaussian overlap[4] for the norm kernel

$$\langle\phi_q|\phi_{q'}\rangle = \exp\left(-(q - q')^2/4\mu\,\frac{q + q'}{2}\right) \tag{3a}$$

where

$$\mu^{-1}(q) = 2\langle\phi_q|\overleftarrow{\partial}_q\,\overrightarrow{\partial}_q|\phi_q\rangle \tag{3b}$$

and analogously

$$\langle\phi_q|\bar{O}p|\phi_{q'}\rangle = \exp\left(-(q-q')^2/4\mu\right)\cdot\left[Op_o + Op_1(q-q') + Op_2(q-q')^2\right] \tag{4}$$

with coefficients Op_i determined as expectation values of $\hat{Op}$ with some derivatives ∂_q^o, ∂_q^1, ∂_q^2.

Working out the Gaussian overlap approximation (GOA) we obtain for $\hat{Op} = H-i\partial_t$,

$$\begin{aligned}<\psi|H-i\partial_t|\psi>_{GOA} = \int dq\, g^*(q,t)\{&<\phi_q(t)|H-i\partial_t|\phi_q(t)> - i\frac{d}{dt} \\ &- \frac{\mu}{4}\partial_q^2 <\phi_q(t)\, H|- i\partial_t|\phi_q(t)> \\ &+ :(-i\frac{d}{dq})<\phi_q(t)|\overleftarrow{\partial}_q\overline{H} - \overline{H}\overrightarrow{\partial}_q|\phi_q(t)>: \\ &+ \frac{1}{2}:|(-i\frac{d}{dq})^2 - \frac{1}{2\mu}|<\phi_q(t)|\overleftarrow{\partial}_q^2\overline{H}-2\overleftarrow{\partial}_q\overline{H}\overrightarrow{\partial}_q+\overline{H}\overrightarrow{\partial}_q^2|\phi_q(t)n^2:\}g(q,t)\end{aligned} \tag{5}$$

where $\frac{d}{dq}$ is to act on everything, but ∂_q only on $|\phi_q>$ or $<\phi q|$. The : ... : express the GOA normal ordering : $\frac{d^n}{dq^n}A(q)$: = $\frac{1}{2^n}\{\frac{d}{dq}\{\frac{d}{dq}...\{\frac{d}{dq}, A\}...\}$. The g(q,t) is the collective wave function, normlaized as $\int dq\, g^*g = 1$; it is related to f(q,t) by

$$g(q,t) = \int dq'(\mu\pi)^{-1/4}\exp(-(q-q')^2/2\mu)f(q',t). \tag{6}$$

Finally, the $\overline{H}$ is $\overline{H} = H - i\partial_t - <\phi_q|H - i\partial_t|\phi_q>$.

Within the approximation (5) we can work out the variational principle (2) varying with respect to both ingredients, g(q,t) as well as $|\phi_q(t)>$; this eventually leads to equations-of-motion for g(q,t) and for $|\phi_q(t)>$. As a further simplification we make the following approximation: we assume the q-motion to be harmonic (which is valid at least for small amplitude motion) and parametrize g as

$$g(q,t) = (\alpha\pi)^{-1/4}\exp(-q^2/2\alpha(t)), \tag{7a}$$

with the correlated state (2) becoming now

$$|\psi(t)> = \int dq\, |\phi_q(t)>\, N\exp(-q^2/2(\alpha-\mu)) \tag{7b}$$

with

$$N = (\alpha/\mu)^{1/4}/(2\pi(\alpha-\mu))^{1/2}.$$

Now, the variation $<\delta\phi_q|$ yields the propagation for the $|\phi_q(t)>$

$$(\hat{W}_q - i\partial_t)|\phi_q(t)> = 0 \tag{8a}$$

where

$$\hat{W}_q = \hat{T} + Tr\{\rho_q V\} \tag{8b}$$

This is nothing else than the TDHF equation for each $|\phi_q(t)>$ separately, moving in its own Hartree-Fock field $\hat{W}_q$. (In fact, in eq. (8) we have neglected the influence of $(\frac{d}{dq})^2$-term in the expansion (5), which means that we have neglected the feedback of the correlations to the mean-field $\hat{W}_q$). The variation δ_{α^*} yields the correlation dynamics

$$\frac{d}{dt}\alpha = 2i\,\mu^2 <\phi_o(t)|\overleftarrow{\partial}_q^2\,\overline{H} + \overline{H}\,\overrightarrow{\partial}_q^2|\phi_o(t)> . \tag{9}$$

It is interesting to note that this correlation dynamics is particularly sensitive to the 2p-2h part of $\overline{H}$; this becomes obvious from eq. (9) keeping in mind that ∂_q is a 1p-1h operator, thus, ∂_q^2 is a 2p-2h operator.

Altogether we have an initial value problem for the correlated state $|\psi(t)>$: We start with a given $|\psi(0)>$ in the form (7b) which means to have a given initial correlation parameter $\alpha(0)$ and an initial deformation basis $\{|\phi_q(0)>\}$; we then propagate $|\phi_q(t)>$ according to TDHF eq. (8) and $\alpha(t)$ according to correlation dynamics (9); this defines $|\psi(t)>$ of eq. (7b) uniquely at all later times. The initialisation, of course, is crucial. By choosing a set $\{|\phi_q(0)>\}$ we define the collective mode for the correlation. In TDHF calculations one usually takes as initial state $|\phi_o(0)>$ Hartree-Fock states for the two separate clusters and boosts them with some relative velocity. Accordingly we have to choose the $\{|\phi_q(0)>\}$ and the $\alpha(0)$ to represent RPA vibrations of the two cluster system.

Finally, we want to evaluate expectation values of observables for the state $|\psi(t)>$. This can also be worked out within the GOA. We obtain, together with the harmonic approximation (7), the following expression for the spreading width of a one-body observable $\hat{A}$:

$$\begin{aligned}<\psi|\Delta^2 A|\psi> = <\phi_o|\Delta^2 A|\phi_o> &+ \frac{1}{2}\left[\frac{\alpha^*\alpha}{\alpha_r} - \mu\right] |<\phi_o|\overleftarrow{\partial}_q\,\overline{A} + \overline{A}\,\overrightarrow{\partial}_q|\phi_o>|^2 \\ &+ i\,\frac{\alpha_i}{\alpha_r}\,\mu <\phi_o|\overleftarrow{\partial}_q\,\overline{A} + \overline{A}\,\overrightarrow{\partial}_q|\phi_o><\phi_o|\overleftarrow{\partial}_q\,\overline{A} - \overline{A}\,\overrightarrow{\partial}_q|\phi_o> \\ &+ \frac{1}{2}\left|\frac{1}{\alpha_r} - \frac{1}{\mu}\right|\mu^2|<\phi_o|\overleftarrow{\partial}_q\,A - A\,\overrightarrow{\partial}_q|\phi_o>|^2\end{aligned} \tag{10}$$

where $\alpha_i = \text{Re}(\alpha)$, $\alpha_i = \text{Im}(\alpha)$ and all quantities depend on t.

Altogether we see that in the harmonic approximation all information on $|\psi(t)>$ is carried by the three quantities $|\phi_o(t)>$, $\partial_q|\phi_o(t)>$ and $\alpha(t)$. In practice, we handle the $\partial_q|\phi(t)>$ as a finite difference $\partial_q|\phi_o(t)> = (|\phi_{\delta q}(t)> - |\phi_o(t)>)/\delta q$. Thus, it remains to propagate two neighboured TDHF trajections $|\phi_o(t)>$ and $|\phi_{\delta q}(t)>$ and to work out additionally the correlation dynamics for $\alpha(t)$. Obviously, this procedure can easily be implemented into any existing TDHF code. The extra expense of TDGCM consists mainly in carrying TDHF several times, namely for each $|\phi_{\delta q}(t)>$ separately.

Of course, the TDGCM theory is not restricted to one correlation channel q only. It can easily be extended, formally and practically, to a multiple superposition $|\psi(t)> = \int d^F q\; |\phi_{q_1 \ldots q_F}(t)>\, f(q_1 \ldots q_F\, , t)$ carrying various collective correlations. In fact, this has been done in the following applications. It is, however, not very enlightening to write down the formula for the general case, since they remain essentially the same except for a confusing multitude of indexes.

3. One-dimensional dynamics

In order to test the TDGCM and to estimate its predictions we have applied the theory to the example of one-dimensional nuclear dynamics (SLAB's). We use a microscopic Hamiltonian similar to the Borche-Koonin-Negele force, consisting in kinetic energy, a folded Gaussian two-body force and a zero-range density dependent force. The Hartree-Fock Hamiltonian to a given density $\rho(x)$ reads then

$$\hat{W} = \frac{h^2}{2m} \partial_x^2 - V_o \int dx' \, \pi^{-1/2} \, \delta r^{-1} \exp\left(-(x - x')^2/ \, \delta r^2\right)\rho(x') + t_3 \, \rho^2(x) \tag{11}$$

where we used $V_o = 50$ MeV, $\delta r = 2$ fm and $t_3 = 8.8$ MeV fm^2. We use a space-grid representation of the single-particle wave function with grid size .3 fm and 140 grid points. The standard initial condition is two four-particle systems placed at a distance of 7.8 fm and having some relative momentum. We adopt quartett symmetry, i.e., the four particles all occupy the 1s state. This system has a binding energy of 11.6 MeV/particle. The compound system, consisting of 8 particles (one quartett in the 1s and the other quartett in the 1p state), has a binding of 15.6 MeV/particle. The TDHF solutions show fusion up to an impact momentum of $p = .28$ fm^{-1}. Altogether these are not precisely the numbers for realistic nuclei, but in orders of magnitude it describes the nuclear situation.

Since we have only one spacial dimension, the variety of collective modes in such a system is rather overseeable. For one single nucleus, we have essentially center-of-mass motion and one compression mode, representing all the $T = 0$ giant resonances in real nuclei. The isovector modes do not occur due to quartett symmetry and a surface mode does not occur because we have a closed-shell system. From the two modes in each cluster we can combine four collective modes in the scattering system. One of them is the total c.m. motion, it is omitted as a trivial mode. There remain three modes; each of them can be excited in two wyas, either by a static deformation or by a dynamic push. Thus, altogether we consider 6 different initialisations, as visualized in fig. 1.

Corresponding to the three modes, we have six observables. We consider the following three of them

$$\hat{P}_{rel} = \int_0^\infty dx \, \hat{j}(x) - \int_{-\infty}^0 dx \, \hat{j}(x) \tag{12a}$$

$$\hat{R}_+^2 = \int_0^\infty dx \, x^2 \, \hat{\rho}(x) + \int_{-\infty}^0 dx \, x^2 \, \hat{\rho}(x) \tag{12b}$$

$$\hat{R}_-^2 = \int_0^\infty dx \, x^2 \, \rho^1(x) - \int_{-\infty}^0 dx \, x^2 \, \hat{\rho}(x) \tag{12c}$$

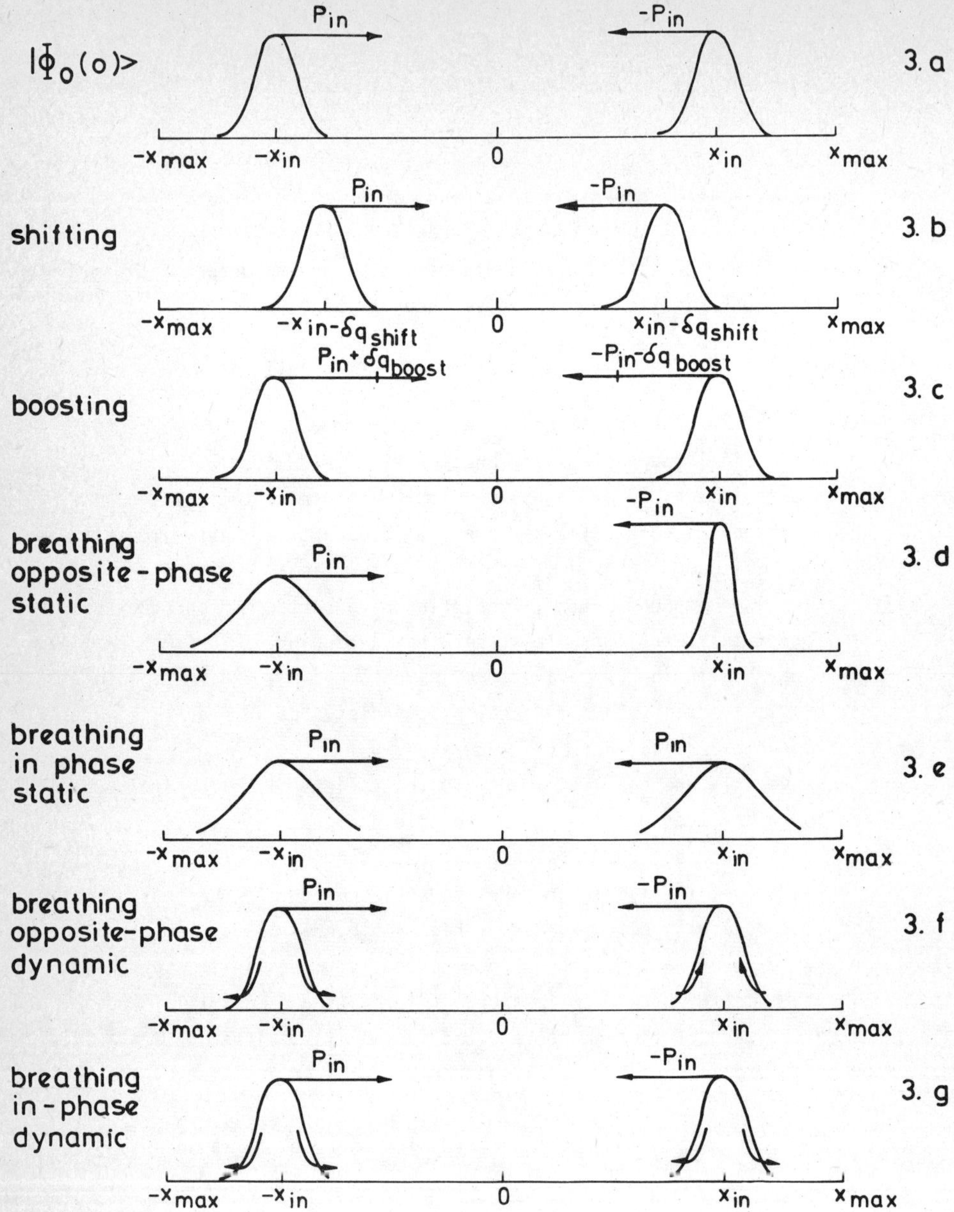

Fig. 1: The various initialization modes eximplified by drawing the initial density profile as a function of coordinate x. The arrows with p_{in} indicate the initial momentum of the clusters. The little arrows indicate the breathing flow of matter inside the clusters.

and in addition the difference in particle number

$$\hat{n}_{rel} = \int_0^{\infty} dx\ \hat{\rho}(x) - \int_{-\infty}^{0} dx\ \hat{\rho}(x) \ . \tag{12d}$$

These are the operators to be inserted in place of $\hat{A}$ in eq. (10).

4. Results and discussion

The first test for the theory is, of course, that it should reproduce the known features of center-of-mass motion. This case can be worked out analytically. We will not give the details of the evaluation here. We only want to quote the result, namely that TDGCM stands this test and shows aptly the spreading of wave packets in coordinate space, but no spreading in momentum space. The study of free c.m. motion, however, has some more importance since in the initial and in the final phase of heavy ion scattering the two clusters move freely. This allows to predict the asymptotic behaviour, for $t \to \infty$, from knowing a few quantities "measured" shortly after the reaction has ended. In that way we obtain, e.g., the very useful relation for the speading width of $\hat{p}_{rel}$

$$\frac{<\psi(+\infty)|\Delta^2\ p_{rel}|\psi(+\infty)>}{<\phi(-\infty)|\Delta^2\ p_{rel}|\phi(-\infty)>} = \frac{\partial p_{out}}{\partial p_{in}} \tag{13}$$

where $p_{out} = <\psi(t)|\ \hat{p}_{rel}\ |\psi(t)>$ for any t after the reaction time and p_{in} is the same for any t before the reaction starts. (To obtain that formula, we have neglected the contribution from the Im(α); it could be shown to be negligible).

Another interesting test of the theory is the study of a RPA vibration within one single cluster. We take $|\phi_0(t)> = |\phi_0>$ as the stationary Hartree-Fock solution and for $|\phi_{\delta q}(0)>$ we start with a static compression of this Hartree-Fock state. The results are shown in fig. 2. We observe harmonic oscillations for the input quantities $(2\mu)^{-1} = <\phi_0|\partial_q\ \partial_q|\phi_0>$ and α_i = Im(α). For the "measurement", namely the spreading width $\Delta^2\ \hat{R}^2$ we have three cases: first, the pure TDHF state (dotted line) produces a constant, but too small width. Second, we have initialized TDGCM with the pure TDHF state (dashed line); as a result the correlated ground state oscillates about the true RPA width. Third, we start TDGCM with the true RPA width (full line) and indeed we obtain a nice stationary line over many periods of oscillation. This proves the physical relevance and the numerical stability of the TDGCM.

As we see in fig. 1, we can initiate each mode in various phases, statically or dynamically or any phase in between. If we start the TDGCM initial condition $|\phi_{\delta q}(0)>$ in different phases the clusters will reach the interaction region in different stages of the internal oscillations and, thus, the coupling will behave quite different. This is shown in fig. 3, where we compare the spreading width $\Delta^2\ R_-^2$ after collision for initialization with the "breathing out-of-phase" mode in various phases. The t_0 means that we have allowed the oscillations to run for the time t_0 be-

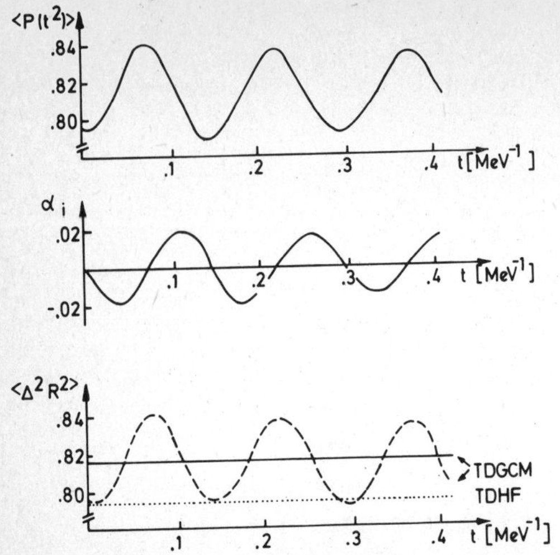

Fig. 2: Breathing oscillations in one nucleus at rest. Initialization with static deformation (analogous to breathing modes in fig. 1. We show $\mu(t)$ according to eq.(3b), $\mathrm{Im}(\alpha(t))$ and $<\Delta^2 \hat{R}^2>$. For $<\Delta^2 \hat{R}^2>$ we give the TDHF results (dotted line) the TDGCM result from pure TDHF start (dashed line) and the TDGCM result for properly correlated start (full line).

fore starting the relative motion of the clusters (thus, producing a phase of 60^0 for curve 1, 0^0 for 2, 150^0 for 3 and 90^0 for 4). As we see, the resulting coupling depends strongly on the initial phase. An obvious but expensive solution of this problem would be to average over many phases. In fact, it turns out that it is sufficient to use a two-channel TDGCM where both channels are initiated into the same mode but with 90^0 phase shift (e.g. channel 1 with static deformation and channel 2 with the corresponding dynamic excitation). This is exemplified in fig. 4, where we show the result of such a two conjugate channel TDGCM. For different initial phase, namely 0^0 and 60^0,

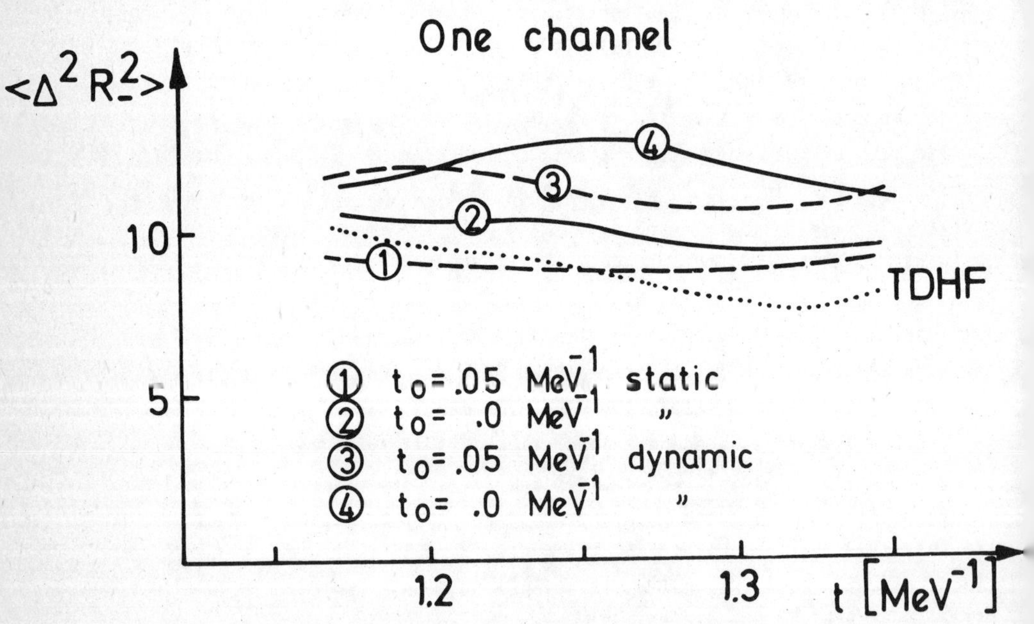

Fig. 3: The $<\Delta^2 \hat{R}^2>$ from the one channel TDGCM for the initial mode "breathing opposite-phase" and various initial phases.

we obtain the same $\Delta^2 R_-^2$, thus, we observe sufficient phase stability.

Now, having shown the numerical reliability of TDGCM and knowing how to produce phase stability, we can evaluate every combination of initialization and final measurement (after the reaction, i.e., $t \to \infty$). The results are given, for various impact momenta p_{in}, in table 1.

First, we observe generally strong contributions from the collective correlations to the spreading widths. Moreover, we see that the dominant contribution always comes from that initial mode which is related to the measurement (see the encircled number in table 1); this feature could be called "channel-memory". The cross channel contributions are much smaller. It may be that many of such small contributions add up to a further substantial enhancement. But this effect is hard to treat in TDGCM; a larger manifold of channels can only be treated statistically.

In table 1 we see also that the correlation effects are strongest for low momenta, near the fusion window at $p_{in} = .28\ fm^{-1}$, and tend to decrease for higher impact momenta. In fig. 5 we have plotted the TDGCM result for the spreading width of E*, the internal excitation energy after reaction, as a function of p_{in}. We see that the correlations indeed produce a significant enhancemant of Δ^2 E*, particularly for

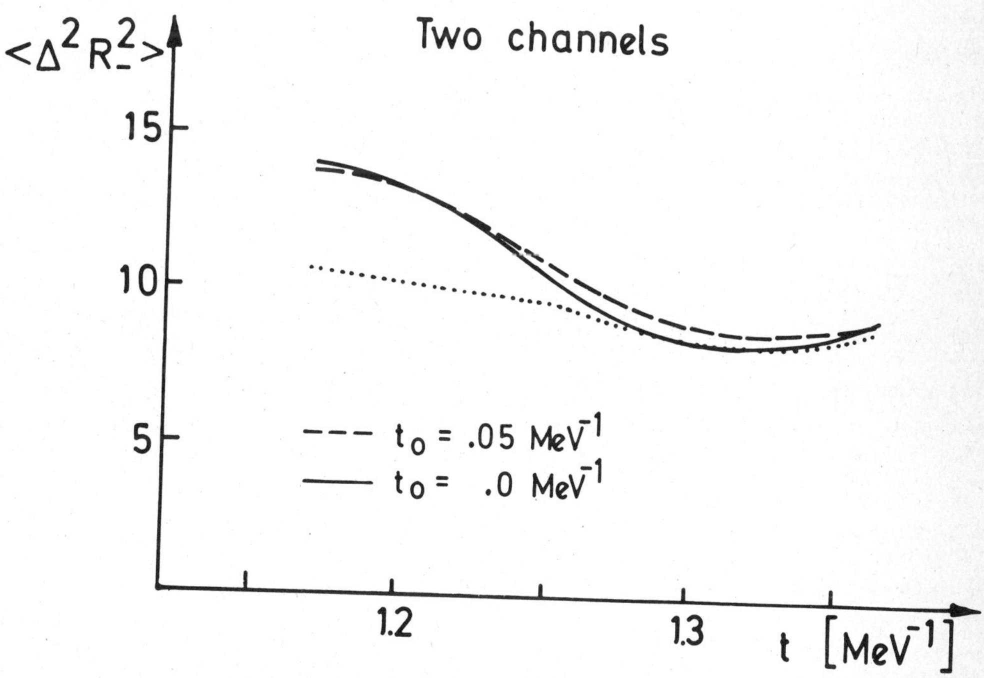

Fig. 4: The $<\Delta^2 \hat{R}^2>$ from two-conjugate-channel TDGCM for the initial mode "breathing-

Initialization	Measurement $\Delta^2 p_{rel}$ [fm^{-2}]	$\Delta^2 R_+^2$ [$f,^4$]	$\Delta^2 R_-^2$ [fm^4]	$\Delta^2 N_{rel}$
$p_{in} = 0.30\ fm^{-1}$				
TDGCM:				
shift + boost	2.40	18.03	10.01	0.51
breathing (+)	0.42	35.60	10.01	0.51
breathing (-)	0.29	16.51	13.14	0.52
TDHF	0.28	16.50	10.00	0.50
$p_{in} = 0.35\ fm^{-1}$				
T.GCM:				
shift + boost	0.88	7.81	6.43	0.36
breathing (+)	0.31	9.60	6.42	0.36
breathing (-)	0.30	5.74	7.86	0.33
TDHF	0.29	5.73	6.42	0.36
$p_{in} = 0.40\ fm^{-1}$				
TDGCM:				
shift + boost	0.63	5.79	4.84	0.24
breathing (+)	0.31	6.04	4.84	0.24
breathing (-)	0.30	4.86	5.02	0.24
TDHF	0.30	4.85	4.84	0.24

Table 1: The spreading width of various measuring operators is given for TDHF and for a couple of explicitly considered correlated time dependent states treated by means of TDGCM.

low p_{in} near the fusion window. For larger p_{in} the contribution decreases and seems to level off at a ratio of about 2, leaving a strong effect also for fast processes.

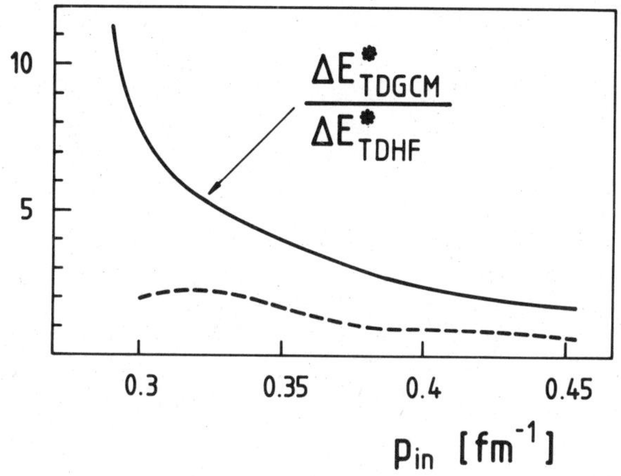

Fig. 5: The enhancement factor for the internal excitation energy E* as function of the initial momentum p_{in} (full line). The dashed line is the contribution due to the imaginary part of the correlation width α, amplified by a factor 100.

5. Conclusion

Time-dependent collective ground state correlations are supposed to describe an important part of the quantum fluctuations about a TDHF trajectory. Formulating the many-body dynamics with a time-dependent generator-coordinate method, we have finally worked out a simple formalism for treating those collective correlations which can easily be implemented on existing TDHF codes. We have applied the method to a one-dimensional model of nuclear dynamics. We observe the following interesting features:

- The method is numerically stable and reliable; this has been shown by tracing stationary RPA oscillations over many periods.
- For each collective mode we need a two-conjugate-channel TDGCM; this is necessary and sufficient to reach phase stability of the results in a heavy ion collision.
- There is a strong "channel-memory", i.e., for a given measuring operator the related initial excitation mode produces the dominant contribution to the spreading width.

- Altogether the collective correlation produce a significant enhancement of spreading widths of one-body observables, particularly for energies near the fusion window.

References

1. H.S. Köhler, Proceedings of the Conference on "TDHF and beyond", Bad Honnef 1982, Springer Tracts in Modern Physics
C.Y. Wong, as above
G. Wolschin, as above
2. S.E. Koonin, G. Sugiyama and H. Friedrich, as ref. 1
J.W. Negele, as ref. 1
3. P.-G. Reinhard, K. Goeke and R.Y. Cusson, Z. Phys. A295 (1980) 45
P.-G. Reinhard, K. Goeke and R.Y. Cusson, preprint 1982
4. P.-G. Reinhard, Nucl. Phys. A261 (1976) 291

A VARIATIONAL PRINCIPLE FOR THE AVERAGE VALUE AND THE DISPERSION OF AN OPERATOR ; APPLICATION TO MEAN FIELD THEORY

H. FLOCARD

Division de Physique Théorique[+], Institut de Physique Nucléaire, F-91406 ORSAY

This communication was prepared in collaboration with R. Balian, P. Bonche and M. Vénéroni.

I- INTRODUCTION

In a recent letter[1)] Balian and Vénéroni proposed a variational principle for the measure of observables in evolving systems. In this section we list some properties which, in our opinion, should be satisfied by any such principle.

a) A measurement involves two objects : an operator which corresponds to the observable of interest, and a density matrix which describes the system under consideration. An operator and a density matrix appear therefore as logical ingredients for a variational principle, the goal of which is the description of a measurement.

b) In an actual experiment the system is prepared at some initial time t_0, i.e. its density matrix D is known at that time. The trial density matrix $\mathcal{D}(t)$ which is to enter the variational principle should therefore satisfy

$$\mathcal{D}(t_0) = D .$$

The actual density matrix $\mathcal{D}(t)$ evolves then according to the Von-Neumann equation up to some time $t_1 > t_0$, at which the measurement is performed by using a specific apparatus which is described by an operator A . If we note $\mathcal{A}(t)$ the trial operator at any time we must have

$$\mathcal{A}(t_1) = A .$$

In other words, any variation on $\mathcal{A}$ and $\mathcal{D}$ consistent with the nature of a measurement at t_1, on a system prepared at t_0, should satisfy

$$\delta \mathcal{D}\big|_{t_0} = 0 \qquad ; \qquad \delta \mathcal{A}\big|_{t_1} = 0 \quad . \tag{1}$$

c) Although the usefulness of the principle will show up mainly when approximations will be introduced, it should also provide the exact equations of motion for the most general variation of the trial operator and density, $\mathcal{A}$ and $\mathcal{D}$.

d) It appears also desirable that the stationary "action" should possess the physical meaning of the quantity for which the principle has been built, i.e. the result of the measurement. This requirement that the action should have physical significance is for example not satisfied by the canonical Schrödinger variational principle for the wave-function $|\Psi\rangle$

$$\delta I_s = 0 \; ; \; I_s = \int_{t_0}^{t_1} dt \, \langle \Psi | \frac{d}{dt} + i H | \Psi \rangle \quad .$$

Indeed when solutions of this variational principle are inserted into the action I_s, this quantity does not acquire any particular significance : it vanishes for the actual solution.

II- A VARIATIONAL PRINCIPLE FOR THE AVERAGE VALUE OF AN OPERATOR

From the considerations of the preceding section, Balian and Vénéroni were led to propose the following principle

$$\delta I = 0 \quad ; \quad I = \mathrm{Tr}\mathcal{A}\mathcal{D}\big|_{t_1} - \int_{t_0}^{t_1} dt \, \mathrm{Tr}\, \mathcal{A}\left(\frac{d\mathcal{D}}{dt} + i[H, \mathcal{D}]\right) \tag{2}$$

Although, several mean field approximations can be derived from this variational principle, the trial density $\mathcal{D}$ is not constrained to be normalized. This feature turns out to be cumbersome for the application we are interested in (evaluation of dispersions). This is why we start anew with the following slightly modified principle.

$$\delta I_a = 0 \; ; \; I_a = \frac{\mathrm{Tr}\mathcal{A}\mathcal{D}}{\mathrm{Tr}\mathcal{D}}\bigg|_{t_1} - \int_{t_0}^{t_1} dt \, \mathrm{Tr}\, \mathcal{A}\left(\frac{d}{dt} \frac{\mathcal{D}}{\mathrm{Tr}\mathcal{D}} + i\left[H, \frac{\mathcal{D}}{\mathrm{Tr}\mathcal{D}}\right]\right)$$
$$\delta \mathcal{A}\big|_{t_1} = 0 \qquad ; \qquad \delta \mathcal{D}\big|_{t_0} = 0 \tag{3}$$

In both definition (2) and (3) of an action, H is the hamiltonian and Tr stands for the trace in the Fock space in which the operators $\mathcal{A}(t)$ and $\mathcal{D}(t)$ act.

For any variational subspaces the variations with respect to $\mathcal{A}$ and $\mathcal{D}$ lead to the equations

$$\left\{ \begin{array}{l} \mathrm{Tr}\ \delta\mathcal{A}\left(\frac{d\tilde{\mathcal{D}}}{dt} + i\,[H,\tilde{\mathcal{D}}]\right) = 0 \quad ; \quad \tilde{\mathcal{D}} \equiv \frac{\mathcal{D}}{\mathrm{Tr}\mathcal{D}} \\ \mathrm{Tr}\ \frac{\delta\mathcal{D}}{\mathrm{Tr}\mathcal{D}} \left\{ \frac{d\mathcal{A}}{dt} + i\,[H,\mathcal{A}] - \mathrm{Tr}\tilde{\mathcal{D}}\left(\frac{d\mathcal{A}}{dt} + i[H,\mathcal{A}]\right) \right\} \end{array} \right. {}^{*} \tag{4.a}$$

These equations can be equivalently written as

$$\left\{ \begin{array}{l} \delta_{\mathcal{A}}\ \mathrm{Tr}\ \mathcal{A}\left(\frac{d\tilde{\mathcal{D}}}{dt} + i\,[H,\tilde{\mathcal{D}}]\right) = 0 \\ \delta_{\mathcal{D}}\ \mathrm{Tr}\ \tilde{\mathcal{D}}\left(\frac{d\mathcal{A}}{dt} + i\,[H,\mathcal{A}]\right) = 0 \end{array} \right. \tag{4.b}$$

where $\delta_{\mathcal{A}}$ and $\delta_{\mathcal{D}}$ stand for the variations with respect to the parameters which define $\mathcal{A}$ and $\mathcal{D}$ in the particular subspaces considered. For an arbitrary choice of these subspaces the existence of solutions to the equations (4) is not guaranteed. It will have to be checked for every particular choice.

If the equations (4) have a solution, then along a trajectory

$$\delta\mathcal{A} = \frac{d\mathcal{A}}{dt}\,\delta t \qquad\qquad \delta\mathcal{D} = \frac{d\mathcal{D}}{dt}\,\delta t$$

are certainly allowed variations. Inserting them into (4) and subtracting the equations, we find that the quantity

$$\mathcal{E}_a = -i\ \mathrm{Tr}\ \mathcal{A}\,[H,\tilde{\mathcal{D}}] \tag{5}$$

is conserved along any trajectory. $\mathcal{E}_a$ is the invariant associated with the time translational invariance of the principle (3), and as such can be called a pseudo-energy.

Let us now find a condition on the variational space for $\mathcal{A}$ which will ensure the property d) presented in the first section, namely that the stationary

* Note the distinct appearance of $\mathcal{D}$ and $\tilde{\mathcal{D}}$ in this second equation.

value of the action I_a is equal to the average value of the operator A at time t_1 (in the restricted space chosen for the density matrix). If a variation $\delta\mathcal{A}$ proportional to $\mathcal{A}$ is possible, equation (4.a) tells us that any solution of the equation of motion satisfies

$$\mathrm{Tr}\,\mathcal{A}\left(\frac{d\tilde{\mathcal{D}}}{dt} + i[H,\tilde{\mathcal{D}}]\right) = 0 \tag{6}$$

and the action (3) is indeed equal to the average value of A :

$$I_a = \mathrm{Tr}\,\mathcal{A}\tilde{\mathcal{D}}\Big|_{t_1} = \mathrm{Tr}\,A\tilde{\mathcal{D}}\Big|_{t_1} \quad . \tag{7}$$

From now on,we assume that the above condition ($\delta\mathcal{A} \propto \mathcal{A}$ allowed) will be satisfied by the variational space for the operator $\mathcal{A}(t)$.

We have now to check the point c) of the introduction by considering the most general variation for the operator $\mathcal{A}$ and the density matrix $\mathcal{D}$. We find first an equation for the normalized density

$$\frac{d\tilde{\mathcal{D}}}{dt} + i[H,\tilde{\mathcal{D}}] = 0 \quad . \tag{8}$$

This is the Von Neumann equation for $\tilde{\mathcal{D}}$. We note that due to the invariance of the action I_a under the gauge transformation

$$\mathcal{D}(t) \rightarrow (\exp f(t)).\,\mathcal{D}(t) \quad ; \quad f(t_0) = 0 \tag{9}$$

we have no information on the evolution of the norm of $\mathcal{D}$. This however does not affect the value of the action. The second equation, once we take eq.(6) into account, writes

$$\frac{d\tilde{\mathcal{A}}}{dt} + i[H,\tilde{\mathcal{A}}] = 0 \tag{10}$$

with

$$\tilde{\mathcal{A}} = \mathcal{A} - \mathrm{Tr}\,\mathcal{A}\mathcal{D} \quad . \tag{11}$$

In other words, up to an unspecified constant, the time evolution of $\mathcal{A}$ follows the Schrödinger equation *. This undeterminacy on $\mathcal{A}$, which has no effect whatsoever on the result of the measurement since $\mathcal{A}$ is fully specified at the time of interest t_1 ($\mathcal{A}(t_1) = A$) , results also from the invariance of the action under a gauge transformation :

$$\mathcal{A}(t) \rightarrow \mathcal{A}(t) + g(t) \quad ; \quad g(t_1) = 0 \tag{12}$$

The simplest way to remove the uncertainty on $\mathcal{A}$ is to impose

$$\frac{d}{dt} \operatorname{Tr} \mathcal{A} \tilde{\mathcal{D}} = 0$$

which implies that the time evolution of $\mathcal{A}$ is governed by the equation

$$\frac{d\mathcal{A}}{dt} + i \, [H, \mathcal{A}] = 0$$

We postpone to section IV the application of eq.(4) to mean field theory and turn now to the extension of the variational principle to the case of the dispersion of an operator.

III- EXTENSION OF THE PRINCIPLE TO MEASURE THE DISPERSION OF AN OPERATOR

Given variational subspaces for $\mathcal{A}$ and $\mathcal{D}$, we have determined an optimal answer to the problem of measuring, at time t_1 , the expectation value $\langle A \rangle$ of the operator A and we know that

$$I_a = \langle A \rangle \tag{13}$$

In a next step we would like to use the same variational spaces to obtain an optimal value for the dispersion Δ of the operator A (at the same time t_1) around its average value $\langle A \rangle$. To do so, we consider the action

---

* Note that $\mathcal{A}(t)$ cannot be identified with an operator in the Schrödinger representation (which is a constant). Rather if we denote it : $\mathcal{A}(t_1, t)$, it obeys the Heisenberg equation with respect to t_1 and the Schrödinger eq.(10) with respect to t .

$$I_d = \mathrm{Tr}\,\mathcal{A}^2\tilde{\mathcal{D}}\Big|_{t_1} - \int_{t_0}^{t_1} dt\,\mathrm{Tr}\left\{\mathcal{A}^2\left(\frac{d\tilde{\mathcal{D}}}{dt} + i\,[H,\tilde{\mathcal{D}}]\right) - \lambda\mathcal{A}\tilde{\mathcal{D}}\right\} \tag{14.a}$$

which is to be extremized with respect to $\mathcal{A}(t)$ and $\mathcal{D}(t)$ subject to the conditions

$$\delta\mathcal{A}\Big|_{t_1} = 0 \qquad \mathcal{A}(t_1) = A - \langle A\rangle$$
$$\delta\mathcal{D}\Big|_{t_0} = 0 \tag{14.b}$$

and where the Lagrange parameter λ is adjusted so as to ensure

$$\mathrm{Tr}\,\mathcal{A}\tilde{\mathcal{D}} = 0 \tag{15}$$

Due to the conditions (14.b), (15) the quantity $\mathrm{Tr}\,\mathcal{A}^2\tilde{\mathcal{D}}\big|_{t_1}$, which, as we shall see , is the stationary value of the action I_d , turns out to be the dispersion of A measured at time t_1 (within the restricted supspace adopted for the variation of $\mathcal{D}$).

$$\mathrm{Tr}\,\mathcal{A}^2\mathcal{D}\Big|_{t_1} \equiv \Delta = \mathrm{Tr}\,A^2\tilde{\mathcal{D}} - (\mathrm{Tr}\,A\tilde{\mathcal{D}})^2\Big|_{t_1} \tag{16}$$

The principles (3) and (14) which have the same structure appear thus as the first two steps towards an utilization of the variational spaces for $\mathcal{A}$ and $\mathcal{D}$ in order to achieve an optimal knowledge of the results of the interaction of the system with the operator A . Indeed generalizations of (14) to higher moments of A are simple.

Let us now derive the equations of motion associated with the action (14). One finds

$$\mathrm{Tr}\left\{\delta(\mathcal{A}^2)\left(\frac{d\tilde{\mathcal{D}}}{dt} + i\,[H,\tilde{\mathcal{D}}]\right) - \lambda\,\delta\mathcal{A}\,\tilde{\mathcal{D}}\right\} = 0 \; ; \; \delta(\mathcal{A}^2) = \delta\mathcal{A}.\mathcal{A} + \mathcal{A}.\delta\mathcal{A}$$
$$\mathrm{Tr}\,\delta\tilde{\mathcal{D}}\left(\frac{d\mathcal{A}^2}{dt} + i\,[H,\mathcal{A}^2] + \lambda\mathcal{A}\right) = 0 \; ; \quad \tilde{\mathcal{D}} = \mathcal{D}/\mathrm{Tr}\,\mathcal{D} \tag{17}$$

An other useful formulation is

$$\delta_{\mathcal{A}} \,\mathrm{Tr} \left\{ \mathcal{A}^2 \left(\frac{d\tilde{\mathcal{D}}}{dt} + i\,[H, \tilde{\mathcal{D}}] \right) - \lambda \mathcal{A} \tilde{\mathcal{D}} \right\} = 0 \tag{18}$$

$$\delta_{\mathcal{D}} \,\mathrm{Tr}\; \tilde{\mathcal{D}} \left(\frac{d\mathcal{A}^2}{dt} + i\,[H, \mathcal{A}^2] + \lambda \mathcal{A} \right) = 0$$

As in the preceding section we find a constant of motion $\mathcal{E}_d$ associated with the time translational invariance of the action (14).

$$\mathcal{E}_d = -i\, \mathrm{Tr}\; \mathcal{A}^2 [H, \tilde{\mathcal{D}}] \tag{19}$$

Similarly the restriction to subspaces for $\mathcal{A}$ which allow variations $\delta \mathcal{A}$ proportional to $\mathcal{A}$ ensures the desired result that the action I_d verifies

$$I_d = \mathrm{Tr}\; \mathcal{A}^2 \tilde{\mathcal{D}} \big|_{t_1} = \Delta \tag{20}$$

The most general variation for $\mathcal{A}$ and $\mathcal{D}$ leads to the equations.

$$\left\{ \mathcal{A},\; \frac{d\tilde{\mathcal{D}}}{dt} + i\,[H, \tilde{\mathcal{D}}] \right\}_+ = \lambda \tilde{\mathcal{D}}$$

$$\left\{ \mathcal{A},\; \frac{d\mathcal{A}}{dt} + i\,[H, \mathcal{A}] \right\}_+ = -\lambda \mathcal{A} + \frac{d}{dt} \mathrm{Tr}\; \mathcal{A}^2 \tilde{\mathcal{D}} \tag{21}$$

where the curly brackets stand for anticommutators. The exact equations of motion

$$\frac{d\tilde{\mathcal{D}}}{dt} + i\,[H, \tilde{\mathcal{D}}] = 0 \;;\quad \frac{d\mathcal{A}}{dt} + i\,[H, \mathcal{A}] = 0 \;;\quad \lambda = 0 \tag{22}$$

are evidently solutions of equations (21). By considering the matrix equations obtained by writing eq.(21) in the basis which diagonalizes $\mathcal{A}(t)$ one sees that most of the matrix elements are identical to those obtained from the exact equations (22). In other words, it is only in some subspaces (in particular the kernel of $\mathcal{A}$) that spurious solutions may arise.

IV- APPLICATION TO THE MEASURE OF A ONE-BODY OPERATOR

Let us now derive the equations for a particular choice of the variational subspaces. We shall restrict $\mathcal{A}(t)$ (and thus A) to be a one-body operator of the type

$$\mathcal{A} = \frac{a}{2} + \sum_{ij} c_i^+ B_{ij} c_j \tag{23}$$

where the set of operators $\{c_i^+\}$, $i = 1..N$ generates the Fock space, the number a and the matrix elements B_{ij} being $\mathbb{C}$-numbers* . For $\mathcal{D}$ we adopt the variational space of uncorrelated density matrices :

$$\mathcal{D} = p \cdot \exp \sum_{ij} c_i^+ R_{ij} c_j \tag{24}$$

where again p and R_{ij} are $\mathbb{C}$ -numbers. These uncorrelated density operators are in one to one correspondance with the set of numbers formed by the partition function Z plus the one-body density matrix $\{\rho_{ij}\}$ defined by

$$Z = \mathrm{Tr}\, \mathcal{D}$$

$$\rho_{ij} = \mathrm{Tr}\, c_j^+ c_i \mathcal{D} / \mathrm{Tr}\, \mathcal{D} \equiv \frac{1}{2} (\delta_{ij} + M_{ij}) \tag{25}$$

(We introduce the NxN matrix M which allows a simpler form for the equations of motion).

A) Equations of motion for the average value

We first consider the equations associated with the action I_a . To derive the explicit form of the equations of motion (4.b) we must evaluate the quantities $\mathrm{Tr}\, \mathcal{A} \frac{d\tilde{\mathcal{D}}}{dt}$, $\mathrm{Tr}\, \tilde{\mathcal{D}} \frac{d\mathcal{A}}{dt}$ and $\mathrm{Tr}\, \mathcal{A} [H, \tilde{\mathcal{D}}]$. This can be done easily by means of the Wick theorem which applies for density matrices of the form (24). One finds

---

*In ref.(1) similar equations were derived (although not explicitly written) for the slightly more general case of operators which do not commute with the particle number operator.

$$2\,\mathrm{Tr}\,\mathcal{A}\tilde{\mathcal{D}} = a + \mathrm{tr}\,B(1+M)$$

$$2\,\mathrm{Tr}\,\mathcal{A}\frac{d\tilde{\mathcal{D}}}{dt} = \mathrm{tr}\,B\dot{M} \tag{26}$$

$$2\,\mathrm{Tr}\,\tilde{\mathcal{D}}\frac{d\mathcal{A}}{dt} = \dot{a} + \mathrm{tr}\,\dot{B}(1+M)$$

In the above equations the notation tr indicates a trace in the one-body space labelled by the index i = 1 ... N whereas Tr denotes a trace in the Fock-space ; the dots indicate a time derivative. For the Hamiltonian H of the many fermions system , we choose

$$H = \sum_{ij} t_{ij}\, c_i^+ c_j + \frac{1}{4}\sum_{ijkl} V_{ijkl}\, c_i^+ c_j^+ c_l c_k \tag{27}$$

with

$$V_{ijkl} = -V_{jikl} = -V_{ijlk} \tag{28}$$

Then the pseudo-energy (5) writes :

$$\mathcal{E}_a = -i\,\mathrm{Tr}\,\mathcal{A}[H,\tilde{\mathcal{D}}] = -\frac{i}{2}\,\mathrm{tr}\,B[W,M] \tag{29}$$

where the Hartree-Fock hamiltonian W is given by

$$W_{ij} = t_{ij} + \frac{1}{2}\sum_{kl} V_{iljk}\,(1+M)_{kl}$$

It is convenient to introduce the notation

$$\sum_{kl} V_{iljk}\, C_{kl} = {}^{tr}V^{C}_{ij}$$

so that the NxN matrix W takes the form

$$W = t + \frac{1}{2}\,{}^{tr}V^{(1+M)} \tag{30}$$

Let us now perform the variations implied by equation (4.b), where $\delta_{\mathcal{A}}$ stands for δ_a and δB_{ij} whereas $\delta_{\mathcal{D}}$ stands for δ_z and δM_{ij}. The variations with respect to a and z lead to two identities 0 = 0. This results from the two gauge invariances already noted ((9), (12)) which are still present in the restricted variational spaces used in this section. We recall that the undeterminacy associated with these gauge transformations does not affect the final result. The variations with respect to B_{ij} and M_{ij} give the two matrix equations

$$\dot{M} + i\,[W, M] = 0 \tag{31.a}$$

$$\dot{B} + i\,[W, B] + i\,T = 0 \tag{31.b}$$

where we introduced the additional mean field T

$$T = \frac{1}{2}\, {}^{\mathrm{tr}}\mathcal{V}^{[B,M]} \tag{32}$$

Eq.(31.a) is the usual T.D.H.F. equation. It is completely decoupled from the time evolution of B. This is the "happy accident" noted in ref.(1) : the T.D.H.F evolution in the space of uncorrelated density matrices is optimal whatever one-body operator is to be measured at the final time t_1. On the other hand, the time evolution of B, described by eq.(31.b) is coupled to that of M through the action of the two mean fields W and T. It is not however necessary to solve eq.(31.b) since, the operator $\mathcal{A}(t_1) = A$ being known at the final time, its expectation value can be computed once the T.D.H.F. equation has been solved. As a last remark we note that the T.D.H.F. equations (31) are not restricted by the condition $M^2 = 1$ (or $\rho^2 = \rho$) which states that the density matrix corresponds to a pure state with a determinantal wave function.

B) Equations of motion for the dispersion

We shall now derive the explicit form of the equations (18) associated with the action I_d. To do so, we must evaluate the quantities $\mathrm{Tr}\, \mathcal{A}^2 \frac{d\tilde{\mathcal{D}}}{dt}$, $\mathrm{Tr}\, \tilde{\mathcal{D}} \frac{d\mathcal{A}^2}{dt}$ and $\mathrm{Tr}\, \mathcal{A}^2 [H, \tilde{\mathcal{D}}]$. Using again the Wick theorem one finds

$$
\begin{aligned}
2\,\mathrm{Tr}\,\mathcal{A}\tilde{\mathcal{D}} &= a + \mathrm{tr}\,B(1+M) \equiv 2\bar{A} \\
4\,\mathrm{Tr}\,\mathcal{A}^2\tilde{\mathcal{D}} &= 4\bar{A}^2 + \mathrm{tr}\,B^2 - BMBM \\
2\,\mathrm{Tr}\,\mathcal{A}^2\frac{d\tilde{\mathcal{D}}}{dt} &= 2\bar{A}\,\mathrm{tr}\,B\dot{M} - \mathrm{tr}\,BMB\dot{M} \qquad (33)\\
2\,\mathrm{Tr}\,\tilde{\mathcal{D}}\frac{d\mathcal{A}^2}{dt} &= 2\bar{A}\,(\dot{a} + \mathrm{tr}\,\dot{B}(1+M)) + \mathrm{tr}\,\dot{B}(B - MBM) \\
2\,\mathrm{Tr}\,\mathcal{A}^2[H,\tilde{\mathcal{D}}] &= -\,\mathrm{tr}\,([W,B]+T)(B-MBM) - 2\bar{A}\,\mathrm{tr}([W,B]M)
\end{aligned}
$$

In constrast with I_a, the action I_d is not invariant under the gauge transformation (12). Indeed shifting $\mathcal{A}^2$ by a constant will in general destroy the property of being the square of an operator. The practical implication is that the variation (18) with respect to a does not lead to an identity 0 = 0 (as does the variation with respect to Z). The variation with a provides a scalar equation which can be used to determine the Lagrange parameter λ . After some tedious but straightforward algebra, one ends up with the equations of motion, which write more conveniently in terms of two matrices X and Y measuring the deviations from the T.D.H.F. equations (31)

$$
\begin{aligned}
X &= \dot{M} + i\,[W,M] \\
Y &= \dot{B} + i\,([W,B] + T)
\end{aligned}
\qquad (34)
$$

In turn the matrices X and Y are obtained in terms of M and B by solving the following matrix equations

$$
\begin{aligned}
MBX + XBM &= i\,([S,M] - T + MTM) \\
BMY + YMB &= i\,([S,B] - U)
\end{aligned}
\qquad (35)
$$

where two new mean fields

$$
S = \frac{1}{2}\,\mathrm{tr}\,V^{B-MBM} \quad ; \quad U = \frac{1}{2}\,\mathrm{tr}\,V^{(BM)^2-(MB)^2} \qquad (36)
$$

have been introduced. From the knowledge of X and Y, we can evolve M and B by equation (34) and proceed further in time. In contrast with eq.(31), eqs.(34-35) describe a fully coupled motion of M and B , which makes the solution more difficult. In addition we recall that the operator B is not known at the initial time t_0 . An iterative process should thus be used (guessing some trial initial value $B(t_0)$) in order to ensure that $B(t_1)$ is indeed the operator chosen at the final time. The variation with respect to a along with the constraint $\mathrm{Tr}\, ct\tilde{\mathcal{D}} = 0$ provide us with two decoupled equations which give the time evolution of the Lagrange parameter λ and the operator constant a , once equations (34-35) have been solved

$$\lambda = \mathrm{tr}\, BX$$
$$a = -\mathrm{tr}\, B(1+M) \qquad (37)$$

V- CONCLUSION

We propose a variational principle which can be used to extract successively the optimal average value and dispersion associated with the measure of a given operator at a final time t_1 on a system whose density matrix is known at some initial time t_0 . For the most general variations the exact equations of motion are recovered. In addition, the stationary values of the actions are equal to the quantities of interest, namely the average value and the dispersion measured in the variational space of the trial density matrices. We have derived the equations for the case of one-body operators and uncorrelated density matrices and showed how the time-dependent Hartree-Fock equations should be modified in order to evaluate a dispersion. Application to the Lipkin model is in progress.

Ref.(1) R. BALIAN, M. VENERONI, Phys. Rev. Let. 47(1981) 1353.

NOTE : Page 1355, 1st Col umn, 3nd Paragraph, the sentence "the best choice for $\underline{D}(t)$ does not depend ...", should be replaced by "the best choice for $\underline{D}(t)$ does depend ...".

On two-body collisions in an equilibrating Fermi gas$^+$

H. Feldmeier* and P. Buck

Institut für Kernphysik, Technische Hochschule,

D-6100 Darmstadt, FRG

1. Introduction

The deeply inelastic scattering of two heavy ions exhibits dissipative phenomena very similar to those of a Brownian motion [WE 80, GR 80]. The stochastic nature of the interaction between the two nuclei results in the drift and diffusion of the mass number and other macroscopic quantities. The intrinsic degrees of freedom serve as a heat bath for the collective degrees of freedom. In contrast to the number of gas particles interacting with the Brownian particle, the number of nucleons involved in a heavy-ion reaction is rather small. Besides this, the collision of two nuclei is extraordinary in the sense that a closed quantum system undergoes a transition from a well ordered pure state to a very complicated state with high excitation energy.

In chapter 2 we investigate a model system of interacting fermions for which we can solve the time-dependent many-body Schrödinger equation numerically. Its dynamical behavior will be illustrated in terms of the one-body Wigner function and the particle number in one half of the volume, serving as the macroscopic variable.

Among the theories used in heavy-ion physics the time-dependent Hartree-Fock theory (TDHF) is the only one which describes the deterministic evolution of a wavefunction. Hence there is no production of

---

+) Work supported by Gesellschaft für Schwerionenforschung mbH

*) Heisenberg fellow; present address: MPI für Kernphysik, Heidelberg

entropy and the system is not dissipative. However, the detailed information about the intrinsic state which is also contained in the wavefunction is of no practical use. It is even not expected to be verified if a very fancy experiment would be performed to measure it. There are only few quantities which can be compared with experimental data (e.g. relative momentum, relative mass), and they show clearly dissipative behavior. The inclusion of collision terms in TDHF [WT 78, OS 79, AY 80, KÖ 80, GW 81] (in the following denoted by TDHF + C) has two aspects. First, the theory becomes dissipative and the infinite mean free path of TDHF becomes finite, and second, the theory provides only a non idempotent one-body density which gives no information about two-body quantities like variances without further assumptions. The first aspect is desired, whereas the second one is a side effect which is of no consequence, since TDHF predictions for variances are known to be wrong anyhow [KD 77, DM 78]. Since a mean-field theory with collision terms should be applicable to our model system we expect systematics seen here to apply to TDHF + C as well. In chapter 3 we draw conclusions about the effect of the residual two-body interaction on the mass variance.

2. Study of a simple model system

The model system [BF 82, FE 80] we want to study has a one dimensional infinite square-well potential, in which A identical fermions (same spin and isospin) are moving. The particles are interacting via a two-body interaction V_{12} which depends on the relative distance between the particles (cf. illustration in fig. 1). The interaction has a repulsive core and an attractive tail. The initial state of the system is chosen such that we start with two gases separated by a wall, having a sharp particle number in each half of the box. At time t = 0 the wall is removed instantaneously and the state of the system evolves according to the Hamiltonian

$$H = \sum_{i=1}^{A} \left\{ \frac{p_i^2}{2m} + V_{box}(x_i) \right\} + \sum_{i<j} V_{12}(|x_i - x_j|) \tag{1}$$

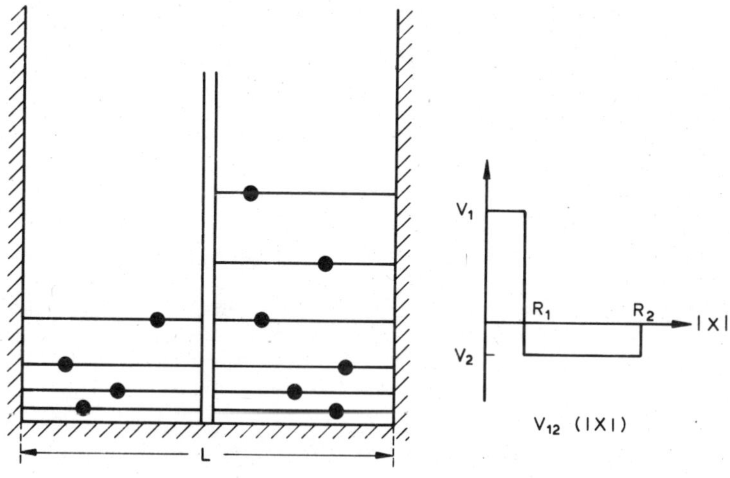

Fig. 1 Schematic illustration of the model system. The two-body interaction $V_{12}(|x|)$ has the parameters V_1 = 63 MeV, V_2 = -7 MeV, R_1 = 2fm, R_2 = 8 fm.

The time-dependent many-body Schrödinger equation

$$i\hbar \frac{\partial}{\partial t} \Psi(x_1, \ldots, x_A; t) = H \Psi(x_1, \ldots, x_A; t) \tag{2}$$

is solved numerically.

In figs. 2 and 3 we illustrate the dynamical evolution of the system by displaying the Wigner function f(x,k,t) at different times.

$$f(x,k,t) = \frac{1}{2\pi} \int e^{-ikz} \langle x - \tfrac{z}{2} | \rho^{(1)}(t) | x + \tfrac{z}{2} \rangle \, dz \tag{3}$$

The Wigner function contains the same information as the one-body density $\rho^{(1)}(t)$ and is the quantum analogue to the classical phase-space distribution. The time-dependent one-body density $\rho^{(1)}(t)$ is calculated from the many-body state $\Psi(x_1, \ldots, x_A; t)$ which is the solution of the Schrödinger equation (2). Figure 2 shows contour maps of the Wigner function for the ideal gas case, where the two-body interaction is turned off. Initially we have four particles in the left half of the box and six particles in the right. After removing the wall at t = 0 the surfaces of the two distributions move more or less parallel without penetrating. The gap survives for a long time and the two gases start mixing only after the fastest particles have hit the opposite wall ($t \approx 5 \cdot 10^{-22}$ s). At a time of about $9 \cdot 10^{-22}$ most of the particles are on the left side. The gas swings back and forth for a long time. Looking at the system for times greater than $40 \cdot 10^{-22}$ s, when the number of particles on each side has settled to the mean value of 5, one realizes a very fissured distribution with an ongoing rapid internal motion. The reason for this lies in the fact that in the ideal gas case without two-body collisions the wavefunction remains a single slater determinant for all times. Thus the whole information about the many-body system is contained in the Wigner function and does not get lost during the course of time.

The time evolution of the Wigner function for the real gas is shown in figure 3. The initial state has again four particles on the left and six on the right. It includes already the two-body interaction but is nevertheless almost the same state as for the ideal gas. However, the dynamical evolution is strongly influenced by the two-body collisions. Already at $t = 1.75 \cdot 10^{-22}$s the two gases begin to mix. The distribution does not show much structure for times greater than $16 \cdot 10^{-22}$s. Also negative values (hatched areas) which existed in the ideal gas case have disappeared.

The operator N_L which counts the number of particles in the left half of the box may serve as a macroscopic variable in our model.

$$N_L = \sum_{i=1}^{A} \tilde{\Theta}(x_i) \quad ; \quad \tilde{\Theta}(x_i) = \begin{cases} 1 & \text{for } 0 \leq x \leq L/2 \\ 0 & \text{for } L/2 \leq x \leq L \end{cases} \tag{4}$$

The time dependence for the expectation value $\langle N_L \rangle$ and the variance

$$\sigma^2 = \langle N_L^2 \rangle - \langle N_L \rangle^2 \tag{5}$$

for the previously discussed cases are shown in figure 4. The fast damping of $\langle N_L \rangle$ in the real gas case reflects the viscosity induced by the collisions. The reason for the higher frequency of the oscillations is that a particle needs not pass all the way to the opposite wall to reverse its velocity. Collisions with other particles have the same effect.

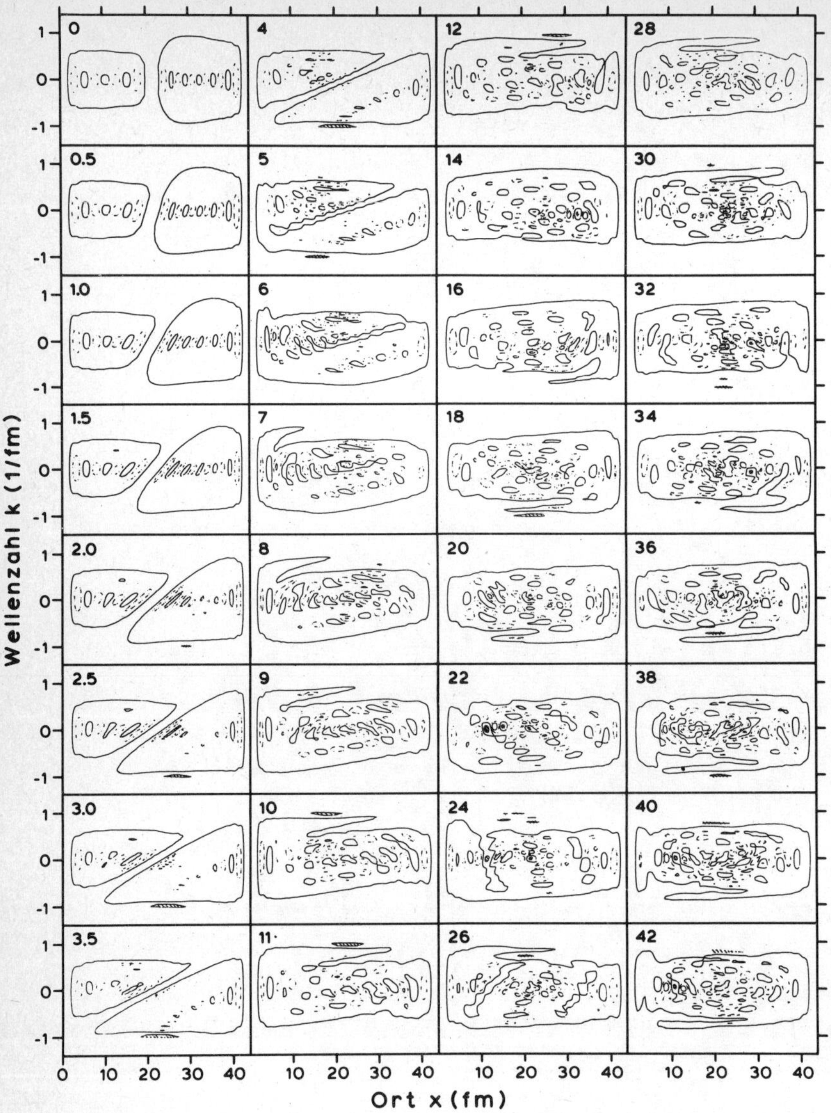

Fig. 2 Contour plot of the Wigner function for the ideal gas. Numbers in the upper left corner of each box denote the time in 10^{-22}s. Contour lines: —— $f = \frac{1}{3\pi}$, --- $f = \frac{2}{3\pi}$, —— $f = \frac{1}{\pi}$; hatched areas: $f < -\frac{1}{6\pi}$.

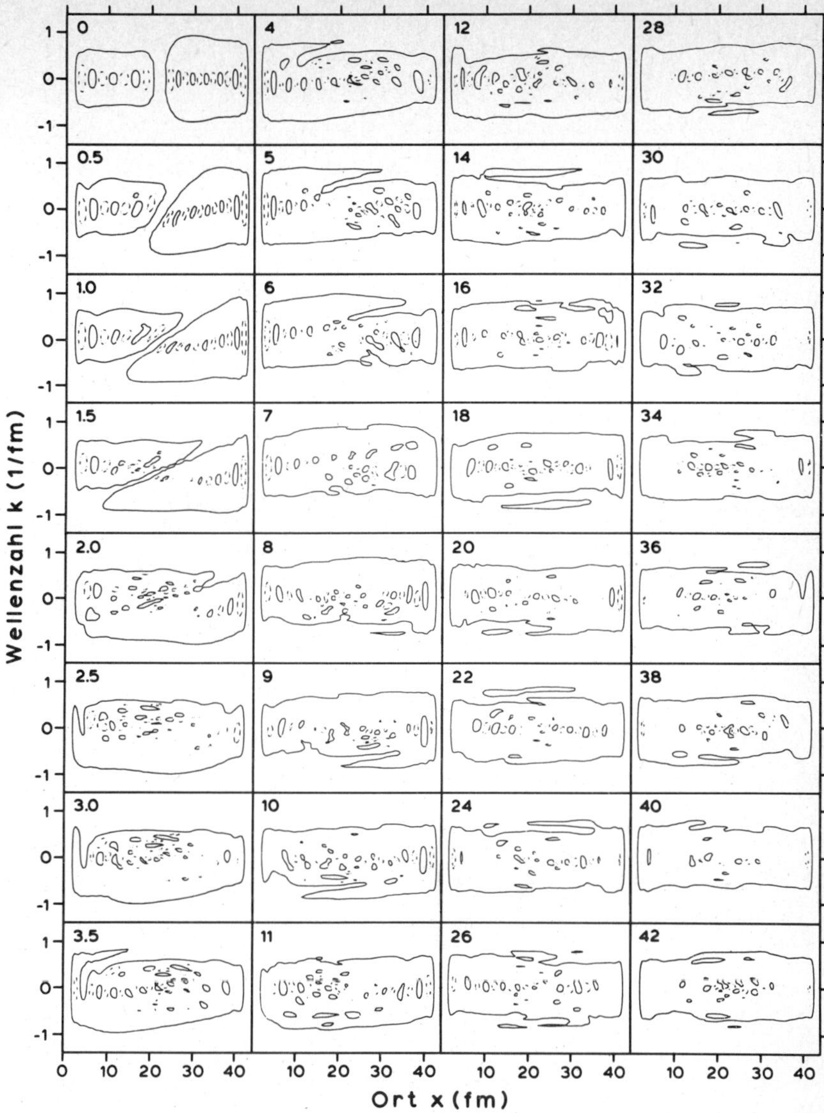

Fig. 3 Contour plot of the Wigner function for the real gas. Otherwise same as fig. 2.

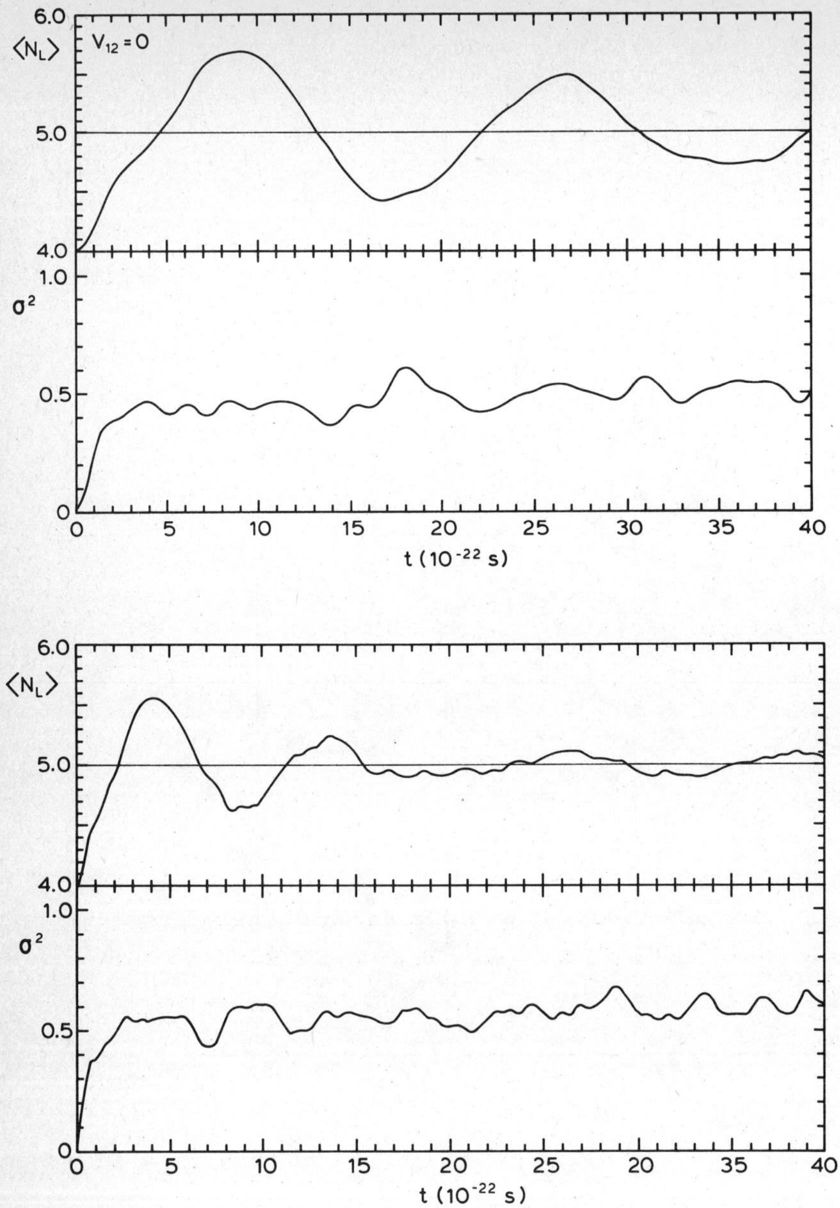

Fig. 4 Mean value $\langle N_L \rangle$ and variance σ^2 of the particle number in the left hand side of the box. Upper part for the ideal gas, lower part for the real gas.

3. Small variances

A remarkable result of the calculation is that the mass variance σ^2 (fig. 4) does not increase significantly if two-body collisions are included. In both cases it saturates rather fast at a value around 0.5 which is small compared to the classical value of $\sigma^2_{cl} = A/4 = 2.5$. This behavior is a consequence of the Pauli exclusion principle together with the fact that the particles are enclosed in a mean potential. To explain this we imagine the state of the system $|\Psi\rangle$ represented in eigenstates $|i, n_L\rangle$ of N_L

$$N_L \, |i, n_L\rangle = n_L \, |i, n_L\rangle \tag{6}$$

$$|\Psi\rangle = \sum_{n_L=0}^{A} \sum_i c_{in_L} \, |i, n_L\rangle \tag{7}$$

In order to achieve a large variance σ^2, the probabilities W_{n_L} to find n_L particles on the left must have a wide distribution (eq. 8).

$$\sigma^2 = \sum_{n_L=0}^{A} W_{n_L} \, (n_L - \langle N_L\rangle)^2 \; ; \quad W_{n_L} = \sum_i |c_{in_L}|^2 \tag{8}$$

Due to the Pauli-principle states with $n_L \neq \langle N_L \rangle$ have on the average $(n_L - \langle N_L \rangle)$ particle-hole excitations compared to states with $n_L \approx \langle N_L \rangle$, if the mean potential is the same for all channels. This causes the occupation probabilities W_{n_L} for $n_L \neq \langle N_L \rangle$ to stay rather small compared to $W_{n_L \approx \langle N_L \rangle}$, since they belong to states with high excitation energy. This case is illustrated in the upper half of fig. 5. Effects like this are seen in heavy-ion data for small reaction times, if the nuclei have a pronounced shell structure [DG 82]. If the particles are not forced to move within one mean potential but rather each component $|i, n_L\rangle$ of the wavefunction is allowed to establish its own selfconsistent potential, then states with $n_L \neq \langle N_L \rangle$ would have a wider or narrower potential adjusted to accomodate n_L nucleons with averaged nuclear matter density, thus avoiding high excitations. This picture is illustrated in the lower half of fig. 5. It corresponds to a nuclid distribution which is governed by the liquid-drop energy-surface [FH 79, SW 80, SS 81] and leads to much larger variances.

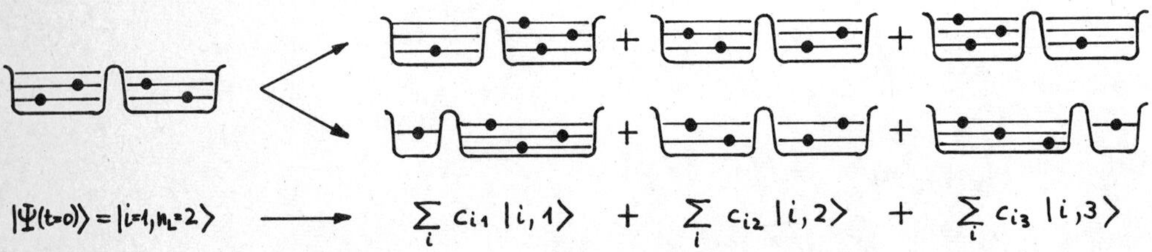

Fig. 5 Illustration for fluctuating mass numbers in a mean potential like TDHF (upper part) and in channel specific potentials (lower part).

Since our model calculation has much similarity to TDHF + C, we expect the small variances to be of general nature for any model which works with one mean single-particle potential. The collision terms will help to equilibrate and increase dissipation but will not lead to large fluctuations of the macroscopic variables as seen in the experimental data. The selfconsistent potential is a fluctuating quantity itself, and replacing it by its mean results in the above described shortcomings.

4. Summary

Although the model we studied is oversimplified, it illuminates the dynamical evolution of two Fermi gases which are brought into contact. The dissipation introduced by a residual two-body interaction was illustrated with the Wigner function and the macroscopic variable N_L which counts the number of particles in one half of the system.

The origin of the small variances in TDHF was discussed, and the results of the model study lead to the prediction that inclusion of short-ranged collision terms in mean-field theories will not increase essentially the fluctuations in the macroscopic variables.

References

AY 80 S. Ayik; Z. Physik A298, 83 (1980).

BF 82 P. Buck and H. Feldmeier; to be published.

DM 78 K.T.R. Davies, V. Maruhn-Rezwani, S.E. Koonin and J.W. Negele; Phys. Rev. Lett. 41, 632 (1978).

DG 82 M. Dakowski, A. Gobbi, W. Nörenberg; Nucl. Phys. A378, 189 (1982).

FE 80 H. Feldmeier; Habilitation, IKDA 80/17, unpublished.

FH 79 H. Freiesleben, K.D. Hildenbrand, F. Pühlhofer, W.F.W. Schneider, R. Bock, D. v. Harrach and H.J. Specht; Z. Physik A292, 171 (1979).

GR 80 D.H.E. Gross; Proc. Int. Workshop on Gross Properties of Nuclei and Nuclear Excitations VIII (1980), ed. H. Feldmeier, Institut für Kernphysik, Darmstadt.

GW 81 P. Grangé, H.A. Weidenmüller and G. Wolschin; Ann. Phys. 136, 190 (1981).

KÖ 80 H.S. Köhler; Nucl. Phys. A343, 315 (1980).

KD 77 S.E. Koonin, K.T.R. Davies, V. Maruhn-Rezwani, H. Feldmeier, S.J. Krieger and J.W. Negele; Phys. Rev. C15, 1359 (1977).

OS 79 H. Orland and R. Schaeffer; Z. Physik A290, 191 (1979).

SS 81 D. Schüll, W.C. Shen, H. Freiesleben, R. Bock, F. Busch, D. Bangert, W. Pfeffer and F. Pühlhofer; Phys. Lett. 102B, 116 (1981).

SW 80 R. Schmidt and G. Wolschin; Z. Physik A296, 215 (1980).

WE 80 H.A. Weidenmüller; in Progress in Particle and Nuclear Physics, Vol. 3, ed. D. Wilkinson, Pergamon Press (1980).

WT 78 C.-Y. Wong and H.H. Tang; Phys. Rev. Lett. 40, 1070 (1978).

FROM TDHF TO NUCLEAR HYDRODYNAMICS+

H. S. Köhler
The Niels Bohr Institute, University of Copenhagen,
DK-2100 Copenhagen Ø, Denmark++

Important developments in the nuclear many-body problem took place in the 1950's. Brueckner formulated a many-body theory based on the assumption that nuclear interactions are basically due to two-nucleon forces[1]. The theory can treat these very strong forces including the strong repulsion at short distances by partial summation of series expansions. The ensuing effective interaction, the Brueckner Reaction-matrix includes in particular many-body effects due to the Pauli exclusion principle and due to the nucleon propagation in a mean field of the other nucleons. These are effects that we now know are essential for the understanding of nuclear saturation and binding.

It is generally accepted that the basic assumption of two-body interactions in nuclei is essentially correct. A critical study does however indicate that three-body forces cannot be neglected[2]. This can also be formulated by saying that mesonic degrees of freedom are not negligible in ground-state nuclei. So, although the picture of a nucleus as consisting of individual nucleons (or quasi-nucleons) moving about independently and interacting through effective interactions is essentially correct it may not be completely so. However it is the picture we shall assume for the time being. An alternative picture would for example be a relativistic mean field theory that explicitly involves a coupling to the mesonic fields[3].

In the 60's and 70's Brueckner theory was applied extensively in Nuclear (or Brueckner) Hartree-Fock calculations[4]. One can state as a summary that these calculations were successful in that nuclear sizes, binding-energies and level-spectra are in fair agreement with experimental data. A partially parallell development took place at this time. The Brueckner reaction-matrix or effective force can be parametrized by a velocity and density-dependent force; usually referred to as a Skyrme-force. Such a parametrization has been rather successful[5]. This approach has the advantage that, because it is fitted to particular data one may increase the predictive power as regards other data or nuclei. Because of the rather lengthy calculations involving several approximations and truncations as well as starting from a semi-phenomenological rather than the "true" force the Brueckner Hartree-Fock calculations falls rather short of a precision fitting. An accurate fit is on the other hand possible by using a simple parametrized force such as the Skyrme-force.

That it is indeed possible to describe nuclei by a simple two-body force is perhaps the most interesting and important finding resulting from this work in the 60's and 70's. The relative simplicity of the force has made it possible to extend the Hartree-Fock ground-state calculations to TDHF-calculations[6]. Although the extension of nuclear ground-state calculations to nuclear collisions may seem straight-forward and require no further justification some caution is required and important corrections may be necessary. It is this latter point that this talk will be concerned with.

In the ground-state the quasi-nucleons interacting via a Brueckner reaction-matrix (or Skyrme-force) have an infinite mean free path. This corresponds to an independent particle picture. This also follows quite generally from the properties of a fermion-system for which the mean-free path $\lambda \sim 1/T^2$ where T is the temperature. This result implies that the transport-coefficients are strongly temperature-dependent near absolute zero temperature. This result was obtained for nuclear matter by Tomonaga in 1938[7].

It is also well known that a nucleon above the fermisurface has a finite mean free path. This fact is contained in the absorptive part of the optical potential for nucleon-nucleus interactions.

The mean free path depends therefore both on the temperature of the fermion-system and on the momentum of the particle. This dependence is found to be[8]

$$\frac{1}{\tau_p(\varepsilon_p,T)} = \frac{(m^*)^3}{16\pi^4\hbar^6} W \frac{(\pi T)^2 + (\varepsilon_p - \mu)^2}{1+\exp((\mu-\varepsilon_p)/T)} \tag{1}$$

τ_p being the lifetime of a quasiparticle of energy ε_p, μ the chemical potential and W a transition rate. It is indicated explicitly that τ_p is a function of ε_p and T. It is also, of course, a function of the density.

Our discussion above suggests a possible correction to TDHF. A finite quasi-particle life-time or "relaxation"-time (finite mean free path) will modify the transport-properties. The TDHF- (or Vlasov) equation that includes one-body collisions only (with the mean field) should be modified or extended to include also two-body collisions. For this modification to play a role the typical relaxation-time should of course be short and at least comparable with typical reaction times.

One should also be aware of another correction. It was pointed out above that the effective interaction in itself depends on the "medium" both through the exclusion-principle and through the mean field. These effects are contained in Brueckner's reaction matrix. When working with a Skyrme-force these effects are approximated by a density-dependence. This may be sufficient for fitting ground-state properties but it may not be adequate for the more complicated excitations reached in heavy-ion collisions. Some information on these modifications of the force is obtained through the work of Faessler et al[9] and by Buchler et al[10].

There are consequently two modifications that we should be concerned about when extending the low-energy TDHF-formalism to higher energies.

1. The modification of the two-body force.
2. A modification of the transport-equation to include two-body collisions.

I shall during the rest of this talk only concern myself with the second modification. (These effects are actually not that separate from each other but it depends on the approach one is taking to these problems).

To implicate the effect of two-body collisions we shall rely on traditional non-equilibrium statistical mechanics methods. Two-body collisions will be assumed to affect transport-properties in the medium; i.e. heat-conduction and viscosity. The effect of these collisions will be a randomnization, i.e. an increase of entropy. Collective, coherent energy will be transferred to heat because of these collisions. In TDHF entropy is conserved; in the fine grained scale.

In previous work we have included this effect of two-body collissions by the time-relaxation method[11,12,13,14]. This implies replacing the collision-term in the transport-equation (K is the two-body interaction)

$$\left(\frac{\partial f_1}{\partial t}\right)_{coll} = \sum f_1 f_2 (1-f_3)(1-f_4)\, K^2\, \delta(\varepsilon_1+\varepsilon_2-\varepsilon_3-\varepsilon_4) \qquad (2)$$

by

$$\left(\frac{\partial f_1}{\partial t}\right)_{coll} = \frac{f_1 - f_0}{\tau} \qquad (3)$$

This approximation is usually referred to as the time-relaxation-method. (In some cases it is exact). Here f_0 is a distribution obtained by locally equilibrating the distribution f. The density, current-, and energy-density are conserved quantities in the two-body collision and this suffices to define f_0.

The relaxation-time would in the case of one particle outside a fermi-sphere be obtained from eq. (1). In a heavy ion collision the distribution of momenta is more complicated. One is dealing with strong deformations of the ground state fermispheres. In general will the relaxation-time be a functional of the distribution-function f. Eq. (1) could however be used as a guide-line as to functional dependence. Information (at T = 0) is also obtainable from the imaginary part of the optical model potential. I shall here also, briefly, review some calculations of the collision term in nuclei. The collision-term is shown graphically in fig. 1. It shows explicitly that one is concerned with a 1h-2h1p coupling. In Brueckner-theory this diagram is referred to as a second order rearrangement term. More specifically we are dealing with the imaginary part of the energy-

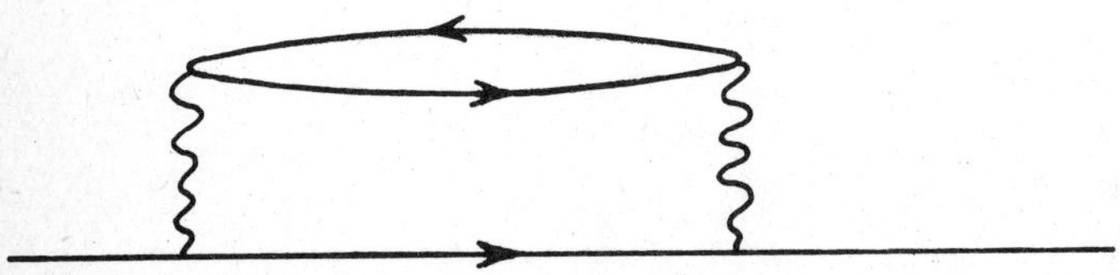

Fig. 1 The two-body collision diagram.

conserving part of this diagram.

This diagram has been calculated with a K-matrix calculated from realistic forces, with the aim of obtaining the optical model. Good agreement with optical model parameters is obtained[15]. Of particular interest is for our discussion of course the agreement with the absorptive part. This same diagram has also been calculated for hole-states in nuclei; the result agrees with experimental widths obtained in knock-out reactions. The interaction used in this calculation[16] is a simple approximation to the K-matrix based on the Moszkowski-Scott separation method. This same interaction was used in a calculation appropriate for heavy-ion collisions: the lifetime of a configuration consisting of two fermi-spheres separated by a momentum P; the relative momentum between two ions. The result is shown in fig. 8 of ref.[11]. The two fermi-spheres had sharp surfaces corresponding to zero temperature. A characteristic of this calculation is the strong P-dependence. As P increases the available phase-space increases fast.

The major difficulty in a detailed calculation of this type is the computation of the effective interaction K in eq. (2) or the diagram in Fig. 1. (or transition rate in eq. (1)). The strong interactions in the nucleus makes a detailed calculation important. The interaction K is a functional of the distribution in momentum-space. We are here actually concerned with the previously modification "1". This may not be so important when calculating the mean field but is probably important for estimating the relaxation-time.

The most detailed calculation from which we can deduce relaxation-times seems to date be that of Izumoto et al.[17]. The purpose of their calculation was actually to calculate the heavy-ion optical potential. As one step in their calculation the authors calculate the imaginary part of single hole states, from which we find the relaxation times shown in Fig. 2. The authors only calculate lifetimes of holes. Their results can probably rather safely be extrapolated into

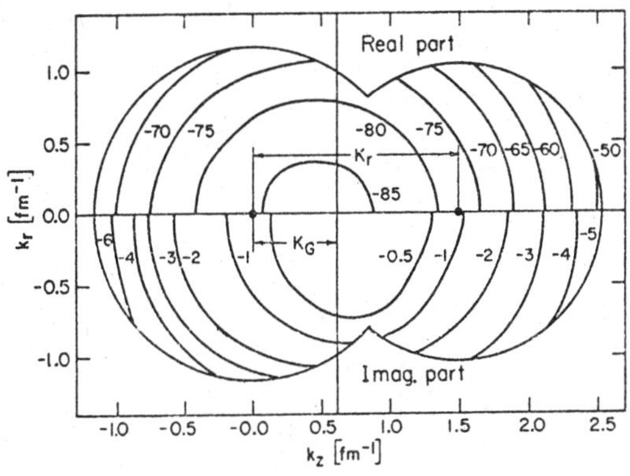

Fig. 2 Single-hole potentials as a function of momentum of a hole-state. The configuration is two zero-temperature fermi-spheres, with radii 1.163 and 1.040 fm^{-1} respectively, separated by a relative momentum K_r = 1.5 fm^{-1}. Relaxation time is $\tau = .32 \cdot V_{im}^{-1} \cdot 10^{-21}$s from Ref.[17].

particle-states. One should note the strong state-dependence. The lifetime varies from $\sim 10^{-21}$ to $\sim 10^{-22}$s from the small to the high momenta. The lifetimes also vary with the separation of the fermi-spheres i.e. with excitation. From eq. (1) we find that a temperature-increase, smearing of the fermi-surfaces would also decrease the relaxation-time. Results obtained by Bertsch[18] suggest however that for small excitations, whether the energy goes into heating or deformation of the fermi-surface is irrelevant; the lifetime depends only on the total excitation energy. This is an approximation that we shall adopt instead of improved estimates.

Although we find that the relaxation of a nucleon-state depends on both the temperature of the medium and the momentum of the nucleon we have only included the temperature-dependence in our calculations. For technical reasons it has so far been prohibitive to include the momentum-dependence. Following Bertsch's result we have also equated temperature with excitation energy. The definition of a temperature (even local temperature) is in a non-equilibrium system strictly not allowed. Our "definition" will simply be in terms of the excitation energy[12].

At low temperature we have used the functional form obtained by Tomonaga[7]. (His is a low temperature expansion). At high temperatures we have used data described in detail in ref.[11]. Two functions have been used with units in MeV and 10^{-21} seconds.

$$\tau_R = \tau(T) = T^{-2} + T^{-1} \tag{4a}$$

$$\tau_M = \tau(T) = T^{-2} + .0058 \tag{4b}$$

The transport-equation is written in terms of the density-matrix and is now:

$$i\hbar \frac{\partial \rho}{\partial t} + \frac{\hbar^2}{m} \nabla_r \cdot \nabla_x \rho - D\rho = -i\hbar(\rho-\rho_0)/\tau(T) \tag{5}$$

with $\rho \equiv \rho(\underline{r},\underline{x},t)$ and $D = U(\underline{r}+\frac{1}{2}\underline{x})-U(\underline{r}-\frac{1}{2}\underline{x})$ where $\underline{x}$ is a non-locality coordinate.

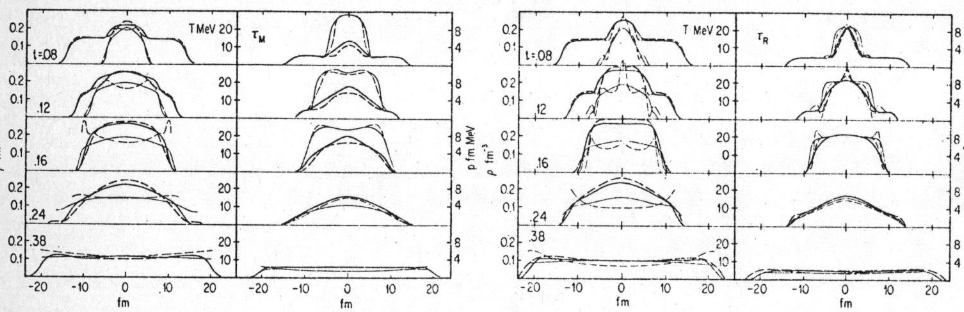

Fig. 3 Comparison between fluid results (full curves) and TDHF (broken curves) for $\tau = \tau_M$, left figure, and for $\tau = \tau_R$, right figure. The set of curves to the left in each figure show densities (heavy lines with scale to the left) and temperatures (thin lines with scale to the right). The set of curves to the right show parallel (thin lines) and perpendicular (full lines) internal pressures. Reference axis is the beam direction.

Fig. 3 show results for collisions between two nuclear slabs at 20 MeV/A CM energy. The broken lines are the results obtained calculating density and temperature as well as pressure, both transverse and longitudinal. Details of this calculation is given in ref.[13]. It should be observed that for the shorter relaxation time, τ_R, the pressure is isotropic while for the longer, τ_M, it is highly unisotropic.

Directly related to this is the more or less obvious fact that for small relaxation-times the properties of the system approaches that of an ideal fluid, i.e. viscosity and heat-conduction goes to zero. It is of interest to find out how the nuclear matter would approach a fluid from a gas state as the temperature (excitation) is increased.

A microscopic derivation of fluid-equations starts from a transport-equation. Fluid-equations have been derived[14] from eq. (5) by a moment-expansion of the Wigner-function, which is the Fourier-transform of $\rho(\underline{r},\underline{x},t)$ with respect to $\underline{x}$. A truncation is made including only the third moment and expressing heat-conduction in terms of $\kappa \cdot \nabla T$, where κ is the heat conductivity. The ensuing equations have been solved (see ref.[14] for details) and the result is shown in Fig. 3 (full lines) for comparison with the results from the transport-equation). It is seen that the fluid approximation to the transport-equation is indeed quite good for these relaxation-times at 20 MeV/A CM. It should be pointed out however that these fluid-equations have a highly unisotropic pressure-tensor. It is probably not approximable by a viscosity term.

A summary of our findings can be given as follows. Nuclear matter behaves at low temperatures $T < 5$ MeV like a gas with long mean free paths. TDHF is applicable for collisions up to ~ 5 MeV/nucleon. As the temperature is increased the importance of two-body collisions increases; the relaxation time decreases. For $T \gtrsim 10$ MeV (for collisions above 20 MeV/nucleon in CM nuclear matter behaves as an ideal fluid. In the intermediate region nuclear matter has to be treated by a transport-equation like eq. (5) with a finite relaxation time. The behaviour is therefore more complicated in this intermediate region. The energy and temperature limits given above are highly approximate. It is rather the goal of heavy-ion experimental and many-body theoretical work to describe this transition from gas to fluid. The property of nuclear matter described say in terms of heat-conductivity and viscosity is largely unknown at these energies and temperatures.

+ Supported in part by US National Science Foundation grant No. PHY-7902654.

++ Permanent address: Physics Department, University of Arizona, Tucson, Arizona 85721 USA.

1. H. S. Köhler, Physics Reports 18C (1975) 217
2. S. A. Coon, Proceedings, International Many Body Conference, Oaxtapec, Mexico, Jan. 1981.
3. J. D. Walecka, Ann. of Phys. 83 (1974) 491
4. H. A. Bethe, Ann. Rev. Nucl. Sci. 21 (1971) 93
5. P. Quentin and H. Flocard, Ann. Rev. of Nucl. Sci. 28 (1978)523
6. J. W. Negele, Rev. Modern Phys. to be published
7. S. Tomonaga, Z. Phys. 110 (1938) 573
8. D. Pines and P. Nozières, The Theory of Quantum Liquids (Benjamin N. Y. 1966) p. 63
9. Amand Faessler, T. Izumoto, S. Krewald and R. Sartor, Nucl. Phys. A359 (1981) 509
10. J. Robert Buchler and Bhaskar Datta, Phys. Rev. C19 (1979) 494
11. H. S. Köhler, Nucl. Phys. A343 (1980) 315
12. H. S. Köhler, Nucl. Phys. A378 (1982) 159
13. H. S. Köhler, Nucl. Phys. A378 (1982) 181
14. H. S. Köhler, Physica Scripta, to be published
15. J. P. Jeukenne, A. Lejeune and C. Mahaux, Phys. Rev. C10(1974)1391
F. A. Brieva and J. R. Rook, Nucl. Phys. A291 (1977) 299
16. H. S. Köhler, Nucl. Phys. 88 (1966) 529
17. T. Izumoto, S. Krewald and Amand Faessler, Nucl.Phys.A357(1981)471
18. G. Bertsch, Z. Phys. A289 (1978) 103

THE INCORPORATION OF PARTICLE COLLISIONS IN THE TIME-DEPENDENT HARTREE-FOCK APPROXIMATION

Cheuk-Yin Wong
Oak Ridge National Laboratory*
Oak Ridge, Tennessee 37830, U.S.A.

Abstract

In the time-dependent Hartree-Fock (TDHF) approximation, particles interact only through the mean field, and the collisions between particles are not included. Previously, we formulated the extended time-dependent Hartree-Fock (ETDHF) approximation to include particle collisions in terms of a temporal variation of the occupation probability n_λ for the single-particle states. In the simplest approximation, the single-particle potential is modified only through the particle density which depends on n_λ. We wish to refine the extended TDHF approximation by studying how particle collisions affect the single-particle potential. We find that it acquires two second-order contributions which are state-dependent and are the generalization of the "core polarization" and "correlation" contributions one encounters in the study of the nucleon-nucleus optical potentials. In consequence, concepts such as energy-dependent single-particle potentials and effective masses may be properly introduced in the extended TDHF approximation. We also wish to review the conservation of energy in the ETDHF approximation. We find that the total energy should include a second-order contribution due to correlations arising from particle collisions. The proper choice of single-particle basis states in the extended TDHF approximation is also discussed.

I. INTRODUCTION

In the time-dependent Hartree-Fock approximation, particles interact only through the mean field, and the collisions between particles are not taken into account.[1] It is a good description for many low-energy phenomena, where the Pauli exclusion principle inhibits collisions between particles. There are, however, problems in nuclear physics in which particle collisions are important. For example, in heavy-ion collisions much above a few MeV per nucleon, the Pauli exclusion principle is not effective in preventing particle collisions. In the problem of nucleon-nucleus reactions, the incident nucleon can excite nucleons in the nucleus from states below the Fermi energy to states above the Fermi energy. Particle collision is also very important in the search for a proper description for the approach to thermal equilibrium.[2,3,6-9]

Previously, we formulated the extended time-dependent Hartree-Fock (ETDHF) approximation to incorporate particle collisions in the mean-field theory. Besides the temporal variation of the single-particle states, there is also a temporal variation of the occupation numbers n_λ governed by a master equation.[2,3] Subsequent studies also lead to similar results.[4,5] The collision term has also been examined from a random-matrix model,[10] and the projection method.[11] Particle collisions are

*Research sponsored by the Division of Basic Energy Sciences, U.S. Department of Energy, under contract W-7405-eng-26 with the Union Carbide Corporation.

conjectured as possible agents of the iterative time-smoothing leading to dissipation in the time-dependent mean-field theory.[12] Other phenomenological ways to introduce the collision term have been presented.[13,14] Numerical ETDHF calculations were performed for simple systems.[15] The complete set of non-Markovian equations involving the one-body Green's function has been recently solved for non-equilibrium nuclear matter, with and without initial-state correlations.[16]

We wish to examine here other refinements of the ETDHF approximation. In the simplest approximation formulated so far, particle collisions affect the single-particle (SP) potential only through the particle density which depends on the occupation numbers. As is well known in nucleon-nucleus reactions, particle collisions give rise to polarization and correlation effects which lead to energy-dependent SP potentials.[17-19] So, in the ETDHF approximation, one expects that particle collisions lead to second-order corrections of the SP potential in a similar manner.

Many questions have been raised about the conservation of energy in the ETDHF approximation.[3-5,10,20] We shall review how the conservation of energy should be formulated and how the total energy should be defined in the presence of particle collisions.

In the ETDHF approximation, it is desirable to work with a set of basis states which give a diagonal occupation matrix so that a description in terms of occupation numbers suffices. However, even though the occupation matrix may be diagonal initially, it may not keep its diagonal property as time evolves. We shall discuss the necessary procedures to maintain the occupation matrix diagonal.

II. EQUATION OF MOTION FOR ψ_λ AND n_λ

For a quantal many-body system much of the dynamical information is contained in the Green's functions introduced by Martin and Schwinger.[21] They are defined by

$$-i\, g^<(1\,1') = \langle \hat{\psi}^+(1')\hat{\psi}(1)\rangle \qquad \text{(II.1)}$$

$$i\, g^>(1\,1') = \langle \hat{\psi}(1)\hat{\psi}^+(1')\rangle \qquad \text{(II.2)}$$

where the expectation values of the field operators are taken with respect to the dynamical wave-packet-like state in question and the argument 1 (and similarly 1′) represents the collection of space and spin coordinates x_1 and time coordinate t_1. The Green's function $-ig^<(1\,1')$ provide information about the occupation of single-particle states, while $ig^>(1\,1')$ the non-occupancy of the single-particle states. The density matrix is the special equal-time case of the Green's function $g^<$.

We assume that the mean-field theory of TDHF, as obtained by using an effective density-dependent interaction $v(x_1x_2)\delta(t_1-t_2)$ and neglecting particle collisions, is a good and approximate representation of the dynamics. It is now necessary to include the residual interaction $v'(12) = v'(x_1x_2)\delta(t_1-t_2)$ between the fermions only in the Born scattering approximation. The equation of motion for $g^<$ in perturbation theory is[22]

$$\left\{ i\hbar \frac{\partial}{\partial t_1} + \frac{\hbar^2}{2m} \nabla_1{}^2 - U_{HF}[\rho(1)] \right\} g^<(1\ 1') \tag{II.3}$$
$$= \int_{-\infty}^{t_1} d\bar{1}\ \Sigma\ (1\ \bar{1}) g^<(\bar{1}\ 1') - \int_{-\infty}^{t_1'} d\bar{1}\ \Sigma^<\ (1\ \bar{1}) g(\bar{1}\ 1')$$

where $U_{HF}[\rho(1)]$ is the density-dependent single-particle potential obtained in the usual way.[1] The mass operators $\Sigma^>$ and $\Sigma^<$ are defined by

$$\Sigma^{\gtrless}(1\ \bar{1}) = -i^2 \int d2d\bar{2}\ v'(1\ 2)\ v'(\bar{1}\ \bar{2}) \mathcal{A}[g^{\gtrless}(1\ \bar{1}) g^{\gtrless}(2\ \bar{2})] g^{\lessgtr}(\bar{2}\ 2), \tag{II.4}$$

where

$$\mathcal{A}[g^{\gtrless}(1\ \bar{1}) g^{\gtrless}(2\ \bar{2})] = g^{\gtrless}(1\ \bar{1}) g^{\gtrless}(2\ \bar{2}) - g^{\gtrless}(1\ \bar{2}) g^{\gtrless}(2\ \bar{1}). \tag{II.5}$$

The quantities Σ and g in Eq. (II.3) are

$$\Sigma(1\ \bar{1}) = \Sigma^>(1\ \bar{1}) - \Sigma^<(1\ \bar{1}), \tag{II.6}$$

and $g(1\ \bar{1}) = g^>(1\ \bar{1}) - g^<(1\ \bar{1})$. (II.7)

From the equations of motion for the Green's functions, we can derive the equations of motion for the single-particle states ψ_λ and the occupation probability n_λ by following the procedures similar to those presented previously.[3] We specialize to an approximate representation of the Green's function in the form

$$-\ ig^<(1\ 1') = \sum_\lambda^\infty n_\lambda^<(t_1 t_1')\psi_\lambda(1)\psi_\lambda^*(1'), \tag{II.8}$$

and $ig^>(1\ 1') = \sum_\lambda^\infty n_\lambda^>(t_1 t_1')\psi_\lambda(1)\psi_\lambda^*(1')$, (II.9)

where

$$n_\lambda^>(t_1 t_1) = 1 - n_\lambda^<(t_1 t_1). \tag{II.10}$$

We also specialize to an approximate representation of the mass operator in the form

$$-i\ \Sigma^<(1\ 1') = \sum_\lambda^\infty \sigma_\lambda^<(t_1 t_1')\psi_\lambda(1)\psi_\lambda^*(1'), \tag{II.11}$$

and $i\ \Sigma^>(1\ 1') = \sum_\lambda^\infty \sigma_\lambda^>(t_1 t_1')\psi_\lambda(1)\psi_\lambda^*(1')$. (II.12)

The consequences of using non-diagonal $n_{\lambda\lambda'}^{\gtrless}$ and $\sigma_{\lambda\lambda'}^{\gtrless}$ will be discussed in Section IV.

From Eq. (II.4), we can relate $\sigma^{\gtrless}$ with $n^{\gtrless}$ by

$$\sigma_\lambda^{\gtrless}(t_1t_1') = \frac{1}{2}\sum_{234} n_3^{\gtrless}(t_1t_1')n_4^{\gtrless}(t_1t_1')n_2^{\lessgtr}(t_1't_1)\ \langle\lambda 2|v'|34\text{-}43\rangle_{t_1}\ \langle 34\text{-}43|v'|\lambda 2\rangle_{t_1'} \tag{II.13}$$

where the symbols 2, 3, and 4 stand for states λ_2, λ_3, and λ_4, respectively, and the subscripts for the matrix elements specify the times at which the matrix elements are taken.

In order to have a Markovian equation of motion, it is necessary to parametrize the time dependence of $n_\lambda^{\gtrless}$ and $\langle\lambda 2|v'|34\rangle_{t_1}$ in simpler forms so that the integration over history can be carried out. The quantity $n_\lambda^{<}(t_1t_1')$ is the probability that a particle at state λ is destroyed at time t_1 and emerges at time t_1'. Anticipating the results of the master equation, we know that a particle or a hole has a lifetime and decays exponentially[3] as $\exp(-\Gamma_\lambda t/\hbar)$. It is reasonable to parametrize $n_\lambda^{<}$ and $n_\lambda^{>}$ by

$$n_\lambda^{\gtrless}(t_1t_1') = n_\lambda^{\gtrless}(T)\ e^{-\Gamma_\lambda|t_1'-t_1|/2} \tag{II.14}$$

where $T = (t_1 + t_1')/2$ and Γ_λ is [see Eqs. (9.11)-(9.17) of Ref. 3]

$$\Gamma_\lambda = \hbar(P_\lambda^{(-)} + P_\lambda^{(+)}), \tag{II.15}$$

with $P_\lambda^{(-)}$ and $P_\lambda^{(+)}$ defined by the master equation (to be obtained below):

$$\frac{\partial n_\lambda^{<}(T)}{\partial T} = -P_\lambda^{(-)}n_\lambda^{<} + P_\lambda^{(+)}(1-n_\lambda^{<}). \tag{II.16}$$

To parametrize the time-dependence of the matrix element $\langle\lambda 2|v'|34\rangle$, we note that the overlap of the single-particle state ψ_λ at two different times is

$$\langle\psi_\lambda(t_1)|\psi_\lambda(t_1')\rangle \cong \exp\left\{-i\int_{t_1}^{t_1'}\varepsilon_\lambda(t'')dt''\right\} \tag{II.17}$$

where

$$\varepsilon_\lambda(t) = (\psi_\lambda(t),\ h\ \psi_\lambda(t)), \tag{II.18}$$

and h is the single-particle hamiltonian.

Equation (II.17) suggests that it is reasonable to approximate the time-dependence of the matrix element by

$$\langle\lambda 2|v'|34\rangle_{t_1} \simeq e^{-i\varepsilon(t_1-T)}\ \langle\lambda 2|v'|34\rangle_T, \tag{II.19}$$

where

$$\varepsilon = \varepsilon_3 + \varepsilon_4 - \varepsilon_\lambda - \varepsilon_2, \tag{II.20}$$

We shall assume that the life-time $\hbar/\Gamma_\lambda$ of the colliding nucleons is much shorter than the mean-free time between collisions, a condition which is perhaps met in heavy-ion and nucleon-nucleus collisions of moderate energies. In that case, during the mean-free time between collisions, the occupation numbers are essentially unchanged. Hence, in the integration of $\overline{t_1}$ on the righthand side of Eq. (II.3), we can take the occupation numbers to depend only on $T = (t_1 + t_1')/2$ and factor them out from the integration in $\overline{t_1}$. We then obtain:

$$(\psi_\lambda(1), \left[i \frac{\partial}{\partial t_1} + \frac{\hbar^2\nabla^2}{2m} - U_{HF}(1)\right] n_\lambda^<(t_1 t_1^+)\psi_\lambda(1))$$

$$= \frac{1}{2} \sum_{234} [n_3^< n_4^< n_2^> n_\lambda^> - n_3^> n_4^> n_\lambda^< n_2^<] \frac{\varepsilon + i\,\Gamma/2}{\varepsilon^2 + (\Gamma/2)^2} \left| \langle\lambda 2| v' |34-43\rangle_{t_1} \right|^2 \tag{II.21}$$

where

$$\Gamma = (\Gamma_\lambda + \Gamma_2 + \Gamma_3 + \Gamma_4). \tag{II.22}$$

We can separate out the proper terms of the above equation (II.21). We get the master equation

$$\frac{\partial n_\lambda^<(T)}{\partial T} = \sum_{234} [n_3^<(T)n_4^<(T)n_\lambda^>(T)n_2^>(T) - n_3^>(T)n_4^>(T)n_\lambda^<(T)n_2^<(T)]$$

$$\times \frac{\Gamma/2}{\varepsilon^2 + (\Gamma/2)^2} \left| \langle\lambda 2|v|34-43\rangle_T \right|^2 . \tag{II.23}$$

We also get an equation involving $n_\lambda^<(t_1, t_1^+)(i\partial_\lambda \psi(1)/\partial t_1)$. The single-particle potential for state λ is obtained by putting a test particle at state λ with varying degrees of occupation. We make a variation with respect to $n_\lambda^<(t_1, t_1^+)$ in the equation containing $n_\lambda^<(t_1, t_1^+)(i\partial\psi_\lambda(1)/\partial t_1)$, and we get the single-particle equation

$$\left[i\hbar \frac{\partial}{\partial t_1} + \frac{\hbar^2}{2m} \nabla_1{}^2 - U_{HF}(1) - \Delta U_\lambda\right]\psi_\lambda(1) = 0 \tag{II.24}$$

where

$$\Delta U_\lambda \psi_\lambda(1) = - \sum_{234} [n_3^> n_4^> n_2^< + n_3^< n_4^< n_2^>] \frac{\varepsilon}{\varepsilon^2 + (\Gamma/2)^2} \langle 34-43| v' |\lambda 2\rangle_{t_1}$$

$$\int dx_2 \psi_2{}^*(x_2) v'(x_1 x_2) \psi_4(x_2) \psi_3(x_1). \tag{II.25}$$

The potential ΔU_λ is spatially non-local and state-dependent. An approximate representation of $\Delta U_\lambda \psi_\lambda(1)$ is

$$\Delta U_\lambda \psi_\lambda(1) \sim - \frac{1}{2} \sum_{234} [n_3^> n_4^> n_2^< + n_3^< n_4^< n_2^>] \frac{\varepsilon}{\varepsilon^2 + (\Gamma/2)^2} \left| \langle\lambda 2| v' |34-43\rangle_{t_1} \right|^2 \psi_\lambda(1) \tag{II.26}$$

Equation (II.23) is the same as what was derived earlier. The SP potential in Eq. (II.24) acquires two new second-order terms which are analogous to the polarization and correlation terms one encounters in the theory of nucleon-nucleus optical-model potentials[17-19] [cf. Eqs. (2.159-2.160) of Ref. 18]. They have the same physical origin as in the optical potential case and lead to a non-local and energy-dependent SP potential. The two contributions have a tendency to have opposite signs. As in the nucleon-nucleus case, the SP potential may be described in terms of an energy-dependent effective mass m^*. It is known that the effective mass m^* of a nucleon is about the nucleon mass m at ε_f but decreases to about 0.7m for $|\varepsilon_\lambda - \varepsilon_f| \gtrsim 2\hbar\omega$.[18,19] So, one should use an effective mass which depends on the energy in a similar manner.

It is worth remarking that in our previous analysis of the equation of motion for ψ_λ and n_λ, the correction term ΔU_λ makes no appearance. This is because the starting point of the previous investigation was not Eq. (II.3) but the difference of Eq. (II.3) with its complex conjugate. In consequence, ΔU_λ was subtracted out. It is by investigation of Eq. (II.3) (or its complex conjugate) separately that the correction term ΔU_λ is obtained.

III. CONSERVATION OF ENERGY IN THE ETDHF APPROXIMATION

There have been many questions raised concerning the conservation of energy in in the ETDHF approximation. At the root of the problem is the confusion as to how the total energy quantity should be defined so that it is conserved as a function of time. Clearly, the additional degree of freedom introduced by particle collisions requires that there is an energy contribution arising from this degree of freedom. Energy quantities calculated without this correlation will not, in general, be conserved.

Kadanoff and Baym obtained the conservation laws in the perturbation theory.[22] In particular, when the Born scattering term is considered, the total energy defined in the following way is a conserved quantity:

$$E_{tot} = \int dx_1\ T(x_1) + (i^2/2) \int dx_1 dx_2\ v(x_1 x_2) g_2(12,1^+2^+) \tag{III.1}$$

where $T(x_1)$ is the kinetic energy density

$$T(x_1) = \frac{\hbar^2}{2im} \lim_{1'\to 1} \nabla_1 \cdot \nabla_{1'}\ g^<(1,\bar{1}'), \tag{III.2}$$

$$g_2(12,1^+2^+) = \mathcal{A}[g^<(1\ 1^+)g^<(2\ 2^+)] + g_{2c}(12,1^+2^+), \tag{III.3}$$

and in the Born scattering approximation

$$\begin{aligned} g_{2c}(12,1^+2^+) = \frac{i}{\hbar} \int d\bar{1}d\bar{2}\ v(\bar{1}\ \bar{2}) \int_{-\infty}^{t_1} d\bar{t} \Big\{ & g^>(1\ \bar{1})g^>(2\ \bar{2}) \mathcal{A}[g^<(\bar{1}\ 1')g^<(\bar{2}\ 2^+)] \\ & - g^<(1\ \bar{1})g^<(2\ \bar{2}) \mathcal{A}[g^>(\bar{1}\ 1')g^>(\bar{2}\ 2^+)] \Big\}. \end{aligned} \tag{III.4}$$

For a system where the effective mean-field interaction is different from the effective residual interaction, the total energy is modified to be[3]

$$E_{tot} = \int dx_1 T(x_1) + \frac{i^2}{2} \int dx_1 dx_2\ v(x_1 x_2) \mathcal{A}[g^<(1\ 1^+) g^<(2\ 2^+)]$$
$$+ \frac{i^2}{2} \int dx_1 dx_2\ v'(x_1 x_2) g_{2c}(12, 1^+ 2^+). \tag{III.5}$$

The first two terms give the Hartree-Fock energy E_{HF}, while the last term is the energy due to correlations arising from particle collisions. With the parametrization of the Green's function as given previoulsy, we find that for slowly varying mean field, the total energy is

$$E_{tot} = E_{HF} + E_{coll} \tag{III.6}$$

where

$$E_{coll} = \frac{1}{4} \sum_{\lambda 234} \left| \langle \lambda 2 | v' | 34-43 \rangle \right|^2 [n_\lambda^< n_2^< n_3^> n_4^> - n_\lambda^> n_2^> n_3^< n_4^<] \frac{\varepsilon}{\varepsilon^2 + (\Gamma/2)^2}. \tag{III.7}$$

The energy E_{coll} due to particle collisions contains two contributions which have a tendency to have the same positive signs. The quantity E_{tot} is conserved if the slowly varying mean field is a good approximation.

IV. CHOICE OF BASIS STATES

For simplicity, the occupation matrix and the mass operator have been assumed to take the diagonal form. We obtain then the simple results as discussed in the previous sections. Will the occupation matrix be always diagonal if it is not constrained to be so?

It is clear that the initial occupation matrix at equal time can be diagonalized with the proper choice of basis states. The non-diagonal elements of the initial occupation matrix at unequal times are expected to be small and are further reduced by the exponential decay factor of Eq. (II.14) so that they can be taken to be zero. As the dynamics proceed, the nondiagonal matrix elements will grow. We shall outline below the necessary procedures to maintain a diagonal occupation matrix.[23]

Previously, an equation of motion for the occupation matrix was obtained as a seven-fold summation of products of occupation matrices.[3] The initial occupation matrix being diagonal, it is reasonable to take $n_{\lambda\lambda'}$ as diagonal in the evaluation of the time derivative of the nondiagonal matrix elements. The seven-fold summation is reduced to a three-fold summation, and we obtain

$$\frac{\partial n_{\lambda\lambda'}}{\partial T} = \sum_{234} \left\{ n_3^<(T) n_4^<(T) n_2^>(T) \frac{n_\lambda^>(T) + n_{\lambda'}^>(T)}{2} - n_3^>(T) n_4^>(T) n_2^<(T) \frac{n_\lambda^<(T) + n_{\lambda'}^<(T)}{2} \right\}$$
$$\times \frac{\bar{\Gamma}/2}{(\bar{\varepsilon})^2 + (\bar{\Gamma}/2)^2} \langle \lambda 2 | v' | 34-43 \rangle_T \langle 34-43 | v' | \lambda' 2 \rangle_T \tag{IV.1}$$

where

$$\bar{\varepsilon} = \varepsilon_3 + \varepsilon_4 - \varepsilon_2 - (\varepsilon_\lambda + \varepsilon_{\lambda'})/2$$

and

$$\bar{\Gamma} = \Gamma_3 + \Gamma_4 + \Gamma_2 + (\Gamma_\lambda + \Gamma_{\lambda'})/2.$$

From the above equation, we see that a good set of basis states is one in which the off-diagonal occupation matrix elements do not grow significantly. Thus, in the collision of two nuclei, the set of initial basis states centered in each of the colliding nuclei is a better set than the set of "gerade" and the "ungerade" states which have components in both centers. The off-diagonal element will become nonzero as time proceeds. With the equation of motion for $n_{\lambda\lambda'}$, as given by Eq. (IV.1), we can evaluate $n_{\lambda\lambda'}$ at every time step. After obtaining $n_{\lambda\lambda'}$, we can diagonalize $n_{\lambda\lambda'}$ to obtain new basis states for which the occupation matrix is diagonal. The procedure can be repeated for subsequent time steps so that a diagonal occupation matrix can be maintained in the study of the dynamics of the system.

V. SUMMARY AND DISCUSSIONS

We have considered three of the outstanding problems in the ETDHF approximation. We found that particle collisions off the energy shell contribute to nonlocal state-dependent corrections to the mean field. In the form of the approximation Eq. (II.26), these corrections correspond to a shift in the single-particle energy. The shift is small for states near the Fermi energy, large and negative for deep-hole states, and large and positive for high-lying particle states.[18] In the optical model, the energy dependence in the potential is an important consideration. It should also be an important factor in the discussion of heavy-ion collisions.

With regard to the question of energy conservation, a total energy quantity including the contributions from particle collision is conserved if the mean-field varies slowly. It is an approximately conserved quantity when the mean field varies rapidly. Energy quantities calculated with only the Hartree-Fock approximation are not conserved in the presence of particle collisions.

The ETDHF makes use of the diagonal nature of the occupation matrix as much as possible. Even though the occupation matrix may be initially diagonal, the off-diagonal matrix element will grow. Knowing the equation of motion of the off-diagonal elements, we can make a unitary transformation to render the occupation matrix in a diagonal form. In a stepwise manner, we can keep the occupation matrix diagonal. Thus, Eqs. (II.23), (II.24), and (IV.1), together with a unitary transformation to diagonalize $n_{\lambda\lambda'}$, constitute the main results of the present refinement of the ETDHF approximation.

There remains an unresolved problem which deserves attention. As it is presented, the ETDHF approximation depends on the expectation value ε_λ of the single-particle hamiltonian h for the basis states ψ_λ. Although this expectation value contains the kinetic energy for collective motion and the single-particle intrinsic static energy, it is not basis independent. It has its origin in the phase factor in the time-dependence of the matrix element $\langle\lambda 2|v'|34\rangle_{t_1}$. It will be of interest to investigate in the future how such a phase factor can be written in terms of some other basis-independent quantities.

ACKNOWLEDGEMENT

The author wishes to thank Dr. C. Mahaux, Dr. P. J. Siemens, Dr. H. H. K. Tang, and Dr. WANG Shun-jin for helpful discussions.

REFERENCES

1. P. Bonche, S. Koonin, and J. Negele, Phys. Rev. C 13, 1226 (1976); for a review of TDHF, see J. Negele (to be published in Reviews of Modern Physics), and K. T. R. Davies, K. R. S. Devi, S. E. Koonin, and M. Strayer, in Heavy Ion Sciences, Vols. I and II - Nuclear Science (ed. by D. A. Bromley), Plenum Press (in press).

2. C. Y. Wong and H. H. K. Tang, Phys. Rev. Lett. 40, 1070 (1978).

3. C. Y. Wong and H. H. K. Tang, Phys. Rev. C 20, 1419 (1979).

4. C. M. Shakin and M. S. Weiss, UCRL Report No. 08500 (unpublished).

5. H. Orland and R. Schaeffer, Z. fuer Phys. A290, 191 (1978).

6. D. Glas and U. Mosel, Nucl. Phys. A264, 2681 (1976); G. Shütte, and L. Wilets, Nucl. Phys. A252, 21 (1975).

7. C. Y. Wong and C. Toepffer, Proceedings of the EPS Topical Conference on Large Amplitude Collective Nuclear Motions (Keszthely, Hungary, 10-16 June, 1979) edited by A. Kiss, J. Nemeth, and J. Zimanyi, p. 400; C. Toepffer and C. Y. Wong, Phys. Rev. C 25, 1018 (1982).

8. G. Wolschin, Phys. Rev. Lett. 48, 1004 (1982).

9. P. Mädler and R. Reif, Nucl. Phys. A273, 27 (1982).

10. P. Grangé, H. A. Weidenmüller, and G. Wolschin, Ann. Phys. (N.Y.) 136, 190 (1981); G. Mantzouranis and H. C. Pauli, Z. fuer Phys. A281, 165 (1977); G. Mantzouranis and H. C. Pauli, Phys. Rev. C 22, 1550 (1980).

11. ZHUO Yi-zhong, ZHANG Jing-shang, WU Xi-zhen, and MA Zhong-yu, Phys. Energ. Fortis, Phys. Nucl. 4, 675 (1980), (English translation Chinese Physics 2, 166 (1982)); ZHUO Yi-zhong, MA Zhong-yu, FENG Ren-fa, and ZHANG Jing-shang, Kexue Tongbao 26, 596 (1981); S. Ayik, Z. fuer Phys. A298, 83 (1980); S. Ayik, Nucl. Phys. A370, 317 (1981).

12. R. Balian and M. Vенéroni, Ann. Phys. (N.Y.) 135, 270 (1981).

13. H. S. Köhler, Nucl. Phys. A343, 315 (1980); H. S. Köhler, Nucl. Phys. A378, 181 (1982).

14. J. Richert, D. M. Brink, and H. A. Weidenmüller, Phys. Lett. 87B, 6 (1979); J. Richert, P. Grange, G. Wolschin, and H. A. Weidenmüller, Nucl. Phys. A356, 260 (1981).

15. C. Y. Wong and K. T. R. Davies, Phys. Lett. 96B, 258 (1980).

16. P. Danielewicz, Ph.D. Thesis, Warsaw University, 1981 (to be published); P. Danielewicz, Workshop on Nuclear Dynamics (Granlibakken, California, 1982) LBL-14138, p. 93.

17. G. F. Bertsch and T. T. S. Kuo, Nucl. Phys. A112, 204 (1968).

18. J. P. Jeukenne, A. Lejeune, and C. Mahaux, Phys. Reports 25C, 85 (1976).

19. C. Mahaux and H. Ngô, Nucl. Phys. A378, 205 (1982); P. F. Bortignon, R. A. Broglia, C. H. Dasso, and FU De-Ji, Phys. Lett. 108B, 24 (1982).

20. G. Schütte and L. Wilets, Phys. Rev. C 25, 673 (1982).

21. P. C. Martin and J. Schwinger, Phys. Rev. 115, 1342 (1959).

22. L. P. Kadanoff and G. Baym, Quantum Statistical Mechanics (Benjamin, New York, 1976).

23. C. Y. Wong and S. J. Wang (to be published).

MEAN-FIELD THEORY AND RANDOM TWO-BODY COLLISIONS

Georg Wolschin
Max-Planck-Institut für Kernphysik, Heidelberg, F.R.G.

A microscopic theory to derive the collision term extending time-dependent mean-field theories for finite systems is discussed. It is based on a random-matrix model for the residual interaction in a diabatic basis of single-particle states. The structure of the energy-conserving function in a finite system with discrete s.p. levels is investigated. The effect of statistical two-body collisions on the equilibration of the occupation numbers is shown analytically in a schematic model. A numerical calculation demonstrates the influence of a collision term on TDHF results.

1. Introduction

The extension of time-dependent mean-field theory to include incoherent effects of the residual interaction leads to the derivation of collision terms. The consideration of incoherent processes that cause a destruction of phase relations in the interacting many-body system is believed to be important in particular in view of the dissipative phenomena that have emerged in the study of heavy-ion collisions[1-3]. Whereas the time-dependent Hartree-Fock approximation incorporates one-body damping, truly irreversible processes provided by random two-body collisions are not accounted for.

Different approaches to derive the collision term in the context of nuclear physics have been proposed. They exploit either Green's function methods[4] of quantum statistical mechanics[5], a perturbation expansion for the density operator[6,7], a time-averaging procedure of the TDHF-equations[8], or stochastic properties of the residual interaction[9,10] as in statistical spectroscopy. These methods lead to a collision term in the equation of motion for the one-body density matrix that is similar to the one in the Boltzmann equation for infinite, homogeneous systems, properly modified[11] to include Fermi statistics. A major difference occurs in finite systems because the single-particle energy levels are discrete, and change in time as the mean field gets distorted. As a consequence the energy-conserving δ-function of the Boltzmann collision term opens up in a finite system, and collisions between nucleons in single-particle levels that are several MeV apart become possible. Overall energy is still conserved through a balance of the energy content of both mean-field and collision term. This has been discussed in detail in the theory of Grangé et al.[10] and will be sketched in the review of that theory in section 2 .

In any of the theories[5-10] that lead to a collision term coherent effects of the residual interaction are not considered. Since they are likely to be unimportant at sufficiently high excitation energy this approximation appears reasonable if the emphasis is on the understanding of dissipative effects based on a time-dependent mean-field theory such as TDHF.

The theory[10] is irreversible already on the A-body level, as a consequence of the ensemble-averaging. The reduction of the equation of motion for the average A-body density matrix to the two- und further to the one-body level gives then rise to the appearance of the collision term. The hierarchy for the ensemble-averaged density matrices that emerges in the reduction process differs from the usual hierarchy obtained when taking subsequent traces of the Liouville equation.

In section 2, the theory is first outlined on the A-body level, the conservation laws are discussed, the reduction to a set of coupled equations for the ensemble-averaged one- and two-body density matrices is performed, and the collision term in the one-body equation of motion is investigated. This part is to a large extent based on Ref. 10 where details can be found.

The role of the collision term in dissipative processes in finite nuclei is discussed in section 3 . The structure of the energy-conserving function is investigated and the time scales that are relevant for the equilibration are presented. The analytical model[12] describes schematically the equilibration in the single-particle occupation numbers on the basis of a nonlinear partial differential equation. Whereas it has not yet been coupled to the time-evolution of the mean field, a numerical investigation[13] with a relaxation ansatz for the collision term exhibits how the Wigner transform of the one-body density approaches statistical equilibrium as a consequence of random two-body collisions which are absent in a pure mean-field approach. Several conclusions are drawn in section 4 .

2. A Random-Matrix Model to derive the Collision Term

In the mean-field approximation, the equations of motion for the A-body density matrix $\rho^{(A)}(\vec{x}_1,..,\vec{x}_A;\vec{x}_1',..,\vec{x}_A';t)$

can be written as

$$i\dot{\varrho}^{(A)}_{HF} = \left[H_{HF}, \varrho^{(A)}_{HF} \right] \tag{2.1}$$

with the self-consistent mean-field Hamiltonian

$$H_{HF} = \sum_{i=1}^{A} h^{(i)}_{HF} = \sum_{i=1}^{A} t_i + \sum_{i=1}^{A} u_i(\varrho^{(1)}) \quad , \tag{2.2}$$

the one-body mean-field

$$u_1 \varrho^{(1)} = \mathop{Tr}_2 \{ v(1,2) \left(\varrho^{(1)}(2,2;t) \varrho^{(1)}(1,1';t) - \varrho^{(1)}(1,2;t) \varrho^{(1)}(2,1';t) \right) \} \tag{2.3}$$

and $v(1,2)$ the two-body interaction. Here, $\varrho^{(A)}_{HF}$ is a product of the $\varrho^{(1)}$, and by taking the trace over (A-1) particles in Eq. (2.1) the time-dependent Hartree-Fock equation

$$i\dot{\varrho}^{(1)} = \left[h_{HF}(\varrho^{(1)}), \varrho^{(1)} \right] \tag{2.4}$$

is recovered.

To go beyond this mean-field approximation in order to include the incoherent effects of the residual interaction, we have formulated in [10] a random-matrix model. As discussed in the introduction, we thus restrict ourselves to effects of the residual force which lead to irreversible behaviour. The matrix elements of the full two-body interaction

$$\mathcal{V} = \sum_{i<j}^{A} v(i,j) \tag{2.5}$$

(which is usually replaced by an effective or density-dependent interaction in the construction of the mean field) are assumed to exhibit stochastic behaviour in a suitably chosen basis. Technically, the matrix $\hat{\mathcal{V}}$ is replaced by an ensemble of statistically distributed matrices. To determine the ensemble, we require in [10] that the mean field (2.3) is the same for all its members and does not fluctuate. This leads us to construct an ensemble of Hamiltonians

$$H = T + \mathcal{V} + V \tag{2.6}$$

with the actual effective operator $\mathcal{V}$ that serves only to determine the mean field, and V an ensemble of operators with mean value zero. The random force V is taken to be a sum of two-body operators V(i,j).

To formulate the statistical properties of V in analogy with methods known from statistical spectroscopy [14] and nuclear reactions [15] a suitable basis of single-particle states has to be chosen. At any time t we consider the antisymmetric matrix elements of V in a basis of single-particle states $|\alpha(t)\rangle$, $|\beta(t)\rangle$ defined by

$$H_{HF}(t)\,|\alpha(t)\rangle = \varepsilon_\alpha(t)\,|\alpha(t)\rangle \tag{2.7}$$

and written as $\langle\alpha\beta|V|\gamma\delta\rangle_t$. These matrix elements have a Gaussian distribution with mean value zero (V describes the statistical fluctuations around the mean field), and second moment

$$\overline{\langle\alpha\beta|V|\gamma\delta\rangle_t \langle\gamma'\delta'|V|\alpha'\beta'\rangle_{t'}} = \overline{V^2_{\alpha\beta\gamma\delta}}\left(\delta_{\alpha\alpha'}\delta_{\beta\beta'} - \delta_{\alpha\beta'}\delta_{\beta\alpha'}\right)\cdot$$
$$\cdot\left(\delta_{\gamma\gamma'}\delta_{\delta\delta'} - \delta_{\gamma\delta'}\delta_{\gamma'\delta}\right)\exp\left[-\frac{(t-t')^2}{2\tau^2_{corr}}\right]. \tag{2.8}$$

The bar denotes the ensemble average. The strength $\overline{V}^2$ of the residual force is independent of time. To connect the states $|\alpha(t)\rangle$ at different times a diabatic basis of single-particle states is chosen[10)], thus ensuring maximum overlap $\langle\alpha(t)|\alpha(t')\rangle$ for $t \neq t'$ at Landau-Zener crossings[16)]. In this basis, the nondiagonal elements $\langle\alpha(t)|\beta(t')\rangle$ $(\alpha\neq\beta)$ never attain large values, in contrast to an adiabatic basis. The statistical assumptions (2.8) are suitable for the diabatic states. The price to be paid is a rather large value $\tau_{corr} = 2\cdot 10^{-21}$ s of the correlation time that determines the decrease in the overlap of the wavefunctions. This, however, will not cause serious problems in the further development of the theory because the rapid change of the s.p. energies with time in the diabatic basis results in an additional, shorter time scale that provides the justification of a weak-coupling approximation. The technical problem of the construction of diabatic states from Eq. (2.7) will not be considered here.

Because of the stochastic nature of V, the A-body density matrix becomes a random variable. Extending the mean-field equation (2.1), its equation of motion becomes

$$i\dot{\rho}^{(A)} = \left[H_{HF}, \rho^{(A)}\right] + \left[V, \rho^{(A)}\right]. \tag{2.9}$$

The ensemble-average of this equation can now be taken. The result will be seen to obey an irreversible equation which, upon taking the trace, essentially reduces to a mean-field equation with a collision term. This program is described in the following parts of this section.

2.1. The equation of motion for the ensemble-averaged A-body density matrix

To obtain the formal solution of Eq. (2.9) it is convenient to transform to the interaction representation

$$\rho^{I} = \mathcal{O}_t\, \rho^{(A)}\, \mathcal{O}_t^{-1} \quad , \quad V^{I} = \mathcal{O}_t\, V\, \mathcal{O}_t^{-1} \tag{2.10}$$

with the time-ordered mean-field propagators

$$\begin{aligned} \mathcal{O}_t &= \underset{\rightarrow}{\exp}\left[i \int_{t_0}^{t} ds\; H_{HF}(s) \right] \\ \mathcal{O}_t^{-1} &= \underset{\leftarrow}{\exp}\left[-i \int_{t_0}^{t} ds\; H_{HF}(s) \right] \end{aligned} \tag{2.11}$$

so that

$$i\dot{\rho}^{I} = [V^{I}, \rho^{I}] \tag{2.12}$$

with the formal solution

$$\rho^{I} = \underset{\leftarrow}{\exp}\left\{ -i \int_{t_0}^{t} [\, V^{I}(s), \cdot\,]\, ds \right\} \rho^{I}(t_0). \tag{2.13}$$

To perform the ensemble-average of this equation, it is convenient to use a cumulant expansion[10,17] for the ensemble-average of the time-ordered exponential, to obtain (after transforming back to Schrödinger representation)

$$i\dot{\overline{\rho^{(A)}}} = \left[H_{HF}, \overline{\rho^{(A)}} \right] - i\, \mathcal{O}_t^{-1} \left[\overline{V^{I}(t), \int_{t_0}^{t} \left[V_{SC}^{I}(s), \overline{\rho^{I}(s)} \right] ds} \right] \mathcal{O}_t \,. \tag{2.14}$$

Given the statistical assumptions, this result is correct for any strength of V . However, V_{SC}^{I} involves time-ordered cumulants of arbitrary order and therefore the result is not yet very useful. The derivation of an integral equation for V_{SC}^{I} is extremely tedious. The lowest-order nonvanishing contributions are

$$\begin{aligned} V_{SC}^{I}(s) \simeq V^{I}(s) + \int_{t_0}^{s} dt_3 \int_{t_0}^{t_3} dt_4 \Big\{ & [\overline{V^{I}(s), [V^{I}(t_3), [V^{I}}(t_4), \cdot]]] \\ & + [\overline{V^{I}(s), [V^{I}}(t_3), [V^{I}(t_4), \cdot]]] - [V^{I}(t_3), [\overline{V^{I}(s), [V^{I}}(t_4), \cdot]]] \\ & - [V^{I}(t_4), [\overline{V^{I}(s), [V^{I}}(t_3), \cdot]]] \Big\} . \end{aligned} \tag{2.15}$$

A physically reasonable approximation for low bombarding energies is the weak-coupling limit formulated in [10]: Only those pairs of V^{I} are averaged that are time-ordered nearest neighbours. Then, $V_{SC}^{I}(s) = V^{I}(s)$ in Eq. (2.14), and using the weak-coupling limit once again to pull out $\overline{\rho^{I}}$ from under the time integral the result[10] is recovered

$$i\dot{\overline{\rho^{(A)}}} = \left[H_{HF}, \overline{\rho^{(A)}}\right] - i\,\overline{\left[V(t), \left[W(t), \overline{\rho^{(A)}}(t)\right]\right]} \tag{2.16}$$

where $W(t) \equiv \int_{t_0}^{t} ds\, O_t^{-1} O_s V(s) O_s^{-1} O_t$. This equation represents the desired transport equation that describes the irreversible approach of the ensemble-averaged A-body density matrix towards statistical equilibrium. It is consistent with the conservation laws of particle number and energy. Taking the trace of Eq. (2.16) over all variables and using the cyclic invariance of the trace, we find immediately that the norm of $\overline{\rho^{(A)}}$ and hence the particle number is conserved.

The proof that the first moment of energy is also conserved has been given in [10]. It requires to show that

$$\frac{d}{dt}\overline{\mathop{Tr}_{1\ldots A}\left\{H\rho^{(A)}\right\}} = \mathop{Tr}_{1\ldots A}\left\{H_{HF}\dot{\overline{\rho^{(A)}}} + \overline{V\dot{\rho}^{(A)}}\right\} \tag{2.17}$$

vanishes. The first term on the r.h.s. is derived as in the usual mean-field theory[18]. Its trace would vanish for $\overline{\rho^{(A)}}$ replaced by $\rho_{HF}^{(A)}$. Here it is rewritten using Eq. (2.16) to become

$$-\mathop{Tr}_{1\ldots A}\left\{H_{HF}\,\overline{\left[V(t), \left[W(t), \overline{\rho^{(A)}}(t)\right]\right]}\right\}$$

whereas the second term becomes using Eq. (2.9)

$$i^{-1}\mathop{Tr}_{1\ldots A}\left\{\overline{V\left[H_{HF}, \rho^{(A)}\right]}\right\}.$$

The two terms cancel because a comparison between Eqs. (2.9) and (2.16) shows that

$$\overline{V\rho^{(A)}} = -i\,\overline{V\left[W, \overline{\rho^{(A)}}\right]}.$$

Overall mean-energy conservation is thus due to a balance of the energy content of the time-dependent mean field, and of the collision term. This is in contrast to both conventional mean-field theory (where the energy contained in the mean field only is conserved), and collision terms in Boltzmann-type equations. A reduction of the equation for the average A-particle density matrix to the one-particle level as described in the next section has to preserve overall energy conservation[10].

2.2 Reduction to the one- and two-body level

A hierarchy of coupled equations for the average density matrices $\overline{\rho^{(k)}}$ (k = 1...A) can now be obtained by taking subsequent traces over the corresponding variables in the A-body equation (2.16). Due to the ensemble-averaging which accounts for the statistical effect of the residual force, this hierarchy differs from the usual one obtained by taking traces over the Liouville equation. There, the one-body density matrix is coupled to the two-body level only (and factorization of $\overline{\rho^{(2)}}$ yields the Hartree-Fock equations). Now the ensemble-averaged $\overline{\rho^{(1)}}$ will be coupled to the two- and three-body level.

To perform the trace over the collisional part of Eq. (2.16), a diabatic approximation to the matrix elements of the single-particle propagator is used,

$$\langle\alpha(t)|\exp\Big\{-i\int_s^t d\tau\, h_{HF}(\tau)\Big\}|\beta(s)\rangle \simeq \delta_{\alpha\beta}\cdot\exp\Big\{-i\int_s^t d\tau\, \varepsilon_\alpha(\tau)\Big\} . \tag{2.18}$$

Hence, the collision term is evaluated under the assumption that in the mean-field propagation of any single-particle state $|\beta(s)\rangle$ no occupation probability is redistributed into other states. The neglect of this redistribution due to terms of the type $\langle\alpha(t)|\partial/\partial t|\beta(t)\rangle,\ \alpha\neq\beta$ puts an upper limit on the collective velocity[10] which may be smaller than the one imposed by the weak-coupling limit used to derive Eq.(2.16). Both approximations are believed to be valid for typical collective velocities of shape deformations in deeply inelastic heavy-ion reactions at several MeV per nucleon above the Coulomb barrier.

With the approximation (2.18) the reduction to equations for the average one- and two-body density matrices can be performed. The result is written as

$$i\dot{\overline{\rho}}^{(1)}_{\mu_1\nu_1} = \left[h^{(1)}_{HF}, \overline{\rho^{(1)}}\right]_{\mu_1\nu_1} + i\, C^{(1)}_{\mu_1\nu_1}\left(\overline{\rho^{(2)}}, \overline{\rho^{(3)}}\right) \tag{2.19}$$

$$i\dot{\overline{\rho}}^{(2)}_{\mu_1\mu_2\nu_1\nu_2} = \left[h^{(1)}_{HF} + h^{(2)}_{HF}, \overline{\rho^{(2)}}\right]_{\mu_1\mu_2\nu_1\nu_2} + i\, C^{(2)}_{\mu_1\mu_2\nu_1\nu_2}\left(\overline{\rho^{(2)}}, \overline{\rho^{(3)}}\right). \tag{2.20}$$

Both the collision term $C^{(1)}$ on the one-body level, and $C^{(2)}$ on the two-body level are functions of $\overline{\rho^{(2)}}$ and $\overline{\rho^{(3)}}$ which survive the reduction of the second term in Eq. (2.16). They contain the effect of the residual interaction V to any order in the weak-coupling limit and lead to a violation of time-reversal invariance in both equations.

The first equation follows from the second one by tracing over $\mu_2 = \nu_2$ with the collision term on the one-body level given by

$$C^{(1)}_{\mu_1\nu_1} = \sum_{\gamma} C^{(2)}_{\mu_1\gamma\nu_1\gamma} \; . \tag{2.21}$$

For simplicity I restrict now the discussion of the collision term to its diagonal elements and obtain

$$\begin{aligned} C^{(2)}_{\mu\gamma\mu\gamma} &= \sum_{\alpha\beta} \overline{V^2_{\mu\gamma\alpha\beta}} \cdot G(\varepsilon_\mu + \varepsilon_\gamma - \varepsilon_\alpha - \varepsilon_\beta) \, \cdot \\ &\cdot \Big[\Big\{ \overline{\rho^{(2)}_{\alpha\beta\alpha\beta}} - \overline{\rho^{(3)}_{\mu\alpha\beta\mu\alpha\beta}} - \overline{\rho^{(3)}_{\gamma\alpha\beta\gamma\alpha\beta}} \Big\} \\ &- \Big\{ \overline{\rho^{(2)}_{\mu\gamma\mu\gamma}} - \overline{\rho^{(3)}_{\mu\gamma\alpha\mu\gamma\alpha}} - \overline{\rho^{(3)}_{\mu\gamma\beta\mu\gamma\beta}} \Big\} \Big] \end{aligned} \tag{2.22}$$

where the normalization

$$\sum_{\alpha_2 \ldots \alpha_A, \beta_2 \ldots \beta_A} \overline{\rho^{(A)}_{\alpha_1 \ldots \alpha_A ; \beta_1 \ldots \beta_A}} = (A-1)! \; \overline{\rho^{(1)}_{\alpha_1\beta_1}} \tag{2.23}$$

has been used. On the one-body level, the off-diagonal elements $C^{(1)}_{\mu\nu}$, $\mu \neq \nu$ can be found in Eq. (8) of Ref. 10. The off-diagonal elements on the two-body level have not yet been derived.

In Eq. (2.22), the strength of the residual interaction is $\bar{V}^2$, and the energy-conserving function G is obtained as

$$\begin{aligned} G(\varepsilon_\mu + \varepsilon_\gamma - \varepsilon_\alpha - \varepsilon_\beta) = \mathrm{Re}\Big[\int_{t_0}^{t} ds \, \exp\Big\{ -i \int_{s}^{t} d\tau \big(\varepsilon_\alpha(\tau) + \varepsilon_\beta(\tau) \\ - \varepsilon_\mu(\tau) - \varepsilon_\gamma(\tau) \big) - \frac{(t-s)^2}{2\tau^2_{corr}} \Big\} \Big] . \end{aligned} \tag{2.24}$$

The discussion of G is deferred to the next section. Here let us proceed with the equations of motion for $\overline{\rho^{(1)}}$ and $\overline{\rho^{(2)}}$. The coupled set of equations (2.19), (2.20) is not closed. It is evident that the natural way to obtain a closed set consists in a factorization of $\overline{\rho^{(3)}}$ in both equations with direct and exchange contributions:

$$\begin{aligned} \overline{\rho^{(3)}_{\alpha_1\alpha_2\alpha_3\beta_1\beta_2\beta_3}} &= \overline{\rho^{(1)}_{\alpha_1\beta_1}} \; \overline{\rho^{(2)}_{\alpha_2\alpha_3\beta_2\beta_3}} - \overline{\rho^{(1)}_{\alpha_2\beta_1}} \; \overline{\rho^{(2)}_{\alpha_1\alpha_3\beta_2\beta_3}} + \\ &+ \overline{\rho^{(1)}_{\alpha_3\beta_1}} \; \overline{\rho^{(2)}_{\alpha_1\alpha_2\beta_2\beta_3}} \; . \end{aligned} \tag{2.25}$$

The resulting equations for $\overline{\rho^{(1)}}$ and $\overline{\rho^{(2)}}$ provide a basis for the calculation of expectation values of one- and two-body operators which is thought to be rather complete. The only feature that is missing, but may be important in low-energy heavy-ion collisions are fluctuations of the mean field.

In the next section, we consider the equation of motion for $\overline{\rho^{(1)}}$ only. Hence, also $\overline{\rho^{(2)}}$ is written as an antisymmetric product of $\overline{\rho^{(1)}}$. Various simplifications will be used in order to discuss the effect of the collision term on dissipative processes in finite fermion systems.

3. Role of the Collision Term in a Finite System

The equation of motion for the average one-body density matrix is written as

$$i \langle \mu | \dot{\overline{\rho^{(1)}}} | \nu \rangle = \langle \mu | [h_{HF}, \overline{\rho^{(1)}}] | \nu \rangle + i\, C_{\mu\nu} (\overline{\rho^{(1)}}) \tag{3.1}$$

The matrix elements of $\dot{\overline{\rho^{(1)}}}$ are given by

$$\langle \mu | \dot{\overline{\rho^{(1)}}} | \nu \rangle = \frac{d}{dt} \overline{\rho^{(1)}_{\mu\nu}} - \sum_{\gamma} \langle \dot{\mu} | \gamma \rangle \overline{\rho^{(1)}_{\gamma\nu}} - \sum_{\gamma} \overline{\rho^{(1)}_{\mu\gamma}} \langle \gamma | \dot{\nu} \rangle \tag{3.2}$$

In occupation-number representation the diagonal elements

$$\overline{\rho^{(1)}_{\mu\mu}} = n_{\mu}(t)\, \varphi_{\mu}(\vec{x},t)\, \varphi^{*}_{\mu}(\vec{x}',t) \tag{3.3}$$

with single-particle occupation numbers $n_{\mu}(t) \equiv n(\varepsilon_{\mu},t)$, $0 \le n_{\mu} \le 1$ are governed by

$$\begin{aligned} \frac{\partial n_{\mu}}{\partial t} = \sum_{\alpha\beta\gamma} \overline{V^{2}_{\mu\gamma\alpha\beta}}\, G(\varepsilon_{\mu}+\varepsilon_{\gamma}-\varepsilon_{\alpha}-\varepsilon_{\beta}) \big[&(1-n_{\mu})(1-n_{\gamma}) n_{\alpha} n_{\beta} \\ &- (1-n_{\alpha})(1-n_{\beta}) n_{\mu} n_{\gamma} \big] . \end{aligned} \tag{3.4}$$

This is similar to the collision term in the Boltzmann equation modified for Fermi statics. It is a balance equation with a gain- and a loss term for the occupation numbers of the time-dependent single-particle levels. The energy-conserving function G, however, has a more complicated structure to be discussed below. The nondiagonal elements of the density matrix obey

$$i \frac{d\overline{\rho^{(1)}_{\mu\nu}}}{dt} = (\varepsilon_{\mu}-\varepsilon_{\nu}) \overline{\rho^{(1)}_{\mu\nu}} + i \langle \dot{\mu} | \nu \rangle \overline{\rho^{(1)}_{\nu\nu}} + i \langle \mu | \dot{\nu} \rangle \overline{\rho^{(1)}_{\mu\mu}} + i \Gamma_{\mu\nu} \overline{\rho^{(1)}_{\mu\nu}}. \tag{3.5}$$

The $\Gamma_{\mu\nu}$-term is the collisional contribution to the time evolution of $\overline{\rho^{(1)}_{\mu\nu}}$. From the structure of the collision term, it is expected that these nondiagonal contributions $\Gamma_{\mu\nu}$ equilibrate faster than the diagonal ones which refer to the s.p. occupation numbers. If the latter reach their equilibrium (the Fermi distribution) within τ_{equ}, the inequality

$$\frac{1}{\Gamma_{\mu\nu}} < \tau_{equ} \tag{3.6}$$

holds. We now proceed to investigate the relaxation in the occupation numbers, and defer the coupling to the time-evolution of the mean field to the last section.

3.1. Structure of the energy-conserving function

The energy-conserving function in the equation (3.4) for the single-particle occupation numbers has been given in Eq. (2.24). If the mean field and the single-particle energies were independent of time as in an infinite homogeneous system, the limit correlation time $\tau_{corr} \to \infty$ can be taken. Then, with $t_o \to -\infty$, $t \to \infty$, one obtains

$$G(\varepsilon_\mu + \varepsilon_\gamma - \varepsilon_\alpha - \varepsilon_\beta) \longrightarrow \pi\,\delta(\varepsilon_\mu + \varepsilon_\gamma - \varepsilon_\alpha - \varepsilon_\beta) \tag{3.7}$$

and the collision term coincides with the one obtained usually[4]. It leads to equilibration in an infinitely extended system, with a Fermi-Dirac distribution for the occupation probabilities in the equilibrium limit.

In a finite system, the single-particle energies depend upon the mean field, and thus upon time. The associated time scale τ_{sp} is shorter than the correlation time τ_{corr} for the single-particle wavefunctions:

$$\tau_{sp} < \tau_{corr} \tag{3.8}$$

whereas the latter is of the order of typical interaction times ($2\cdot 10^{-21}$s), τ_{sp} is of the order of 10^{-22}s [19]. As a consequence, the width of the energy-conserving function is not determined by τ_{corr} but instead by the rapid change of the single-particle energies as the system deforms. Hence, we evaluate G neglecting the effect of τ_{corr}:

$$G(\varepsilon_\mu + \varepsilon_\gamma - \varepsilon_\alpha - \varepsilon_\beta) \simeq \mathrm{Re}\left[\int_{t_o}^{t} ds \exp\left\{-i\int_{s}^{t} d\tau\, \Delta\varepsilon(\tau)\right\}\right] \tag{3.9}$$

where

$$\Delta\varepsilon(\tau) \equiv \varepsilon_\alpha(\tau) + \varepsilon_\beta(\tau) - \varepsilon_\gamma(\tau) - \varepsilon_\mu(\tau).$$

In a first-order expansion of $\Delta\varepsilon(\tau)$,

$$\Delta\varepsilon(\tau) \simeq \Delta\varepsilon(t) + (\tau - t)\,\Delta\dot{\varepsilon}(t) \tag{3.10}$$

the G-function becomes

$$G_{\mu\gamma\alpha\beta} = \int_0^{t-t_0} du \cos\left[\Delta\varepsilon\cdot u - \frac{1}{2}\Delta\dot{\varepsilon}\cdot u^2\right] \tag{3.11}$$

which is integrated to obtain

$$G_{\mu\gamma\alpha\beta} = \left[\frac{\pi}{|\Delta\dot{\varepsilon}|}\right]^{1/2} \left\{\left[\frac{1}{2} - C(z)\right]\cos\left(\frac{\pi}{2}z^2\right) + \left[\frac{1}{2} - S(z)\right]\sin\left(\frac{\pi}{2}z^2\right)\right\} \tag{3.12}$$

with the Fresnel integrals

$$S(z) = \int_0^z \sin\left(\frac{\pi}{2}u^2\right)du\ , \quad C(z) = \int_0^z \cos\left(\frac{\pi}{2}u^2\right)du$$

and $z = |\Delta\varepsilon| / \sqrt{\pi|\Delta\dot{\varepsilon}|}$.

A rather accurate approximation to this result is written as [20]

$$G_{\mu\gamma\alpha\beta} \simeq \left[\frac{\pi}{|\Delta\dot{\varepsilon}|}\right]^{1/2} \cdot \left[2 + 4.142\,z + 3.492\,z^2 + 6.670\,z^3\right]^{-1} . \tag{3.13}$$

The result is shown in Figure 1 .

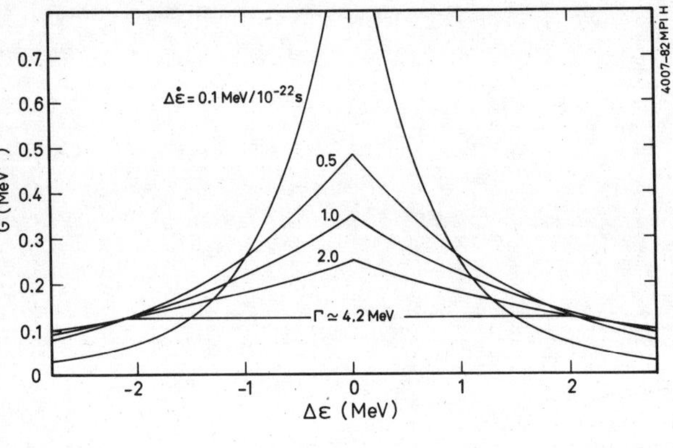

Fig. 1 : The energy-conserving function $G(\Delta\varepsilon)$ in a finite fermion system for different velocities $\Delta\dot{\varepsilon}$ (Ref.21).

The width of the energy-conserving function is seen to increase in proportion with $(\Delta\dot{\varepsilon})^{1/2}$. For heavy-ion collisions, the difference $\Delta\dot{\varepsilon} = (\dot{\varepsilon}_\alpha + \dot{\varepsilon}_\beta - \dot{\varepsilon}_\gamma - \dot{\varepsilon}_\mu)$ has typically the value

$$\Delta\dot{\varepsilon} \simeq 1\,MeV/10^{-22}s. \tag{3.14}$$

This value is obtained using

$$\frac{d(\Delta\varepsilon)}{dt} = \frac{d(\Delta\varepsilon)}{dQ}\frac{dQ}{dt}$$

and a collective speed of $\dot{Q} = Q_{max}/(4\cdot 10^{-21}s)$ where Q_{max} is the deformation at the scission point. The associated characteristic time scale τ_{sp} becomes

$$\tau_{sp} = (\Delta\dot{\varepsilon})^{-1/2} \simeq 2.5\cdot 10^{-22}s \tag{3.15}$$

which is indeed an order of magnitude shorter than the correlation time. This short value also provides the posteriori justification of the weak-coupling limit that has been used in the derivation of the previous section. Due to the finite correlation time $\tau_{corr} > \tau_{sp}$, the energy-conserving function will have exponential tails which are not included in Fig. 1 .

The collision term in a finite system is thus linked to the time-dependence of the mean field: Fast changes of the mean field result in a large width of the energy window provided by G , and a rapid action of the collision term which converts ordered nucleonic motion coherent with the mean field into incoherent motion, until the single-particle occupation numbers have acquired a Fermi distribution and the mean field does not change any further.

3.2. A schematic model for the action of the collision term:

Whereas TDHF corresponds to constant single-particle occupation numbers $n_\mu = 1$ or 0 , the collision term describes the equilibration in the fermion system due to the action of the residual force. To emphasize its effect on the occupation numbers according to the reduced equation (3.4), let us outline a schematic model that leads to analytical solutions[12] and still preserves the essential physical ingredients such as Fermi statistics, particle-number and energy conservation.

The collision term is written in the form of a master equation with gain and loss term, respectively

$$\frac{\partial n_\mu}{\partial t} = (1-n_\mu)\sum_\beta W_{\beta\to\mu}\, n_\beta - n_\mu \sum_\beta W_{\mu\to\beta}\,(1-n_\beta) \tag{3.16}$$

where

$$W_{\mu\to\beta} = \sum_{\alpha\gamma} \overline{V^2_{\mu\gamma\alpha\beta}}\; G(\varepsilon_\mu+\varepsilon_\gamma-\varepsilon_\alpha-\varepsilon_\beta)\,(1-n_\alpha)\,n_\gamma \tag{3.17}$$

and $W_{\beta\to\mu}$ accordingly. The aim is to transform the collision term into a tractable differential equation which contains the essential physical ingredients. The nonlinearity in the transition function W is disregarded; improvement with an iteration scheme appears possible. With the replacement (g_β the s.p. level density)

$$\begin{aligned}\sum_\beta W_{\beta\to\mu}\, n_\beta &\longrightarrow \int d\beta\, W_{\beta\mu}\, g_\mu \cdot n_\beta \\ \sum_\beta W_{\mu\to\beta}\,(1-n_\beta) &\longrightarrow \int d\beta\, W_{\mu\beta}\, g_\beta\,(1-n_\beta),\end{aligned} \tag{3.18}$$

the additional assumption $W_{\beta\mu} = W_{\mu\beta}$, and a Taylor expansion of n_β and $g_\beta \cdot n_\beta$ around $\varepsilon_\beta = \varepsilon_\mu$ the nonlinear partial differential equ.[12] for $n \equiv n(\varepsilon_\mu, t)$ is obtained

$$\frac{\partial n}{\partial t} = -\frac{\partial}{\partial \varepsilon}\left[v\cdot n(1-n) + n^2\,\frac{\partial D}{\partial \varepsilon}\right] + \frac{\partial^2}{\partial \varepsilon^2}\left[D\cdot n\right]. \tag{3.19}$$

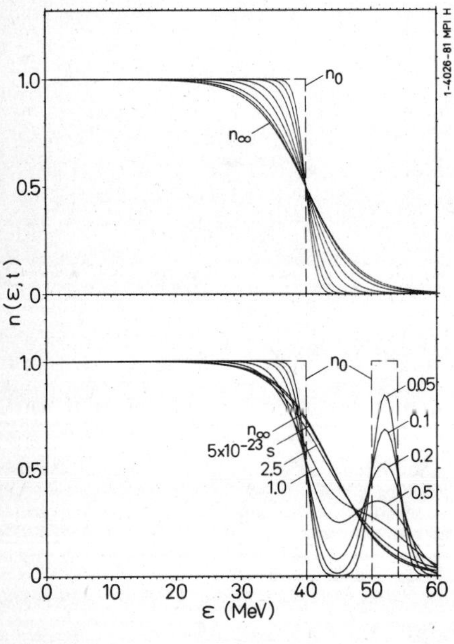

Fig. 2: Equilibration in a finite fermion system due to the action of the residual force, modelled by the analytical solution of Eq. (3.19) (Ref.12).

Transport coefficients have been introduced via moments of the transition function ($x = \varepsilon_\beta - \varepsilon_\mu$)

$$D = \frac{1}{2} g_\mu \int dx\, W(\mu,x) x^2$$

$$v = g_\mu^{-1} \frac{d}{d\varepsilon_\mu} (g_\mu \cdot D). \tag{3.20}$$

In Eq. (3.19), I have expressed the dissipative effects in the s.p. occupation through the drift term v, the diffusive effects through the diffusion term D. Both transport coefficients combine to determine the speed of the relaxation process as well as the equilibrium distribution. Unlike in case of linear diffusion equations of the Fokker-Planck type[1-3], solutions of Eq. (3.19) are rather difficult to obtain. This is a consequence of the nonlinear terms which are due to the exclusion principle. They are essential for the proper description of the time evolution.

The Eq. (3.19) has been solved analytically in Ref. 12 through a nonlinear transformation in the limit of constant transport coefficients. As is to be expected, it has a Fermi-type equilibrium limit

$$n_\infty(\varepsilon) = \left[1 + \exp\left(-\frac{v}{D}(\varepsilon - \varepsilon_F)\right)\right]^{-1}. \tag{3.21}$$

Analytical solutions for any initial distribution $n_o(\varepsilon)$ (dashed curve in Fig. 2) are obtained as[12)]

$$n(\varepsilon,t) = \int_{-\infty}^{\infty} n_o(x)\, f(x,\varepsilon;t)\, dx \Big/ \left[\int_{-\infty}^{\infty} f(x,\varepsilon;t)\, dx\right] \tag{3.22}$$

where

$$f(x,\varepsilon;t) = \exp\left\{-\frac{1}{2} D^{-1}\left[vx - 2v\int_0^x n_o(y)\,dy\right]\right\} \exp\left[-(\varepsilon-x)^2/(4Dt)\right].$$

These solutions are shown in Fig. 2 for schematic initial conditions. In the lower part, projectile particles can be viewed as occupying states in the continuum region of the potential, and the residual force acts to equilibrate the system towards the equilibrium limit (3.21). Here we have not coupled the time evolution of the occupation numbers to the mean-field evolution.

In the model, the equilibration time is obtained as a ratio of the transport coefficients

$$\tau_{equ} = 4D/v^2 \tag{3.23}$$

and thus, since $\tau_{equ} \propto D/v^2 \propto 1/\overline{V^2}$, essentially by the strength of the residual interaction that enforces the equilibration process. A microscopic calculation of the transport coefficients and the investigation of their dependence on the collective velocity (through the function G) is desirable. In Fig.2, I have chosen $D = 20 \cdot 10^{23}\,\text{MeV}^2\text{s}^{-1}$

and $v=-5\cdot 10^{23}$ MeV s^{-1} such that $-D/v$ = 4 MeV (to be compared with the temperature) and τ_{equ} = $3.2\cdot 10^{-23}$s .

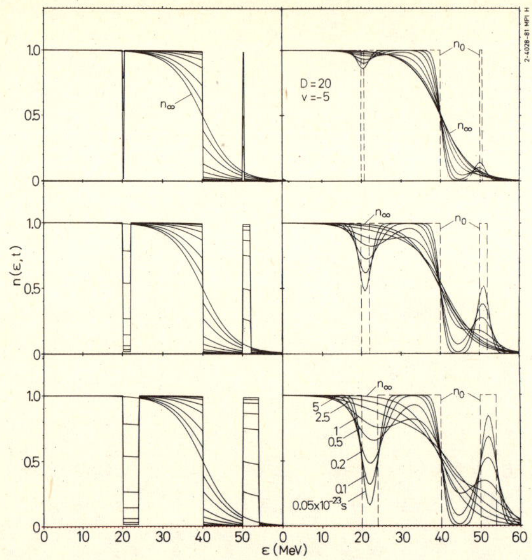

Fig. 3 : Comparison of the results for the collision term (r.h.s.) with the relaxation ansatz (l.h.s.) for various initial distributions (Ref. 12).

In Fig. 3 the results of the collision term are compared to the corresponding relaxation ansatz (l.h.s.)[22]

$$\frac{\partial n}{\partial t} = (n_\infty - n)/\tau_{equ} \tag{3.24}$$

which enforces equilibration towards n_∞ . The sequence of time is the same as on the r.h.s. For small times, and especially for small initial disturbances, the relaxation ansatz leads to a slower equilibration. Numerical solutions of the full occupation-number equation obtained by Toepffer and Wong for an infinite system[23] exhibit a similar behaviour. It may be worthwhile to replace the relaxation ansatz by the solution of Eq. (3.19) in actual extended mean-field calculations. This has not yet been done. In the next section, the modification of numerical TDHF-results through the action of a phenomenological collision term[13] is reported.

3.3. Effect of a Collision Term on TDHF-Results

Following the work of Richert et al.[13)] we have studied the effect of a collision term on one-dimensional TDHF-results in a reaction between two slabs. The TDHF equations have been rewritten in terms of the Wigner transform

$$f(x,k;t) = (2\pi)^{-1}\int_{-\infty}^{+\infty} ds\, \exp(-iks)\,\rho(x+\tfrac{1}{2}s, x-\tfrac{1}{2}s) \tag{3.25}$$

of the one-body density matrix as

$$\left(\frac{\partial}{\partial t} + \frac{k}{m}\frac{\partial}{\partial x}\right) f(x,k;t) + F_{pot}(f) = 0 \tag{3.26}$$

with the Wigner transform $F_{pot}(f)$ of the self-consistent single-particle potential. The effective interaction of Ref. 24 with parameters adapted[13)] to a genuine one-dimensional system is used. A phenomenological collision term which has not yet been related to the theory presented above is added to the r.h.s. of Eq. (3.26)

$$I_{coll} = (2\tau)^{-1}\left[f(x, 2\bar{k}-k;t) - f(x,k;t)\right]. \tag{3.27}$$

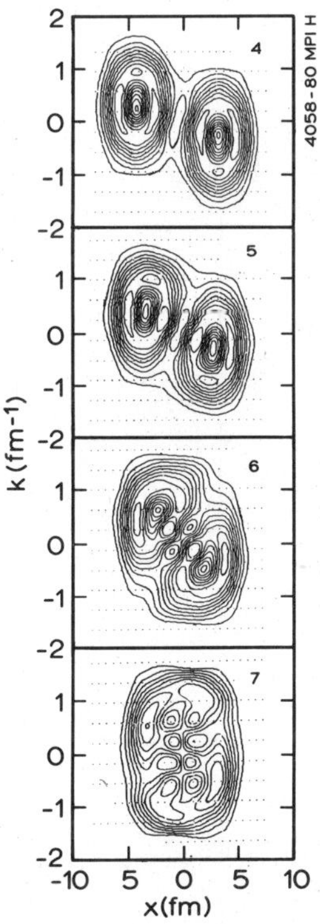

Fig. 4 : Time evolution of the Wigner transform f(x,k;t) of the single-particle density matrix in the entrance phase of the collision. Times in 10 fm/c (Ref.13).

It imposes an equilibration in momentum space towards the local mean momentum

$$\bar{k}(x,t) = \int k\, f(x,k;t)\, dk \,/ \int f(x,k;t)\, dk \tag{3.28}$$

with a relaxation time τ . Through the value of τ a connection with microscopic derivation of the collision term is feasible. In [13] we have treated it as a parameter.

To solve the equation for the Wigner function supplemented with the collision term, $f(x,k;t)$ has been expanded on a basis of harmonic functions, and the truncated set of coupled differential equations for the expansion coefficients has been solved numerically. Results for a collision of two one-dimensional slabs containing 8 nucleons each are shown in Figs. 4 and 5 for a bombarding energy of 7.5 MeV/nucleon. At this energy three-dimensional TDHF-calculations for $^{16}O + ^{16}O$ and similar systems are known to exhibit a low-ℓ cutoff for fusions ("fusion window")[25] .

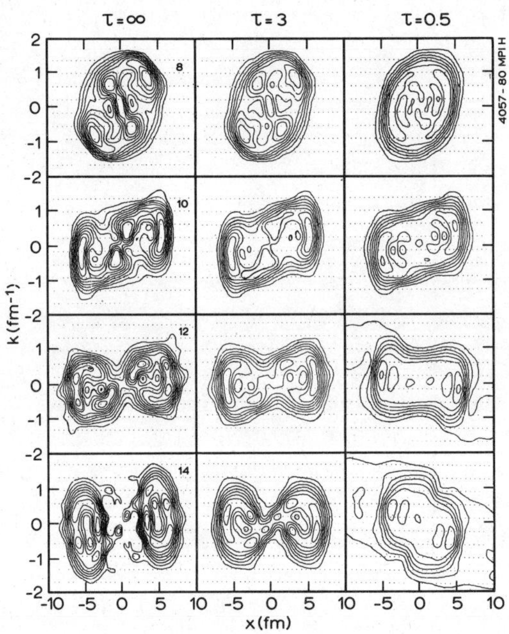

Fig. 5 : Time evolution of the Wigner function in the exit phase of the collision for three different strengths τ (in 10 fm/c) in the relaxation ansatz for the collision term (Ref. 13).

For reasonable strengths of the collision term as expressed by the equilibration time (τ = 5 ... 30 fm/c) very little modification of the TDHF results is found in the initial phase of the collision (Fig. 4). The effect of the residual two-body forces becomes apparent during the final stages where they lead to a slowing-down of the translational motion beyond what is found from the effects of the mean field, and eventually induce fusion.

This is illustrated in the contour plots of Fig. 5 for the Wigner function in the exit phase of the collision where the TDHF result is compared with results for two different strengths of the collision term. For small τ the collisions drive the system towards the equilibrated compound nucleus. The equilibration cannot be described in a pure mean-field calculation without the effect of two-body collisions. Similar results have been obtained by Köhler[26]. Our results are also likely to affect predictions of two- and three-dimensional TDHF-calculations for the occurrence of the fusion window. Experimental evidence for the fusion window has not been found[27], and it may well turn out to be an artifact of the TDHF-approximation.

4. Conclusions

We have been concerned with the extension of time-dependent mean-field theory in order to include the incoherent effect of the residual interaction. The consideration of random two-body collisions appears as a mandate if the goal is a microscopic understanding of dissipative processes starting from a mean-field approach. In the initial stages of a collision problem, TDHF may be a good approximation to the time-evolution of the many-body system, but at later stages the collision term becomes important, leads to a steady increase of entropy, and may eventually induce fusion (that is statistical equilibrium), unless the collision partners separate before due to the dynamical evolution of the process.

The random-matrix model of Grangé et al.[10] has been formulated in section 2 in such a way as to include the residual interaction for arbitrary strength. The corresponding equation for the average A-body density matrix has a memory kernel which is extremely difficult to handle. The subsequent restriction of the theory to the tractable weak-coupling limit is sufficiently accurate for nuclear collisions at energies of several MeV/nucleon above the Coulomb barrier.

The results of the theory describe dissipative behaviour. In actual calculations, they should yield expectation values of one-body operators somewhat different from the ones obtained in TDHF: The mean energy loss in a collision will be larger, the mass drift at high excitation energies is expected to increase, predictions for a fusion window will be modified. A better microscopic understanding of the measured mean values in nuclear collisions appears to be possible, beyond the theoretical interest in the origin of dissipation.

The collision term will certainly also contribute to the expectation values of two-body operators. Coupled equations for the ensemble-averaged one- and two-body density matrix have been derived to treat this problem. However, statistical fluctuations of the mean field itself have not been considered in the theory for reasons of technical feasibility, and this may, indeed, limit the microscopic understanding of the large fluctuations observed in dissipative nuclear collisions on the basis of the present theory.

I am indebted to several colleagues for their support of the theoretical work, and I thank P. Grangé, J. Richert and H.A. Weidenmüller for the intense collaboration. This work was sponsored in part by the CRN, Strasbourg and the Deutsche Forschungsgemeinschaft, Bonn.

References:

1. H.A. Weidenmüller. Prog. Part. and Nucl. Phys. Vol. 3, p.49 (1980), Pergamon Press, and references therein.
2. A. Gobbi and W. Nörenberg, in: Heavy-Ion Collisions, Ed. R. Bock, North-Holland Publ. Comp., Amsterdam - New York - Oxford, 1980, Vol. 2
3. G. Wolschin, in: Nuclear Structure and Heavy-Ion Collisions, LXXVII Corso Varenna, Soc. Italiana di Fisica, p. 507 (1981)
4. L.P. Kadanoff and G. Baym, Quantum Statistical Mechanics, Benjamin, New York (1962)
5. C.Y. Wong and H.H.K. Tang, Phys.Rev.Lett. 40 (1978) 1070, and Phys. Rev. C20 (1979) 1419;
C.M. Shakin and M.S. Weiss, preprint UCRL-80500 (1977, unpublished)
H. Orland and R. Schaeffer, Z. Physik A290 (1978) 191;
G. Schütte and L. Wilets, Phys. Rev. C25 (1982) 673
6. G. Mantzouranis and H.C. Pauli, Z.Physik A281 (1977) 165, and Phys. Rev. C22 (1980) 1550
7. H.C. Pauli, preprint MPI-H-1982-V7
8. R. Balian and M. Véneroni, Ann. Phys. (N.Y.) 135 (1981) 270
9. S. Ayik, Z. Physik A298 (1980) 83
10. P. Grangé, H.A. Weidenmüller and G. Wolschin, Ann.Phys. (N.Y.) 136 (1981) 190
11. E.A. Uehling and G.E. Uhlenbeck, Phys. Rev. 43 (1933) 352
12. G. Wolschin, XIV. Masurian Summer School in Nuclear Physics, Mikolajki, Poland, Sept. 1981, in press, and Phys.Rev.Lett. 48 (1982) 1004
13. J. Richert, D.M. Brink and H.A. Weidenmüller, Phys.Lett. 87B (1979) 6;
P. Grangé, J. Richert, G. Wolschin and H.A. Weidenmüller, Nucl. Phys. A356 (1981) 260
14. T.A. Brody, J. Flores, J.B. French, P.E. Mello, A. Pandey and S.S.M. Wong, Rev.Mod.Phys. 53 (1981) 385
15. C. Mahaux and H.A. Weidenmüller. Ann.Rev.Nucl.Particle Sci. 29 (1979) 1
16. D.L. Hill and J.A. Wheeler, Phys.Rev. 89 (1953) 1102
17. R.F. Fox, Phys. Lett. C48 (1978) 179
18. J.W. Negele, in: Theoretical Methods in Medium-Energy and Heavy-Ion Physics (K.W. McVoy and W.A. Friedman, Eds.), p.235, Plenum Press New York / London, 1978
19. M.C. Nemes and H.A. Weidenmüller, Phys. Rev. C24 (1981) 450
20. M. Abramowitz and J.A. Stegun, in: Handbook of Mathematical Functions, Dover Publications, Inc., New York, 7.3.33
21. P. Grangé and G. Wolschin, in preparation
22. W. Nörenberg, Proc. Int. Workshop on Gross Properties of Nuclei Nuclear Excitations IX, Hirschegg 1981 (Ed. H. Feldmeier); preprint GSI-81-14(1981); Phys.Lett. 104B (1981) 107
23. C. Toepffer and C.Y. Wong, Phys. Rev. C 25 (1981) 1018
C. Toepffer, Z. Physik A 305 (1982) 263

24. P. Bonche, S.E. Koonin and J.W. Negele, Phys.Rev. C13 (1976) 1226
25. H. Flocard, S.E. Koonin and M.S. Weiss, Phys.Rev. C17 (1978) 1682
26. H.S. Köhler, Nucl.Phys. A343 (1980) 315 and these proceedings
27. A. Lazzarini et al., Phys.Rev. C24 (1981) 309

HOW COLLISION TERMS RELATE TO THE DYNAMICS OF QUANTUM CORRELATIONS*

A.F.R. de Toledo Piza
Inst. Física, USP, CP 20516, 01000 São Paulo, Brasil

1. INTRODUCTION

Time dependent, self consistent mean field theories approximate the time evolution of interacting quantum subsystems in terms of a set of effective, coupled unitary time displacement laws for the subsystems themselves. This offers the enormous simplification of avoiding the dynamic entanglement of the individual subsystems that would in general result from the complete dynamical law, due to time evolving correlations among them. It implies also, however, that the approximate state-representatives (density matrices) become inadequate at least in so far as the effects of such a dynamical entanglement are lost: the degree of quantum coherence of state-representatives of subsystems is preserved by the unitary mean field law but is in general altered by the exact evolution. Collision terms are non-unitary corrections to time dependent mean field theories which attempt at redressing this defect while still retaining as much as possible of the simplification offered by these theories. In the following development, based on refs. (1) and (2), I will describe a way of scrutinizing the works of the exact dynamical law so as to isolate unitary (mean field) and non-unitary (collision) effects in the effective time-displacement laws for subsystems. This will also throw some light on the general connection between non-unitary effects and the dynamics of intersubsystem correlations. After that, I will indicate how one can proceed in order to obtain a description of effective dynamical laws for subsystems, including both unitary and non-unitary effects, which is as much as possible closed in terms of the disentangled state-representatives themselves, and finally comment on possible further reductions of the theory aimed at making it fit for consumption.

2. COLLECTIVE/INTRINSIC SUBSYSTEMS

Even though I will eventually turn to nuclear one-body mean fields, it is easier to begin with the consideration of two sub-systems as in the case of some damped collective mode of a nuclear system. Here the subsystems are characterized respectively in terms of distinct (commuting) collective (typically few, symmetric) and intrinsic (typically many) degrees of freedom; and the Hilbert space for the system may be naturally considered as a product of a collective and an intrinsic space:

$$H = H_C \otimes H_I \ .$$

The quantum state of the composite system will be described by a normalized vector $|\psi\rangle$ of this product space, or alternatively by a projection operator of trace one, $F = |\psi\rangle\langle\psi|$. (This implies that I am adopting a restriction to pure quantum states of the composite system. While it is possible to work with more general states, this simplifies the discussion and illustrates all the relevant points). Furthermore, given the state F, one can obtain the corresponding state-representatives for each of the two subsystems by taking partial traces of this object. Thus

$$\rho = Tr_I F \qquad \text{and} \qquad R = Tr_C F$$

are the appropriate subsystem state representatives in H_C and in H_I respectively. These are hermitean operators of trace one, but they are not in general projectors, which means that the subsystem states are in general not pure states, but mixtures. Choosing the representations which diagonalize these operators, one can write them as

$$(1) \quad \rho = \sum_k |c_k\rangle p_k \langle c_k| \qquad \text{and} \qquad (2) \quad R = \sum_k |I_k\rangle p_k \langle I_k|$$

where the $|c_k\rangle$ and the $|I_k\rangle$ are orthornormal "natural states", defined as the eigenvectors of ρ and R respectively. The fact that these operators have the same spectrum is a result of the theory of Hilbert-Schmidt kernels, as shown in detail by von Neumann[3]. Associated with it is the existence of a priviledged decomposition of the state vector $|\psi\rangle$ as

$$(3) \quad |\psi\rangle = \sum_k \alpha_k |c_k\rangle \otimes |I_k\rangle \qquad \text{with} \qquad p_k = \alpha_k^2 \ .$$

The common eigenvalues p_k reveal, therefore, the degree of entanglement of the two subsystems in the state $|\psi\rangle$: one single nonvanishing eigenvalue implies uncorrelated subsystems (factored state $|\psi\rangle$), while the existence of a large number of nonvanishing eigenvalues implies strongly intercorrelated subsystems.

I will now state how dynamics appears in terms of this representation. As F evolves in time in accordance with the Liouville-von Neumann equation

(4) $$i\dot{F} = [H,F] \equiv LF$$

where H is the Hamiltonian of the <u>composite</u> system, both the natural states, $|c_k>$ and $|I_k>$, and the eigenvalues p_k will also evolve in time. Since at any time the natural states constitute orthonormal sets, <u>their</u> time evolution can be described in terms of hermitean time-displacement generators $h^{(c)}(t)$ and $h^{(I)}(t)$ respectively. Thus, in particular

$$i \frac{d}{dt}\left[|c_k><c_k|\right] = \left[h^{(c)}(t), |c_k><c_k|\right]$$

which allows one to write the time derivative of eq. (1) as

(5) $$i\dot{\rho} = \sum_{k\ell} |c_k>(p_\ell - p_k)\, h^{(c)}_{k\ell} <c_\ell| + i \sum_k |c_k> \dot{p}_k <c_k| \quad .$$

Here $h^{(c)}_{k\ell} \equiv <c_k|h^{(c)}|c_\ell>$. A similar form results of course for the time-derivative of eq. (2), involving $h^{(I)}$ and states $|I_k>$. By inserting (3) into eq. (4) one can moreover deduce algebraically that

(6) $$(p_\ell - p_k)h^{(c)}_{k\ell} = \alpha_\ell <c_k I_\ell|H|\psi> - <\psi|H|c_\ell I_k> \alpha_k \qquad (k \neq \ell)$$

(7) $$(p_\ell - p_k)h^{(I)}_{k\ell} = \alpha_\ell <c_\ell I_k|H|\psi> - <\psi|H|c_k I_\ell> \alpha_k \qquad (k \neq \ell)$$

and (8) $$\dot{\alpha}_k = \mathrm{Im}\, <c_k I_k|H|\psi> \quad ,$$

relating the dynamical ingredients of eq. (5) to the complete dynamics of the composite system.

The effective dynamics of the subsystem described by ρ (the "collective" subsystem) is clearly splitt into two parts of diverse nature in eq. (5). The first part comes from the time evolution of the natural states and is fully unitary. Let me call this the "mean field" part. It is governed by the effective generator $h^{(c)}$. This, in the special correlation free case in which $F = \rho R$ (i.e., the state vector $|\psi>$ is a simple product state), reduces just to $Tr_I HR$, but otherwise is more complicated: the mean field part will depend, in general, on correlation effects. The second part involves time rates of change of <u>eigenvalues</u> of the hermitean ρ, and is thus essentially <u>non unitary</u>. Let me call it the "collision" part. Its form, taken in conjunction with eqs. (8) and (3), shows that it arises out of the time evolution of the degree of entanglement of the subsystems, this being characterized by the set of weights entering into the priviledged expansion (3).

It will have been noticed that the basic trick allowing for the separation of a unitary "mean field" part from a non-unitary,"collision" part on the r.h.s. of eq. (5) resides in the use made of the natural states. Thus, except for the all-important "external" ingredient consisting in the adopted initial factorization of the Hilbert space H (i.e., the "a priori" characterization of each of the two subsystems), it is an "intrinsic" trick, in the sense that it operates in terms of properties of the state vector itself, rather than of properties depending on some base in which it may find itself expanded. This trick is quite general, and has in fact been used as such in the work of Wong and Tang on collision terms[4] but occurs also in a work of 1935 by Schrödinger[5].

Next for consideration is the fact that eq. (5), together with the corresponding equation for $\dot{R}$, are not closed equations. Their essential dynamical ingredients involve again all the information present in the fully correlated, time dependent state $|\psi\rangle$ through eqs.(6), (7) and (8), so that the question concerning the possibility of effectively closing them in terms of ρ and R arises naturally. This can be acomplished by writing the density F as $F=\rho R+F'$ and then eliminating formally the traceless correlation part F' after substituting this form in eq. (4). In this way one can replace eq. (5) by[1,6]

$$(9)\quad i\dot{\rho}(t) = \mathrm{Tr}_I\Big[L(t,0)F'(0)\Big] + \mathrm{Tr}_I\Big[LR(t)\Big]\rho(t) - i\int_0^t dt'\,\mathrm{Tr}_I\Big[K(t,t')R(t')\rho(t')\Big]$$

and a similar equation for $\dot{R}(t)$. The Liouvillian memory kernel $K(t,t')$ describes the buildup of correlations out of the uncorrelated part ρR of F; and the first term on the r.h.s. describes the effects at time t of the existing correlations at time zero, F'(0). Thus, to the extent that these effects make themselves felt at time t, one cannot eliminate completely the information contained in the correlation part F' of F.

Rather than elaborate on the content of this type of equation and of its ingredients in the present context, I will now swich to the consideration of a one-body subdynamics of a many-body system. The question of closing the description will then lead to contact with one-body mean field theories and "Time-Dependent Hartree-Fock collision terms".

3. ONE-BODY SUBSYSTEM

Switching to the case of the one-body subdynamics is actually straightforward, implying a deep structural afinity between the treatment of collective and single nucleon subsystems from the present point of view. The important differences that will eventually emerge will all hinge on physical rather than structural properties of the two types of subsystem. One may, in fact, directly take ρ of eq. (1) to be the one-body density associated with the fully correlated A-body state $|\psi\rangle$. At the same time, R becomes the corresponding (A-1)-body density, while eqs. (5)-(8) translate the dynamical law, eq. (4), into this language. Eq. (3) also exists in this case. Adopting an explicit configuration space representation it becomes, in fact

$$(10) \qquad \langle x_1 \ldots x_A|\psi\rangle = \sum_k \alpha_k \langle x_1|c_k\rangle\langle x_2 \ldots x_A|I_k\rangle \quad .$$

It should be observed that Pauli correlations are treated in eq. (10) on the same footing with all other correlations: if $|\psi\rangle$ is a determinantal state, eq. (10) reduces to a row expansion of the appropriate normalized determinant of rank A. Moreover, in eqs. (5)-(8) the product states are not antisymmetrized, but the appropriate projection results automatically from the antisymmetry of $|\psi\rangle$ and from the symmetry of H. It is actually easy to check that eqs. (6) and (8) can be rewritten in second quantized form as

$$(11) \qquad (p_\ell - p_k)h^{(c)}_{k\ell} = \frac{1}{A}\langle\psi|\left[c^+_\ell c_k , H\right]|\psi\rangle \qquad (k \neq \ell)$$

and

$$(12) \qquad i\,\dot{p}_k = \frac{1}{A}\langle\psi|\left[c^+_k c_k , H\right]|\psi\rangle \quad ,$$

where the fermion creation operators $c^+_k(t)$ are associated with the one-body natural orbitals $|c_k\rangle$ of eq. (1), and $\sum_k p_k(t) = 1$. The reduced one-body subdynamics associated with eq. (4) appears thus in the guise of eq. (5), the two terms on its r.h.s. giving respectively the mean field part and the collision part of an effective one-body Liouvillian. These objects are determined by eqs. (6) and (8), or by eqs. (11) and (12), in terms of the complete information contained in the fully correlated state $|\psi\rangle$.

It is of course possible to carry the paralelism of the present one-body case with the collective/intrinsic case further and obtain a pair of formal coupled equations for ρ and R by eliminating a traceless correlation part F' defined as before by $F = \rho R + F'$. This singles out a given particle and treats Pauli and dynamical correlations on the same level, a somewhat unphysical option which may be particularly ill-suited for approximate treatments. It is in fact preferable in this case to give a privileged status to Pauli correlations and to treat all particles in a symmetric fashion. This is accomplished by defining a different traceless correlation part F_1' of the full density through $F = F_0 + F_1'$ where F_0 is best written in Fock space as[7]

$$F_0 = \prod_k \left[(1-Ap_k)\, c_k c_k^+ + Ap_k\, c_k^+ c_k\right] \quad . \tag{13}$$

The formal elimination of the correlation part F_1' can be carried out essentially as in the collective/intrinsic case yielding for eqs. (11) and (12) respectively

$$(p_\ell - p_k) h_{k\ell}^{(c)} = \frac{1}{A}\,\mathrm{Tr}\left[c_\ell^+ c_k L_1(t,0) F_1'(0)\right] + \\ + \frac{1}{A}\,\mathrm{Tr}\left[c_\ell^+ c_k \left(LF_0(t) - i\int_0^t dt' K_1(t,t') F_0(t')\right)\right] \;, \quad k \neq \ell \tag{14}$$

and

$$i\,\dot{p}_k = \frac{1}{A}\,\mathrm{Tr}\left[c_k^+ c_k L_1(t,0) F_1'(0)\right] - \frac{i}{A}\,\mathrm{Tr}\left[c_k^+ c_k \int_0^t dt' K_1(t,t') F_0(t')\right] \quad . \tag{15}$$

As was the case with eq. (9), these expressions carry memory effects involving both the "uncorrelated" density F_0 and initial correlations $F_1'(0)$. The cooresponding memory kernels ar given by

(16) $K_1(t,t') = LG(t,t')Q(t')L$ and (17) $L_1(t,0) = LG(t,0)$

where

$$G(t,t') = T\exp\left(-i\int_{t'}^{t} d\tau\, Q(\tau) L\right) \tag{18}$$

and $Q(t)$ is a Liouville operator with the property $\dot{F}_1' = Q\dot{F}$. It can be constructed entirely out of ingredients occuring in F_0, namely the occupation probabilities $p_k(t)$ and the associated states $c_k^+(t)$, and is given explicitely in ref. (2). I will refrain from giving more details here, and just state that what it does essentially is to select only relevant many-particle correlation parts from the objects upon which it acts.

4. DISCUSSION: ONE-BODY MEAN FIELD AND COLLISION EFFECTS

Together with eq. (5), eqs. (14) and (15) perform the decomposition of the effective one body subdynamics into unitary and non-unitary parts in as closed a form as possible. The big breach is again the unavoidable involvement of the initial correlations $F_1'(0)$ in both parts. The term involving LF_0 in eq. (14) gives rise to a unitary contribution of standard Hartree-Fock form, involving the one body density ρ of eq. (1). A corresponding term is absent from eq. (15). Finally, terms involving K_1 represent contributions due to dynamically generated many-body correlations.

Before any further commitment on the significance or use of these equations it is crucial to recall what are the physical facts standing behind them in the context of nuclear dynamics: in few words, single-particle, mean field approximations to nuclear dynamics only became successfull and respectable when carried considerably beyond the limits of a straight Hartree-Fock context. Because of radical properties of nucleon-nucleon forces, adequate success required the consideradion of effective, density dependent interactions designed so as to account for the effects of essential correlations induced by these forces. It must certainly be considered a remarkable fact that it has been possible to include enough correlations via the effective forces with completely coherent (determinantal) one body densities in a closed theory for such densities in the static case[8]. Extentions to dynamical situations ("Time-Dependent Hartree-Fock") involve certainly a good deal of extrapolation of the more carefully controlable applications to nuclear ground states, but quantitative successes are still impressive.

In such a framework, the correlation dependent terms of eq. (14) acquire a special significance if one intends to keep some connection to more fundamental levels of description. In fact, at least part of the correlation effects become essential ingredients to produce a sufficiently healthy effective mean field: it is now well known that the bare LF_0 term of eq. (14) is not good enough for that. Actually, current TDHF calculations with effective forces should rather be interpreted in terms of the replacement of the full r.h.s. of that equation by the effective Hartree-Fock field under the assumption of a determinantal one body density and frozen one body occupation probabilities, $\dot{p}_k = 0$, instead of eq. (15). This implies of course a truncation "by ansatz" of all non-unitary effects and furthermore restricts in principle dynamical correlation corrections to the mean field by forcing them through the gauge of the effective two body interaction.

The collision part of the effective dynamical law as given by eq. (15), on the other hand, can be seen to depend in an essential way on objects which are rigorously the "diagonal" (i.e., $k = \ell$) counterparts of the correlation corrections to the time-dependent effective mean field part, eq. (14) (cf. also eqs. (11) and (12)). These are however just the corrections to the mean field which can at best be described as being obtained, in the dynamical case, by extrapolation of results obtained in stationary contexts. It is clear moreover that the "diagonal" contributions of eq. (15) must vanish in the stationary case. Present knowledge of these objects is therefore extremely meager.

Even so, and in order to have an exploratory idea of the possible content of eq. (15), one may look into the structure of the second term on the r.h.s. of eq. (15) assuming that, as in the case of the mean field part, the remaining initial correlation term has been sufficiently taken into account in terms of a suitable effective interaction in the memory kernel K_1, eq. (16). The relevant object is then

$$\mathrm{Tr}\; c_k^+ c_k \int_0^t dt' \Big[H, G(t,t') Q(t') \left[H, F_0(t') \right] \Big]$$

from which, under customary weak-coupling|Markof approximations[9,10] one can readily obtain a Boltzmann-like collision term including the appropriate occupation factors which account for the exclusion principle. It looks thus as this form constitutes a stable structural element, at least in the weak-coupling limit, in the sense that it does not appear to be sensitive to the meanders of particular derivations[9-11]. Details are however overshadowed by the uncertainties in the effective interactions.

I cannot sufficiently acknowledge two rather short periods of intense work in São Paulo and complementary periods of stimulating correspondence with M.C. Nemes. As the references indicate, this was crucial for the elaboration of this report.

REFERENCES

* Work supported in part by CNPq and FAPESP (Brasil)

1) M.C. Nemes and A.F.R. de Toledo Piza, Preprint IFUSP/P-287, August 1981

2) M.C. Nemes and A.F.R. de Toledo Piza, Preprint IFUSP/P-317, February 1982
3) J. von Neumann, Mathematical Foundations of Quantum Mechanics, Princeton U. Press 1955, chap. VI
4) C.Y. Wong and H.H.K. Tang, Phys. Rev. Lett. 40, 1070 (1978)
5) E. Schrödinger, Proc. Camb. Phil. Soc. 31, 555 (1935); ibid. 32, 446 (1936). See also O. Kübler and H.D. Zeh, Ann. Phys. (N.Y.) 76, 405 (1973)
6) C.R. Willis and R.H. Picard, Phys. Rev. A9, 1343 (1974)
7) J. Des Cloiseaux, in Many Body Physics, C. De Witt and R. Balian, eds., Gordon and Breach, N.Y. 1968, pp. 5-36.
8) J.W. Negele, MIT Preprint CTP#898, October 1981 (to be published in Revs. Mod. Phys.)
9) P. Grangé, H.A. Weidenmüller and G. Wolschin, Preprint MPIH-1981 v.8, to be published in Ann. Phys. (N.Y.) (and references therein)
10) S. Ayik, Zeit. für Physik A298, 83 (1980); Preprint GSI-81-12 (May 1981).
11) H.C. Pauli, Heidelberg preprint MPI H-1982-V7.

BEYOND THE TIME-DEPENDENT HARTREE-FOCK:

The collision terms in a mean-field theory.

H.C. Pauli

Max-Planck-Institut für Kernphysik, Heidelberg, FRG

Abstract: (1) A many-body Hamiltonian is proposed, for which TDHF constitutes a rigorous solution. This allows to define a residual interaction appropriate for TDHF. (2) The impact of the residual interaction is studied in a non-perturbative approximation and without ensemble averages. (3) The further reduction to the one-body level is possible without ad hoc assumptions, and produces a closed equation of motion for the one-body density. (4) The residual interaction causes a collision term which acts simultaneously on intrinsic and on collective motion.

At this conference, we have heard on the many facets of the time-dependent Hartree-Fock equations, on their beauty of producing almost parameter-free results, and on their success in comparison with the experimental data. But even more than forty years after their first formulation by Dirac |1|, it is not perfectly clear - at least not to me - what type of approximation they stand for vis à vis the full quantum-mechanical many-body problem, and because of this, it is unclear, how and in which sense one could go "beyond" this formalism.

In this short lecture I cannot give a review or even comment on the numerous attempts of this kind |2 - 14|, rather shall I concentrate on some aspects of the problem, as they were published by Mantzouranis and myself |11| or were laid down in a recent series of preprints |12 - 14|.

In presenting my arguments I shall put weight on the following four questions. 1. Does a physical problem, does a Hamiltonian exist, to which TDHF is a rigorous solution ? 2. Can one treat a suitably defined residual interaction without perturbation theory of finite order or without using the method of ensemble averages ? 3. Can one reduce the equations of motion for the A-body density to those for the one-body density without using the idempotency condition at any stage? and 4. Are the modifications of the TDHF-equations - the terms beyond

TDHF - are they relevant in any physical context ?

For to be specific, I discuss a closed system of A fermions which move subject to their pairwise interaction $\tilde{V}$. This system is defined by the Hamiltonian $\tilde{H}$

$$\tilde{H} = \sum_{m_1,m_2} \langle m_1|K|m_2\rangle a^+_{m_1} a_{m_2} + \frac{1}{4}\sum_{m_1,m_2..m_4} \langle m_1,m_2|\tilde{V}|m_3,m_4\rangle a^+_{m_2} a^+_{m_1} a_{m_3} a_{m_4} . \quad (1)$$

I aim at the solution of the time-dependent state $|t\rangle$, i.e. $(\tilde{H} - i\hbar \frac{d}{dt})|t\rangle = 0$, or more specifically at an appropriate solution of the A-body density operator $\tilde{M}(t) = |t\rangle\langle t|$ by means of

$$i\hbar\, d/dt\, \tilde{M}(t) = [\tilde{H}, \tilde{M}(t)]. \quad (2)$$

In the lack of a rigorous solution, one must approach this problem by a kind of perturbative approach. One defines a soluble Hamiltonian H , whose time-dependent amplitudes $|\mu(t)\rangle$,

$$(H - i\hbar\, d/dt)|\mu(t)\rangle = 0 \quad (3)$$

provide us with a complete set of states (the Hilbert space) for a systematic discussion of the "residual interaction V_S" which is defined by

$$\tilde{H} = H + V_S \quad (4)$$

One is tempted to identify the unperturbed states $|\mu(t)\rangle$ with the TDHF solutions. This, at least, would be intuitively appealing. However, one realises immediately that TDHF is defined by a one-body operator $h_{HF}(t)$; the Hamiltonian on the A-body level is largely unknown. The most straightforward identification, $H_W = \sum_i h_{HF}(i)$ cannot be correct by two reasons: 1) H_W and H do not have the same energy, they differ by half the interaction energy; 2) The so defined H_W has <u>only</u> <u>one</u> many-body state as a rigorous solution $(H_W - i\hbar\, d/dt)\,|\mu_o(t)\rangle = 0$, and <u>not</u> a complete set.

Some time ago, Mantzouranis and myself |10| have proposed another unperturbed Hamiltonian H

$$\begin{aligned} H = & \sum_{m_1,m_2} \langle m_1(t)|K|m_2(t)\rangle a^+_{m_1} a_{m_2} \\ & - \frac{1}{2}\sum_{m_1,m_2} \langle m_1(t),m_2(t)|\tilde{V}|m_1(t),m_2(t)\rangle a^+_{m_1} a_{m_1} a^+_{m_2} a_{m_2} \\ & + \sum_{m_1,m_2,m} \langle m_1(t), m(t)|\tilde{V}|m_2(t),m(t)\rangle a^+_{m_1} a_{m_2} a^+_m a_m . \end{aligned} \quad (5)$$

By definition, this genuine two-body operator cannot cause more than zero or one-particle-one-hole excitations, when acting on an arbitrary state $|\mu(t)\rangle$. The one-body amplitudes can be set zero by properly choosing the time-derivative of single-particle states $|m(t)\rangle$. This condition leads straightforward to

$$i\hbar \, d/dt \, |m(t)\rangle = h^{(\mu)}(t)|m(t)\rangle. \tag{6}$$

The time-dependent one-body operator $h^{(\mu)}(t)$,

$$\langle m_i|h^{(\mu)}|m_j\rangle = \langle m_i(t)|K|m_j(t)\rangle + \sum_m{}' p_m(t)\langle m_i(t),m(t)|V|m_j(t),m(t)\rangle \tag{7}$$

looks formally like the TDHF Hamiltonian, but the state-dependent occupation numbers

$$p_m^{(\mu)}(t) = \langle\mu(t)|a_m^+(t)\ a_m(t)|\mu(t)\rangle \tag{8}$$

defined as the probability to find state $|m(t)\rangle$ occupied in the many-body state $|\mu(t)\rangle$ need not be zero or one by necessity, as in the familiar theories.

Thus, one has found an answer to the first question: The TDHF equations are a rigorous solution to the Hamiltonian H as given by Eq. (5).- Simultaneously one has found that part of the full Hamiltonian H which by definition cannot be included in TDHF, the residual interaction V_S. It is defined rigorously by $V_S = \tilde{H} - H$, and thus by

$$\begin{aligned} V_S = \frac{1}{4} & \sum_{m_1,m_2,m_3,m_4} \langle m_1,m_2|\tilde{V}|m_3,m_4\rangle a_{m_2}^+ a_{m_1}^+ a_{m_3} a_{m_4} \\ - & \sum_{m_1,m_2,m_3} \langle m_1,m_2|\tilde{V}|m_3,m_2\rangle a_{m_2}^+ a_{m_1}^+ a_{m_3} a_{m_2} \\ + \frac{1}{2} & \sum_{m_1,m_2} \langle m_1,m_2|\tilde{V}|m_1,m_2\rangle a_{m_2}^+ a_{m_1}^+ a_{m_1} a_{m_2} \end{aligned} \tag{9}$$

The operator V_S has no diagonal elements in between the Hartree-Fock states, i.e. $\langle\mu(t)|V_S|\mu(t)\rangle = 0$, and can generate only two-particle-two-hole amplitudes when applied to any $|\mu(t)\rangle$.

We proceed now to the second question: How can we incorporate the residual interaction - i.e. those terms beyond TDHF - at least approximately ? - As usually, the problem can be treated further in interaction representation. The solutions to the unperturbed problem Eq. (3), defines a set of complete states $|\mu(t)\rangle$, which can be understood as a unitary transformation of the initial states $|\mu(0)\rangle$, i.e. $|\mu(t)\rangle = U(t)|\mu(0)\rangle$. The unitary operators can be used to define $V(t) = U^+(t)\ V_S(t)\ U(t)$ and $N(t) = U^+(t)M(t)U(t)$.

Then, Eq. (2) is replaced quite straightforwardly by

$$i\hbar\, d/dt\, N(t) = [V(t),\, N(t)]\,. \tag{10}$$

Its formal solution

$$N(t) = N(0) + \frac{1}{i\hbar}\int_0^t ds\,[V(s),\, N(s)] \tag{11}$$

is treated in a somewhat unconventional way. The Hilbert space is defined by the unperturbed problem, and therefore it makes sense to divide $N(t)$ into its diagonal part $N_D(t)$ and its off-diagonal parts $N_S(t)$, which obey a set of coupled equations

$$N_D(t) = N_D(0) + \frac{1}{i\hbar}\int_0^t ds\,[V(s),\, N_S(s)]\,, \text{ and}$$

$$N_S(t) = N_S(0) + \frac{1}{i\hbar}\int_0^t ds\,[V(s),\, N_D(s)] + \frac{1}{i\hbar}\int_0^t ds\,[V(s), N_S(s)]\,. \tag{12}$$

This couple of equations can be analysed in the following way. For a given operator function $N_S(t)$, the diagonal elements are determined as a simple quadrature and this way are fixed. Opposed to this, for a given $N_D(t)$, the second equations remains an integral equation for $N_S(t)$ and can be formally solved by an infinite series. Retaining the lowest non-trivial terms (we call it the approximation of rank zero) one has for a diagonal initial state $N(t_o) = N_D(t_o)$

$$d/dt\, N_S(t) = \frac{1}{i\hbar}[V(t),\, N_D(t)] \tag{13}$$

$$d/dt\, N_D(t) = -\frac{1}{\hbar^2}\int_0^t ds\,[V(t), [V(s),\, N_D(s)]]\,. \tag{14}$$

In other words, the diagonal elements are determined from a closed (integral) equation; once $N_D(t)$ is known, one can determine the off-diagonal $N_S(t)$. - In a series of preprints |12-14|, I have carefully investigated whether the restriction to rank zero can do any harm ot introduce inconsistencies; it does not. Needless to say, that particle number and energy are conserved.

The procedure indicated above corresponds to an infinite series in conventional perturbation theory. Rewriting equation as an integral equation, one has for example

$$N_D(t) = N_D(t_o) + \left(\frac{1}{i\hbar}\right)^2 \int_o^t ds_1 \int_o^{s_1} ds_2\, [\overline{V(s_1), [V}(s_2), N_D(t_o)]]$$

$$+ \left(\frac{1}{i\hbar}\right)^4 \int_o^t ds_1 \int_o ds_2 \int_o ds_3 \int_o^{s_3} ds_4\, [\overline{V(s_1), [V}(s_2), [\overline{V_S(s_3), [V}(s_4) N(t_o)]]]]$$

$$+ \ldots \tag{15}$$

for the diagonal elements, and a similar equation for the off-diagonal elements. In other words: Out of the infinite series of conventional perturbation theory, the above procedure picks out those graphs which contain an even power of interactions $V(t)$, and among them only those which are contracted pairwise and sequentially in time. This is the same series as one would have obtained with a suitably defined ensemble average over a Gaussian distribution of matrix elements.

I should conclude this section with the remark that I consider Eq.(14) as an approximation to the exact equation for the actual system of particles rather than an exact solution for an ensemble of systems. Conceptually, both concepts are very different from each other, but I do not intend to deepen this point in the present context.

Last but not least, we can assume that the diagonal elements $P_\mu(t) = \langle\mu|N_D(t)|\mu\rangle$ change with time slower than the rapidly oscillating functions $\langle\mu|V(t)|\mu'\rangle$. The argument of N_D in Eq. (14) under the integral can thus be replaced by its value at the upper integration limit, i.e.

$$N_D(t) = -\frac{1}{\hbar^2}\int_0^t ds\left[V(t),\left[V(s),\, N_D(t)\right]\right] \quad . \tag{16}$$

I have verified elsewhere |12| that this substitution is consistent within the lowest rank of approximation. - We have thus found an answer to question number 2 and can proceed with the reduction to the one-body problem.

The reduction to the one-body level is more or less straightforward and yields in normal representation

$$\begin{aligned} d/dt\, p_{m_1} = -\sum_{m_2,m_3,m_4} w(m_1,m_2,m_3,m_4;t)\{&p_{m_1}p_{m_2}(1-p_{m_3})(1-p_{m_4}) \\ &-(1-p_{m_1})(1-p_{m_2})\,p_{m_3}\,p_{m_4}\} \end{aligned} \tag{17}$$

This equation associates the change per unit time of the one-body occupation numbers, as defined by Eq. (8), with their value at the actual time t in terms of the matrix elements of the residual interaction, which are condensed into the time-dependent transition rates w, i.e.

$$\begin{aligned} w(m_1,m_2,m_3,m_4;t) = \frac{1}{\hbar^2}\int_0^t ds\, &\langle m_1(t),\, m_2(t)|V|\, m_3(t)m_4(t)\rangle \cdot \\ &\cdot \langle m_3(s),m_4(s)|V|m_1(s)m_2(s)\rangle \quad + \text{c.c.} \end{aligned} \tag{18}$$

Together with the generalised TDHF-equations Eq. (6) and (7), the latter represent a set of coupled integro-differential equations for the self-consistent determination of the amplitudes $|m(t)>$ and their occupation $p_m(t)$. All these equations, Eqs. (6-8) and (17-18), can be cast into the compact form

$$\dot{\rho} = \frac{1}{i\hbar}\left[h_{HF}(t), \rho(t)\right] + C[\rho] \tag{19}$$

which related the change per unit time of the one-body density operator $\rho(t)$ to the density at the actual time. The first of the two terms on the r.h.s. represents the usual TDHF commutator, the second term $C[\rho]$,

$$C[\rho] = \sum_m |m(t)> \dot{p}_m(t) <m(t)| \tag{20}$$

represents all what is"beyond TDHF".In the present approximation, the operator C has a structure like the "collisions terms in the Boltzmann equation".

The reduction to the one-body level involves one type of approximation which should be outlined in short. Primarily, in the reduction, $\dot{p}_m(t)$ is a functional of the diagonal elements of the two- and of the three-body density, $p_{m_1,m_2}(t)$ and $p_{m_1,m_2,m_3}(t)$, respectively. They appear in characteristic combinations with the transition rates $w(m_1m_2m_3m_4,t)$. These transition rates are kind of absolute squares with respect to the quantum numbers m_i, and therefore vary smoothly with any of the four, the other three being fixed. One is thus entitled to the following type of substitution

$$\sum_{m_3} w(m_1,m_2,m_3,m_4;t)\; p_{m_1m_2m_3}(t)$$
$$\sim \frac{1}{6}\, p_{m_1m_2}(t) \sum_{m_3} w(m_1,m_2,m_3,m_4;t)\quad p_{m_3}(t) \tag{21}$$

and this gives directly Eq. (17). The structure of Eq. (17) is caused by the nature of the residual interaction and the way they enter into the formalism. It is not a property of the density matrices. In fact, a factorisation of the three-body density into a product of one-body matrices would be inconsistent within the formalism.

So far we have concentrated on the formal aspects of the problem. In view of the rather weak assumptions made - restriction to the lowest rank (I) and smoothness of transition rates (II) - we must conclude that terms beyond TDHF are doubtlessly present and in their crudest approximation can be formulated by collision terms in a self-consistent fashion.

We arrive thus at our fourth and last question: "Collision terms are present, but are they relevant in any physical context?" - The answer to this question cannot be given in generality, because the unperturbed TDHF states depend drastically on the initial condition. However we can discuss two particular, realistic cases.

(1) <u>The quasistationary case</u>: Suppose, the initial configuration corresponds to a stationary Hartree-Fock solution with eigenstates $|m\rangle$ and single-particle states ε_m. The time-dependent amplitudes are then given by $|m(t)\rangle = |m\rangle \exp(-i\varepsilon_m t/\hbar)$. In this case the time-dependent transition rates can be easily evaluated and yield

$$w(m_1,m_2,m_3,m_4;t) = \frac{1}{\hbar}\,\delta(\varepsilon_{m_1}+\varepsilon_{m_2}-(\varepsilon_{m_3}+\varepsilon_{m_4}))\,|\langle m_1,m_2|V_S|m_3,m_4\rangle|^2$$

After some 10^{-24} seconds the time-dependent phases settle into a constant, producing essentially the energy conserving δ-function. Thus, by virtue of the residual interaction, transitions between any pair of single-particle states are possible only if the single-particle energies match: $\varepsilon_{m_1}+\varepsilon_{m_2}=\varepsilon_{m_3}+\varepsilon_{m_4}$. If the initial occupation has been chosen such that all energy levels up to a certain energy, (the Fermi energy ε_F), are occupied and all others are empty ($p_m = 0$), the residual interaction cannot cause transitions by lack of phase space. In other words: The non-degenerate Hartree-Fock ground state is stationary and not affected by the residual interaction. If, however, the initial occupation allows for transition at the same energy, the system will equilibrate to an equilibrium occupation

$$p_m = e^{-(\varepsilon_m-\varepsilon_F)/kT}$$

within the equilibration time. Both, very rough estimates as well as the more detailed calculation of Wolschin |8| produce equilibration times of a few times 10^{-23} seconds for realistic cases. -

(2) <u>The heavy-ion collision or coherent motion</u>:
The above considerations disregard like most of the actual calculations |2-11| the impact of collective motion. The latter is of particular importance in a heavy-ion collision, which shall be sketched in short. Prior to the collision, the TDHF solutions for a time-dependent single-particle state $|m(t)\rangle$ have the structure

$$\langle x|\,m(t)\rangle = \langle x-\xi(t)|m\rangle\; e^{i(k\cdot x\,-\,\omega_m t)}$$

in configuration space with

$$\hbar\omega_m - \varepsilon_m - (\hbar k)^2/2m$$

These wave functions reflect the fact that the TDHF-nucleus travels through space on the collective trajectory $\xi(t) = \hbar kt/m + \xi(0)$ with the collective momentum $\hbar k$. Assuming for the moment a straightline trajectory for all times, the time-dependent transition rate w can be evaluated, with the result

$$w(m_1,m_2,m_3,m_4;t) = \delta(\varepsilon) \cdot F(t) \ .$$

The function $F(t)$ is given elsewhere in detail |13| and changes slowly in time; it differs from zero essentially only when the ions are in contact. More important is the δ-function in front of F. Its argument has the structure

$$\varepsilon = (\varepsilon_{m_1} + \varepsilon_{m_2} - \varepsilon_{m_3} - \varepsilon_{m_4}) + \frac{\hbar^2}{4m}(k_1-k_2)^2 - \frac{\hbar^2}{4m}(k_3-k_4)^2$$

This displays visibly the important and novel property of the present rather rigorous approach. Transitions are possible not only between different intrinsic states $|m\rangle$ with the same collective momentum, but also between states of different collective momentum. With other words: The residual interaction can cause intrinsic excitation with its characteristic spread over many single-particle states, but simultaneously it causes collective excitations with a characteristic spread over many trajectories. The initial, sharp TDHF trajectory gets transformed during the time of contact into a bundle of TDHF trajectories. A formalism "Beyond TDHF" can thus eventually account for the large variances of collective variables like c.m. position, angle or mass, which seem to appear in the experimental cross section and which so far were impossible to obtain in TDHF calculations without "collision terms".

References

|1| P.A.M. Dirac, Proc. Cambridge Phil.Soc. 26 (1930) 376, 27 (1931) 240.

|2| L.P. Kadanoff and G. Baym, Quantum Statistical Mechanics, W.A. Benjamin, New York (1962).

|3| H. Orland and R. Schaeffer, Z. Phys. A290 (1978) 191.

|4| C.Y. Wong and H.H. Tang, Phys.Rev.Lett. 40 (1978) 1070 and Phys. Rev. 20 (1979) 1420.

|5| S. Ayik, Z. Phys. A298 (1980) 83.

|6| J. Richert, D.M. Brink, and H.A. Weidenmüller, Phys.Lett. 87B (1979) 6 .

|7| P. Grangé, H.A. Weidenmüller and G. Wolschin, Ann.Phys.(N.Y.) 136 (1981) 190

|8| G. Wolschin, "Equilibration in finite Fermion Systems". Lectures at the XIV. Masuranian Summer School in Mikolajki, Poland, Sept. 1981 (in press); MPI H-1981-V34.

|9| G. Mantzouranis and H.C. Pauli, Z. Phys. A278 (1976) 373.

|10| G. Mantzouranis and H.C. Pauli, Z. Phys. A281 (1977) 165.

|11| G. Mantzouranis and H.C. Pauli, Phys. Rev. C22 (1980) 1550.

|12| H.C. Pauli, "Ensemble Averages versus Perturbation Theory". Preprint, MPI-H-1982-V6, Heidelberg 1982.

|13| H.C. Pauli, "Beyond Time-Dependent Hartree-Fock: The collision Term in Mean-Field Theory". Preprint, MPI-H-1982-V7

|14| H.C. Pauli, Preprint MPI-H-1982-V5

QUANTUM STATISTICAL APPROACH TO EXTENDED MEAN-FIELD THEORY

S. Ayik
Physik-Department, TU München
8046 Garching

The time-dependent Hartree-Fock theory is generalized in order to include two-body collisions. Using the projection operator techniques of statistical mechanics a transport equation is derived for the uncorrelated part of the full A-particle density matrix. This leads to an equation of motion for the one-particle density matrix which includes a collision term. The collision term is considered in the weak-coupling limit and a master equation is obtained for the occupation probabilities of the single-particle levels.

1. INTRODUCTION

Inclusion of two-body collisions in a self-consistent mean-field description is a natural step in developing a unique understanding of low-energy nuclear dynamics. For this purpose, recently, several attempts have been made to derive the extended time-dependent Hartree-Fock equations /1-6/. In the present work this problem is studied in the frame of statistical mechanics. Using the well known Projection Operator Techniques we derive a transport equation for the single-particle density matrix.

The use of projection operator techniques to derive transport equations has become increasingly more common in recent years /7,8/. Although these techniques are very general they have been especially useful in problems where systems are in contact with reservoirs or where one is interested only in the diagonal matrix elements of the density matrix in some representation. In these problems the degrees of freedom of the reservoirs or the off-diagonal part of the density matrix are considered to be irrelevant and the irrelevant part of the density operator is formally eliminated. One is then left with an exact equation for the remaining relevant part of the density operator. However, in many problems involving coupled dynamical systems it is necessary to have a description where each degree of freedom is considered to be on an equal footing. In this work we derive a transport equation for the coupled A-particle system where the irrelevant degrees of freedom are not the dynamical degrees of freedom of the system but rather the correlations between particles produced by the interactions. The correlations are formally eliminated and we obtain an exact equation for the relevant part of the A-particle density operator. This yields, for the one-particle density matrix, an equation motion which includes a collision term /4/.

2. EXTENDED MEAN-FIELD EQUATIONS

We consider an interacting A-particle system with the total Hamiltonian

$$H = \sum_{i}^{A} \frac{1}{2m} p_i^2 + \sum_{i<j}^{A} v(i,j) \qquad (2.1)$$

where the first and second terms denote the kinetic energy and the two-body interactions of the particles. The time evolution of the system is determined by the Liouville equation,

$$i \frac{\partial}{\partial t} F(t) = [H, F(t)] \equiv LF(t) \qquad (2.2)$$

where $F(t) = F(12\cdot\cdot A, 1'2'\cdot\cdot A';t)$ is the exact A-particle density matrix and $1\cdots A$ abbreviates the coordinates including spin-isospin of the particles. The one-particle density matrix $\rho(t)$ is obtained from A-particle density matrix by taking trace over the subspace corresponding to A-1 particles,

$$\begin{aligned} \rho(11';t) &= \mathrm{Tr}\,\{F(t)\}_{A-1} \\ &= \int\cdots\int d2\cdot\cdot dA\; F(12\cdot\cdot A,\ 1'2\cdot\cdot A;t). \end{aligned} \qquad (2.3)$$

In general the total density matrix can be written as the sum of two terms,

$$F(t) = F_0(t) + F_{cor}(t) \qquad (2.4)$$

where $F_0(t)$ is the uncorrelated component and it is defined in terms of one-particle density matrices as

$$F_0(t) = \mathcal{A}\,\rho(11';t)\cdots\rho(AA';t) \qquad (2.5)$$

with $\mathcal{A}$ is the antisymmetrization operator.

The second term in the r.h.s. of (2.4) as the correlated component of the total density matrix and it is given by the difference between the exact density matrix and the uncorrelated component.

The TDHF approximation to the many-body dynamics is obtained if the correlated component in (2.4) is neglected completely / 9/. This means that one assumes the factorization structure (2.5) for the total density matrix at all times. As a result, inserting (2.5) into (2.2) and taking trace over (A-1)-particle space we obtain TDHF equations

$$i \frac{\partial}{\partial t} \rho_{HF}(t) = [\mathcal{H}_{HF}(t), \rho_{HF}(t)], \qquad (2.6)$$

for the one-particle density matrix. The self-consistent Hartree-Fock Hamiltonian is determined by

$$\mathcal{H}_{HF}(t) = \frac{p^2}{2m} + \mathrm{Tr}\, v\, \rho_{HF}(t). \tag{2.7}$$

In general the many-body density matrix deviates from the factorized form (2.5) as a result of the correlations. In the following, using the projection operator techniques of statistical mechanics, we derive a generalized equation for the one-particle density matrix, in which the correlations are taken into account.

In the usual applications of the projection operator techniques in deriving transport equations, the degrees of freedom which corresponds to a part of the system or the off-diagonal part of the density matrix in some representation are regarded irrelevant. Then one obtains for the remaining relevant degrees of freedom an exact equation of motion by eliminating the irrelevant degrees of freedom with the use of properly defined projection operators /7,8/. In the problem of an interacting A-identical particles that we consider, all the dynamical degrees of freedom should be treated on an equal footing and consequently they cannot be seperated into relevant and irrelevant parts. However, the structure of (2.4) suggests that we can regard the uncorrelated component $F_0(t)$ as the relevant part and the correlated component $F_{cor}(t)$ as the irrelevant part of the total density matrix. Then in order to eliminate correlations we should define an operator C(t) which projects the total density matrix on to the uncorrelated part. It can be seen that the integral operator which is defined as

$$C(t) = \mathcal{A} \sum_{J=1}^{A} \prod_{K\neq J}^{A} \rho(KK';t) \int dK-(A-1)\, \mathcal{A} \prod_{K=1}^{A} \rho(KK';t) \int dK \tag{2.8}$$

has the right property and the application of C(t) on the total density matrix

$$C(t)\, F(t) = F_0(t), \tag{2.9}$$

gives the relevant part of the density matrix. Therefore C(t) when it is applied to F(t) behaves like a time dependent projection operator on the uncorrelated component. Similarly Q(t)=1-C(t) acts like a projection operator on to the uncorrelated component. In addition C(t) and Q(t) have the property that although they are time dependent, they effectively commute with the time derivative,

$$C(t) \frac{\partial}{\partial t} F(t) = \frac{\partial}{\partial t} \left[C(t)F(t)\right] = \frac{\partial}{\partial t} F_0(t), \tag{2.10}$$

and similarly for Q(t).

Applying C(t) and Q(t) to both sides of (2.2) and using the property (2.10) we obtain two coupled equations for the correlated (irrelevant) and the uncorrelated (relevant) parts of the density matrix,

$$i \frac{\partial}{\partial t} F_o(t) = C(t)L \{F_o(t) + F_{cor}(t)\} \tag{2.11}$$

$$i \frac{\partial}{\partial t} F_{cor}(t) = Q(t)L \{F_o(t) + F_{cor}(t)\} \quad . \tag{2.12}$$

The second equation can be solved formally as

$$F_{cor}(t) = G(t,t_o)F_{cor}(t_o) - i \int_{t_o}^{t} dt' G(t,t')L_1(t')F_o(t') \tag{2.13}$$

where $L_1(t)$ denotes the Liouville operator corresponding to the residual interactions $V(t) = v - U(\rho)$, and $U(\rho)$ is the meanfield potential. The propagator $G(t,t')$ is given by

$$G(t,t') = T\cdot\exp\{-i \int_{t'}^{t} du\, Q(u)\, L\} \tag{2.14}$$

where T denotes the time-ordered integration due to the time dependence of Q. As seen from (2.13) the correlated part of density matrix consists of two contributions. The first term on the r.h.s. describes the propagation of initial correlations whereas the second term determines dynamical correlations produced by the residual interactions. Even initially there are no correlations, in this case the first term vanishes, the correlated part of the density matrix does not vanish as a result of the dynamical correlations. Eliminating the correlated component from the equation (2.11) we obtain a transport equation

$$i \frac{\partial}{\partial t} F_o(t) = CLF_o(t) - i \int_{t_o}^{t} dt' CL_1G(t,t')L_1F_o(t') \tag{2.15}$$

for the relevant part of the density matrix, where a term due to the initial correlations is omitted. We can immediately derive an equation for the one-particle density matrix by taking trace over the sub-space of (A-1) particles on both sides of (2.15). As a result we obtain a generalized equation for the one-particle density matrix as

$$i \frac{\partial}{\partial t} \rho(t) = [\mathcal{H}(t),\rho(t)] - iK(\rho) \tag{2.16}$$

which describes the single-particle motion under the influence of the mean field $\mathcal{H}(t)$ and the collision term $K(\rho)$. The collision term is given by

$$K(\rho) = \int_{t_o}^{t} dt' \, Tr\{L_1G(t,t')L_1F_o(t')\}_{A-1} \tag{2.17}$$

where Tr denotes integrations over the coordinates of (A-1)-particles. The equations (2.15) and (2.17) are in principle exact and they are obtained from the Liouville equation (2.2) by formal manipulations. However, they provide a general basis for including correlations and for going beyond the self-consisent meanfield description of the many-body dynamics. The collision term can be studied in different approximations. In low energy, the dynamic is mainly determined by the mean-field and the collision term can be considered in a weak coupling approximation. In high energy limit the collision term plays a dominant role. In fact for classical systems it can be shown that (2.17) reproduces the collisoin term of the Boltzmann equation /10/.

Before we go into the approximate evaluation of the collision term in the weak-coupling limit, we consider the expectation values of operators, in particular two-body operators, in the present formulation. The expectation value of an operator $\mathcal{O}$ by definition is given by

$$\langle \mathcal{O} \rangle (t) = \mathrm{Tr}\, \mathcal{O}\, F(t) \tag{2.18}$$

Using the decomposition (2.4), the expectation value can be expressed in terms of sums of two terms,

$$\langle \mathcal{O} \rangle\ (t) = \langle \mathcal{O} \rangle_0 (t) - i \int_{t_0}^{t} dt' \ \langle \mathcal{O}\, G(t,t') L_1 \rangle_0 (t') \tag{2.19}$$

where the result (2.13) is used for $F_{cor}(t)$ and $\langle \cdots \rangle_0 = \mathrm{Tr}(\cdots)F_0$ denotes the expectation value over the relevant part of the density matrix. As seen from (2.19) the expectation value of an operator can be given as the sum of the mean field contribution, which is the first term on the r.h.s. of (2.19), and the contribution arising from the correlations.

For one-body operators the contributions due to the correlations vanishes due to the fact that L_1 in (2.19) is a traceless two-body operator and a self-contraction does not contribute. But for a two-body operator, the contributions due to the correlations do not vanish in general. For example if the two-body interactions are taken as the operator $\mathcal{O} \to v$, the second term in (2.19) gives the correlation energy. In many cases, it may be more convenient to convert (2.19) into a differential equation. Taking time derivative and using (2.11) and (2.12) we obtain a differential equation

$$\frac{d}{dt} \langle \mathcal{O} \rangle = \langle -i \mathcal{O} \mathcal{L} \rangle_0 - \int_{t_0}^{t} dt' \langle \mathcal{O}\, L_1 G(t,t') L_1 \rangle_0 \tag{2.20}$$

for the rate of expectation value, where $\mathcal{L}$ denotes the Liouville operator corresponding to the mean-field Hamiltonian. This equation can be regarded as a

Langevin equation for the collective operator $\mathcal{O}$, in which the first term on the r.h.s. describes the driving force on $\mathcal{O}$ due to the mean field and the second term determines the dissipation. In fact, under certain approximations (2.20) can be converted into a classical equation of motion for the collective variables /11/.

3. WEAK-COUPLING LIMIT AND MASTER EQUATION

The collision term (2.17) provides a basis for introducing different approximations Here we consider the weak-coupling limit and evaluate the collision term by assuming some general statistical properties for the residual interactions (random two-body collisions). The collision term in general involves the two different characteristic times, (i) the decay time τ_λ, of the intermediate propagator G, which measures the average time interval between subsequent two-body collisions, (ii) the correlation time, τ_{cor}, of the residual interactions which corresponds to the duration time of a two-body collision. In low energy nuclear reactions, due to large mean free path, λ, of nucleons the time spent between subsequent collisions is much larger than the duration time of a collision. Consequently the weak coupling limit is defined by $\tau_{cor} \ll \tau_\lambda$ /12/. Classically this condition corresponds to a dilute gas limit and it can be expressed in terms of the range, σ, of the interactions and the mean free path as $\sigma \ll \lambda$. In the weak coupling limit the leading order contribution to the collision term is obtained by replacing the full propagator in (2.7) with the mean- field propagator,

$$K(\rho) = \int_{t_0}^{t} dt' \, Tr\left[V, g_0\left[V, F_0^{(4)}(t') \, g_0^+\right]\right], \tag{3.1}$$

where the mean -field propagator is given by

$$g_0(t,t') = T\cdot\exp\{-i \int_{t'}^{t} du \, \mathcal{H}(u)\} \tag{3.2}$$

and the uncorrelated part of A-particle density matrix $F_0 = F_0^{(A)}$ is replaced by the uncorrelated part of 4-particle density matrix $F_0^{(4)}$, as a result of two-body character of the residual interactions. The approximate expression (3.1) for the collision term is still time reversible and it carries memory effects. However, inconsistent with the weak-coupling limit, it can be further simplified. We consider the correlation function,

$$\chi(t,t') = \langle V \, g_0(t,t') V \, g_0^+(t,t') \rangle_0 \tag{3.3}$$

of the residual interactions. As a result of the random properties of the residual interactions, the correlation function decays over a time interval

$t-t' \sim \tau_{cor}$, which is defined as the correlation time /12/. The collision term has essentially the same structure as the correlation function and consequently the decay time of the collision kernel (memory time) is determined by the correlation time. The characteristic variations in the one-particle density matrix $\rho(t)$ take place over time intervals of order τ_λ. However, the variations of $\rho(t)$ during time intervals of order τ_{cor} and hence the memory effects in the collision term can be neglected in the weak-coupling limit. As a result, within the Markov approximation the collision term becomes as

$$K(\rho) = \int_0^\infty d\tau \quad Tr\left[V, \left[g_o V g_o^+, F_o^{(4)}(t)\right]\right] \tag{3.4}$$

where the upper limit of the τ-integration is extended to infinity. Furthermore, if the variations of the mean field during time intervals of order τ_{cor} is neglected, the meanfield propagator is approximately given by

$$g_o(t,t-\tau) \cong \exp\{-i\tau\mathcal{H}(t)\}. \tag{3.5}$$

Introducing some random properties for the residual interaction and neglecting the memory effects in the collision term the equation (2.16) with $K(\rho)$ given by (3.4) becomes time irreversible and it describes the evolution of the system towards statistical equilibrium. In the derivation of collision term (3.4) the decay of the intermediate propagator G in (2.17) is neglected. However, it is possible to take the decay of the propagator into account and one obtains a mean-field propagator in(3.4) which is modified by a decay matrix /4/.

In the remaining part of this sectiion we use the extended mean-field equation (2.16) with the collision term (3.4) to derive a quantum mechanical master equation for the occupation probabilities of the single-particle states. We expand the one-body density matrix in terms of the complete set of TDHF single-particle states,

$$\rho(t) = \sum_{\alpha\beta} |\alpha(t)\rangle \rho_{\alpha\beta}(t) \langle\beta(t)| \tag{3.6}$$

where the single-particle states are determined by

$$i \frac{\partial}{\partial t} |\alpha(t)\rangle = \mathcal{H}(t)|\alpha(t)\rangle \tag{3.7}$$

The collision term in general is not diagonal in the TDHF representation. Hence the off-diagonal elements $\rho_{\alpha\beta}, \alpha \neq \beta$ of the density matrix do not vanish and they describe the non-statistical aspects of two-body collisions. We assume that the mean-field describes the essential coherent aspect of the single-particle motion and neglect the additional coherence introduced by the residual interactions.

Therefore in the following we neglect the off-diagonal elements. Inserting the expansion (3.6) into (2.16) we obtain a master equation for the occupation probabilities $\rho_{\alpha\alpha}(t)\equiv\rho_{\alpha}(t)$ of single-particle states,

$$\frac{\partial}{\partial t}\rho_{\alpha} = \sum W_{\alpha\beta,\gamma\delta}(t)\left[\rho_{\gamma}\rho_{\delta}(1-\rho_{\alpha})(1-\rho_{\beta})-\rho_{\alpha}\rho_{\beta}(1-\rho_{\gamma})(1-\rho_{\delta})\right] \qquad (3.8)$$

where the transition probabilities are given by

$$W_{\alpha\beta,\gamma\delta}(t) = |V_{\alpha\beta,\gamma\delta}(t)|^{2}\pi\ \delta\left[\varepsilon_{\alpha}(t)+\varepsilon_{\beta}(t)-\varepsilon_{\gamma}(t)-\varepsilon_{\delta}(t)\right] \qquad (3.9)$$

and $\varepsilon_{\alpha}(t)=\langle\alpha(t)|\mathcal{H}(t)|\alpha(t)\rangle$ denotes the diagonal elements of the mean-field Hamiltonian. If we include the decay of the intermediate propagator G in (2.17) into the calculation, as a result of finite width of single-particle states, we get a soft δ-function instead of δ-function in the transition probabilities.

4. CONCLUSIONS

We have formulated a theory in order to generalize the TDHF description by including consistently two-body collisions. In usual applications of the projection operator techniques of statistical mechanics, one derives transport equations for the relevant degrees of freedom by formally eliminating the degrees of freedom which are regarded irrelevant. Here we use these techniques to derive an exact equation for the uncorrelated component of the A-particle density matrix by eliminating the correlations from the equation of motion. By reducing the resulting equation to the single-particle degree of freedom, we obtain for the one-particle density matrix a generalized equation which includes a collision term. The exact collision term provides a basis for introducing approximations and for studying the equilibration processes in quantum and classical systems. For example for dilute classical systems one obtains the Boltzmann collision term. In the present work we consider the collision term in the weak-coupling limit and derive a quantal master equation for the single-particle occupation probabilities.

REFERENCES

1. C.Y. Wong and H.F. Tang, Phys.Rev.Lett. 40 (1978) 1070
2. H. Orland and R. Schaeffer, Z.Phys. A290 (1979) 191
3. G. Mantzouranis and H.C. Pauli, Z.Phys. A281 (1977) 165
4. S. Ayik, Z.Phys. A298 (1980) 83
5. P. Grangé, H.A. Weidenmüller and G. Wolschin, Ann.Phys. 136 (1981) 190
6. R. Balian and M. Veneroni, preprint, Orsay (1981)
7. S. Nakajima, Progr. Theor.Phys. 20 (1958) 948
8. R.W. Zwanzig, Quantum Statistical Mechanics. P.H.E. Meijer, (ed.), p. 139,

New York:Gordon and Breach 1966 and Physica 30, (1964) 1109
9. P. Bonche, S. Koonin and J. Negele, Phys.Rev. C13 (1976) 226
10. C.R. Willis and R.H. Picard, Phys.Rev. A9 (1974) 1343
11. S. Ayik, Nucl.Phys. A370 (1981) 317
12. H.A. Weidenmüller, in Progress in Particle and Nuclear Physics, Vol. 3, ed. D.H. Wilkinson, Pergamon, Oxford 1980.

A BOLTZMANN EQUATION APPROACH TO THE DAMPING OF GIANT RESONANCES IN NUCLEI

P. Schuck
Institut Laue-Langevin, 156X, 38042 Grenoble Cedex, France

J. Winter
University of Munich, Sektion Physik, Garching, Germany

ABSTRACT

The Vlasov equation plus collision term (Boltzmann equation) represents an appropriate frame for the treatment of giant resonance (zero sound modes) in nuclei. With no adjustable parameters we obtain correct positions and widths for the giant quadrupole resonances.

— — — — — —

It is well known [1] that schematic forces of the multipole-multipole type give a surprisingly good description of giant resonances within the RPA - or equivalently the TDHF-approach; this concerns for instance the position of energies and to less extent transition densities but it is of course impossible within this formalism to account for the (spreading) width of the giant resonances (we treat here intermediate to heavy nuclei for which the decay or escape width should be negligible).

We will be concerned with the giant quadrupole resonance only but our theory is applicable for other multipoles as well [2]. For our model (harmonic oscillator plus quadrupole-quadropole force) the Vlasov equation becomes identical to the TDHF-equation and moments of the Vlasov equation with respect to powers of the momentum break off after the second. This means that the coupled fluid dynamic equations for density, velocity and pressure tensor are still exact.

To account for damping we include a collision term [3] to the Vlasov equation but still break off the moments after the second, certainly a very good approximation in view of what we said above. For the collision integral we take the form of Uehling-Uhlenbeck [4] which has certain approximative features as for instance energy conservation of the quasiparticle energies during the collision process and its validity for zero excitation only [4]. This however does not seem to be a serious drawback since we are able to explain the width of the giant quadrupole resonance quantitatively.

For the two body collisions we take for convenience only the s-wave part of the Gogny force as an average representation of the scattering in all channels and we neglect exchange. The Gogny force [5] has been chosen because it is a good phenomenological representation of a microscopic G-matrix which should be used in the collision integral [4]. The fact that we use different effective forces in the collision integral and for the mean field is not contradictory since they have to describe very different processes.

In this way we arrive at a model [6] for the damping of the giant quadrupole resonance which is free of parameters (the strength of the quadrupole-quadrupole force is as usual determined from the self-consistency condition of the harmonic oscillator).

Our basic equation is the Boltzmann equation [4] :

$$\frac{\partial f}{\partial t} + \frac{\vec{p}}{m}\frac{\partial f}{\partial \vec{R}} - \frac{V}{\partial \vec{R}}\frac{f}{\partial \vec{p}} = I[f]$$

$$I[f] = \frac{2\pi}{\hbar}\int \frac{d^3p'}{(2\pi\hbar)^3}\frac{d^3p_1}{(2\pi\hbar)^3}\frac{d^3p_1'}{(2\pi\hbar)^3} v^2(\vec{p} - \vec{p}_1)\,\delta(\vec{p} + \vec{p}\,' - \vec{p}_1 - \vec{p}_1{}') \tag{1}$$

$$\delta(\varepsilon_p + \varepsilon_{p'} - \varepsilon_{p_1} - \varepsilon_{p_1'})\ \ [\bar{f}\bar{f}'\, f_1 f_{1'} - f f' \bar{f}_1 \bar{f}_{1'}]$$

where $\bar{f} = 1 - f$ and $f' = f(\vec{R},\vec{p}\,',t)$ etc., $\varepsilon_p = p^2/2m$, and $v(p)$ is the Fourier transform of the two-body interaction; the mean field is given by

$$V(\vec{R},t) = \frac{1}{2}\, m\, \omega_o^2 R^2 - \chi q(t)\, \hat{Q} = :\ \frac{1}{2}\, m\, (\Omega_x^2(t)\, X^2 + \Omega_y^2(t) Y^2 + \Omega_z^2(t) Z^2)$$

$$q(t) = \mathrm{Tr}(\hat{Q}\rho(t));\ \hat{Q} = 2Z^2 - X^2 - Y^2. \tag{2}$$

For this potential the left hand side of (1) is the exact transcription into Wigner space of $i\hbar\,\dot{\rho} = [h,\rho]$ and in general it is the $\hbar = 0$ limit of TDHF. It should be realized that the collision term in (1) represents the $\hbar \to 0$ limit of its quantum mechanical counterpart; this corresponds to treating collisions in the local density approximation, i.e. the nucleus is considered locally as a piece of nuclear matter. In this sense we think that the approximation of energy conservation during the scattering process is less important than in the discrete case. The approximation that I[f] or (1) is strictly only valid for zero excitation [4,7] is certainly less justified since we here are concerned with giant resonances. This point needs further studies in the future but at the moment we take the classical form of the collision term as it can be found in the literature [4]. Finally we want to point out that in the samll amplitude limit studied here the collision term (1) can be derived from the sum of 2 particle-2 hole Feynman graphs shown in Fig.1 [4].

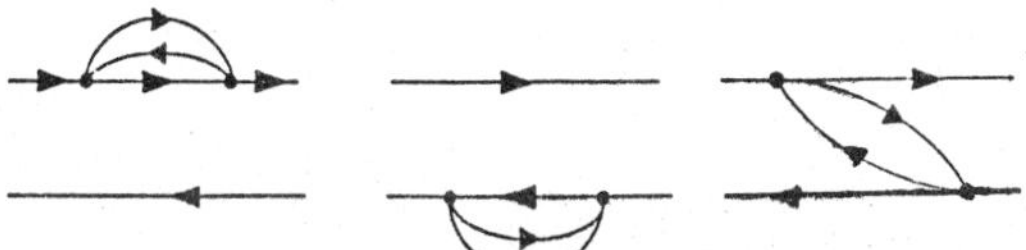

Fig. 1 - Feynman graphs contributing to the collision integral in the linearised form.

Without collision term (1) can be solved analytically (up to the solution of classical eqs. of motion). The result is :

$$f(\vec{R},\vec{p},t) = F(\lambda - \tilde{H}) \tag{3}$$

where F(x) is an arbitrary function and

$$\bar{p}_i = \xi_i(p_i - m\,u_i) \;;\; \bar{X}_i = \frac{X_i}{\xi_i} \;;\; u_i = \dot{\alpha}_i X_i \;;\; \alpha_i = \ln \xi_i \tag{4}$$

The parameters ξ_i obey the (coupled) set of classical equations :

$$\ddot{\xi}_i - \frac{\Omega_i^2(0)}{\xi_i^3} + \Omega_i^2(t)\,\xi_i = 0 \qquad \xi_i(0) = 1;\; \dot{\xi}_i(0) = 0$$

In the small amplitude limit, χ being determined from the usual oscillator self-consistency condition, this yields the well known result [1] $\omega_{2^+} = \sqrt{2}\,\omega_o \simeq 60\ A^{-1/3}$ MeV in very good agreement with experiment.

For small amplitudes it is possible to develop (3) around equilibrium $F_o(\lambda - H)$ where F_o is now the Wigner transform of the groundstate density matrix. For what follows it is very important not to choose for F_o a step function as would be suggested by lowest order Thomas Fermi theory [8]. This would correspond to an infinite Fermi system at zero temperature (T = 0) and consequently no two-body collisions can take place since the collision integral is proportional to T^2 [9]. Nuclei are at zero temperature but even there a smearing out of the Fermi step function is present due to either two-body correlations or, most importantly the finiteness of the system. For a harmonic potential one can deduce such a finite diffusivity T from a Strutinsky averaged Wigner transform of the density matrix [10]. One obtains a value constant over the nucleus of T ≃ 4 MeV (A = 224) which is a rather high value indeed. More realistic is a Woods Saxon potential for which we show in Fig.2 the phase space distribution [11] for A = 184 and in Fig.3 the corresponding diffusivity is extracted in fitting to the curves in Fig.2 a Fermi distribution of the type

$$f(R,p) = \left[1 + \exp\frac{\varepsilon_p - \mu(R)}{T(R)}\right]^{-1} \tag{6}$$

We see that here the diffusivity is strongly varying from small values in the interior to rather high values in the surface. In spite of this fact the Fermi function (6) can be quite effectively represented by distributions

$$f \simeq \theta(\mu - \varepsilon_p) - \frac{\pi^2}{6}\,T^2\delta'\,(\mu - \varepsilon_p) \tag{7}$$

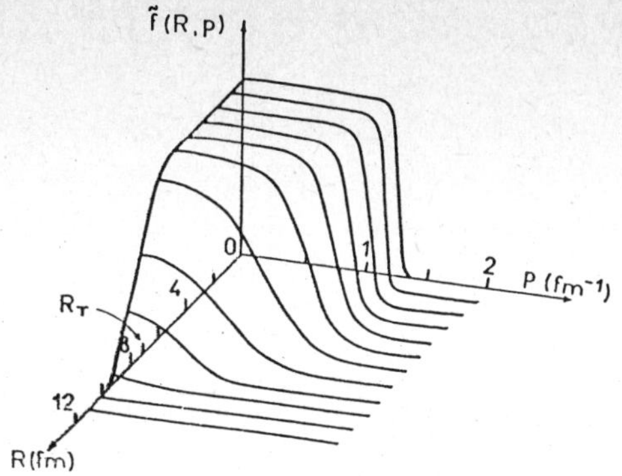

Fig. 2 - Averaged phase space distribution for A = 184 in a Woods Saxon potential. For details see ref. [11].

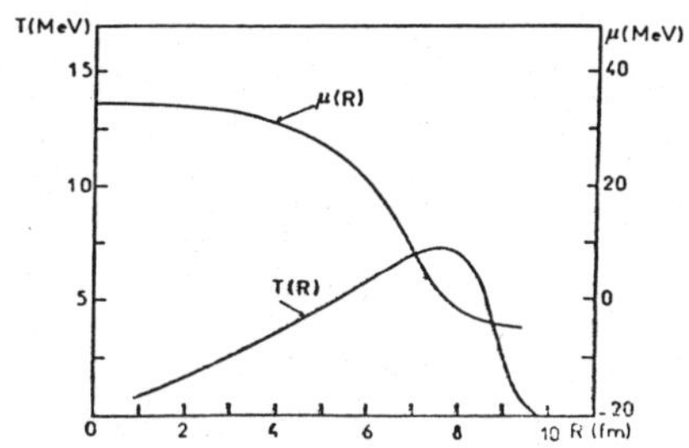

Fig. 3 - Local diffusivity and Fermi energy as extracted from the distribution in Fig. 2.

which then helps very much to perform the integrals in (1). Taking in addition moments of (1) with respect to $\vec{p}$ and truncating after second order we obtain [12] (we use summation convention) :

$$\dot{\rho} + \mathrm{div}\,(\rho\,\vec{u}) = 0$$

$$m\left(\frac{\partial}{\partial t} + u_k \frac{\partial}{\partial X_k}\right) u_i = - \frac{\partial V}{\partial X_i} - \frac{1}{\rho}\frac{\partial}{\partial X_k} P_{ki} \qquad 8(a\text{-}c)$$

$$\left(\frac{\partial}{\partial t} + u_k \frac{\partial}{\partial X_k}\right) P_{ij} + P_{ik}\frac{\partial}{\partial X_k} u_j + (i \leftrightarrow j) = \frac{1}{m}\int d^3p\; p_i p_j I$$

as a coupled system for ρ, $\vec{u}$, $\overleftrightarrow{P}$, the zeroth, first and second moment of f. As mentioned above these equations are exact in our model for vanishing collision term, which can be verified with (2) and (4).

Evaluating the second moment of the collision integral in (8c) one obtains in the small amplitude limit

$$d^3p\ p_i p_j I = \frac{mp^{id}}{\tau}\ \pi_{ij} \tag{9}$$

where $\overleftrightarrow{\pi}$ is a traceless tensor defined by

$$P_{ij} = p^{id}\ (\delta_{ij} + \pi_{ij}),\quad p^{id} = \frac{1}{5}\ (3\ \pi^2)^{2/3}\quad \rho^{5/3} \tag{10}$$

which describes the deformation of the Fermi surface, and τ is the local collision time. In the same limit, and in first order of the expansion (7), the inverse collision-time can be calculated analytically and turns out to be

$$\frac{1}{\tau(R)} = \frac{m^3}{16\pi h^7}\ T^2 \cdot \frac{1}{p_F^5} \int_o^{2p_F} dp\ p^2\ (4p_F^2 - p^2)\ v^2(p) \tag{11}$$

$$p_F(R) = (\frac{3\pi^2}{2}\ \rho)^{1/3}.$$

In the case of a superposition of gaussians like the Gogny-force even the last integral in (11) can be performed, and we therefore get an explicit expression for the local collision time.

Going over from eq. (8c) for the pressure tensor $\overleftrightarrow{P}$ to an equation for the deformation tensor $\overleftrightarrow{\pi}$ and making further use of the restriction to small amplitudes, we obtain the approximate solution

$$\pi_{ij} = \delta_{ij} - \int_{-\infty}^{t} dt'\ e^{-\frac{t-t'}{\tau}} \left(\frac{\partial u_i}{\partial X_j} + (i \leftrightarrow j) - \frac{2}{3}\frac{\partial u_i}{\partial X_i}\ \delta_{ij} \right)_{t'} \tag{12}$$

In this limit we have in our model

$$\vec{u}(t) = \alpha(t)\vec{u}\ ;\qquad \vec{u} = (-X, -Y, 2Z) \tag{13}$$

and therefore we obtain a position independent π_{ij} if we introduce an effective inverse collision time averaged over the pressure which turns out to be the result of the expansion of (8) to first order in $(\Omega_\infty\ \tau)^{-1}$ (see (16)).

$$\frac{1}{\tau_{eff}} = \int d^3R\ \frac{p^{id}}{\tau} \Big/ \int d^3R\ p^{id} \tag{14}$$

From (8(a-c), 12-14) one then obtains [12] :

$$B\ddot{\alpha} + C_o\alpha + (C_\infty - C_o)\dot{\beta} = 0; \quad \dot{\beta} = \int_{-\infty}^{t} dt' e^{-\frac{t-t'}{\tau_{eff}}} \dot{\alpha} \tag{15}$$

with

$$B = m\int d^3r \; \rho|\vec{u}|^2 \qquad C_o = B\,\Omega_o^2 \; ; \; C_\infty = B\,\Omega_\infty^2 \tag{16}$$

where Ω_o and Ω_∞ are the well known [8,9] frequencies of normal sound ($\tau_{eff} = 0$) and zero sound ($\tau_{eff} = \infty$) respectively. Equation (15) is the one of a damped harmonic oscillator with a friction kernel which is non local in time reflecting the fact that an appreciable elapse of time can occur during successive collisions. This memory effect was discussed qualitatively by Nörenberg in a recent work [13] but no determination of τ_{eff} for zero sound modes was attempted. Here we give for the first time a completely self contained theory with no adjustable parameters for giant resonances in nuclei.

In order to discuss more specifically the influence of damping it is convenient to transform (15) to a third order differential equation [12] :

$$B\,\dddot{\beta} + \frac{B}{\tau_{eff}}\,\ddot{\beta} + C_\infty\dot{\beta} + \frac{C_o}{\tau_{eff}}\,\beta = 0 \tag{17}$$

This equation immediately reproduces the well-known limiting frequencies of ordinary and zero sound but for finite τ_{eff}, as in our case, the characteristic equation of (16) has in general complex frequencies Ω as solution. In the limit Re $\Omega\,\tau_{eff} \to \infty$ one obtains [12] :

$$\Omega = \Omega_\infty\left[1 + \frac{1}{4}\left(-\frac{1}{2} - \left(\frac{\Omega_0}{\Omega_\infty}\right)^2 + \frac{3}{2}\left(\frac{\Omega_0}{\Omega_\infty}\right)^4\right)\left(\frac{\Omega_\infty}{\tau_{eff}}\right)^2\right] - \frac{i}{2\tau_{eff}}\left(1 - \left(\frac{\Omega_0}{\Omega_\infty}\right)^2\right) \tag{18}$$

From (14) and (9) we can calculate τ_{eff} with the s-wave part of the Gogny force [5] or more conveniently with a one Gaussian equivalent to the Gogny force : $v(r) = v_o\, e^{-r^2/r_o^2}$ with $v_o = 26.5$ MeV and $r_o = 2.25$ fm (this force yields the same pairing properties as the original Gogny force [14]). For the model nucleus A = 224 and the corresponding harmonic oscillator diffusivity [10] $T \simeq 4$ MeV we then obtain a width of

$$\Gamma \simeq 3 \text{ MeV} \tag{19}$$

The only A-dependence of the width in our model comes from the A-dependence of T. As a matter of fact the distribution function $f(\varepsilon)$ for a harmonic oscillator [10] becomes (not unexpectedly) more and more infinite matter (i.e. step function) like the larger the nucleus and a detailed investigation [15] shows that the h.o. - diffusivity can be represented very accurately by the law (for a position dependent diffusivity in a Woods Saxon like potential (Fig. 2) this A-dependence could be changed somewhat)

$$T^2 \sim A^{-1/3} \tag{20}$$

Together with (19) we thus have the smooth trend of the width for all nuclei which is represented together with the experimental values in Fig. 4.

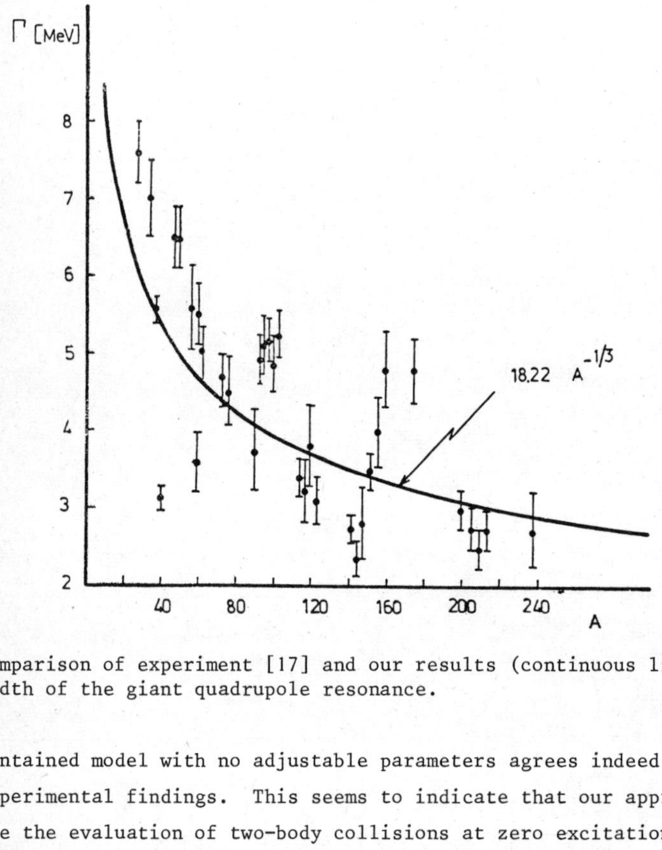

Fig. 4 - Comparison of experiment [17] and our results (continuous line) for the width of the giant quadrupole resonance.

Our self-contained model with no adjustable parameters agrees indeed very nicely with the experimental findings. This seems to indicate that our approximations, for instance the evaluation of two-body collisions at zero excitation energy, are not too drastic. Further studies have however to be performed to see in detail how excitation energy influences the rate of two-body collisions. A preliminary account of how this can be done is given in Ref. [7]. We also would like to stress at this point that our theory is a linear one. A classical estimate for the quadrupole giant resonance in heavy nuclei yields roughly a 8 % change of nuclear half axes compared to their spherical equilibrium values. This seems to justify the linearization though second and higher order contributions like those shown in Fig. 5 [16] cannot be ruled out completely.

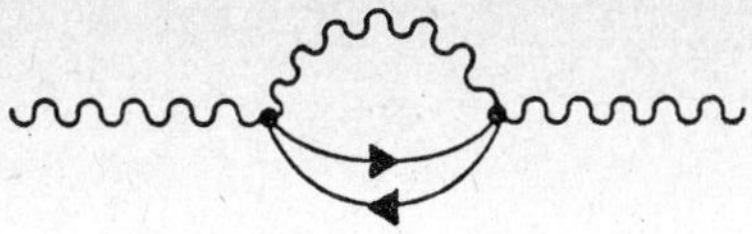

Fig. 5 - Second order effect in the change of the density leading to incoherent particle scattering from the surface oscillation.

In spite of these restrictions it seems evident from this work that the surprisingly strong diffusivity of the Wigner distribution and its A-dependence play an important role in the two-body scattering mechanism.

A more elaborate version of this work will be given elsewhere.

Acknowledgements

It is a pleasure for us to thank R. Hasse for comments. One of us (P.S.) is also grateful to W. Götze, P. Wölfle for discussions and to M. Durand and V.S. Ramamurthy for their collaboration on the Wigner transform of the density matrix. J. Winter would like to thank the Institut Laue-Langevin for its hospitality accorded to him during several visits.

References

[1] A. Bohr and B.R. Mottelson, (1975) Nuclear Structure, Benjamin (New York) Vol. II
[2] P. Schuck, J. Winter, to be published
[3] J. Winter, to be published
P. Schuck, (1977) Proceedings of the Topical Conference on Heavy Ion Collisions Fall Creek Falls, Tennessee (Oak Ridge Lab.)
[4] E.A. Uehling, G.E. Uhlenbeck, (1933) Phys. Rev. 43, 552
L.P. Kadanoff and G. Baym, (1962) Quantum Statistical Mechanics, Benjamin, New York
[5] J. Déchargé, D. Gogny, (1980) Phys. Rev. 1568
[6] P. Schuck, June 1981, unpublished Lectures notes, GSI
[7] P. Schuck, (1972) J. of Low Temp. Phys. 7, 459
[8] P. Ring, P. Schuck, (1980) the Nuclear Many Body Problem, Springer
[9] G. Baym, C. Pethik, (1978) Monographs and Texts in Physics and Astronomy, Vol. XXIX, part II, p.1
[10] M. Prakash, S. Shlomo, V.M. Kolomietz, (1981) Nucl. Phys. A370, 30
[11] M. Durand, V.S. Ramamurphy, P. Schuck, (1982) Phys. Lett. 113B, 116
[12] J. Winter, to be published
[13] W. Nörenberg, (1981) Phys. Lett. 104B, 107
[14] P. Schuck et al. to be published
[15] M. Durand, V.S. Ramamurphy, P. Schuck, to be published
[16] P. Schuck, (1976) Z. Phys. A279, 31
[17] F.E. Bertrand, (1981) Nucl. Phys. A354, 129c.

TUNNELING OF A MANY-FERMION SYSTEM IN ONE DIMENSION

J. W. NEGELE
Center for Theoretical Physics
Massachusetts Institute of Technology
Cambridge, Massachusetts 02139

ABSTRACT

A method is presented for stochastic evaluation of ground state properties and tunneling decay of many-fermion systems in one spatial dimension. Application to a model problem defined to produce saturation, strong short-range repulsive-core correlations, and spontaneous fission provides the opportunity to test the stationary-phase approximation and assumptions concerning the effective interaction presently utilized in the mean-field theory of spontaneous fission.

I. INTRODUCTION

Since the first application to the nuclear many-body problem at the Saclay TDHF workshop three years ago, significant progress has been made in developing functional integral techniques for nuclear physics.(1) Despite new insight provided by alternative formulations of functional integral representations of the evolution operator, fuller understanding of the freedom in defining the mean field, and experience obtained from a variety of applications, two crucial, fundamental questions still remain unanswered: What is the validity of the stationary-phase approximation (SPA) and how does one define an appropriate effective interaction to use in the resulting mean-field equations? These questions are obviously interrelated and would be resolved simultaneously by a fundamental treatment of many-body correlations.

In the absence of a practical general theory for correlations, it would be of significant value to test both the SPA and the treatment of the effective interaction by comparison with the exact solution for a model which incorporates the essential features of the nuclear many-body problem. Thus, in this work, Monte Carlo techniques will be described for the exact evaluation, subject only to controllable statistical sampling errors, of functional integrals specifying ground state observables and tunneling decay of many-fermion systems. Much of this work is based upon the approach to Monte Carlo developed by S. Koonin(3) and extensive discussions with H. Orland and the application to tunneling is being pursued in collaboration with G. Bertsch.

Given that one of the primary objectives is accurate treatment of many-body correlations, I have chosen to utilize a coordinate-space path integral representation which is well-suited to forces with strongly repulsive cores. Unfortunately, as will become evident subsequently, Monte Carlo techniques for such path integrals are presently only practical in one spatial dimension, so this convenience in treating repulsive cores comes at a high price. Nevertheless, it is instructive to begin with a simple, well-controlled problem and I believe there is considerable relevant many-body physics to be learned in one-dimension. Furthermore, there is reason to hope that the use of

Monte Carlo evaluation of other functional integral representations, such as the auxiliary field described at this meeting by Steve Koonin, or the Monte Carlo methods of Kalos or Zabolitzky may ultimately allow one to extend the present investigation to higher dimensions.

II. QUANTUM MECHANICS FOR ONE PARTICLE IN A POTENTIAL WELL

It is useful to tersely review the fundamental functional integral for the quantum mechanics of a single particle, and its evaluation by the SPA and Monte Carlo methods. The real-time evolution operator may be evaluated by dividing the time continuum into discrete steps of size ε, inserting complete sets of position space and momentum space eigenstates at each step, replacing $\hat{x}$ and $\hat{p}$ by their eigenvalues and integrating out the momenta to obtain[2]

$$<x_{N+1}|e^{-iHT}|x_0> = \int (x_1 \dots x_N)\, e^{i\varepsilon \sum_n [\frac{m}{2}\left(\frac{x_{n+1}-x_n}{\varepsilon}\right)^2 - V(x_n)]} \tag{1}$$

The exponent is recognized as a discrete approximation to the classical action S(x).

A) The Stationary Phase Approximation

To calculate eigenstates of H it is convenient to evaluate the Fourier transform of the trace of the evolution operator:

$$\begin{aligned} i\int_0^\infty dT e^{iET} \mathrm{tr}\, e^{-HT} &= \sum_n \frac{<n|\;|n>}{E_n - E - i\eta} \\ &= i\int dT e^{iET} \int d\bar{q} \int \mathfrak{D}[q(t)] e^{iS[q(t)]}\Big|_{q(T)=q(0)=\bar{q}} \end{aligned} \tag{2}$$

For a particle in a single well, sequential application of the SPA to the three integrals yields a sum over all periodic classical trajectories in the well with energy E, and the resulting geometric series yields poles when the Bohr-Sommerfeld quantization condition is satisfied. Corrections to the lowest SPA systematically sum the fluctuations of the coordinates about the classical trajectory at each time step. For the double well the new feature arises that trajectories with imaginary time provide the leading SPA contribution in classically forbidden regions, corresponding to classical motion in the inverted potential. Summation of the multiple geometric series of all classical trajectories in the three allowed and forbidden regions yields an expression for the level density for which a WKB formula for the level splitting between the two lowest eigenstates of a degenerate well or the lifetime of a metastable state may be calculated. In the case of a double well with a high barrier, the classical solution may be picturesquely visualized in terms of instantons, classical solutions in the inverted well connecting arbitrary numbers of periodic solutions in the two classically allowed wells. Corrections to the SPA again build up all the fluctuations about these classical paths required to obtain the full quantum mechanics of the exact functional integral.

B) Monte Carlo Evaluation of Functional Integrals

An integral with an integrand expressed as the product of a function f(x) times a probability distribution P(x) may be evaluated stochastically

$$\int f(x)\ P(x) = \lim_{N\to\infty} \frac{1}{N} \sum_{i=1}^{N} f(x_i) \tag{3}$$

where the x_i's are randomly distributed according to the distribution P(x). Although this formula is very general, it is only useful when the variance may be made sufficiently small, which in practice requires that the integrand be positive or at least dominantly positive. This requirement forces one to deal with the Euclidean, or imaginary time, evolution operator to eliminate cancelling phases with the corollary that the exponential weighting of eigenstates restricts practical calculations to low-lying states. Nearly cancelling positive and negative contributions arising from antisymmetry pose additional problems for fermions which will be apparent below.

Even when one has selected a functional integral representation for the evolution operator, considerable freedom exists in the choice of the physical quantity to be calculated. The fundamental trade-off is building into the calculation as much of the physics of the solution as one understands a priori versus unduely prejudicing the results by misconceptions.

1) Initial Value Form

A physically transparent way to formulate a random walk for the ground state introduced by Koonin[3] is to use the imaginary-time counterpart of eq.(1) to generate a population of points distributed according to $\Psi(x)$. The matrix element

$$\begin{aligned} <\Phi_2|e^{-\beta(H-E)}|\Phi_1> &= \int dx_1 \ldots dx_N \; <\Phi_2|x_N><x_N|e^{-\varepsilon(H-E)}|x_{N-1}> \times \\ &\times \; <x_{N-1}|e^{-\varepsilon(H-E)} \ldots |x_2><x_2|e^{-\varepsilon(H-E)}|x_1><x_1|\Phi_1> \end{aligned} \tag{4}$$

may be evaluated as a multidimensional integral using the Monte Carlo method as follows, where for the moment E should be regarded as an arbitrary constant. An ensemble of points $\{x_1\}$ is selected with probability distribution $\phi(x_1)$ (which is necessarily assumed positive). For each x_{n-1}, x_n is generated according to the distribution

$$<x_n|e^{-\varepsilon(H-E)}|x_{n-1}> = \frac{2m\pi}{\varepsilon} e^{-\frac{m}{2\varepsilon}(x_n - x_{n-1})^2 - (\frac{V(x_n)+V(x_{n-1})}{2} - E)\varepsilon} \tag{5}$$

The kinetic energy alone, represented by the gaussian factor, would distribute x_n within in a region of the order $\sqrt{\varepsilon}$ of x_{n-1}, leading to the physical spreading of a free wave packet. Depending upon whether V-E is positive or negative, the distribution is normalized less than or greater than one, resulting in the deletion of points in classically forbidden regions and the replication of points where the potential is most attractive. Finally the matrix element in eq.(4) is calculated by evaluating $\phi_2(x_N)$ for the ensemble of values $\{x_N\}$ as in eq.(3).

The ground state energy may be calculated two ways. Evaluating the ratio of two matrix elements of the form of eq.(4) yields the trial energy

$$E_0 = \lim_{\beta\to\infty} \frac{\langle H\Phi| e^{-\beta(H-E)} |\Phi\rangle}{\langle\Phi| e^{-\beta(H-E)} |\Phi\rangle} \tag{6}$$

In practice, once β is sufficiently large one can generate independent populations of $\{x\}$ distributed according to the ground state wave function by continuing iteration of eq.(5) and thereby improve statistics to any required accuracy. Thus far E has been an arbitrary parameter which controls the overall growth or decrease of the population $\{x\}$. The value of E which keeps the population stable provides an independent determination of the ground state energy which will be denoted here as the normalization energy.

Importance sampling may be introduced to decrease the variance of observables of interest.[4] Essentially, one writes the evolution operator for $\phi(x)\psi(x)$ rather than $\psi(x)$ by multiplying and dividing by a physically motivated trial function $\phi(x)$ at each step. For example, the numerator of eq.(5) is evaluated

$$\begin{aligned}\langle\Phi|He^{-\beta(H-E)}|\Phi\rangle &= \int dx_1\ldots dx_N \langle\Phi|H\tfrac{1}{\phi}|x_N\rangle \times \\ &\times \langle x_N|\phi e^{-\varepsilon(H-E)}\tfrac{1}{\phi}|x_{N-1}\rangle\ldots\langle x_2|\phi e^{-\varepsilon(H-E)}\tfrac{1}{\phi}|x_1\rangle\langle x_1|\phi|\Phi\rangle\end{aligned} \tag{7}$$

where

$$\begin{aligned}&\langle x_n|\phi e^{-\varepsilon(H-E)}\tfrac{1}{\phi}|x_{n-1}\rangle \\ &= \sqrt{\frac{2m\pi}{\varepsilon}}\; e^{-\frac{m}{2\varepsilon}\left(x_n - x_{n-1} - \varepsilon\frac{\frac{\partial}{\partial x}\phi(x_{n-1})}{m\,\phi(x_{n-1})}\right)^2 - \varepsilon\frac{(H-E)\phi(x_{n-1})}{\phi(x_{n-1})}}\end{aligned} \tag{8}$$

Evolution of random walks using eq.(8) differs from eq.(5) in two respects. The gaussian is now shifted by a drift term proportional to ϕ'/ϕ which improves statistics by moving points toward regions where the wave function is largest and points are added and deleted only to the extent to which $H\phi$ differs from $E\phi$. If $\phi(x)$ were the ground state wave function, the energy estimate would be exact and in practice a reasonable trial function significantly increases the accuracy of the Monte Carlo approximation.

Whereas the initial value form offers the advantage of utilizing physical insight in the selection of $|\Phi\rangle$ to obtain an excellent approximation to the energy, other observables are either inconvenient to calculate or require the introduction of additional approximations. The energy splitting of the two lowest eigenstates in a double well, and thus the penetrability of a potential barrier, may be approximately calculated by the following method.[5] Let ψ_L and ψ_R denote approximate eigenstates localized in the left and right wells respectively. To within errors which are exponentially small for a high barrier, the two lowest eigenstates are $\psi_+ \sim \frac{1}{\sqrt{2}}(\psi_L+\psi_R)$ with energy E_0 and

$\psi_- \sim -\frac{1}{\sqrt{2}}(\psi_L - \psi_R)$, with energy E_0+2V, where $V = \langle\psi_L|H|\psi_R\rangle$ and the next eigenstate is separated by energy $\Delta \gg 2V$. Taking $\beta \gg \frac{1}{\Delta}$, the expansion for $e^{-\beta(H-E_0)}$ may be approximated by the lowest two terms

$$e^{-\beta(H-E_0)} \approx |\psi_+\rangle\langle\psi_+| + |\psi_-\rangle\, e^{-2V\beta}\langle\psi_-| + \mathcal{O}e^{-\beta\Delta} \tag{9}$$

so that defining trial functions Φ_L and Φ_R localized in the left and right wells respectively

$$\frac{\langle\Phi_L|e^{-\beta(H-E_0)}|\Phi_R\rangle}{[\langle\Phi_L|e^{-\beta(H-E_0)}|\Phi_L\rangle\langle\Phi_R|e^{-\beta(H-E_0)}|\Phi_R\rangle]^{1/2}}$$

$$\approx \frac{\langle\Phi_L|\psi_L\rangle\langle\Phi_R|\psi_R\rangle}{[|\langle\Phi_L|\psi_L\rangle|^2|\langle\Phi_R|\psi_R\rangle|^2]^{1/2}} \quad \frac{(1-e^{-2V\beta})}{(1+e^{-2V\beta})} = \tanh(V\beta) \tag{10}$$

independent of these trial functions. If one wishes to calculate the lifetime of a metastable state, one may do so by erecting a barrier to convert the potential into a degenerate double well, calculating the matrix element V, and applying the Fermi golden rule with the calculated final density of states. Although this latter procedure has been checked for a representative test case[(5)], all the relevant physics of barrier penetration is already contained in the splitting of the double well, and we shall henceforth use this quantity to test the validity of the SPA and the effective interaction.

2) Evaluation of Trace

A convenient method for the evaluation of ground state expectation values of operators is calculation of an appropriate trace. A local one-body operator, for example, may be calculated as follows

$$\langle 0|\mathcal{O}|0\rangle = \lim_{\beta\to\infty} \frac{\sum_\alpha \langle\alpha|\mathcal{O}|\alpha\rangle e^{-\beta E_\alpha}}{\sum_\alpha e^{-\beta E_\alpha}} = \lim_{\beta\to\infty} \frac{\mathrm{Tr}\,\mathcal{O}e^{-\beta H}}{\mathrm{Tr}\,e^{-\beta H}} \tag{11a}$$

$$= \lim_{\beta\to\infty} \frac{\int dx \langle x|\mathcal{O}e^{-\beta H}|x\rangle}{\int dx\langle x|e^{-\beta H}|x\rangle} = \lim_{N\to\infty} \int dx_1 \dots dx_N \left(\sum_{s=1}^{N} \frac{\mathcal{O}(x_s)}{N}\right) \frac{e^{-S(x_1\dots x_N)}}{\int dx_1\dots dx_N\, e^{-S(x_1\dots x_N)}}$$

$$= \lim_{N\to\infty} \frac{1}{M}\sum_{i=1}^{M} \frac{1}{N}\sum_{s=1}^{N} \mathcal{O}(x_s^{(i)})$$

where the action is given by

$$S(x_1\dots x_N) = \sum_{n=1}^{N} \left[\frac{m}{2\varepsilon}(x_{n+1} - x_n)^2 + \varepsilon V(x_n)\right] \tag{11b}$$

with the periodic condition $x_{N+1} = x_1$ or by the corresponding importance sampled expression in eq.(8) and $\{x_s^{(i)}\}$ denotes the ith set of variables $x_1 \dots x_N$ distributed according to the probability distribution

$$P = \frac{e^{-S(x_1 \ldots x_N)}}{\int dx_1 \ldots dx_N \; e^{-S(x_1 \ldots x_N)}} \tag{11c}$$

Note that since $\mathcal{O}$ may be evaluated at any time step, statistical errors have been minimized in eq.(11) by averaging over all times.

Evaluation of non-local operators such as the kinetic energy is somewhat more difficult. Placing $\frac{p^2}{2m}$ at the nth time step in the imaginary time counterpart to eq.(1) yields the results

$$\langle T\rangle = \frac{1}{M}\sum_{i=1}^{M}\sum_{n} -\frac{m}{2}\frac{(x_{n+1}^{(i)} - x_n^{(i)})(x_n^{(i)} - x_{n-1}^{(i)})}{\varepsilon^2} \tag{12a}$$

or

$$T = \frac{1}{M}\sum_{i=1}^{M}\sum_{n} \left(\frac{1}{2\varepsilon} - \frac{m}{2}\frac{(x_n^{(i)} - x_{n-1}^{(i)})^2}{\varepsilon^2}\right) \tag{12b}$$

depending upon whether the two p's are grouped with adjacent steps or both p's are associated with the same step. Equation (12b) is particularly unsuitable since the $\frac{1}{\varepsilon}$ singularity in the first term must be cancelled numerically by the second, but even the alternative (12a) is generally not sufficiently accurate. Hence, it is convenient to use the virial theorem,

$$\langle T\rangle = \frac{1}{2}\langle xV'(x)\rangle \tag{12c}$$

and to evaluate the ground state energy by calculating the expectation value of the local operator $\tilde{V}(x)=V(x) + \frac{1}{2}xV'(x)$. The only practical disadvantage of this method is that step function potentials and potentials with more than one infinite barrier cannot be treated at all and potentials with very large gradients may yield undesirably large variance.

The energy of the first excited state, and in particular the splitting in a double well, may be calculated by evaluating the correlation function

$$\Gamma(\tau) = \frac{\mathrm{tr}\,\mathcal{O}(t+\tau)\mathcal{O}(t)e^{-\beta H}}{\mathrm{tr}\,e^{-\beta H}} = \frac{1}{MN}\sum_{i=1}^{M}\sum_{s=1}^{N}\mathcal{O}(x_s^{(i)})\,\mathcal{O}(x_{s+r}^{(i)}) \tag{13}$$

where $\tau = r\varepsilon$, $s+r$ is defined mod N by periodicity, and $\{x_s^{(i)}\}$ is distributed as before according to eq.(12). If $\beta \gg \tau \gg \Delta$, where Δ is the energy of the second excited state as before, then

$$\frac{\mathrm{tr}\,\mathcal{O}(t+\tau)\,\mathcal{O}(t)e^{-\beta H}}{\mathrm{tr}\,e^{-\beta H}} = \frac{\sum_\alpha \langle\alpha|\mathcal{O}|\gamma\rangle\langle\gamma|\mathcal{O}|\alpha\rangle\, e^{-E_\alpha\beta}\, e^{-(E_\gamma - E_\alpha)\tau}}{\sum_\alpha e^{-E_\alpha\beta}}$$
$$\rightarrow \sum_\gamma e^{-(E_\gamma - E_0)\tau}|\langle 0|\mathcal{O}|\gamma\rangle|^2 \rightarrow e^{-(E_1-E_0)\tau}|\langle 0|\mathcal{O}|1\rangle|^2 - |\langle 0|\mathcal{O}|0\rangle|^2 \tag{14}$$

and the splitting (E_1-E_0) may be determined by the decay of the correlation function.

Exact evaluation of operators of interest thus reduces to the problem of sampling the distribution P in eq.(12), which may be accomplished by means of the Metropolis algorithm.[6] Considering first the case of a single variable, a Markov random walk to generate a population distributed according to $e^{-S(x)}$ may be generated as follows. Given a value $x^{(i)}$, a new value $x^{(i+1)}$ is defined by selecting a tentative value x_t distributed about $x^{(i)}$ according to a function $f(|x_t-x^{(i)}|)$ and accepting or rejecting this tentative choice on the basis of the change in action, $\Delta S = S(x_t)-S(x^{(i)})$. Specifically, the algorithm proceeds by the following three steps:

a) Define

$$x_t = x^{(i)} + \xi\Delta x \tag{15}$$

where ξ is a random variable uniformly distributed on the interval $[-1/2,1/2]$ and Δx is a parameter selected to optimize convergence by yielding an acceptance ratio of the order of 50%. Thus, $f(x)$ is $\theta(\Delta x/2 - |x|)$.

b) If $S(x_t) \leq S(x^{(i)})$, then $x^{(i+1)} = x_t$.

c) If $S(x_t) > S(x^{(i)})$, select a random variable η uniformly distributed on the unit interval $[0,1]$

$$\text{If } e^{-[S(x_t)-S(x^{(i)})]} > \eta, \text{ then } x^{(i+1)} = x_t$$
$$\text{If } e^{-[S(x_t)-S(x^{(i)})]} \leq \eta, \text{ then } x^{(i+1)} = x^{(i)}$$

Thus, if the action is decreased, the tentative value is accepted. If the action is increased, the tentative value is accepted with probability $e^{-\Delta S}$ and rejected otherwise.

It is simple to verify that $e^{-S(x)}$ is an equilibrium solution to the random walk defined by this rule. Consider the probabilities that $x\to y$ and that $y\to x$. In the case $S(y) < S(x)$, straightforward application of the rule yields the ratio

$$\frac{P(x\to y)}{P(y\to x)} = \frac{f(|x-y|)}{f(|y-x|)} \quad \frac{1}{e^{-[S(x)-S(y)]}} = \frac{e^{-S(y)}}{e^{-S(x)}} \tag{16}$$

The same ratio is obtained for the case $S(x) < S(y)$ so that summing over all x and using the fact $\sum_x P(y\to x) = 1$,

$$\sum_x e^{-S(x)} P(x\to y) = \sum_x e^{-S(y)} P(y\to x) = e^{-S(y)} \tag{17}$$

Thus, the rule evolves the distribution $e^{-S(x)}$ into itself so it is indeed an equilibrium solution.

Generalization to distributions of many variables is straightforward. One may sweep through the time lattice varying each x_n independently in turn, and accepting or rejecting the tentative change on the basis of $\Delta S = S(x_1^{(i)}\ldots x_{n-1}^{(i)}, x_t, x_{n+1}^{(i)}\ldots x_N^{(i)}) - S(x_1^{(i)}\ldots x_N^{(i)})$. Since each step in the sweep satisfies the microreversibility condition eq. (16), $e^{S(x_1\ldots x_N)}$ is an equilibrium solution as before. For subsequent application to collective motion, it is useful to note that the argument also applies to variation of any combination of the x's as well. The question of convergence of the random walk is much more complicated. An especially simple proof may be constructed in the special case that each step has a non-vanishing probability for reaching every configuration[7] and a proof for the method described here can be constructed using the Perron-Frobenius Theorem.[8]

A pedagogical application of the Metropolis method to the calculation of the trace for the harmonic and anharmonic oscillators is given by Creutz and Freedman.[7] A particularly instructive result is the graph of a typical trajectory in a double well in which one clearly observes random fluctuations about the classical solutions in the minimum in each well connected by instanton-like paths in the classically forbidden region.

III. THE MANY-FERMION PROBLEM.

A number of alternative functional integral representations exist for many-fermion systems, including the many-particle generalization of the path integral, eq. (1), an integral over an auxiliary field, and integrals over overcomplete sets of boson coherent states, determinants, and Grassman variables. These formulations thus offer a number of different possibilities for application of the SPA and Monte Carlo techniques.

A) SPA Bounce Equations

Application of the SPA to the many-particle Feynman path integral yields classical equations of motion which are not a particularly useful starting point for the nuclear many-body problem. All the remaining formulations yield similar equations at the SPA level, corresponding to self-consistent Hartree Fock(or in some cases Hartree) equations in real or imaginary time for solutions in classically allowed or forbidden regions respectively.[1] Including exchange terms for convenience, the bounce equation for spontaneous fission is

$$\left[-\frac{\partial}{\partial\tau} + \frac{1}{2m}\nabla^2 - \int dr'\, v(r-r') \sum_k \phi_k^c(r',-\tau)\phi_k^c(r',\tau)\right] \phi_\ell^c(r,\tau) = -\frac{\alpha_\ell^c}{T}\, \phi_\ell^c(r,\tau) \tag{18a}$$

yielding a total width equal to the sum over partial widths for each distinct solution $\{\phi_i^c\}$ to eq.(18a)

$$\Gamma = \sum_c K^c e^{-\int d\tau\, dr \sum_k \phi_k^c(r,-\tau) \frac{\partial}{\partial\tau} \phi_k^c(r,\tau)} \tag{18b}$$

The salient uncertainties in this result discussed in the introduction are the absence of an explicit small parameter for the SPA and the justification for replacing the bare interaction by an effective interaction.

B) Monte Carlo Solution of the Many-Fermion Problem

The alternative functional integral representations of the many-fermion evolution operator discussed above also yield significantly different Monte Carlo problems. In the Feynman path integral, the number of stochastic variables equals the number of particles. All the other forms involve fields which must be defined on a spatial mesh and hence necessarily involve many more variables than particles. In the model problem discussed subsequently for example, one needs on the order of 20 mesh points in an interval occupied by a single particle for adequate accuracy, as well as many additional points at the edge of the nucleus to generate the exponential tail of the wave function, so even a ten-particle problem in one dimension involves over 200 degrees of freedom. A comparable problem in three dimensions involves 10^7 degrees of freedom rendering the study of hard core potentials impractical with present computers.

A second crucial consideration is the necessity of casting the integrand of the integral to be performed by Monte Carlo sampling into non-negative or at least dominantly positive form. This appears to be possible for the auxiliary field form in three dimensions, whereas at present no corresponding coordinate-space form has been developed which is suitable in more than one dimension.

Insertion of a complete set of antisymmetrized many-particle states $|x_1 \ldots x_A\rangle$ at each time step yields the following generalization of eq.(5) for the matrix element of the evolution operator for infinitesimal time ε

$$\begin{aligned}
&\langle x_1^{\,n} \ldots x_A^{\,n}| e^{-\varepsilon(H-E)} |x_1^{\,n-1} \ldots x_A^{\,n-1}\rangle \\
&= \sum_P \varepsilon^P \, e^{-\frac{m}{2\varepsilon} \sum_i (x^n_{Pi} - x_i^{n-1})^2 - \varepsilon\{\frac{1}{4} \sum_{ij} [v(|x_i^{n-1} - x_j^{n-1}|) + v(|x_i^n - x_j^n|)] - E\}} \\
&= e^{-\frac{m}{2\varepsilon} \sum_i (x_i^n - x_i^{n-1})^2} \det\left| e^{-\frac{m}{2\varepsilon}[(x_i^n - x_j^{n-1})^2 - (x_i^n - x_i^{n-1})^2]} \right| \times \\
&\times e^{-\varepsilon\{\frac{1}{4} \sum_{ij} [v(|x_i^{n-1} - x_j^{n-1}|) + v(|x_i^n - x_j^n|)] - E\}}
\end{aligned} \tag{19}$$

where the superscript on x denotes the time and the subscript indicates the particle. In one dimension, the signs in eq.(19) are totally innocuous. The sign of the matrix element at each step arising from the determinant is just the sign of the permutation required to bring the x^n's into the same order as the x^{n-1}'s. Hence the cummulative sign after any number of steps is path-independent and is the sign of the permutation required to make the final order equal to the initial order. It does not affect matrix elements of the evolution operator with antisymmetrized states or the trace of the evolution operator times a symmetric operator.

In higher dimension the sign is path-dependent. Two particles may interchange in a single step, generating a minus sign, or circle around each other by a series of small steps, each corresponding to a positive matrix element. The net result is nearly complete cancellation of positive and negative contributions from which it becomes exceedingly difficult to extract statistically significant information for large β.

For Monte Carlo calculations in one-dimension, it is convenient to work in the subspace $x_1 < x_2 < x_3 \ldots < x_A$. The determinant in eq.(19) already yields zero when any two coordinates are equal, and one in addition simply includes θ-functions forbidding all interchanges. Physically, this is equivalent to assuming an infinite potential at all boundaries $x_i = x_j$ which generates the nodes required by antisymmetry. In this domain the determinant may be replaced by the following product in the limit $\varepsilon \to 0$:

$$\det \left| e^{-\frac{m}{2\varepsilon}\left[(x_i^n - x_j^{n-1})^2 - (x_i^n - x_i^{n-1})^2\right]} \right| \underset{\varepsilon\to 0}{\longrightarrow} \prod_{i=2}^{A} \left(1 - e^{-\frac{m}{\varepsilon}(x_i^n - x_{i-1}^n)(x_i^{n-1} - x_{i-1}^{n-1})}\right) \tag{20}$$

yielding an action which may be sampled efficiently. As in the single-particle case, importance sampling may be introduced by multiplying and dividing by a trial function $\phi(x_1 \ldots x_A)$ and eq.(19) is replaced by

$$\langle x_1^n \ldots x_A^n | \phi e^{-\varepsilon(H-E)} \frac{1}{\phi} | x_1^{n-1} \ldots x_A^{n-1} \rangle = \sum_P \varepsilon^P e^{-\frac{m}{2\varepsilon} \sum_i \left(x_{Pi}^n - x_i^{n-1} - \frac{\varepsilon \frac{\partial}{\partial x_i^{n-1}} \phi(x_1^{n-1} \ldots x_A^{n-1})}{\phi(x_1^{n-1} \ldots x_A^{n-1})} \right)^2} \tag{21}$$

A convenient form of trial function, both for importance sampling and for the matrix elements arising in the initial value form, is the following

$$\phi(x_1 \ldots x_n) = \prod_{i<j} g(x_i - x_j) \prod_k f(x_k) \tag{22}$$

where g is odd and asymptotically approaches ± 1 and f localizes the density properly. Note this ansatz is different from the usual Jastrow form since the antisymmetry is included in the two-body rather than one-body component. Nuclear matter is calculated by placing A particles in a box of length L with periodic boundary conditions. For this case $g(x) = \sin\left(\frac{x}{L}\right)$ and $f(x) = 1$ which provides an efficient means of representing a Slater determinant. For finite nuclei $g(x) = \tanh(bx)$, $f(x)$ is a fermi function, and the parameters are adjusted to produce a reasonable density and two-body correlation function.

Thus far, the discussion has tacitly assumed a single spin state. For spin-independent forces and spin degeneracy M, the ground state energy and spin-independent observables may be calculated by including determinatal factors and forbidding interchanges only within each of the ordered subsets $\{x_1 \ldots x_{\frac{A}{M}}\}$, $\{x_{\frac{A}{M}+1} \ldots x_{\frac{2A}{M}}\}$, $\ldots \{x_{\frac{(M-1)}{M}A} \ldots x_A\}$. When the spin degeneracy, M, equals the number of particles, A, one obtains the boson ground state.

In order to convert the fission problem into a double-well problem, it is necessary to add a potential which becomes strongly repulsive when a collective variable corresponding to the distance between two fission fragments becomes large. For the symmetric fission of a four-particle system, a symmetric operator which accomplishes this is

$$U = \sum_{ijk\ell} \theta(x_\ell - x_j)\theta(x_\ell - x_i)\theta(x_k - x_j)\theta(x_k - x_i) U\left[\frac{x_k + x_\ell - x_i - x_j}{2}\right]. \tag{23}$$

In the ordered subspace, U is a function of $R = \frac{x_4+x_3-x_2-x_1}{2}$ which is just the desired separation of the two fragments. Equation (23) may be generalized straightforwardly to larger systems and arbitrary numbers and species of fragments by appropriate groupings of θ-functions.

As in the single-particle case, the energy is easily evaluated by a matrix element of the form in eq.(6). The kinetic energy operator acts upon the trial function eq.(22) and the resulting function is sampled stochastically. Center-of-mass motion may be treated either by working in the cm and evaluating the intrinsic hamiltonian, which requires the calculation of mixed derivatives of the form $\sum_{i\neq j} P_i P_j$ acting on Φ, or by adding a one-body harmonic oscillator potential acting on the cm coordinate and explicitly subtracting $\frac{1}{2}\hbar\omega$. When the trace is calculated, the energy is best evaluated using the virial theorem, eq.(12c). It is straightforward to verify for a two-body potential $v(|x_i - x_j|)$ and a potential of the form eq.(23) that the total energy is obtained by replacing $v(x)$ by $\hat{v}(x)=v(x)+\frac{1}{2}xv'(x)$ and $U(R)$ by $\hat{U}(R) \equiv U(R)+\frac{1}{2}RU'(R)$.

C) One-Dimensional Model

In order to construct a one-dimensional model as relevant as possible to nuclear physics, it is desirable to define a saturating system interacting with a two-body potential exhibiting the qualitative behavior of the nuclear force. To define the potential quantitatively, it is useful to require that relevant dimensionless ratios be comparable in one and three dimensions. The fundamental length scale in a saturating system is specified by the saturation density ρ_0, so we define

$$\ell_0 = \rho_0^{-1/D} \tag{24a}$$

where D is the dimension. To within uninteresting geometrical constants, ℓ_0 specifies the characteristic distance between particles. The Schrodinger equation may then be reduced to dimensionless form by measuring all lengths in units of ℓ_0 and energies in units of

$$E_0 = \frac{\hbar^2}{m\ell_0^2} \tag{24b}$$

The potential used in this work is graphed in Figure 1 and was defined such that the scaled binding energy per particle, $\frac{E/A}{E_0}$, fermi gas kinetic energy per particle $\frac{T/A}{E_0}$, core radius $\frac{r_0}{\ell_0}$, and maximum attraction $\frac{V_{max}}{E_0}$ are roughly comparable to the three-dimensional values shown in Table 1. The explicit form of the potential is

$$V(x) = \sum_{i=1}^{2} \frac{V_i}{\sigma_i\sqrt{\pi}} e^{-\frac{x^2}{\sigma_i^2}} \tag{25}$$

with $V_1=12$, $V_2=-12$, $\sigma_1=0.2$ and $\sigma_2=0.8$ in units such that $\hbar=m=1$ and $\rho_0 \sim 1.25$.

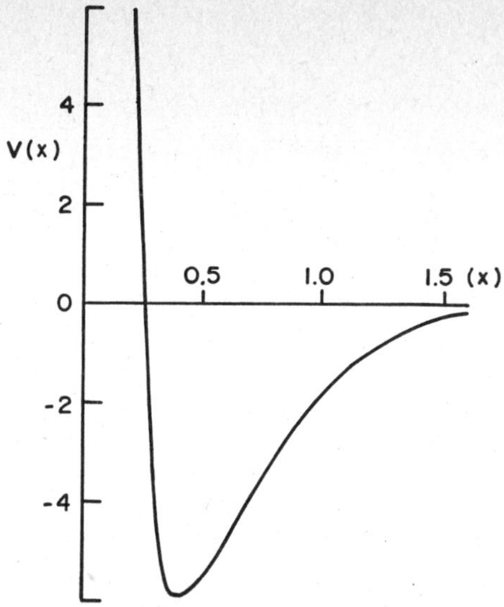

Fig. 1. Two-body potential, eq.(25), used in the one-dimensional model.

Table 1

Comparison of Characteristic Scaled Quantities in Nuclear Matter and the One-Dimensional Model

	D=3	D=1
$\frac{E/A}{E_0}$	-0.09	-0.07
$\frac{T/A}{E_0}$	0.13	0.17
$\frac{r_0}{\ell_0}$	1.25	0.98
$\frac{V_{max}}{E_0}$	0.56	0.39

A crucial consideration in practical calculations is extrapolation of observables as the time step $\varepsilon\to 0$. One can show that the symmetrized expression, eq.(5), used in this work is formally accurate to order ε^2, but one still needs to know how small ε must be in order for quadratic convergence to set in. Heuristically, since the gaussian factor in eq.(5) confines $\Delta x \sim \sqrt{\varepsilon}$ and the smallest characteristic scale of variation of the potential is 0.2, one would expect to see convergence at $\varepsilon \sim (0.2)^2 = 0.04$. Monte Carlo calculations of the binding energy of nuclear matter, obtained by placing 6 particles in a periodic cell, are shown in Figure 2 as a function of ε. One observes that the heuristic estimate $\varepsilon \sim 0.04$ is quite reasonable and that both the normalization energy and trial energy in the initial-value formulation yield consistent and suffici-

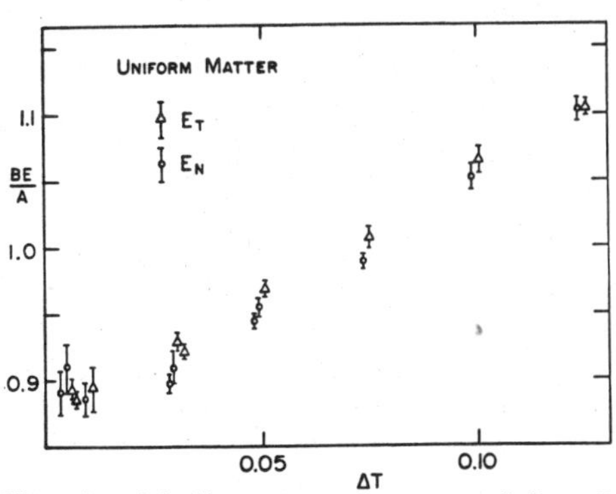

Fig. 2. Binding energy per particle of uniform matter as a function of time mesh size $\varepsilon = \Delta T$. The trial energy, E_T, and normalization energy, E_N, were calculated using 10^4 to 10^5 events.

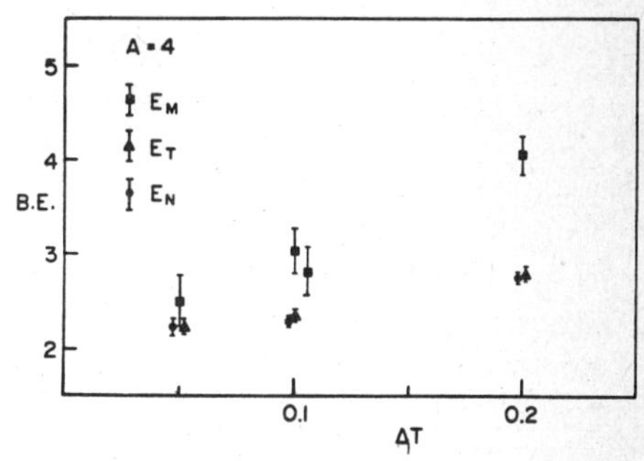

Fig. 3. Binding energy of 4-particle bound state as a function of time mesh size $\varepsilon = \Delta T$ using 10^4 events. The trial energy, E_T, and normalization energy, E_N, were calculated using the initial value form and the energy E_M was obtained using the Metropolis algorithm to evaluate the trace.

ently accurate results with of the order of 10^4 to 10^5 independent events. An analogous calculation of the binding energy of a four-particle bound state is shown in Figure 3 on a compressed scale using roughly 10^4 events. Whereas the variance and convergence with ε of the energies computed from the initial value form are comparable to those in Figure 2, one observes that evaluation of the trace by the Metropolis algorithm using the virial theorem energy expression yields variances roughly four times larger and somewhat less accuracy for large ε. The superior accuracy and variance of the initial value form is obvious from eq.(6). If Φ were the exact ground state, sampling the numerator and denominator with any set of x's would yield the exact energy with zero variance. To the extent to which Φ is a good approximation, it will thus compensate for error in the x's resulting either from statistical fluctuations or from a large value of ε. The importance-sampled version of the trace calculation would presumably yield comparable precision. The nuclear matter saturation curve obtained from a sequence of calculations analogous to that of Figure 2 is shown in Figure 4. In all these calculations, six particles were placed in a cell with periodic boundary conditions and the density was varied by changing the length of the cell. When the cell is made sufficiently large, the six particles form a bound state, so the low density limit is simply a constant equal to the binding energy per particle of a six-particle state. When more particles are included in the cell the surface energy contributes less. Hence, the low-density constant region approaches closer to the saturation point and in the infinite volume limit the saturation curve is constant below the saturation point. Calculated binding energies for 2 to 16 particles are well reproduced by a semi-empirical formula of the form $BE(A) \sim E_0 A - E_s$ where E_0 is the binding energy at saturation, -1.06, and the surface energy E_s is $\sim$1.7.

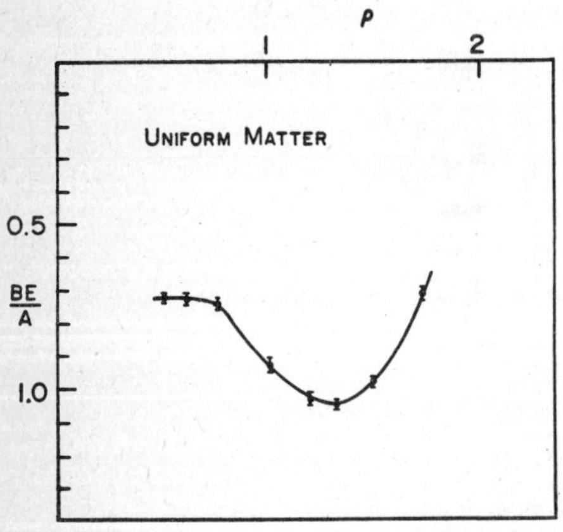

Fig. 4. Saturation curve calculated for six particles in a periodic cell.

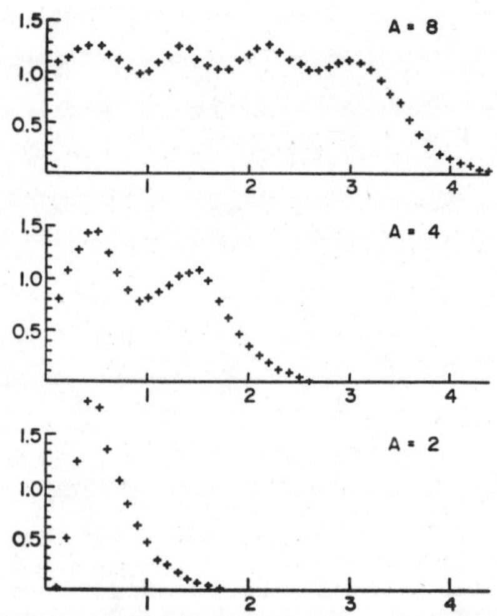

Fig. 5. Ground state density distributions for two, four, and eight-particle bound states.

In order to produce spontaneous fission, a long-range repulsive interaction must be added to the Hamiltonian analogous to the Coulomb interaction. In the present case, the range and strength were adjusted such that a twelve-particle system was stable with respect to single-particle emission and was unstable with respect to fission into two six-particle fragments.

Ground-state density distributions for three nuclei with this force are shown in Figure 5. Note that due to the long-range repulsion, the average interior density is slightly lower than the saturation density $\rho_0 = 1.25$ (defined without this repulsion). The quantum density fluctuations are especially interesting. In the two-body system, antisymmetry forces the density to vanish at the origin, and the size of the low-density region is increased by the repulsive core. In nuclei as large as A=8, the density fluctuations are comparable to those arising in the Hartree-Fock approximation, and a detailed comparison with mean-field theory would be most instructive.

Calculations of barrier penetration are at present still preliminary pending a thorough study of statistical ncertainties. In the case of ground state properties displayed in Figures 2-5, it was extremely valuable to have several independent calculations of the same observable so as to critically assess the accuracy and statistical errors of each separate calculation. The primary problem in the tunneling calculation is the fact that two grossly different energy scales entering into the problem simultaneously place very stringent demands on the calculation. On the one hand, as seen already in Figure 2, the strong repulsive core requires that the time step ε be very small. On the other hand, the collective motion through the barrier involves a large mass moving in a large double well potential so that calculation of the energy splitting requires the total time, $\beta=N\varepsilon$, to be large. Hence, N must be very large, on the order of 10^4 or 10^5. The fundamental problem with such a large time lattice is assuring that true equilibrium has been reached which is genuinely independent of the initial condition. The ultimate physics objective, of course, is to understand just this interplay between two-body correlations involving the large energy scale and the collective dynamics involving the low energy scale, and it is not surprising that this complicated physics question should ultimately translate into a correspondingly complicated stochastic calculation.

Straightforward modifications may be introduced into both the initial-value and trace calculations described above to improve the statistics relevant to the collective dynamics. In both cases, the key is to explicitly introduce the collective variable discussed in connection with eq.(23), which for symmetric fission is $R = \frac{2}{N}[\sum_{i=N/2+1}^{N} x_i - \sum_{i=1}^{N/2} x_i]$. In the initial-value form, it is convenient to introduce importance sampling into the numerator of the many-body generalization of eq.(10) using a function $\phi \sim e^{-\alpha R}$. This function generates a net drift of the collective motion from the right well to the left, thereby significantly improving the statistics.

In using the Metropolis algorithm to evaluate the trace, a convenient means of facilitating equilibration of the collective variable is to explicitly vary the variable R at each time step, as well as each of the individual coordinates x_i. As remarked in the discussion of the Metropolis algorithm, any change in the configuration accepted or rejected on the basis of the change in the action satisfies the reversibility condition, eq.(16), and thus yields the correct equilibrium distribution. Whereas it would take many individual random changes of each coordinate to move two fragments a significant distance from each other, it is clear that a much smaller number of random steps in R will fully explore this degree of freedom. Preliminary results to date show the expected qualitative behavior in the evolution of R. As in the case of a single particle in a double well, the trajectory of R is comprised of regions of fluctuations around the parent ground state and the two fragment scission point, connected by instanton-like paths through the fission barrier. It now remains to obtain a statistically reliable correlation function, eq.(14), from which the energy splitting may be determined, and to confirm the result with the corresponding initial-value evaluation using eq.(10).

D) Future Prospects

On the basis of these initial calculations, it is clear that the model problem described above can be solved to the required accuracy and compared in detail with mean field theory. Comparison of exact ground states with various mean field calculations is already a rich testing ground for approximations presently utilized in nuclear theory. One can define classes of Skyrme forces to reproduce the exact mass table and saturation density and then directly compare a number of observables. This should give significant insight, for example, into the role of the effective mass and compression modulus. More microscopic approximations, such as the Brueckner G-matrix, systematic truncation of the coupled-cluster hierarchy, RPA corrections, etc. may also be definitively explored. Given an acceptable ground state approximation, one may then compare the SPA result using the bounce equations eq.(18) with an exact solution to explore the validity of the SPA and the use of a ground state effective interaction. One may thus expect to obtain insight into a variety of approximations which are presently used but not understood.

Physics in three dimensions, of course, is much different and richer than in one dimension. It is well-known, for example, that one-dimensional systems don't undergo phase transitions, that mean field theory breaks down near the critical point, and that two-body collisions cannot yield dissipation. Given the scale of the computational problem in one dimension, however, it is not likely that treatment of realistic potentials in higher dimensions will be practical in the near future. Exact solutions to interesting problems do not come easily, so at this stage all I can suggest is to be patient, to think hard about the physics contained in the one-dimensional problem which is presently accessible to us, and to savor the insight thus obtained.

REFERENCES

1) A thorough review by the author including extensive references to recent work will appear in Rev. Mod. Phys., October 1982.

2) E. Abers and B. Lee, Phys. Rept. 9, 1 1973.

3) S. E. Koonin, Proceedings of Santa Barbara Workshop on Nuclear Theory, 1981 and private communication.

4) D. M. Ceperly and M. H. Kalos, Monte Carlo Methods in Statistical Physics, Ed. K. Binder, Springer Verlag NY 1979, p. 145.

5) G. Bertsch, Private Communication.

6) N. Metropolis, A. Rosenbluth, M. Rosenbluth, A. Teller and E. Teller, J. Chem. Phys. 21 (1953) 1087.

7) M. Creutz and B. Freedman, Ann. Phys. NY 132, 427 (1981).

8) Karlin and Taylor, Course in Stochastic Processes, Academic Press (1975).

Mean-Field Monte Carlo Method for Many-Body Ground States*

S. E. Koonin, G. Sugiyama, and H. Friedrich†

W. K. Kellogg Radiation Laboratory, Caltech, Pasadena California 91125 USA

ABSTRACT

A method is described for calculating the exact ground state energy of a many-body system, whether fermion or boson. The Hubbard-Stratonovich representation of the imaginary-time many-body propagator is used to express the ground-state energy as a ratio of two functional integrals. When discretized on a space-time mesh, the Monte-Carlo evaluation of this ratio is equivalent to many TDHF evolutions of the system in a random mean-field. The method is illustrated by application to a many-boson system in one dimension with a zero-range two-body interaction.

1. Introduction

Exact solutions of the many-body Schroedinger equation are of interest for several reasons. Among these are their uses as benchmarks against which to test approximation methods and as tests of a given Hamiltonian by comparison with experimental observables. To date, such solutions have been obtained for many-body ground-states using the Green's Function Monte-Carlo [1] (GFMC) or the related Path-Integral Monte-Carlo [2,3] (PIMC) approach. These methods have been applied with considerable success to liquid He [1], the electron gas [2], and three- and four-nucleon systems with state-independent potentials [3,4]. Although the GFMC and PIMC algorithms can provide a restricted description of fermion systems within the fixed-node approximation [1,2], the proper inclusion of the Pauli principle is a major difficulty in these methods and has, in fact, prevented their unrestricted application to nuclear systems with $A > 4$ or with state-dependent potentials. This is basically because the many-body wavefunction is described by a statistically evolving ensemble of configurations (specified by the coordinates of each nucleon). The Pauli principle, which enforces a spatially non-local relation between configurations which differ by the interchange

*Work supported in part by National Science Foundation grants PHY77-21602 and PHY79-23638.

†Fellow of the Heisenberg Program of the Deutsche Forschungsgemeinschaft

of a pair of nucleons, is thus difficult to implement in the simple local algorithms used to evolve the ensemble.

In this note, we discuss an alternative algorithm for finding the exact ground state energy of a many-body system. It is based on a Monte-Carlo evaluation of the Hubbard-Stratonovich representation of the imaginary-time many-body propagator. Bosons and fermions are treated on the same footing. Instead of an ensemble of configurations, the many-body wavefunction is represented by a set of single-particle wavefunctions evolving in a randomly fluctuating one-body potential. As a result, anti-symmetrization for fermions can be enforced exactly.

2. Method

The basic idea behind our method is to use the imaginary-time evolution operator to filter a trial wavefunction to the exact ground state Ψ. Thus, we write the ground-state energy E_0 for a system of A particles as

$$E_0 = \lim_{T\to\infty} \frac{\langle H\Phi | e^{-HT} | \Phi\rangle}{\langle \Phi | e^{-HT} | \Phi\rangle} . \tag{2.1}$$

Here Φ is, in principle, any trial wavefunction not orthogonal to Ψ, although in practice our method is tractable only if Φ is a symmetrized product of single-particle orbitals for bosons or a Slater determinant for fermions. As in the GFMC and PIMC methods, the efficiency of MFMC calculations will be enhanced if Φ closely approximates Ψ. The many-body hamiltonian H is assumed to contain the kinetic energy and a local two-body potential:

$$H = \sum_{i=1}^{A} \frac{p_i^2}{2m} + \tfrac{1}{2} \sum_{i\neq j=1}^{A} v(x_i - x_j). \tag{2.2}$$

(For simplicity, we write our equations assuming only one spatial dimension, x, and omit spin and isospin labels. We also put $\hbar$=1.)

Upon applying the Hubbard-Stratonovich transformation [5] to e^{-HT}, we can rewrite (2.1) as

$$E_0 = \lim_{T\to\infty} \frac{\int D[\sigma(x,t)]\, e^{\tfrac{1}{2}\int_0^T (\sigma,v\sigma)dt} \langle\Phi | U_\sigma | \Phi\rangle \dfrac{\langle H\Phi | U_\sigma | \Phi\rangle}{\langle \Phi | U_\sigma | \Phi\rangle}}{\int D[\sigma(x,t)]\, e^{\tfrac{1}{2}\int_0^T (\sigma,v\sigma)dt} \langle\Phi | U_\sigma | \Phi\rangle} . \tag{2.3}$$

Here, $\sigma(x,t)$ is a real field integration variable whose measure is defined in [5]. The symbol $(\sigma,v\sigma)$ denotes an inner product in which σ is thought of as a vector and v a matrix: $(\sigma,v\sigma) = \int dx\, dx'\sigma(x)v(x-x')\sigma(x')$. Finally, the evolution operator U_σ corresponds to evolution from t=0 to t=T under the time-dependent single-particle hamiltonian (in first-quantized notation) $h_\sigma = \frac{-1}{2m}\partial^2/\partial x^2 \pm \frac{1}{2}v(0) + \int dx'\, v(x-x')\sigma(x',t)$, where the -(+) refers to fermions (bosons). In essence, the Hubbard-Stratonovich transformation expresses the many-body propagator as the coherent superposition of an infinite number of one-body propagators, each involving a different one-body potential.

To evaluate (2.3) numerically, we bound space-time to a region $0 \le t \le T$, $|x| \le L/2$ and define an $(N+1)\times M$ mesh $t_i = (i-1)\Delta t$, $i=1,\cdots,N+1$ with $\Delta t = T/N$ and $x_j = (j-m/2-\frac{1}{2})\Delta x$, $j=1,\cdots,M$ with $\Delta x = L/M$ and M even. Of course, we suppose that Δx and Δt are sufficiently small. With the definition $\sigma_{ij} = \sigma(t_{i-\frac{1}{2}}, x_j)$, (2.3) can be discretized as:

$$E_0(T=N\Delta t) = \frac{\int D[\sigma_{ij}]\, e^{\frac{1}{2}\sum_{i=1}^{N}(\sigma,v\sigma)_i \Delta t} <\Phi|U_\sigma|\Phi> \dfrac{<H\Phi|U_\sigma|\Phi>}{<\Phi|U_\sigma|\Phi>}}{\int D[\sigma_{ij}]\, e^{\frac{1}{2}\sum_{i=1}^{N}(\sigma,v\sigma)_i \Delta t} <\Phi|U_\sigma|\Phi>}, \tag{2.4}$$

where the measure can be taken as $D[\sigma_{ij}] = \prod_{i=1}^{N}\prod_{j=1}^{M} d\sigma_{ij}$ (any overall constants cancel in the ratio), the inner product is $(\sigma,v\sigma)_i = \sum_{j=1}^{M}\sum_{k=1}^{M} \sigma_{ij}\, v_{jk}\, \sigma_{ik}(\Delta x)^2$ with $v_{jk} = v(x_j - x_k)$, and $U_\sigma = \prod_{i=1}^{N} U_i$ with U_i effecting the one-body evolution from t_i to t_{i+1} under the single-particle hamiltonian $h_i(x_j) = -D^2/2m + \sum_{k=1}^{M} v_{jk}\,\sigma_{ik}(\Delta x)$ (D^2 is the usual 3-point discretization of the second-derivative). We may omit the self-energy term, $\pm\frac{1}{2}v(0)$ from h as this results in a constant shift of the energy scale for the time evolution which does not affect the right-hand side of (2.4).

As is clear from the standard derivation of the Hubbard-Stratonovich transformation by discretizing the time evolution [5], any approximation for U_i must be accurate through $O(\Delta t^2)$; we have used the Crank-Nicholson formula familiar from TDHF calculations, $U_i \sim (1-h_i\Delta t/2)(1+h_i\Delta t/2)^{-1}$, whose effect on a single-particle wavefunction can be quickly evaluated by Gaussian elimination [6]. As is also clear from the usual derivation, the integrals over σ in (2.4) will not converge unless $(\sigma,v\sigma)_i$ is negative definite; i. e., the eigenvalues of v_{jk} are all less than zero. While this property is certainly not true in general, it can always be guaranteed to be so for a given v by adding an appropriate two-body interaction term to H. This makes the eigenvalues of the effective v_{jk} negative definite and shifts E_0 in a trivial way [7].

Equation (2.4) is in a form suitable for Monte-Carlo evaluation. To make this explicit, we rewrite it as

$$E_0 = \frac{\int D[\sigma_{ij}]\, W[\sigma_{ij}] \dfrac{<H\Phi | U_\sigma | \Phi>}{<\Phi | U_\sigma | \Phi>}}{\int D[\sigma_{ij}]\, W[\sigma_{ij}]} ; \qquad (2.5a)$$

$$W[\sigma] = e^{\frac{1}{2}\sum_{i=1}^{N} (\sigma, v\sigma)_i \Delta t} <\Phi | U_\sigma | \Phi> . \qquad (2.5b)$$

Thus, E_0 is the average of $<H\Phi | U_\sigma | \Phi>/<\Phi | U_\sigma | \Phi>$ over all field configurations, weighted by $W[\sigma]$. To actually compute E_0, we use the well-known algorithm of Metropolis et al. [1] to generate an uncorrelated sequence of σ- fields distributed according to W and then simply average $<H\Phi | U_\sigma | \Phi>/<\Phi | U_\sigma | \Phi>$ over these configurations. Of course, this requires that $W[\sigma]$ be positive definite, which can be shown to always be the case for a symmetrized product trial boson state or for a spin and/or isospin symmetric deterimental trial state for fermions interacting through a spin-isospin independent potential. If $W[\sigma]$ is not positive definite, $|W|$ can be used as the weight and the sign $W/|W|$ appended to the energy contribution from each configuration. However, even here W must be predominantly of one sign for the denominator in (2.5a) to remain large and good statistical accuracy be achieved. While we have no guarantee that W be well-behaved in the general case, results for fermion systems treated by other methods offer some encouragement on this point [7].

Our general method is in some respects similar to one used in Monte-Carlo simulations of relativistic field theories [8]. There, the fermion degrees of freedom are "integrated out", leaving only a boson theory with an effective action (analogous to our $W[\sigma]$). However, the principal problem in such calculations is the evaluation of the enormous determinant (of dimension equal to the number of lattice sites) appearing in the effective action, essentially due to the presence of the filled Dirac sea. For the non-relativistic fermion systems we propose to treat (such as light nuclei), the dimensionality of the determinant required to compute $<\Phi | U_\sigma | \Phi>$ is relatively small and its direct evaluation is possible. This also emphasizes the advantage of the MFMC method over the GFMC or PIMC methods in that positive and negative contributions to the norm are canceled exactly rather than statistically.

3. Model Calculation

For a first investigation of the MFMC method, we have chosen a system of A bosons of mass m in one-dimension interacting with each other through an attractive zero-range potential: $v(x-x') = -V_0\delta(x-x')$. The hamiltonian is thus

$$H = \sum_{i=1}^{A} \frac{p_i^2}{2m} - \tfrac{1}{2}V_0 \sum_{i\neq j} \delta(x_i - x_j) + \tfrac{1}{2} mA\Omega^2(\sum_i x_i/A)^2, \tag{3.1}$$

where we have confined the center-of-mass motion in an oscillator potential of frequency Ω. This additional term adds both one- and two-body parts to H, and it is easy to show that the effective $(\sigma, v\sigma)$ is negative definite provided $\Omega<\sqrt{6AV_0/mL^3}$ The oscillator frequency is also bounded from below by the requirement that the oscillator length, $\sqrt{1/mA\Omega}$, be smaller than the mesh size, L, so that the mesh boundaries have no affect on the solution. When this is satisfied, we are free to choose zero boundary conditions for the single-particle wavefunctions at $x = \pm L/2$.

The hamiltonian (3.1) is exactly soluble [9]; the eigenvalue of its only bound state is $E_0 = \tfrac{1}{2}\Omega - A(A-1)(A+1)V_0^2m/24$. It can also be solved in the Hartree approximation provided that the oscillator term (which gives only a very small contribution to the Hartree energy for the parameters we have used) is not treated self-consistently. In this latter case, the energy is found to be $E_H = \pi^2 m\Omega^2 b^2/24A - A(A-1)^2V_0^2m/24$ and the Hartree wavefunction is $\Phi(x_1,\cdots,x_A) = \prod_{i=1}^{A}\varphi(x_i)$ with $\varphi(x) = \sqrt{b/2}/\cosh bx$ and $b = mV_0(A-1)/2$. This is the trial function we used in our MFMC calculations. We also chose to work in physical units appropriate to nuclear systems, with $\hbar^2/m = 41.47 MeV{-}fm^2$ and took $V_0 = 41.47 MeV{-}fm$.

We have performed MFMC calculations of $E_0(T)$ for several different numbers of particles. The Metropolis et al. algorithm was implemented so that the entire σ -field over all space at one particular value of time was updated simultaneously; since only one single-particle wavefunction, $\varphi(x,t)$, is involved, an efficient algorithm could then be used for evaluating the change in W due to each update. Care was taken to ensure that the successive field configurations used for the energies $\langle H\Phi|U_\sigma|\Phi\rangle/\langle\Phi|U_\sigma|\Phi\rangle$ were sufficiently uncorrelated with each other so that the energies were statistically independent. (For an acceptance ratio of about 0.5, every 20'th Metropolis sweep of the mesh. The initial field configuration was taken to be the time-independent Hartree field; it was "thermalized" by some 1000 sweeps before the calculation of the energy began.) A mesh of 30 spatial points and up to 160 time points was used; we verified that our results do not depend significantly on either Δt or Δx when these parameters are small enough and that changes in Ω shift the large-T value of E_0 in the expected way. A typical calculation of some 60 time steps took 4 hours of CPU time on a VAX 11/750 without floating point accelerator (about 5 minutes on a CDC 7600).

Results for the A=6, 10, and 20 systems are shown in Figures 1-3, where we plot $E_0(T)$. In all cases, Ω was 25 MeV. The energies shown are the results of averaging over some 200 field configurations. For A=6, we used Δx = 0.15 fm, for A=10, Δx was 0.10 fm, and for A = 20, Δx was 0.04 fm. As can be seen from the figures, reasonable variations in the time step do not change our results. $E_0(T)$ shows an initial relaxation and asymptotically approaches a value which fluctuates around the expected result for each A. The convergence becomes more rapid with increasing A. This is due to the nature of the spectrum of excited states of our model, as can be seen from Figure 4, where we plot for A=10 the logarithm of the difference between $E_0(T)$ and its asymptotic value. Two different relaxation scales are clearly seen. The rapid initial relaxation corresponds to the energy gap between the intrinsic ground state and the excited continuum (at zero intrinsic energy); this is indicated by the dotted line. Note that this energy gap increases with increasing A. The slower relaxation is that associated with the center-of-mass motion in the harmonic oscillator potential.

4. Discussion

A number of considerations are important if more realistic systems are to be treated with the MFMC method. One of these is an adequate but tractable choice for the trial wavefunction since the method clearly becomes more efficient as the trial state approaches the true ground state. A strong repulsive core in the two-body potential (such as that between two He atoms or two nucleons) means that an independent-particle wavefunction is a poor approximation to the exact eigenstate, but our method is computationally feasible only for such trial wavefunctions; precisely how much of a penalty this implies remains to be investigated. Another important consideration is one of sheer computing power. Realistic systems require a spatial grid fine enough to resolve the details of the short-range few-body correlations in the wavefunction yet large enough to adequately enclose the system. Moreover, each single-particle wavefunction must be stored at every point in space time and determinants of moderate size must be evaluated often for fermion systems.

The results presented above are an encouraging demonstration that the MFMC method can be applied to describe the ground state energy of a simple many-boson system; it is particularly noteworthy that the computational effort for such a boson system does not increase with the number of particles. The proper treatment of fermion systems, the primary motivation for studying the MFMC method, entails no foreseeable problems beyond those we have already faced and calculations along these lines for finite systems are in progress. Also of interest would be calculations of infinite systems, such as "nuclear matter", obtained by imposing periodic boundary conditions on a system confined within a fixed region of space.

References

[1] D. M. Ceperley and M. H. Kalos in *Monte-Carlo Methods in Statistical Physics*, K. Binder, ed. (Springer-Verlag, N. Y., 1979) p.145.

[2] D. M. Ceperley and B. Alder, Phys. Rev. Lett. **45**, 566 (1980).

[3] J. G. Zabolitsky and M. H. Kalos, Nucl. Phys. **A356**, 114 (1981).

[4] G. Maddison, Y. Alhassid, K. Chow, and S. E. Koonin, to be published.

[5] S. Levit, Phys. Rev. **C21**, 1594 (1980).

[6] P. Bonche, S. E. Koonin, and J. W. Negele, Phys. Rev. **C13**, 1226 (1976); R. Varga, *Matrix Iterative Analysis* (Prentice-Hall, Englewood Cliffs, 1962) p. 195.

[7] R. Blankenbecler, D. J. Scalapino, and R. L. Sugar, Phys. Rev. **D24**, 2278 (1981).

[8] See, for example, J. Kogut et al., Phys. Rev. Lett. **48**, 1140 (1982) and references cited therein.

[9] B. Yoon and J. W. Negele, Phys. Rev. **A16**, 1451 (1977).

FIGURE 1: ENERGY PLOT A=6

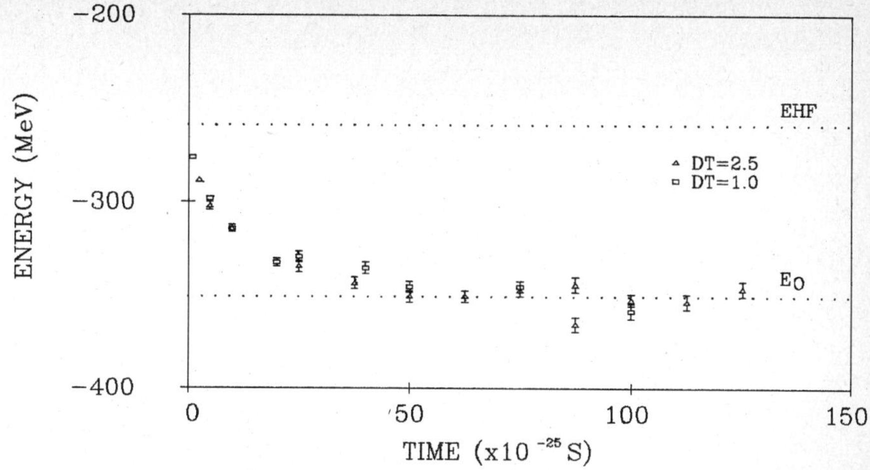

FIGURE 2: ENERGY PLOT A=10

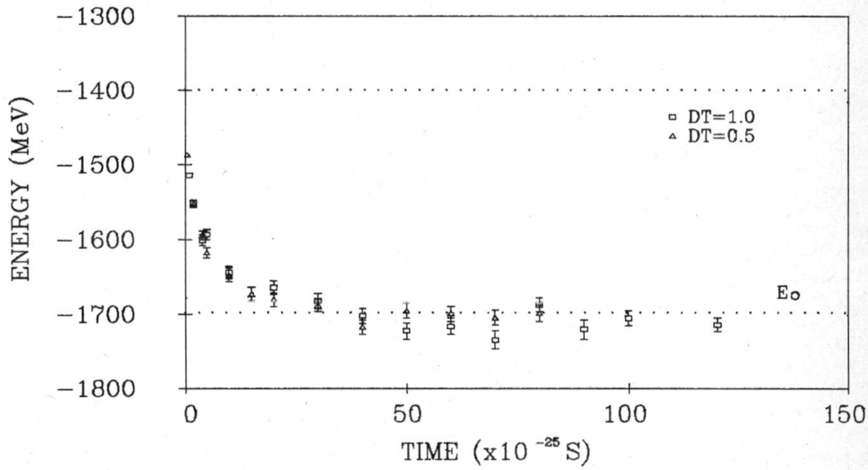

FIGURE 3: ENERGY PLOT A=20

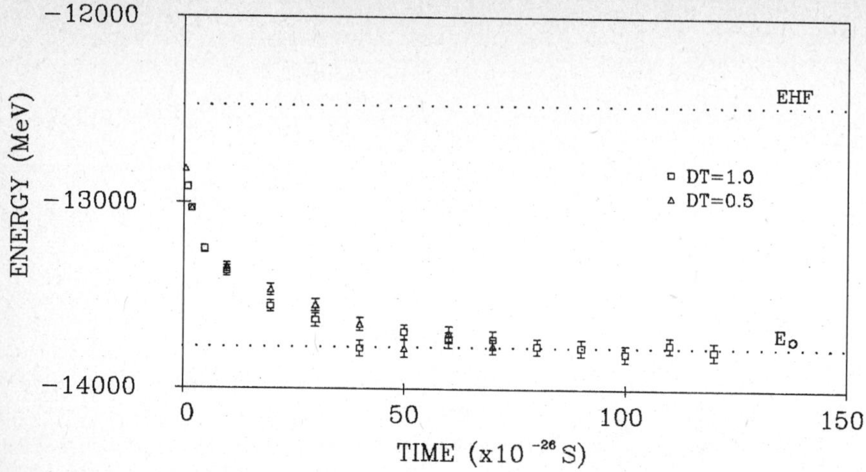

FIGURE 4: ENERGY GAP PLOT A=10

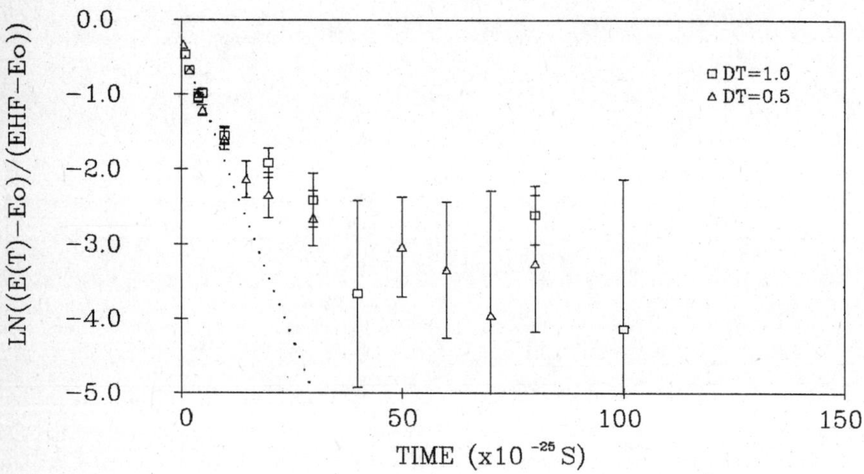

1/Degeneracy Expansion of Collective Quantum Fields
and
Higher Effective Actions

H. Kleinert
Institut für Theoretische Physik
Freie Universität Berlin
1000 Berlin 33, Arnimallee 14

I Introduction

The low lying collective excitations of complicated many-body systems display, to a certain approximation, quasiclassical behavior. This means that the action of "characteristic" collective fluctuations is large with respect to $\hbar$ such that their presence is suppressed by a "quantum Boltzmann factor" $e^{i\mathcal{A}/\hbar}$. I would like to point out two alternative systematic ways of expanding physical amplitudes according to fluctuation sizes. One has a somewhat restricted range of applications and is useful only for model systems with large degeneracy. The other one is quite general and appropriate to the treatment of realistic physical nuclei which are characterized by strong exchange forces as well as pairing and higher cluster formation, for example alpha particle clusters.

If a system is described by a single Bose quantum field $\varphi(x) = \varphi(\mathbf{x},t)$ all properties of the system are contained in the set of all Green's functions[1]

$$G^{(n)}(x_1,\ldots,x_n) \equiv \langle 0|T(\varphi(x_1)\cdot\ldots\cdot\varphi(x_n))|0\rangle \tag{1}$$

where $\varphi(x)$ is the fully interacting Heisenberg field and $|0\rangle$ the ground state. These Green's functions can be studied simultaneously by considering the single generating functional

$$Z[j] = \langle 0|T e^{\frac{i}{\hbar}\int dx\, j(x)\varphi(x)}|0\rangle \tag{2}$$

whose functional derivatives are equal to $G^{(n)}$:

$$G^{(n)}(x_1,\ldots,x_n) = Z[j]^{-1} \frac{\hbar}{i} \frac{\delta}{\delta j(x_1)} \cdots \frac{\hbar}{i} \frac{\delta}{\delta j(x_n)} Z[j] \Big|_{j=0} \tag{3}$$

Such an object has a simple, quasiclassical limit which can efficiently be evaluated by using the path integral representation

$$Z[j] = \int \mathcal{D}\varphi(x)\, e^{\frac{i}{\hbar}\left(\mathcal{A}[\varphi] + \int dx\, j(x)\varphi(x)\right)} \tag{4}$$

where $\mathcal{A}[\varphi]$ is the action and $e^{\frac{i}{\hbar}(\mathcal{A}[\varphi] + \int dx\, j\varphi)}$ the "quantum Boltzmann factor" for each field configuration in the presence of the external current. If one formally takes the limit $\hbar \to 0$, this integral is dominated by the saddle point which leads to the classical field equation

$$\left[\frac{\delta}{\delta\varphi(x)} \mathcal{A}[\varphi] + j(x)\right]_{\varphi=\varphi_{cl}(x)} = 0 \tag{5}$$

and the partition function

$$Z[j] \sim Z_{cl}[j] = e^{\frac{i}{\hbar}\left(\mathcal{A}[\varphi_{cl}] + \int dx\, j(x)\varphi_{cl}(x)\right)} \tag{6}$$

Quantum corrections can be calculated by setting[2]

$$\varphi(x) = \varphi_{cl}(x) + \sqrt{\hbar}\,\varphi'(x) \tag{7}$$

and expanding the action $\mathcal{A}[\varphi]$ in powers of $\varphi'(x)$:

$$Z[j] = Z_{cl}[j] \cdot \int D\varphi' \exp i \left\{ \int dx_1 dx_2 \, \varphi'(x) \frac{\delta^2 A[\varphi_{cl}]}{\delta\varphi_{cl}(x_1)\delta\varphi_{cl}(x_2)} \varphi'(x_2) \right.$$

$$\left. + \hbar^{1/2} \int dx_1 dx_2 dx_3 \frac{\delta^3 A[\varphi_{cl}]}{\delta\varphi_{cl}(x_1)\delta\varphi_{cl}(x_2)\delta\varphi_{cl}(x_3)} \varphi'(x_1)\varphi'(x_2)\varphi'(x_3) + \ldots \right\} \quad (8)$$

$$= Z_{cl}[j] + \hbar \, Z^{(1)}[j] + \hbar^2 \, Z^{(2)}[j] + \ldots$$

The quadratic piece in $\varphi'(x)$ can be considered as a new propagator and the higher powers $\hbar^{\frac{n}{2}-2} \varphi'^{\,n}$ as interaction terms. Calculating the associated Feynman graphs leads to the expansion of $Z[j]$ in powers of $\hbar$.

Moreover, if time dependent oscillating solutions are known, it is straightforward to find the quantized energy level and their quasi-classical corrections. To lowest order in $\hbar$, this goes as follows: Consider a family of such orbits $\varphi^T_{cl}(x)$ differing only by the period T. Denote $\frac{2\pi}{T}$ by ω and define a Lagrangian-like quantity

$$L(\omega) \equiv \mathcal{A}[\varphi_{cl}^T]/T \quad (9)$$

as a function of ω. The variable ω may be considered as the velocity-like Lagrange coordinate $\dot{q}$. Consequently we define

$$p(\omega) \equiv \frac{\partial L(\omega)}{\partial \omega} \;\hat{=}\; \frac{\partial L}{\partial \dot{q}} \quad (10)$$

as the associated canonical momentum and the Legendre transform

$$E(p) = p\omega - L(\omega) \;\hat{=}\; p\dot{q} - L \quad (11)$$

as the energy variable. The quasiclassical quantization rule is then

$$\rho(\omega_n) = 2\pi n \hbar \tag{12}$$

(or $2\pi(n+\frac{1}{2})\hbar$ if the orbits have turning points) and the energy levels are found as

$$E_n = p(\omega_n) \cdot \omega_n - L(\omega_n) \tag{13}$$

For multiply periodic orbits, the other frequencies are treated in complete analogy. These statements can be proved directly from the path integral[3].

In recent years, semiclassical expansion methods around the solution of field equations have been applied also to collective quantum fields[4]. Such fields appear in fermion systems if the fermion degrees of freedom are eliminated in favor of particle-hole or particle-particle composite fields[5] by a simple trick of quadratic completion[6]. For a typical fermion Lagrangian

$$Z = \int D\psi D\psi^{+} \exp\left[i \left\{ \int dx\, \psi^{+}(x)(i\partial_t - \xi)\psi(x) - \frac{1}{2}\int dxdx'\, V(x,x')\psi^{+}(x)\psi(x)\psi^{+}(x')\psi(x') \right\} \right] \tag{14}$$

one may, for example, introduce a trivial multiplicative complete square involving the fermionic density

$$(\det V)^{1/2} \int D\rho\, e^{\frac{i}{\hbar}\frac{1}{2}\int dxdx'\, V(x,x')(\rho(x)-\psi^{+}(x)\psi(x))(\rho(x')-\psi^{+}(x')\psi(x'))} \tag{15}$$

after which the fermion integral becomes quadratic, resulting in the partition function of a fermion system without two-particle interactions but moving in an average external field determined by $\rho(x)$:

$$Z_1[\rho] \equiv e^{\frac{i}{\hbar}\mathcal{A}'[\rho]} = \int \mathcal{D}\psi \mathcal{D}\psi^\dagger \exp\Big\{ i \int dx\, \psi^\dagger(x) \Big(i\partial_t - \xi - \int dx'\, V(x,x')\rho(x') \Big) \psi(x) \Big\} \tag{16}$$

Here, the integral over the Fermi fields can easily be performed, resulting in the fermionic action

$$\mathcal{A}'[\rho] = -i\hbar\, \mathrm{tr} \log \Big(i\partial_t - \xi - \int dx'\, V(x,x')\rho(x') \Big)$$

with the symbol tr denoting the functional trace in all field labels x. What remains is a purely collective formulation of the original fermionic theory in terms of the density field $\rho(x)$:

$$Z_1 = (\det V)^{1/2} \int \mathcal{D}\rho(x)\, e^{\frac{i}{\hbar}\left\{ \mathcal{A}'[\rho] + \frac{1}{2}\int dx\, dx'\, V(x,x')\rho(x)\rho(x') \right\}} \tag{17}$$

This path integral contains the same physical information as (14), albeit in another field language. Alternatively one can complete the square using a bilocal density

$$\exp\Big\{ -\frac{i}{\hbar}\,\frac{1}{2} \int dx\, dx'\, V(x,x') \big| \rho(x,x') - \psi^\dagger(x)\psi(x') \big|^2 \Big\} \tag{18}$$

or a pair field

$$\exp\Big\{ \frac{i}{\hbar}\,\frac{1}{2} \int dx\, dx'\, V(x,x') \big| \Delta(x,x') - \psi(x)\psi(x') \big|^2 \Big\} \tag{19}$$

In this case, the collective action redescribes the many fermion system in terms of either of these two collective fields.

Finally, it is possible to choose an arbitrary complete square[5]

$$
\begin{aligned}
&(\det V_1)^{1/2}(\det V_2)^{1/2} \int D\rho(x_1x_2)D\Delta(x_1x_2)D\Delta^+(x_1x_2) \\
&\exp \frac{i}{2\hbar} \{ \int dx_1dx_2dx_1'dx_2' \, V_1(x_1x_2x_1'x_2')(\rho(x_1x_2)-\psi^+(x_1)\psi(x_2))(\rho^+(x_1'x_2')-\psi^+(x_1')\psi(x_2')) \\
&+ \int dx_1dx_2dx_1'dx_2' \, V_2(x_1x_2x_1'x_2')(\Delta(x_2x_2')-\psi(x_2)\psi(x_2'))(\Delta^+(x_1'x_1)-\psi^+(x_1')\psi^+(x_1))\}
\end{aligned}
\tag{20}
$$

as long as we require $V_1+V_2 = [\alpha\delta(x_1-x_2)\delta(x_1'-x_2')V(x_1,x_2)]-\beta[x_1\leftrightarrow x_1']-(1-\alpha-\beta)[x_2\leftrightarrow x_2']$. such that the quartic four fermion interaction is eliminated in favor of quadratic terms.

Historically, this type of approach has been extremely useful for the study of second order phase transitions in several physical systems, such as superconductors and superfluid ^{3}He.[5] In these systems, an attractive interaction leads to the formation of Cooper pairs and therefore the pair field (19) appears in the collective version

Hand in hand went the application of this approach to many soluble field theoretic models in two space-time dimensions (Schwinger model, Thirring model, degenerate shell model[7]) and collective quantum fields have opened up new ways of describing and understanding their structural properties.

As a matter of fact, there is a whole new class of soluble models which can be constructed by means of collective quantum fields: It consists of all field theories in which the same field occurs many times, say N, with N tending to infinity[8]. In this case the action has the generic form

$$
\begin{aligned}
A &= \int dx \sum_{n=1}^{N} \psi_n^+(x)(i\partial_t - \xi)\psi_n(x) \\
&- \frac{1}{N}\frac{1}{2}\int dx\,dx' \, V(x,x') \sum_{n=1}^{N} \psi_n^+(x)\psi_n(x) \sum_{n=1}^{N} \psi_n^+(x')\psi_n(x')
\end{aligned}
\tag{21}
$$

where the $\frac{1}{N}$ factor ensures the existence of the limit $N \to \infty$. This interaction can be eliminated by the complete square

$$(\det V)^{N/2} \int D\rho(x)\, e^{\frac{i}{\hbar}\frac{N}{2}\int dxdx' V(x,x')(\rho(x)-\frac{1}{N}\sum_{n=1}^{N}\psi_n^+(x)\psi_n(x))(\rho(x')-\frac{1}{N}\sum_{n=1}^{N}\psi_n^+(x')\psi_n(x'))} \tag{22}$$

Due to the complete generacy of the N levels, the fermion integral is the same for everyone of the $n = 1, 2, 3, \ldots$ fields ψ_n and the collective path integral becomes

$$Z_1 = (\det V)^{N/2} \int \mathcal{D}\rho\, e^{N\frac{i}{\hbar}\mathcal{A}_{coll}[\rho]} \tag{23}$$

$$\equiv (\det V)^{N/2} \int \mathcal{D}\rho\, e^{N\frac{i}{\hbar}\left(\mathcal{A}'[\rho] + \frac{1}{2}\int dxdx' V(x,x')\rho(x)\rho(x')\right)}$$

Now we see that the number N appears as an overall factor such that in the limit of large N the path integral is dominated by the saddle point in just the same way as the previous fundamental field theory (4) in the quasiclassical limit $\hbar \to 0$. Thus, in the collective quantum field, the quantity $\frac{1}{N}$ plays the same role as the expansion parameter $\hbar$ in (6). In the limit $N \to \infty$, the partition function is given by the solution of the field equation

$$\frac{\delta}{\delta\rho}\mathcal{A}_{coll}[\rho]\Big|_{\rho=\rho_\infty} = \frac{\delta}{\delta\rho}\Big[-i\hbar\, \mathrm{tr}\log\left(i\partial_t - \xi - \int dx' V(x,x')\rho(x')\right) + \frac{1}{2}\int dxdx' V(x,x')\rho(x)\rho(x')\Big]_{\rho=\rho_\infty} = 0 \tag{24}$$

resulting in

$$Z_1 \underset{N\to\infty}{\longrightarrow} Z_\infty \equiv \det V^{N/2}\, e^{N\frac{i}{\hbar}\mathcal{A}_{coll}[\rho_\infty]} \tag{25}$$

The corrections in $\frac{1}{N}$ can be obtained by expanding the exponent in powers of $\rho' \equiv (\rho - \rho_{00})/\sqrt{N}$ and calculating Feynman graphs using the quadratic part of ρ' as an inverse propagator and the higher powers $\left(\frac{1}{N}\right)^{\frac{n}{2}-2} \rho'^{\,n}$ as vertices.

There can also be the generic interaction forms

$$e^{-\frac{i}{\hbar}\frac{1}{2N}\int dx\,dx'\,V(x,x')\left|\sum_{n=1}^{N}\psi_n^{+}(x)\psi_n(x')\right|^2} \tag{26}$$

$$e^{-\frac{i}{\hbar}\frac{1}{2N}\int dx\,dx'\,V(x,x')\left|\sum_{n=1}^{N}\psi_n(x)\psi_n(x')\right|^2} \tag{27}$$

in which case one obtains collective bilocal density fields

$$\rho(x,x') \sim \frac{1}{N}\sum_{n=1}^{N}\psi_n^{+}(x)\psi_n(x') \tag{28}$$

or collective pair fields

$$\Delta(x,x') \sim \frac{1}{N}\sum_{n=1}^{N}\psi_n(x)\psi_n(x') \tag{29}$$

In each case, the limit $N \to \infty$ leads to an exact solution of this theory via the extremum. Moreover, periodic solutions can be quantized by rules like (12).

Fascinated by the successes[5,7], several groups have applied collective quantum fields to realistic nuclear problems[9-11]. They immediately ran, however, into difficulties. These can be traced back to the amgiguity in the choice of the collective field (15), (18), or (19). In nuclear problems, there are usually important correlations in both $\rho(x)$ and $\rho(x,x')$, due to the presence of exchange forces, and often also in $\Delta(x,x')$ if there are strong pairing forces. In fact, calculations based only on a density field lead, in lowest order, to Hartree equations, which often fail to reproduce experimental results. This is why inclusion of $\rho(x,x')$ becomes necessary, which is achieved via the Hartree-Fock equation. The pairing field, finally, is taken into account by means of the Hartree-Fock-Bogoljubov (HFB) equations. By exploiting the freedom in introducing the collective fields it is in principle possible[5] to choose V_1 and V_2 such that to lowest order in V all three forces are properly taken into account. The special natural role of the HFB equations as a lowest approximation, however, reveals itself only after a study of fluctuations of ρ and Δ fields. These are quite different for different choices of V_1 and V_2. The HFB fields mini- mize the fluctuation corrections to lowest order in V.[10] Within the collective field approach this requirement is quite unnatural. In order to see this consider the collective field theory based on the pure density field $\rho(x)$. Within this theory the natural lowest correction is of the one-ρ loop type which from (17) can be seen to be

$$\left[\det V(1-VG_\rho G_\rho)\right]^{-1/2}$$

where G_ρ are the propagators of the Fermi field moving in the presence of the average background field ρ, as determined by (16). The first factor V cancels the $(\det V)^{1/2}$ in front of (17). Expanding the remainder in powers of V gives

$$\exp\left\{-\frac{1}{2}\sum_{n=1}^{\infty}\frac{1}{n}\left(VG_\rho G_\rho\right)^n\right\} \tag{30}$$

The n=1 term consists of the exchange part of the potential energy V but calculated within the Hartree basis. If V has the property that its exchange part is zero there is no such correction.

Now, the Hartree-Fock-Bogoljubov field is distinct by the fact that n=1 fluctuation correction for the general collective fields (20) vanishes. Those for n $\geq$ 2, however, will in general be non-zero. Since the potential V is in no way small, there is no special reason to select the n=1 term over the others.

It must be stressed that restricting the freedom in the choices of V_1 and V_2 is of crucial importance if one wants to extract physical information from approximations which always begin with mean fields. While mathematically any choice V_1 and V_2 leads to the same theory, this can in general, be guaranteed only due to violent fluctuations of ϱ and Δ fields. Approximating ϱ and Δ by their mean values and small fluctuations around these may lead to results which have nothing to do with physical properties of the system. In general one can say that integrating out, in a partition function, a subset of degrees of freedom (and this is what we have done when going to ϱ and Δ fields) can create completely fictitious spectra in the remaining fluctuating fields. A simple mathematical model can illustrate this point. Consider the partition function of a system with two degrees of freedom, both of energy ε:

$$Z = \int \frac{dx}{\sqrt{2\pi\varepsilon}} \frac{dy}{\sqrt{2\pi\varepsilon}} e^{-\varepsilon \frac{x^2+y^2}{2}} = 1 \tag{31}$$

Integrating out the y variable leaves

$$Z = \int \frac{dx}{\sqrt{2\pi\varepsilon}} e^{-\varepsilon \frac{x^2}{2}} \tag{32}$$

which is a system with only one degree of freedom of energy ε. Thus only one of the physical frequencies can be recovered from the second form of the partition function.

Let us now look at the same partition function but change variables from x, y to ξ, η:

$$x = \mathrm{ch}\,\Theta\, \xi + \mathrm{sh}\,\Theta\, \eta$$

$$y = \mathrm{sh}\,\Theta\, \xi + \mathrm{ch}\,\Theta\, \eta$$

Then it reads

$$Z = \int \frac{d\xi}{\sqrt{2\pi\varepsilon}} \frac{d\eta}{\sqrt{2\pi\varepsilon}} e^{-\frac{\varepsilon}{2} \mathrm{ch}\,\Theta\, (\eta + \mathrm{th}\, 2\Theta\, \xi)^2 - \frac{\varepsilon}{2} \frac{1}{\mathrm{ch}\,\Theta} \xi^2} \tag{33}$$

If we now integrate out the η variable we obtain

$$Z_1 = \int \frac{d\xi}{\sqrt{2\pi\varepsilon}} \frac{1}{\sqrt{\mathrm{ch}\, 2\theta}} e^{-\frac{\varepsilon}{2}\frac{1}{\mathrm{ch}\, 2\theta}\xi^2} \qquad (34)$$

and see that the remaining fluctuation in ξ has frequency $\varepsilon/\mathrm{ch}\, 2\theta$ which has nothing to do with either of the physical frequencies. Certainly, the final integral over ξ leads to the same total partition function (31)

$$Z = 1$$

but the intermediate "action"

$$\mathcal{A}[\xi] = \frac{\varepsilon}{2}\frac{1}{\mathrm{ch}\, 2\theta} - \frac{1}{2}\log \mathrm{ch}\, 2\theta \qquad (35)$$

gives no insight into the physical fluctuations.

From (33) we see that integrating out part of the field variables has the effect of producing constraints on the remaining fields which, in general, create a new spectrum, completely different from the physical one[13].

Thus in realistic nuclear applications, collective quantum fields are generally of little use+. Notice that these difficulties do not occur in the above described successful applications of the method which were selected such that at least one of the following conditions was satisfied:

+Unless one decides to sum over many collective field fluctuations, brute force on the computer, via Monte Carlo calculations, as has been done recently for the ground state energy of exactly soluble models (see S. Koonin's and J. Negele's lectures at this meeting).

1) Close to a second order phase transition, only the collective field which destabilizes at the critical point is a relevant variable. All others are irrelevant.[5]
2) In soluble models, the choice of the collective field is irrelevant since any of the three forms gives an exact reformulation of the same partition function.[7]
3) In the limit $N \to \infty$, one channel is always uniquely selected.[8]

Thus we arrive at the following alternative:

Either we must restrict our attention to model systems which consist of a highly degenerate or at least almost degenerate shell. Then the degeneracy can play the role of N and 1/N=1/degeneracy becomes a good expansion parameter for the collective quantum fields, which leads to a systematic sequence of approximations.

Or we must develop another systematic way of dealing with collective phenomena which is capable of treating the three two-particle correlation on the same footing, while retaining all the other attractive features of the collective quantum field theory. These can be characterized as follows:

1) The quantum problem is reduced to an extremal principle for classical collective variables.
2) The semiclassical quantization rules can directly be imposed.

We shall see that such a method indeed exists.

While this alone would justify its detailed study it has another important feature which, in general, makes it far superior to the collective field approach.

3) There is a straightforward extension which permits the description of collective effects in any higher multiparticle clusters.

The method is based on higher effective actions which is an adaption of thermodynamics to field theory.

Let us first recall the standard effective action as it appears in all textbooks on quantum theory[1] (For brevity, we shall from now on set $\hbar=1$).

II Simple Effective Action

Consider the logarithm of the generating functional

$$W[j] = \frac{1}{i} \log Z[j] \tag{36}$$

It is well-known, and easy to verify,[14] that its functional derivatives are the connected Green's functions, i.e.

$$G_c^{(n)}(x_1,\dots,x_n) = \frac{\delta}{i\delta j(x_1)} \cdots \frac{\delta}{i\delta j(x_n)} W[j] \tag{37}$$

In particular

$$\Phi(x) \equiv G_c^{(1)}(x) = \langle \varphi(x) \rangle = \frac{\delta W[j]}{\delta j(x)} \tag{38}$$

is the vacuum expectation of the field in the presence of the external current j. The effective action is defined as the Legendre transform

$$\begin{aligned} \Gamma[\Phi] &= W[j] - \int dx \, \frac{\delta W[j]}{\delta j(x)} \, j(x) \\ &= W[j] - \int dx \, \Phi(x) \, j(x) \end{aligned} \tag{39}$$

in which j has to be eliminated in favor of Φ by inverting (38). By definition, this effective action is a functional of a classical object which has no more fluctuations since $\Phi(x)$ is the expectation of the fluctuating field $\varphi(x)$. Still, $\Gamma[\Phi]$ contains all quantum mechanical information on the system: First of all, ground state and large amplitude collective oscillations can be found from extremizing $\Gamma[\Phi]$ with time independent or time dependent fields $\Phi(\underline{x},t)$. This directly follows follows (39): By construction, the effective action has the derivative from

$$\frac{\delta \Gamma}{\delta \Phi(x)} = \int dx' \frac{\delta W[j]}{\delta j(x')} \frac{\delta j(x')}{\delta \Phi(x)} - \int dx' \, \Phi(x') \frac{\delta j(x')}{\delta \Phi(x)} - j(x) = -j(x) \tag{40}$$

which vanishes in the absence of external currents. Second, with little effort it can be shown that higher Green's functions can be reconstructed by putting all possible tree diagrams built from vertices and branches where the vertices are the functional derivatives of $\Gamma[\Phi]$ of third and higher order

$$\Gamma^{(n)}(x_1,\dots,x_n) \equiv \frac{\delta}{\delta\Phi(x_1)} \cdot \dots \cdot \frac{\delta}{\delta\Phi(x_n)} \Gamma[\Phi] \quad . \tag{41}$$

which are the one-particle irreducible (OPI) amputated Green's functions, shortly called vertex functions, and where the branches are the full connected propagators[+]

$$G_c(x_1,x_2) = i\,\Gamma^{(2)^{-1}}(x_1,x_2) \equiv G(x_1,x_2) \tag{42}$$

The construction of $\Gamma[\Phi]$ can in fact proceed by calculating G and the vertex functions and composing from these the effective functional[1].

$$\Gamma[\Phi] = \sum_{n=0}^{\infty} \frac{1}{n!} \int dx_1 \dots dx_n \, \Gamma^{(n)}(x_1,\dots,x_n)\, \Phi(x_1) \cdot \dots \cdot \Phi(x_n) \tag{43}$$

This functional can then be extremized to find ground state and collective excitations.

[+]The inverse is understood in the functional matrix sense, i.e.

$$\int dx_2 \, \Gamma^{(2)}(x_1,x_2)\, \Gamma^{(2)^{-1}}(x_2,x_3) = \delta(x_1-x_3)$$

Such an expansion would, however, not be useful in studying quasiclassical limits. For this, $\Gamma[\Phi]$ has to be calculated in the loop expansion which collects, at each loop level, infinitely many powers of Φ such that $\Gamma[\Phi]$ becomes a non-perturbative object. This goes as follows[14]. One inserts into the action

$$\varphi(x) = \Phi(x) + \varphi'(x) \tag{44}$$

and expands

$$\begin{aligned} A[\varphi] &= A[\phi+\varphi'] = A[\phi] + \int dx \frac{\delta A[\phi]}{\delta\phi(x)} \varphi'(x) \\ &+ \frac{1}{2!} \int dx_1 dx_2 \frac{\delta^2 A[\phi]}{\delta\phi(x_1)\delta\phi(x_2)} \varphi'(x_1)\varphi'(x_2) \\ &+ \frac{1}{3!} \int dx_1 dx_2 dx_3 \frac{\delta^3 A[\phi]}{\delta\phi(x_1)\delta\phi(x_2)\delta\phi(x_3)} \varphi'(x_1)\varphi'(x_2)\varphi'(x_3) \\ &+ \ldots \end{aligned} \tag{45}$$

and considers the Feynman graphs composed of the Φ-dependent propagator

$$G_\Phi = i\left[\frac{\delta^2 \mathcal{A}[\Phi]}{\delta\Phi\,\delta\Phi}\right]^{-1} \tag{46}$$

and the Φ-dependent interactions

$$\frac{\delta^{(n)} \mathcal{A}[\Phi]}{\delta\Phi(x_1)\cdot\ldots\cdot\delta\Phi(x_n)}, \quad n \geq 3 \tag{47}$$

If one adds up all vacuum graphs which are one-particle irreducible, then the result is $\Gamma[\Phi]$. The expansion can be ordered by the number of loops which increases with the number of explicit $\hbar$'s in each term. Thus effective action is indeed a non-perturbative object. It has been used efficiently for the study of phase transitions in Bose systems.

Notice that this method is far superior to the original quasiclassical expansion since each loop term with an explicit power in $\hbar$ is equivalent to a whole infinite set of diagrams in the expansion (8).

The effective action can be used to quantize periodic orbits. Let $\Phi^T(x)$ be an oscillating extremum of period T, then we can define the effective Lagrangian

$$L(\omega) \equiv \Gamma[\Phi^T]/T$$

and proceed just as in equs. (9) to (13).

III 1/Degeneracy Collective Effective Actions

This leads us to the first proposal for going beyond time-dependent Hartree-Fock equations. If a many-body system consists of a large number of degenerate or of almost degenerate single particle levels, say N, we can introduce this number explicitly into the interaction (15) and rewrite it in the form (20). The result is a collective partition function of the form (22). There could also exist strong pairing forces, in which case there would be an additional term like (29). This would bring in a collective $\Delta(x,x')$ field which we shall omit, for brevity. It is then possible to perform the limit $N\to\infty$ and obtain a $1/N$ expansion of the partition function analogous to the semiclassical expansion (8).

According to Ch. II, however, a much better non-perturbative approach can be based on the loop expansion of the collective effective action. Just as before in the effective action of the fundamental field φ, each loop sums up a whole infinite set of powers in 1/N. In addition, higher loops are suppressed by higher powers in 1/N. In order to calculate this effective action we introduce a fictitious external collective source $N\, i\int dx\, j(x)\rho(x)$ and calculate

$$Z[j] \equiv e^{iW[j]} = \int \mathcal{D}\rho\; e^{Ni\left[\mathcal{A}_{coll}[\rho] + \int dx\, j(x)\rho(x)\right]} \tag{48}$$

The expectation of the density field is then given by

$$\bar{\rho}(x) = \langle \rho(x) \rangle = \frac{\delta W[j]}{\delta j(x)} \tag{49}$$

and we can define a collective effective action

$$\begin{aligned}\Gamma_{coll}[\bar\rho] &= W[j] - \int dx \frac{\delta W}{\delta j(x)}\, j(x) \\ &= W[j] - \int dx\, \bar{\rho}(x)\, j(x)\end{aligned} \tag{50}$$

This has the same pleasant properties as the previous one defined for the fundamental field. It can be expanded in a series of loops, which are the one-particle irreducible vacuum graphs, with $\bar{\xi}$- dependent propagators $G_{\bar{\xi}}$ and vertices $\delta^n\Gamma/\delta\bar{\xi}\ldots\delta\bar{\xi}$. Moreover, higher numbers of loops are depressed by factors 1/N such that, for high degeneracy, these have a decreasing importance.

The extremum of $\Gamma_{coll}[\bar{\xi}]$ determines non-perturbative collective ground state and large amplitude collective excitations.

For periodic orbits, writing the effective collective action as $L_{coll}(\omega) = \Gamma_{coll}[\bar{\xi}^T]/T$ leads to a quantization just as described before.

IV Two-Particle Effective Action

As discussed in the introduction, for general nuclei the 1/degeneracy limit cannot be used and density, exchange, and pairing correlations have to be considered on the same footing. This can be done by another type of effective action which is constructed directly involving the composite fields $\psi^+(x)\psi(x')$, $\psi(x)\psi(x')$, $\psi^+(x)\psi^+(x')$. For this, one introduces into the generating fermion functional with external bilocal source terms[16,17] and forms

$$Z[\mu,\lambda] \equiv e^{iW[\mu,\lambda]} = \int D\psi D\psi^+ \exp\{i\, A[\psi,\psi^+]$$

$$+ i \int dx dx' [\psi^+(x)\mu(x,x')\psi(x) + \frac{1}{2}(\psi^+(x)\psi^+(x')\lambda(x,x') + \lambda^+(x,x')\psi(x)\psi(x'))]\} \qquad (51)$$

The derivatives with respect to μ and λ lead to the bilocal density and pair expectations

$$\frac{\delta W[\mu,\lambda]}{\delta\mu(x,x')} = \langle T\, \psi^+(x)\psi(x')\rangle \equiv \rho(x,x') \qquad (52)$$

$$\frac{\delta W[\mu,\lambda]}{\delta\lambda^+(x,x')} = \frac{1}{2}\langle T\psi(x)\psi(x')\rangle \equiv \frac{1}{2}\Delta(x,x')$$

$$\frac{\delta W[\mu,\lambda]}{\delta\lambda(x,x')} = \frac{1}{2}\langle T\, \psi^+(x)\psi^+(x')\rangle \equiv \frac{1}{2}\Delta(x',x)^+ \equiv \frac{1}{2}\Delta^+(x,x') \qquad (53)$$

which are just the normal and anomalous Green's functions of the system. The two can be treated simultaneously by employing a bispinor notation for the fields ψ, ψ^+:

$$\varphi \equiv \begin{pmatrix} \psi \\ \psi^+ \end{pmatrix} \qquad (54)$$

and rewriting Z_1 in the form

$$Z_1[K] = e^{iW[K]} = \int \mathcal{D}\varphi \; e^{i\mathcal{A}[\varphi] + \frac{i}{2}\varphi^T K \varphi} \tag{55}$$

where K contains the sources μ and λ in a two-by-two matrix

$$K \equiv \begin{pmatrix} \lambda^+ & -\mu^T \\ \mu & \lambda \end{pmatrix} \tag{56}$$

For brevity, vector notation has been used to denote functional contraction of indices

$$\varphi^T K \varphi \equiv \int dx\,dx' \, \varphi(x)\, K(x,x')\, \varphi(x') \tag{57}$$

Now the derivative of W with respect to K gives

$$\frac{\delta W[K]}{\delta K(x,x')} = \frac{1}{2}\, G(x,x') \tag{58}$$

where the matrix

$$G(x,x') = \begin{pmatrix} \Delta(x,x') & -g(x',x) \\ g(x,x') & \Delta^+(x,x') \end{pmatrix} \tag{59}$$

collects the two types of Green's functions.

Using this functional $W[K]$ one introduces the effective action of bilocal fields as

$$\Gamma[G] \equiv W[K] - W_K \, K \tag{60}$$

and sees again that it is extremal in G :

$$\frac{\delta \Gamma[G]}{\delta G} = -\frac{1}{2} K = 0 \tag{61}$$

It is easy to construct $\Gamma[G]$ which consists of a free piece[16,17]

$$\Gamma^{(0)} = \frac{i}{2} \operatorname{tr}(G_0^{-1} G T) - \frac{i}{2} \operatorname{tr} \log i G^{-1} , \tag{62}$$

where G_0 is the free propagator of the doubled $\varphi = \begin{pmatrix} \psi \\ \psi^+ \end{pmatrix}$ field

$$G_0(x, x') \equiv \langle T \varphi(x) \varphi(x') \rangle \Big|_{V=0} = \begin{pmatrix} 0 & G_{0\,12}(x-x') \\ -G_{0\,12}(x'-x) & 0 \end{pmatrix} \tag{63}$$

with

$$G_{0\,12}(x-x') = \int \frac{dE}{2\pi} \frac{d^3 p}{(2\pi)^3} \, e^{-iE(t-t') + i\mathbf{p}(\mathbf{x}-\mathbf{x}')} \frac{i}{E - \left(\frac{p^2}{2m} - \mu\right)} ,$$

plus interacting pieces $\Gamma^{int}[G]$. The rule of composing these is the following: Consider all vacuum diagrams which do not fall into pieces by cutting two internal lines. These are called two-particle irreducible (TPI). Associate with each leg the full Green's function G and perform the integration just as in an ordinary Feynman diagram. The lowest three terms are[16,17]

$$\Gamma^{int}[G] = -\frac{1}{8} VGG + \frac{i}{48} (VGG)^2 + \frac{1}{48} (VGG)^3 + \dots \tag{64}$$

corresponding to the TPI vacuum graphs

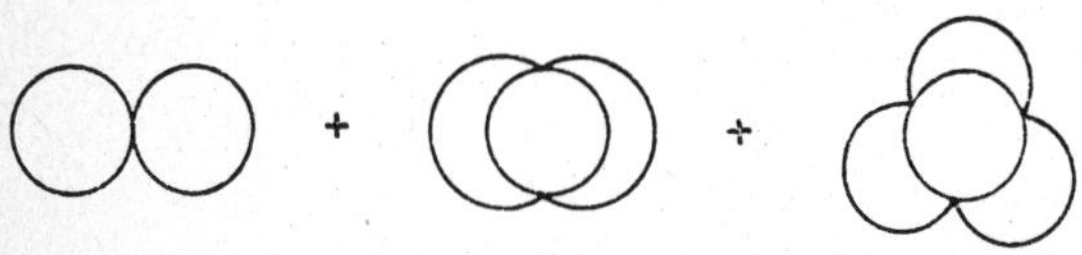

(65)

Keeping only the first graph, extremization $\delta\Gamma/\delta G = 0$ leads to the equation of motion

$$iG^{-1} = iG_0^{-1} - \frac{1}{2} VG \tag{66}$$

which is recognized as the time-dependent Hartree-Fock-Bogoljubov[15] equation. The higher diagrams give specific prescriptions on how to go beyond this if the most important correlations are of the two-particle kind.

V Effective Action with Two and Four Particle Correlations

In nuclear physics, four-particle clusters are known to play an important role. This suggests the introduction of an effective action in which four-particle correlations are taken into account explicitly[17,18]. The consequence is one more variable in the extremal principle and this always improves the quality of the extremum. Thus we introduce, in addition to the bilocal source, another one which couples to four particles and consider

$$Z_1[K,\gamma] = e^{iW[K,\gamma]} = \int \mathcal{D}\varphi \exp\left\{ iA[\varphi] + \frac{i}{2}\varphi^T K \varphi - \frac{i}{4!}\gamma\varphi\varphi\varphi\varphi \right\} \tag{67}$$

The derivative of W with respect to γ gives the four-particle Green's function

$$-\frac{\delta}{\delta\gamma} W[K,\gamma] = \frac{1}{4!}\langle 0|T\varphi(x_1)\cdots\varphi(x_4)|0\rangle = \frac{1}{4!} G^{(4)}(x_1\ldots x_4) \tag{68}$$

This can be separated into connected and disconnected parts as

$$G^{(4)}(x_1\ldots x_4) = G_c^{(4)} + (G\times G + 2\,\text{perm.}) \tag{69}$$

Moreover, the four-particle connected part has singular factors, namely a single particle propagator at each leg. For calculations it is useful to work with the smoothest possible object, which sometimes can be approximated by a constant. Therefore, we remove those factors and introduce the four particle vertex function α via

$$G_c^{(4)}(x_1\ldots x_4) \equiv -i\,\alpha\, GGGG \tag{70}$$

The higher effective action is now introduced as a functional of G and α :

$$\Gamma[G,\alpha] = W[K,\gamma] - W_K K - W_\gamma \gamma \tag{71}$$

where K and γ have to be expressed in terms of G and α by inverting the defining equations

$$W_K = \frac{1}{2} G \tag{72}$$

$$W_\gamma = \frac{i}{4!} \alpha GGGG - \frac{3}{4!} GG \tag{73}$$

By construction, the effective action $\Gamma[G,\alpha]$ is extremal in G and α :

$$\frac{\delta \Gamma[G,\alpha]}{\delta G} = -\frac{1}{2} K + \frac{1}{4} \gamma G - \frac{i}{6} \alpha G^3 \gamma \tag{74}$$

$$\frac{\delta \Gamma[G,\alpha]}{\delta \alpha} = -\frac{i}{4!} \gamma G^4 \tag{75}$$

and can now be used to obtain non-trivial ground state configurations of two- and four-particle distributions as well as quantized large-amplitude periodic orbits.

Notice that equation (75) is essential in calculating the non-perturbative formation of four-particle clusters. As a matter of fact, it can be considered as a "gap equation" for α-particle condensation[19].

VI Ways Beyond Landau's Theory of Fermi Liquids

In nuclear physics, Landau's theory of Fermi liquids has greatly helped studying low lying excited states. Within the present framework, this theory constitutes a certain lowest approximation which arises as follows[17,20]:

Let G_0 be the ground state extremum of $\Gamma[G]$. Then we can write

$$G = G_0 + \delta G \tag{76}$$

and expand, up to quadratic order in δG,

$$\begin{aligned}\Gamma[G] &= \Gamma[G_0] + \delta^2\Gamma[G_0, \delta G] + \dots \\ &= \Gamma[G_0] + \frac{1}{2!}\left.\frac{\delta^2\Gamma[G]}{\delta G\,\delta G}\right|_{G_0} \delta G\,\delta G + \dots\end{aligned} \tag{77}$$

where the linear term is absent due to the extremality of $\Gamma[G_0]$. Inserting for $\Gamma[G]$ the lowest non-trivial approximation to (62) and (64) we find[16]

$$\delta^2\Gamma[G_0, \delta G] = \frac{i}{4}\,\delta G\, G_0^{-1} \times G_0^{-1}\delta G - \frac{1}{8} V\,\delta G\,\delta G \tag{78}$$

For simplicity, let us assume a translationally invariant Fermi liquid where all particles can be labeled by a momentum $\mathbf{p}$ with an energy $\varkappa(\mathbf{p})=$. At low temperature, the particles occupy all levels up to an energy $\varkappa(\mathbf{p}) = \mu \sim E_F$ which is called Fermi energy. In rotational invariant systems, this corresponds to a sphere in momentum space with all $|\mathbf{p}|$ smaller than the Fermi momentum p_F. The free particle propagator is $(p \equiv (E, \mathbf{p}) \equiv (p_0, \mathbf{p}))$

$$G_{0_{12}}(p) = \frac{i}{E - \xi(\mathbf{p})} \qquad \xi(\mathbf{p}) \equiv \varkappa(\mathbf{p}) - \mu \tag{79}$$

Thus we can form a single loop integral

$$-i \int \frac{dp_o}{2\pi} G_{o_{12}}(p + \frac{q}{2}) G_{o_{12}}(p - \frac{q}{2})$$
$$= i \int \frac{dp_o}{2\pi} \frac{1}{p_o+\frac{q_o}{2} - \xi(\underset{\sim}{p}+\frac{\underset{\sim}{q}}{2})} \frac{1}{p_o-\frac{q_o}{2} - \xi(\underset{\sim}{p}-\frac{\underset{\sim}{q}}{2})} \tag{80}$$

where $\underset{\sim}{q}$ is the total and $\underset{\sim}{p}$ the relative momentum of the two particle lines in the loop. This can be rewritten as

$$\frac{1}{q_o-\xi(\underset{\sim}{p}+\frac{\underset{\sim}{q}}{2})+\xi(\underset{\sim}{p}-\frac{\underset{\sim}{q}}{2})} \int \frac{dp_o}{2\pi} \left[- \frac{i}{p_o+\frac{q_o}{2} - \xi(\underset{\sim}{p}+\frac{\underset{\sim}{q}}{2})} + \frac{i}{p_o-\frac{q_o}{2} - \xi(\underset{\sim}{p}-\frac{\underset{\sim}{q}}{2})}\right] \tag{81}$$

Now, the dp_o integral can be performed most conveniently by rotating the contour such that it coincides with the imaginary $p_o = ip_4$ axis. In order to allow for non-zero temperature we may split the integral into a sum over Matsubara frequencies

$$\int \frac{dp_o}{2\pi} \to iT \sum_{p_4 = 2\pi T\cdot(n+\frac{1}{2})} \tag{82}$$

Then (81) becomes

$$\frac{1}{q_o - \xi(\underset{\sim}{p}+\frac{\underset{\sim}{q}}{2})+\xi(\underset{\sim}{p}-\frac{\underset{\sim}{q}}{2})} \left(n(\underset{\sim}{p}+\frac{\underset{\sim}{q}}{2})-n(\underset{\sim}{p}-\frac{\underset{\sim}{q}}{2})\right) \tag{83}$$

where $n(\underset{\sim}{p}) = (e^{\xi(\underset{\sim}{p})/T} + 1)^{-1}$ is the Fermi distribution.

In the long-wavelength the limit $q \to 0$ this becomes

$$(q_0 - \frac{\partial \xi}{\partial p} \cdot q)^{-1} \frac{\partial n}{\partial \xi} \frac{\partial \xi}{\partial p} \cdot q \tag{84}$$

But $\partial \xi / \partial p$ is the group velocity of the particles $v(p)$. Moreover, for low temperature, the derivative $\partial n / \partial \xi$ is strongly peaked at the surface of the Fermi sea $\varkappa \sim E_F$ such that we can write approximately

$$-i G_{012} G_{012} \approx -(q_0 - v \cdot q)^{-1} v \cdot q \, \delta(\xi - E_F) \tag{85}$$

If we neglect pairing effects which would be carried by the diagonal parts of G, equation (85) can be inserted into (78) to write the effective action $\delta^2 \Gamma[G_0, \delta G]$ in the approximate form

$$\delta^2 \Gamma [G_0, \delta G] \approx \frac{1}{4} \delta G \left(\frac{q_0}{v \cdot q} - 1 \right) \left(\frac{\partial n}{\partial \xi} \right)^{-1} \delta G - \frac{1}{8} V \delta G \delta G \tag{86}$$

Extremizing this in δG we obtain the differential equation in momentum space [19]

$$(q_0 - v \cdot q) \delta G = v \cdot q \frac{\partial n}{\partial \xi} \frac{1}{2} V \delta G \tag{87}$$

This is recognized as Landau's equation for quasiparticle densities with a collision term.

The development of the effective action $\Gamma[G]$ to higher orders in δG allows for a straightforward extension of this equation, adding on the right hand side of (87)

$$\underset{\sim}{v}\cdot q\;\frac{\partial n}{\partial \xi}\sum_{n=2}^{\infty}\delta^2\Gamma^{(n)}/\delta G\,\delta G\Big|_{G=G_0}\delta G \tag{88}$$

where $\Gamma^{(n)}$ denotes the graphs (65) with powers $V^2, V^3, \ldots$etc. But there is an even more dramatic way of going beyond Landau's theory. Being in the possession of the higher effective action $\Gamma[G,\alpha]$ involving also the four particle vertices we can take the extremum G_0 and α_0 and expand $\Gamma[G,\alpha]$ quadratically around this. The equations for $\delta G = G - G_0$ and $\delta\alpha = \alpha - \alpha_0$ account for the possibility of small oscillations in densities and vertices, and solving these equations will lead to far better approximations than any finite number of higher corrections (88).

VII Higher Effective Action in Collective Fields

If a system has high degeneracy and strong four-particle clusters, there is the possibility of using higher effective actions in conjunction with the fluctuating collective field theory (48). For this we can add, in the exponent, the bilocal source term

$$N \frac{i}{2} \int dx\, dx' \, \rho(x)\, K(x,x')\, \rho(x') \tag{89}$$

and obtain a generating functional

$$Z[j,K] = e^{Ni\, W[j,K]} \tag{90}$$

which can then be used to calculate an effective action

$$\Gamma[\bar{\rho}, \bar{\alpha}] = W[j,K] - W_j\, j - W_K\, K \tag{91}$$

where

$$W_j \equiv \bar{\rho} \tag{92}$$

and $\bar{\alpha}$ is defined by

$$W_K = \frac{1}{2}\bar{\alpha} + \frac{3}{2}\bar{\rho}\bar{\rho} \tag{93}$$

This effective action can then be treated in close analogy with the functional $\Gamma[G,\alpha]$ discussed in Chapter IV.

VIII Outlook

We have pointed out a variety of possible ways of going beyond time-dependent Hartree-Fock equations. Each of them has a parameter which in some sense is small and systematically organizes an approximation scheme.

In simple models of many-body systems, one possible parameter is the inverse degeneracy of the single particle levels. Then the lowest approximation consists in extremizing the collective action which leads to the time-dependent Hartree equations. The solutions can be corrected by powers 1/degeneracy=1/N. There exist also non-perturbative ways of going beyond the leading order in 1/N. They are based on forming the collective effective action and expanding it in vacuum diagrams organized by the number of loops, which for high N are suppressed by increasing powers in 1/N.

If there are strong multiparticle clusters, higher effective collective actions must be used which give even more dramatic non-perturbative corrections in 1/N.

For small degeneracies, the most efficient way to go beyond time-dependent Hartree is provided by the higher effective actions in the composite fields $\psi^{+}\psi$, $\psi\psi$, $\psi^{+}\psi^{+}$. Here, the lowest approximation always leads to the time-dependent Hartree-Fock-Bogoljubov equation. The expansion in loop diagrams is characterized by increasing leading orders in $\hbar$. If four-particle variables are included, a non-perturbative gap equation is obtained also for α-particle condensation.

In all these approaches, a quantum mechanical many-body system is described in terms of classical c-number variables whose field equations follow from an extremal principle.

The quantization of multiple periodic solutions presents no problem since we are always in possession of an action and there are simple rules of determining the discrete set of allowed orbits.

Much detailed work will be necessary in order to see whether the proposed methods lead to a significant improvement of our understanding of complicated many-nucleon systems.

References

1) See, for example, C. Itzykson and J.B. Zuber, Quantum Field Theory, McGraw-Hill, New York, 1980.

2) R.P. Feynman and A.R. Hibbs, Quantum Mechanics and Path Integrals, New York, 1965. The extension to fluctuating field theory was first used extensively by
S.A. Langer, Ann. Phys. (N.Y.) 41, 108(1967),
later by
A.A. Belavin, A.M. Polyakov, A.S. Schwartz, and Yu.S. Tyupkin, Phys. Lett. 59B, 85(1975).
G. 't Hooft, Phys. Rev. Lett. 37, 8(1976).

3) R.F. Dashen, B. Hasslacher, and A. Neveu, Phys. Rev. D10, 4114(1974).

4) H. Kleinert and H. Reinhardt, Nucl. Phys. A332, 331(1979).

5) H. Kleinert, Fortschr. Phys. 26, 565(1978).

6) R.L. Stratonovich, Dok. Akad. Nauk. SSSR 115, 1097(1957), (Sov. Phys. Dokl. 2, 416(1958).

7) H. Kleinert, Phys. Lett. 69B, 9(1977) and Lectures presented at NATO Advanced Study Institute on Non-Linear Equations in Physics and Mathematics, Istanbul, August 1977, ed. A.O. Barut (Reidel, Dardrecht, Holland, 1978).

8) T.H. Berlin and M. Kac, Phys. Rev. 86, 821(1952).
H.E. Stanley, Phys. Rev. Lett. 20, 589(1968), Phys. Rev. 176, 780 (1968).
E. Helfand, Phys. Rev. 183, 562(1969).
M. Kac and C.J. Thompson, Proc. Norwegian Acad. Science (1971).
D. Gross and A. Neveu, Phys. Rev. D10, 3225(1974).

9) D. Ebert and H. Reinhardt, Nucl. Phys. A298, 60(1978), for recent work see H. Reinhardt's lecture at this meeting.

10) S. Levit, Phys. Rev. C21, 1594(1980), S. Levit et al., Phys. Rev. C21, 1603(1980), C22, 1979(1980), for recent work see J. Negele's lecture at this meeting.

11) Y. Alhassid and S.E. Koonin, Phys. Rev. C23, 1590(1981).

12) The coherent state path integrals, advanced by J.P. Blaizot and H. Orland, J. Physique Lett. 41, 531(1980) and Phys. Rev., Urbana preprint, 1981, are in principle capable of including exchange and pairing forces. Their approach is, however, very complicated to handle in practical calculations. See also
H. Kuratsuji and T. Suzuki, Phys. Lett. 92B, 19(1980), J. Math. Phys. (N.Y.) 21, 472(1980) and T. Suzuki's lecture at this meeting.

13) This has been of importance in calculating the lifetimes of metastable states
I.H. Duru, H. Kleinert, and Y. Ünal, Journ. Low Temp. Phys. 42, 137(1981).

14) For a detailed discussion as well as references to the earlier work see
H. Kleinert, Fortschritt. Phys. 30, 187(1982), Phys. Lett. A84, 259 (1981).

15) H. Kleinert, Lett. Nuovo Cimento 31, 521(1981).

16) H. Kleinert, Fortschr. Phys. 30, (in press).

17) H. Kleinert, Phys. Lett. A84, 199(1981).

18) H. Kleinert, Journ. of Phys. 68, 239(1982).

19) H. Kleinert, Phys. Lett. A84, 202(1981).

SEMICLASSICAL ANALYSIS OF MANY-FERMION SYSTEM IN THE GENERALIZED COHERENT-STATE REPRESENTATION

Toru Suzuki and Hiroshi Kuratsuji[+)]

Niels Bohr Institute, 2100 Copenhagen Ø, Denmark

[+)] Kyoto University, 606 Kyoto, Japan

1. Introduction

Time-dependent Hartree-Fock method is obtained as a restriction of the full, time-dependent Schrödinger dynamics of many-fermion system to a family of determinantal wave functions. Confronted with various facets of this restriction, one may try to extend the method from each point of view. Thus, for instance, shortcomings associated with its single-particle aspect might be remedied by introducing two-body correlation effect explicitly. One of such prominent aspects of the TDHF method is the so-called 'classical' aspect which makes it difficult to extract an unambiguous quantum mechanical information from the outcome of the method: In TDHF, one-body motion in a mean field is treated quantum mechanically, while the time evolution of the field itself is governed by a nonlinear equation which conflicts with the superposition principle. This implies that the concept of wave function is not inherent in the method. Although the TDHF equation contains certain quantum mechanical informations, it is this aspect of a wave function where lies the mission of a standard semiclassical method to 'sew a quantum flesh to classical bones'.[1)] In order to surmount the limitation of TDHF in this respect, therefore, one must develop a relevant semiclassical method of many-fermion system based on classical TDHF bones. A natural framework to accomplish this purpose is provided by a path-integral method applied to many-fermion system, which is known to yield TDHF equation in a classical limit.[2-5)] We adopt here the approach developed in Refs.4,6-8) which utilizes a quantum mechanical representation based on generalized coherent states.

2. Representation Based on Slater Determinants

TDHF equation describes a continuous time-development of a determinant and thus defines a path in a space of a set of parameters which labels determinants. On the other hand, actual time evolution of a full wave function takes place in a quantum mechanical Hilbert space.

Path integral method enables one to find a relationship between the two, by describing the latter in terms of a path concept in a parameter space which fixes a quantum mechanical representation basis. Hence, one is led to adopt a representation in which a whole family of determinants is taken as a basis set.

To safely define the concept of path one must impose the following conditions:

(I) The set of basis states $\{|Z\rangle\}$ which defines a labelling of paths should be complete. This implies that there is a 'resolution of unity' relation in the state space considered:

$$\int d\mu(Z)|Z\rangle\langle Z| = 1 \qquad (1)$$

under a proper integration measure $d\mu(Z)$.

(II) For a description of closed systems, the Hamiltonian should not have a matrix element which makes a wave function escape from the space.

The family of determinants satisfies the condition (I): Actually, it is highly nonorthogonal and overcomplete. The relation (1) reflects the closed property of a Lie algebra of particle-hole operators, and the determinants $\{|Z\rangle\}$ are regarded as generalized coherent states[9] associated with the corresponding Lie group. More specifically, if one takes the Thouless' form of parametrization,

$$|Z\rangle = \mathcal{N}_Z^{-\frac{1}{2}}\exp\{\Sigma_\mu Z_\mu (a^\dagger b^\dagger)_\mu |0\rangle, \quad a_p|0\rangle = b_h|0\rangle = 0 \qquad (2)$$

where $\mathcal{N}_Z$ is a normalization factor and $\{Z_\mu;\ \mu=(ph)=1,2,\ldots,M\}$ are complex numbers, then the integration measure is given by[8]

$$d\mu(Z) = \text{const.}\times\ \det\{\ g\ \}\cdot\Pi_\mu dReZ_\mu dImZ_\mu$$
$$g_{\mu\nu}(Z^*,Z) = \partial^2\log\mathcal{N}_Z/\partial Z_\mu^*\partial Z_\nu \qquad (3)$$

In this case, the r.h.s. of (1) denotes unity in the state space having fixed number of particles determined by $|0\rangle$. The expression (3) is, in fact, quite general. To describe a system having pairing correlations, for instance, one has simply to replace p-h creation operators by pair creation operators and $|0\rangle$ by the fermion vacuum.

The condition (II) is satisfied if the Hamiltonian can be expressed as a polynomial of the elements of the Lie algebra as is usually the case.

3. Path Integral Expression for the Propagator

Once the conditions I,II are established, it is straightforward to express the propagator of the system, $U(t_f,t_i)\equiv\exp\{iH(t_f-t_i)/\hbar\}$ for a time-independent Hamiltonian H, in a path integral form:[4)]

$$U(t_f,t_i) = \lim_{n\to\infty}\int\cdots\int\prod_{k=0}^{n}d\mu(Z_k)\cdot|Z_f\rangle\langle Z_i|\cdot\exp(iS_n\{Z\}/\hbar)$$
$$S_n\{Z\} = -\sum_{k=1}^{n}\varepsilon(i\hbar\log\langle Z_k|Z_{k-1}\rangle+\langle Z_k|H|Z_{k-1}\rangle/\langle Z_k|Z_{k-1}\rangle) \tag{4}$$

where $Z_k=\{(Z_k)_\mu\}$ $(Z_0=Z_i, Z_n=Z_f)$ denotes the k-th meshpoint on the path at time $k\varepsilon+t_i$ and $\varepsilon=(t_f-t_i)/n$. (For a well-behaved Hamiltonian, one can alternatively use a diagonal expression, $H=\int d\mu(Z)|Z\rangle H'(Z^*,Z)\langle Z|$, where H' is a contravariant symbol of H[9)], to obtain a slightly different form of a path integral.)

The integration over Z_k (k=1,...,n-1) may be evaluated in the stationary phase phase approximation (SPA), from which one obtains a variational condition (for $\varepsilon\to 0$),

$$\delta S = 0, \quad S\{Z(\cdot)\} = \int_{t_i}^{t_f}\langle Z(t)|(i\hbar\partial_t-H)|Z(t)\rangle dt \tag{5}$$

leading to the TDHF equation of motion.

4. Formal Properties of the TDHF Equation and Its Solutions

Under the parametrization (2) of determinants, the TDHF equation of motion takes a form

$$dZ_\mu/dt = \{Z_\mu,\mathscr{H}\} = -(i/\hbar)\Sigma_\nu(g^{-1})_{\mu\nu}\partial\mathscr{H}/\partial Z^*_\nu \tag{6}$$

where the Poisson bracket is defined by

$$\{A,B\} \equiv -(i/\hbar)\Sigma_{\mu\nu}(g^{-1})_{\mu\nu}(\partial A/\partial Z_\mu\cdot\partial B/\partial Z^*_\nu-\partial B/\partial Z_\mu\cdot\partial A/\partial Z^*_\nu) \tag{7}$$

and the 'classical' Hamiltonian is given by $\mathscr{H}\equiv\langle Z|H|Z\rangle$. For any quantum mechanical operator F satisfying a similar condition (II) as for H, one can associate a 'classical' counterpart given by the covariant symbol $\mathscr{F}(Z^*,Z)\equiv\langle Z|F|Z\rangle$.

One-body nature of TDHF can be seen in the following identity[8)] for the commutator of any one-body operator F and an arbitrary operator G:

$$\langle Z|[F,G]|Z\rangle = \{\mathscr{F},\mathscr{G}\} \tag{8}$$

Accordingly, if there is a one-body operator F which is commutable with the Hamiltonian, its classical image is also conserved:

$$[F,H] = 0 \;\rightarrow\; d\mathscr{F}/dt = \{\mathscr{F},\mathscr{H}\} = 0 \tag{9}$$

The form of (6) suggests that the structure of the TDHF equation is quite analogous to the classical Hamilton's equation of motion[10)]. Indeed, as summarized below, the TDHF solutions are regarded as classical trajectories in a 2M-dimensional phase space characterized by the metric tensor $g_{\mu\nu}$.(It is not fully classical, however, in the sense of the 'large-N' limit.)

i) Since (6) is a coupled first order differential equation in the time variable, a single point on a path at any fixed time is sufficient to uniquely determine the solution.

ii) The system is conservative so that the classical energy $E_c = \mathcal{H}$ is a constant of motion. Hence the trajectories are confined to a 2M-1 dimensional hypersurface of the phase space. If one can find M-1 additional constants of motion, the system is integrable. This is not generally the case for $M \geq 2$, and the phase space may be separated into regular and irregular regions[11,12].

iii) For a given solution having energy E_c, the classical (TDHF) action is given by

$$S_c = \int_i^f \omega - E_c(t_f - t_i),$$
$$\omega = (i\hbar/2)\cdot(dZ_\mu \partial/\partial Z_\mu - dZ_\mu^* \partial/\partial Z_\mu^*)\log \mathcal{N}_Z \qquad (10)$$

One can utilize an invariance property associated with the 1-form ω-Edt to express (10) in the following form:

$$S_c = S_f - S_i, \quad S_{f,i} \equiv \int_0^{f,i} \omega - E_c dt \qquad (11)$$

The integration in (11) is taken along a path from some fixed point O on the energy surface to $Z_f(Z_i)$. The integration paths $O\to i$, $O\to f$ may be taken arbitrarily as fas as they lie on the same surface of the phase space flow.

iv) The exterior derivative of the 1-form ω,

$$d\omega = -i\hbar g_{\mu\nu} dZ_\mu^* \wedge dZ_\nu \qquad (12)$$

is the absolute integral invariant of Poincaré and defines a symplectic structure of the phase space. Starting with (12) one can construct a series of integral invariants, the highest of which is the phase space volume element

$$\Omega = \text{const.}\times \det\{g\} \Lambda_\mu (dZ_\mu^* \wedge dZ_\mu) \qquad (13)$$

This expression agrees with the invariant measure (3). The invariance of (13) along TDHF paths is the Liouville's theorem.

5. Semiclassical Propagator

In the lowest order SPA the expression (4) reduces to

$$\hat{U}(t_f, t_i) = \int d\mu\{Z\} |Z(t_f)\rangle e^{iS_c\{Z\}/\hbar} \langle Z(t_i)| \qquad (14)$$

where the integration is now taken over the classical(TDHF) solutions. From the property i) above, the integration is replaced by the one on phase space points at some fixed time t_0; $d\mu\{Z\} \to d\mu(Z(t_0))$. Due to the property iv), the time t_0 can be arbitrary chosen.

The semiclassical propagator (14) is also obtained naturally if one tries to recover the superposition principle starting with a (single path) TDHF propagator,

$$U_{TDHF}(t_f,t_i) = |Z(t_f)\rangle e^{iS_c\{Z\}/\hbar}\langle Z_i| \tag{15}$$

which is defined for a given initial determinant $|Z(t_i)\rangle=|Z_i\rangle$. The propagator (15) satisfies the exact Schrödinger equation, $(i\hbar\partial_t - H)U=0$, if the Hamiltonian happens to be a one-body operator. Even in this case, however, U_{TDHF} does not coincide with the exact propagator: The initial ($t_f=t_i$) value of U_{TDHF} does not give the correct expression, unity, but the projector to $|Z_i\rangle$. To obtain an expression which coincides with the exact propagator for one-body Hamiltonians in terms of TDHF solutions, one has simply to take a linear superposition of U_{TDHF} for different initial determinants so as to satisfy the correct initial condition. This linear combination is built in as the resolution-of-unity relation (1). Consequently, one recovers the expression (14).

The propagator U obeys a number of relations which may or may not be related to a symmetry of the Hamiltonian. Similarly, the semiclassical propagator $\tilde{U}$ satisfies such relations as time-translation/reversal symmetry, Hermitian conjugate property, etc.[8] In SPA, $\tilde{U}$ satisfies also a product relation, $\tilde{U}(t_f,t_i) \simeq \tilde{U}(t_f,t_m)\cdot\tilde{U}(t_m,t_i)$, from which follows the unitarity of $\tilde{U}$ within the same approximation.

6.Quantization of Bound States

Stationary states are characterized by periodic time dependence. Since the classical energy is conserved along TDHF trajectories, one can extract time dependence in a similar manner. This enables one to deduce a semiclassical analog of a stationary wave function and to obtain quantization conditions. In the following we discuss integrable classical systems where the trajectories lie on (M-dimensional) invariant tori in 2M-dimensional phase space. For non-integrable systems, there appear chaotic trajectories, the quantization condition of which is yet to be established[12,13].

The invariant tori are determined by M constants of motion, including energy. Therefore, a phase space point can be specified as $Z=\{Z_\mu\} \leftrightarrow \{E,\upsilon,\theta\}$ where the energy E and $\upsilon=\{\upsilon_\alpha;\alpha=1,\dots,M-1\}$ are the 'classical numbers' which define an invariant torus, while $\theta=\{\theta_\beta;\beta=$

$1,\ldots,M\}$ is the coordinate on the torus. In terms of these variables the invariant integration measure (3) may be written as $d\mu(Z)=\tilde{\rho}(E,\upsilon,\theta)$ $\times dEd\upsilon d\theta$ under a suitable transformation of the weight function $\tilde{\rho}$.

The exact time evolution of a wave function is given by

$$|\Psi(t_f)\rangle = U(t_f,t_i)|\Psi(t_i)\rangle = \Sigma_m|m\rangle e^{iE_m(t_f-t_i)/\hbar}\langle m|\Psi(t_i)\rangle \quad (16)$$

where $|m\rangle$ denotes a stationary eigenstate of the Hamiltonian specified by a set of quantum numbers m. One can analogously write down the equation for a semiclassical wave function $\tilde{\Psi}$ as

$$\begin{aligned}|\tilde{\Psi}(t_f)\rangle &= \tilde{U}(t_f,t_i)|\tilde{\Psi}(t_i)\rangle \\ &= \int\tilde{\rho}dEd\upsilon d\theta|E\upsilon\theta_f\rangle e^{iS_f/\hbar}e^{-iS_i/\hbar}\langle E\upsilon\theta_i|\tilde{\Psi}(t_i)\rangle\end{aligned} \quad (17)$$

where $\theta_i(\theta_f)$ denotes the initial(final) point of a path on a torus and $S_i(S_f)$ has been introduced in (11). The equation (17) can be regarded as an integral equation for $\tilde{\Psi}$, which has a solution

$$\langle E\upsilon\theta|\tilde{\Psi}(t)\rangle = \tilde{A}(E,\upsilon)\exp\{i(\int^{\theta}\omega-Et)/\hbar\} \quad (18)$$

By comparing (16) with (17) and (18) one thus obtains $E=E_m$ for stationary states. Quantization conditions for the 'classical numbers' $\{E,\upsilon\}$ are obtained from a requirement of one-valuedness of the wave function (18) on the invariant torus[7]:

$$\oint\omega = 2\pi\hbar\cdot n \quad \text{(n: integer)} \quad (19)$$

where the integrations are over topologically independent closed loops on the torus. This is the quantization condition of Einstein[14].

After the quantization, the m-th semiclassical wave function (18) is defined on the m-th invariant torus having $E=E_m$ and $\upsilon=\upsilon_m$, i.e., $\tilde{\Psi}_m(\theta)=\text{const.}\times\exp(i\int^{\theta}\omega/\hbar)$. Because of the nonorthogonal character of the basis states $|Z\rangle$, one may extend the definition of the semiclassical wave function over the whole phase space as

$$\langle Z|\tilde{m}\rangle = \int\tilde{\rho}(Z_m)d\theta_m\langle Z|Z_m\rangle\tilde{A}(Z_m)\exp\{i\int^{Z_m}\omega/\hbar\} \quad (20)$$

where $Z_m=\{E_m,\upsilon_m,\theta_m\}$. In SPA, the expression (20) reduces to (18) on the torus.

7. Illustrative Examples

<Harmonic Oscillator>

The Hamiltonian of one-dimensional harmonic oscillator is given by

$$H = \hat{p}^2/2m + m\omega^2\hat{x}^2/2 = (\varepsilon/2)(b^{\dagger}b+bb^{\dagger}) + (\kappa/2)(b^{\dagger}b^{\dagger}+bb) \quad (21)$$

where the boson creation/annihilation oberators are defined through

$$\hat{x} = \xi(b^\dagger+b), \quad \hat{p} = i\eta(b^\dagger-b), \quad 2\xi\eta = \hbar \tag{22}$$

The basis vector is the boson coherent state

$$|Z\rangle = \exp(-Z^*Z/2+Zb^\dagger)|0\rangle, \quad b|0\rangle = 0 \tag{23}$$

which describes a wave packet centered at

$$x = \xi(Z^*+Z), \quad p = i\eta(Z^*-Z) \tag{24}$$

with dispersion $\Delta x=\xi$ and $\Delta p=\eta$. The classical equation of motion is given by the ordinary Hamilton's equation. For a given initial condition, $Z(t=0)=Z_0$, the classical solution is

$$Z(t) = Z_0\cos\omega t - i(\hbar\omega)^{-1}(\varepsilon Z_0+\kappa Z_0^*)\sin\omega t \tag{25}$$

having a classical action $S_c=-\varepsilon t/2$. One can show that the semiclassical propagator $\tilde{U}$ coincides with the exact one in this case by substituting the solution (25) into the expression (14) and then performing an gaussian integration over the classical solutions.

In the following we adopt an optimal choice for the parameters in (2), $\xi=\sqrt{\hbar/2m\omega}$, $\eta=\sqrt{m\hbar\omega/2}$, which gives $H=\hbar\omega(b^\dagger b+\frac{1}{2})$. The invariant tori of this system are ellipses in the p-x plane. It is easily seen that the quantization condition (19) reproduces the exact spectrum, $E_n=\hbar\omega(n+\frac{1}{2})$, under the above choice of parameters ξ,η. In order to compare the wave functions, we shall first express the exact n-th eigenstate in the coherent-state representation:

$$\langle Z|\Psi_n\rangle = \langle Z|(b^\dagger)^n/\sqrt{n!}|Z\rangle = (n!)^{-\frac{1}{2}}(Z^*)^n e^{-Z^*Z/2} \tag{26}$$

The peak of the wave function (26) occurs at $|Z|=\sqrt{n}$, i.e., $p^2/2m + m\omega^2x^2/2 = \hbar\omega n$, which coincides with the n-th quantized torus. One can calculate a full semiclassical wave function over the p-x plane from (20). The integration measure on the torus is obtained from $d\mu(Z)=d\mathrm{Re}Z\,d\mathrm{Im}Z/\pi=dE\,d\theta/2\pi\hbar\omega$ where $\tan\theta=m\omega x/p$. By taking a simple choice, $\tilde{A}=1$, one can perform the line integration over the ellipse and obtain

$$\langle Z|\tilde{\Psi}_n\rangle = \mathrm{const.}\times(n!)^{-1}n^{n/2}e^{-n/2}(Z^*)^n e^{-Z^*Z/2} \tag{27}$$

which has the same functional dependence on Z as the exact wave function (26). The factor $(n!)^{-\frac{1}{2}}$ in (26) is also reproduced for large n.

<R(4) model>

The model consists of two orbits specified by a signature $\sigma=\pm1$, each having Ω(=even)-fold degeneracy distinguished by a quantum number $k=\pm1,\pm2,\ldots,\pm\Omega/2$. The Hamiltonian contains a pairing and a particle-hole interaction, the latter of which may be regarded as a prototype of the quadrupole force:

$$H = -(G/8)(P^{\dagger}P+PP^{\dagger}) - (X/4)QQ$$
$$P^{\dagger} = \Sigma_{\sigma}S_{+}(\sigma), \quad Q = \Sigma_{\sigma}\sigma\cdot S_0(\sigma) \tag{28}$$
$$S_{+}(\sigma) = \tfrac{1}{2}\Sigma_k c^{\dagger}_{k\sigma}c^{\dagger}_{\bar{k}\sigma}, \quad S_0(\sigma) = \tfrac{1}{2}\Sigma_k(c^{\dagger}_{k\sigma}c_{k\sigma}-c_{k\sigma}c^{\dagger}_{k\sigma})$$

The operators S_{+}, $S_{-}\equiv(S_{+})^{\dagger}$ and S_0 constitute a quasi-spin for each $\sigma=\pm1$ orbit. Hence the Hamiltonian possesses $R(4)\sim SU(2)\times SU(2)$ symmetry and is easily diagonalized in the basis of either group. The model has been analyzed in connection with the attenuation factors in the interacting boson approximation[15]. A semiclassical quantization problem has been studied in ref.16) for the case of particle number $N=\Omega$. We discuss below a general case based on the framework given above.

Since the model is described by the orientation of two quasi-spins, the number of degrees of freedom M is two and the phase space is four-dimensional. The Hamiltonian is commutable with the number operator, $\hat{N} = \Sigma_{\sigma k}c^{\dagger}_{k\sigma}c_{k\sigma}$, which is a one-body operator. Therefore, even when the basis vectors (determinants) break a number-conservation, the classical (TDHFB) equation of motion allows two constants of motion, energy and the average particle number. This means that the system is integrable and the classical trajectories lie on two-dimensional invariant tori. The basis vectors can be parametrized by the classical orientation ψ_{σ}, ϕ_{σ} $(\sigma=\pm1)$ of quasi-spins as

$$|\{\psi,\phi\}\rangle = \mathcal{N}\exp\{\Sigma_{\sigma}\tan(\psi_{\sigma}/2)e^{-i\phi_{\sigma}}S_{+}(\sigma)\}|0\rangle \tag{29}$$

where $|0\rangle$ denotes a south pole $(S_0=-S=-\Omega/4)$ state. The TDHF equation describes a motion of the quasi-spin vectors on a sphere with radius S. If one introduces four physical variables instead of $\{\psi,\phi\}$ as

$$n = \langle\hat{N}\rangle/\Omega = 1-\tfrac{1}{2}\Sigma_{\sigma}\cos\psi_{\sigma}, \quad \gamma = \Sigma_{\sigma}\phi_{\sigma}$$
$$q = \langle Q\rangle/\Omega = -\tfrac{1}{2}\Sigma_{\sigma}\sigma\cdot\cos\psi_{\sigma}, \quad \delta = \Sigma_{\sigma}\sigma\cdot\phi_{\sigma} \tag{30}$$

then the equations of motion take a simple form:

$$(h\Omega/4)d\gamma/dt = \partial\mathcal{H}/\partial n, \quad (h\Omega/4)dn/dt = -\partial\mathcal{H}/\partial\gamma$$
$$(h\Omega/4)d\delta/dt = \partial\mathcal{H}/\partial q, \quad (h\Omega/4)dq/dt = -\partial\mathcal{H}/\partial\delta \tag{31}$$

The classical Hamiltonian $\mathcal{H}=\langle H\rangle$ is independent of γ, the total polar orientation of quasi-spins. Thus one recovers the conservation of the average number Ωn. In terms of the variables (30), the 1-form ω, (10), is given by

$$\omega = (\hbar\Omega/4)(nd\gamma+qd\delta) \tag{32}$$

in accordance with (31). On the two-dimensional invariant torus determined by a fixed energy and average number, there are two independent closed loops. From the quantization condition in the γ-direction one obtains

$$\langle\hat{N}\rangle = \Omega n = \text{even integer} \tag{33}$$

which quantizes the average number of particles. From the condition on closed loops in the q-δ plane for fixed γ, we obtain the quantization of energy. Explicit dependence of $\mathcal{H}$ on q and δ shows that the period and the canonical action$\sim\oint q d\delta$ are given by the first and the third elliptic integrals, respectively. The quantized spectrum is plotted in Fig.1 together with the exact spectrum. Depending on the relative strengths of pairing (G) and particle-hole (X) forces, there appear four different types of orbits, from which one can classically locate a phase transition in the quantized spectrum.

The validity of the semiclassical approach may be further examined by looking at the wave functions. In the classical limit, the wave function is concentrated on the quantized invariant torus which is shown in the upper part of Fig.2. One can visualize a phase space structure of the exact eigenfunction $|m\rangle$ by calculating

$$\langle\{\psi,\phi\}|m\rangle = A(q,\delta)e^{iW(q,\delta)}e^{iN\gamma/4} \tag{34}$$

The γ-dependence of (34) is trivially reproduced by the semiclassical wave function (18). In the lower part of Fig.2 are shown the amplitudes $A(q,\delta)$ for a few cases. One can see a correspondence between the exact wave function and the classical invariant torus. One can

Fig.1 Exact(solid lines) and semiclassical(+) energy spectra of the R(4) model plotted against the relative strength of the pairing(G) and the particle-hole(X) forces. Dotted lines distinguish regions associated with the four different types of classical trajectories.

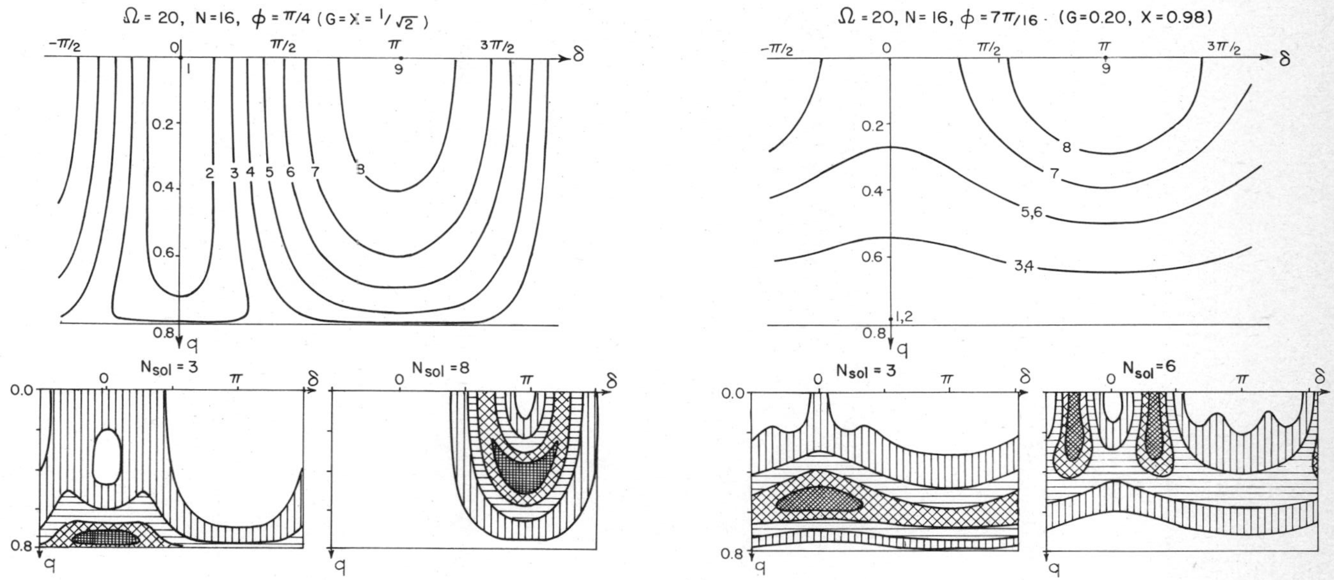

Fig.2 (Upper part) Projection of the classically quantized tori on the q-δ plane for $\phi=\pi/4$ and for $\phi=7\pi/16$ ($G=\cos\phi$, $X=\sin\phi$). On each line is given the solution number N_{sol}(=m) from the lowest one. The figure is symmetrical about the δ-axis and only the positive-q part is shown. (Lower part) Selected phase space picture of the amplitude $A(q,\delta)$ of the exact solution. Darker regions indicate a concentration of the wave function. Wave functions for the other solutions are also peaked roughly around the corresponding quantized tori.

also trace a change of the structure of the exact wave function similar to the phase transition appearing in the classical orbits. This change roughly agrees with the dotted lines in Fig.1.

Another possible check of the semiclassical method may be given by the matrix elements of observables. Because of a symmetry of the Hamiltonian, the 'Q-moment' $\langle m|Q|m\rangle$ of the eigenstates vanishes exactly. One can, instead, calculate the matrix element $\langle m|Q^2|m\rangle$ which gives the sum-rule value of Q-transition for the m-th excited state. In the semiclassical method one can calculate the quantity from the value, $\partial E_\lambda/\partial\lambda|_{\lambda=0}$, where E_λ is the quantized energy of the Hamiltonian $H+\lambda Q^2$. The result is shown in Table I. Except for the quantized states close to a separatrix, the main features of the exact result are reproduced.

Table I

	m	1	2	3	4	5	6	7	8	9
G=0.92	E	26.2	75.9	118.9	152.2	169.4	163.9	134.7	88.1	30.8
X=0.38	S	20	66	108	136	152	140	110	68	20
G=0.71	E	64.0	144.0	161.5	157.5	142.6	120.1	91.8	58.4	20.2
X=0.71	S	20	114	138	136	124	106	82	54	20
G=0.56	E	207.5	216.1	77.8	123.3	105.3	93.6	73.0	47.1	16.5
X=0.83	S	240	240	80	102	100	90	70	46	20

Calculated values of $\langle m|Q^2|m\rangle$ for the m-th quantized state. E=exact and S=semiclassical. $\Omega=20$, N=16.

REFERENCES

1) M.V.Berry and K.E.Mount, Rep.Prog.Phys.35(1972)315.
2) H.Reinhardt, Nucl.Phys.A346(1980)1.
3) S.Levit, J.W.Negele and Z.Paltiel, Phys.Rev.C21(1980)1603.
4) H.Kuratsuji and T.Suzuki, Phys.Lett.92B(1980)19.
5) J.Blaizot and H.Orland, Phys.Rev.C24(1981)1740.
6) H.Kuratsuji and T.Suzuki, J.Math.Phys.21(1980)472.
7) H.Kuratsuji, Phys.Lett.103B(1981)79, Phys.Lett.108B(1982)367.
8) T.Suzuki, preprint NBI-82-22.
9) A.M.Perelomov, Comm.Math.Phys.26(1972)222, Sov.Phys.Uspekhi 20(1977)703.
10) A.K.Kerman and S.E.Koonin, Ann.Phys.100(1976)332.
11) M.V.Berry, in 'Topics in Nonlinear Dynamics', AIP Conf.Proceedings No.46(American Institute of Physics, N.Y., 1978) p.16.
12) I.C.Percival, Adv.Chem.Phys.36(1977)1.
13) M.C.Gutzwiller, in 'Path Integrals', eds. G.J.Papadopoulos and J.J.Devereese (Plenum, N.Y., 1978) p.163.
14) A.Einstein, Verh.Deut.Phys.Ges.19(1917)82.
15) K.Matsuyanagi, Prog.Theor.Phys.67(1982), in press. See also, R.Piepenbring et al., Nucl.Phys.A348(1980)77.
16) Y.Mizobuchi, Prog.Theor.Phys.65(1981)1450.

MEAN-FIELD THEORY OF NUCLEAR DYNAMICS

H. Reinhardt
Zentralinstitut für Kernforschung, Rossendorf
8051 Dresden, PF 19, DDR

1. Introduction

The aim of nuclear theory is to understand the structure and the dynamics of many-nucleon systems in terms of the underlying two-body interaction, while the understanding of the hadronic interaction itself is subject to subnuclear physics. Given the complexity of the nuclear many-body problem rather crude approximations have to be introduced to arrive at practicable equations. Among these the concept of independent nucleons moving in a self-consistent potential has proved successful in the understanding of low-energy nuclear phenomena. For energies up to several MeV per nucleon the Pauli principle essentially suppresses two-body collisions, which results in a long mean-free path of the nucleons. This concept was in particular supported by the success of the static Hartree-Fock calculations, which reproduced astonishingly well the ground state characteristics of the nuclei such as binding energies and charge density distributions as well as single particle energies (see e.g. ref. 1). Furthermore the small amplitude vibrations have been successfully described within the RPA, which is known to arise in the small amplitude limit from the time-dependent Hartree-Fock (TDHF) approximation [2]. Stimulated by the success of the static Hartree-Fock and RPA calculations one has applied the full TDHF theory to the so-called large amplitude collective motion in particular to the description of heavy ion collisions [3]. However, the standard TDHF theory as initial value problem suffers from conceptual ambiguities arising from the fact that the initial condition appropriate to a given physical situations are not specified by the theory itself.
What one would like to have is a theory which incorporates the successful independent particle (mean-field) concept into a complete quantum description. This goal has been recently achieved by exploiting path integral techniques. A quantum mean-field theory could be formulated, which in principle may provide an overall understanding of nuclear dynamics. This talk is intended to survey the quantum mean-field description of bound states, spontaneous decay processes as nuclear fission and reactions between self-bound many-body systems. I shall rather briefly treat the bound state and the fission problem, since

I expect that Prof. Negele will report in detail on these subjects, and shall concentrate on the presentation of the recently developed mean-field S-matrix theory.

2. Path integral approach and mean-field theory

During the last five years there has been a rapid increase in the interest of path integral methods in the study of the nuclear many-body problem. The original interest in these methods arose from their great flexibility in view of which they allow in a very elegant way to derive modified perturbation expansions. By means of these methods one could for the first time give a strict non-perturbative foundation of the so-called nuclear field theory [4]. In the subsequent time the path integral approach has proved as a very powerful and versatile method for deriving appropriate mean-field approximations for many-body systems. In particular they have given new insights into the nature of TDHF and have shown the precise relation of TDHF to the exact Schrödinger theory. In contrast to the familiar initial-value TDHF the path integral approach yields by applying the stationary phase method a mean-field approximation,which is of quantum nature in the sense that the corresponding mean-field is uniquely defined by the quantum process considered.

In what follows I will briefly sketch the formulation of the mean-field approximation in the path integral approach and discuss the different approaches and their mutual relationships. Roughly speaking we can distinguish three approaches: the field-theoretic approach [5-8], the approach of the MIT group [9-11] and the determinantal approach [12,13]. The first approach starts from the functional integral representation of the time-evolution operator

$$\mathcal{U}(t_f, t_i) = \exp\left[-i H(t_f - t_i)\right] \quad , \quad \hbar = 1 \tag{2.1}$$

in terms of classical fields [14], which are ordinary c-number fields in the case of Bose systems and anticommuting (Grassmann) fields in the case of Fermi systems. After a linearization$^{x)}$ of the two-body interaction in the Hamiltonian

$$H = H(\xi, \xi^+) = \int dx\, \xi^+(x) \left(\frac{-\Delta_x}{2m}\right) \xi(x) + \frac{1}{2} \int dx\, dy\, \xi^+(x) \xi^+(y)\, V(x,y)\, \xi(y)\, \xi(x) \tag{2.2}$$

---

x) To avoid confusion let me emphasise that this linearization has not to be understood in the sense of the familiar linearization of the equation of motion (RPA), which contains an approximation, but is exact.

by means of an auxiliary commuting (Bose) field ϱ the original elementary fields can be integrated out and one obtains the following representation of the time-evolution operator

$$\mathcal{U}(t_f,t_i)=(\mathcal{D}et\, V)^{\frac{1}{2}}\int\mathcal{D}\varrho\, \exp\left[\frac{i}{2}\int_{t_i}^{t_f}dt\, \varrho(t)\, V\varrho(t)\right]\, \mathcal{U}_\varrho(t_f,t_i) \tag{2.3}$$

where

$$\mathcal{U}_\varrho(t_f,t_i) = T\, \exp\left[-i\int_{t_i}^{t_f}dt\, \hat{h}_\xi[\varrho(t)]\right] \tag{2.4}$$

is defined by the s.p. Hamiltonian

$$\begin{aligned} \hat{h}_\xi[\varrho] &= \int dx\, dy\, \xi^+(x)\, h[\varrho](x,y)\, \xi(y), \\ h[\varrho](x,y) &= \left[-\frac{\Delta x}{2m} + \int dy'\, V(x,y')\, \varrho(y',y',t)\right]\delta(x,y). \end{aligned} \tag{2.5}$$

According to the relation (2.5) the exact time-evolution operator of an interacting many-body system can be expressed as a superposition of s.p. operators U_ϱ describing the time-evolution of the non-interacting system in an "external" field $\int dy\, V(x,y)\varrho(y,y,t)$. The path integral over the bilocal density field $\varrho(x,x',t)$ forms just the superposition of the infinite number of U_ϱ's corresponding to all possible external fields. The weight of a particular density field $\varrho(t)$ (ϱ-trajectory) is determined by the two-body interaction V and is given by $\exp[\frac{i}{2}\int\varrho(t)\, V\varrho(t)\, dt]$.

The MIT group [9-11] arrives at a similar relation (2.3) by using the Hubbard-Stratonowitch relation [15] except that the s.p. Hamiltonian is here given by

$$h'[\varrho](x,y) = \left[-\frac{\Delta x}{2m} - \frac{1}{2}V(x,x) + \int dy'\, V(x,y')\, \varrho(y',y',t)\right]\delta(x,y). \tag{2.6}$$

It differs by the term $-\frac{1}{2}V(x,x)$ which eliminates the self-interaction present in the third term. As model studies show, for large particle numbers $N \gg 1$ the self-interaction can be discarded in the mean-field description.

In particular situations it is even advantageous to include the self-interaction in the mean-field description. This is not only because of technical reasons ($-\frac{1}{2}V(x,x)$ may be infinite!) but yields also a better approximation.

Let us now consider a particular quantum process leading the system from a state ϕ_i after a time $t_f - t_i$ into a state ϕ_f. Within the laws of quantum mechanics this process is uniquely described by the corresponding matrix element of the time-evolution operator (transition amplitude) $\langle \phi_f | U(t_2,t_1) | \phi_i \rangle$. An appropriate time-dependent mean-field approximation to the transition amplitude can be found by using the path integral formula (2.3) for U and evaluating the resulting functional integral in the SPA. This yields the following temporally non-local mean-field equation

$$\rho(x,x',t) = \frac{\langle \phi_f | U_\rho(t_f,t) \xi^+(x') \xi(x) U_\rho(t,t_i) | \phi_i \rangle}{\langle \phi_f | U_\rho(t_f,t_i) | \phi_i \rangle} \quad . \tag{2.7}$$

The solutions to this equation satisfy the usual TDMF equation for the density matrix

$$i \dot{\rho} = [h[\rho], \rho]. \tag{2.8}$$

Since the mean-field equation is temporally non-local it contains already the necessary boundary condition, which is uniquely determined by the external wave functions ϕ_i and ϕ_f, but has not to be supplied from the outside, which in turn allows an unambiguous quantum mechanically interpretation of the mean-field results (see below). The boundary condition to eq. (2.8) is explicitly obtained by taking the eq. (2.7) at an arbitrary but fixed time $t_o \in [t_i, t_f]$. For a fixed time eq. (2.7) represents a temporally non-local self-consistent condition which couples the mean-field solutions at all times in a non-linear way. (This reflects just the fact that the mean-field solution of eq. (2.7) is not determined by an initial value but simultaneously by the initial and final wave functions ϕ_i and ϕ_f).
If the external wave functions are Slater determinants the mean-field values of symmetry operators (i.e. single particle operators which commute with the exact Hamiltonian) are conserved. The proof proceeds in exactly the same way as in the ordinary TDHF theory [16] except that one has to use the extended form of Wick's theorem [17].
If the mean-field solution ρ to eq. (2.7) is known the mean-field value of the transition amplitude is given by the stationary phase result of the functional integration

$$\langle \phi_f | U(t_f,t_i) | \phi_i \rangle = \left[\mathrm{Det} \frac{\delta^2 \mathcal{A}[\rho]}{\delta\rho \, \delta\rho} \Big/ \mathrm{Det}\, V \right]^{-\frac{1}{2}} \exp\left[i \mathcal{A}[\rho] \right], \tag{2.9}$$

where the effective action $A[\rho]$ is defined by the logarithm of the integrand:

$$\mathcal{A}[\rho] = \frac{1}{2} \int_{t_i}^{t_f} dt \, \rho(t) V \rho(t) - i \log \langle \phi_f | U_\rho(t_f,t_i) | \phi_i \rangle \tag{2.10}$$

The pre-exponential factor gives the effect of the small quantum fluctuations of the (density) field around its mean-field value ϱ. In practice it is usually very hard to evaluate the corresponding functional determinants. This task has been accomplished so far only for periodic TDMF orbits as they arise in the description of bound states [7] and spontaneous fission [18] (see ref. [19]). Following the method of ref. [7] the quantum corrections have been recently evaluated also for the level densities [20]. Due to technical difficulties arising in the evaluation of the quantum corrections in practical applications one will restrict oneself to the mean-field approximation (neglecting the pre-exponential factor in eq. (2.9)).

You will surely have noticed that the exchange term of the mean-field is missing in the s.p. Hamiltonian $h[\varrho]$ (2.5) as well as in $h'[\varrho]$ (2.6). This is a drawback of the two approaches of refs. [5-8] and [9-11]. While they are physically very transparent in the zero'th order SPA the exchange term does not naturally arise at the same time as the direct one, though these two approaches can, in principle, be reformulated so that the exchange term can be included in the mean-field approximation.

The exchange term arises naturally in the determinantal functional integrals of refs. [12,13] but the price is a much more involved formalism. Here the functional integration is performed over complex variables $z_{\alpha\gamma}(t)$ which parametrize via Thouless's theorem $|z\rangle = \exp[\sum_{\alpha,\gamma} z_{\alpha\gamma} a_\alpha^+ a_\gamma] |\phi_0\rangle$ the Slater determinantal basis $|z\rangle$ of the many-body Hilbert space. The $z_{\alpha\gamma}(t)$'s play an analogous role as the auxiliary field $\varrho(x,y,t)$. Since one works here in the determinantal basis the exchange terms are handled from the beginning on the same footing as the direct ones. The mean-field approximation arises again in the SPA to the functional integral over the $z_{\alpha\gamma}(t)$'s.

Though the three approaches differ in technical details the physics involved is essentially the same. In the remaining part of my talk I will discuss results obtained within the first approach.

3. Mean-field description of bound states (Quantization of TDHF)

For the description of bound states it is convenient to start from the (canonical) partition function

$$z^T = tr\, \mathcal{U}(\tfrac{T}{2}, -\tfrac{T}{2}) = \sum_n e^{-iE_n T} \tag{3.1}$$

since the poles of its Fourier transform

$$Z^E = -i\int_0^\infty dT\, e^{iET}\, Z^T = \sum_n \frac{1}{E-E_n} \tag{3.2}$$

determine the quantized energies E_n.
Using the path integral formula (2.3) for $U(\frac{T}{2}, -\frac{T}{2})$ and evaluating the resulting functional integral in the SPA one derives a mean-field approximation to the partition function Z^T completely analogously to the case of a single matrix element of U considered above. The trace fixes via the resulting temporally non-local mean-field equation the boundary condition to the mean-field to be periodic. Using Bloch's theorem the mean-field equation can be cast into the form of an hermitean eigenvalue problem

$$(i\partial_t - h[\rho])|f_\nu(t)\rangle = -\epsilon_\nu |f_\nu(t)\rangle \tag{3.3a}$$

where the f_ν's are periodic functions satisfying the self-consistent condition

$$\rho(t) = \sum_\nu |f_\nu(t)\rangle\, n_\nu \langle f_\nu(t)|\,, \quad f_\nu(\tfrac{T}{2}) = f_\nu(-\tfrac{T}{2}) \tag{3.3b}$$

Evaluating the Fourier integral of the mean-field partition function in the SPA one finds from the pole condition of Z^E the quantization condition x)

$$\sum_\nu n_\nu \int_{-\frac{T}{2}}^{\frac{T}{2}} dt \langle f_\nu(t)|\, i\partial_t\, |f_\nu(t)\rangle = 2\pi n \tag{3.4}$$

which selects out from the branches of periodic TDHF solutions those corresponding to quantized eigenstates.

The self-consistent eigenvalue problem (3.3) is obviously solved by all the static HF solutions, which are supposed to form the "band heads" of the branches of non-trivial periodic TDHF solutions [21]. The existence of such branches is known at least for rotational bands where the periodic mean-field solutions are given by the self-consistent cranking (SCC) model. In fact for the Yrast states the SCC model provides the most general (periodic) TDHF solutions. For the SCC solutions the quantization condition yields just the quantization of the angular momentum [22,23]

$$tr(j_x \rho) = I \tag{3.5}$$

where I is the quantum number of the square of the nuclear spin. If one includes the leading order quantum corrections to the mean-field

x) The same quantization condition was also obtained within the gauge-invariant quantization method [32].

approximation (see ref. [7]) the quantization condition (3.5) is modified to [23]

$$tr(j_x \varrho) = I + \tfrac{1}{2} \tag{3.6}$$

which except for I < 2 agrees very well with the result obtained by the projection method [24]

$$tr(j_x \varrho) = \sqrt{I(I+1)}. \tag{3.7}$$

Unfortunately so far nothing is known for realistic systems about non-cranking periodic TDHF solutions since a numerical solution of the self-consistent eigenvalue problem seems hard to be contrived. The self-consistent eigenvalue problem cannot be solved by the standard iteration procedure used in the static HF problem due to numerical instabilities, which are likely to occur since these equations follow from an extremal principle but not from a minimal principle.

4. Spontaneous fission

The spontaneous decay of a many-body system can be treated in the path integral approach analogously to the bound state problem by considering the evolution of the system for pure imaginary times $t = -i\tau$ [18]. The resulting mean-field equation describing the decaying system follows from eq. (3.3a) after the Wick rotation $\tau = it$ to be

$$(\partial_\tau + h[\varrho])\,|f_\nu(\tau)\rangle = \epsilon_\nu\,|f_\nu(\tau)\rangle \tag{4.1a}$$

where the self-consistent condition reads now

$$\varrho(\tau) = \sum_\nu |f_\nu(\tau)\rangle\, n_\nu \,\langle f_\nu(-\tau)|, \quad f_\nu(\tfrac{T}{2}) = f_\nu(-\tfrac{T}{2}) \tag{4.1b}$$

and the functions $f_\nu(\tau)$ are again periodic.

As in the real time case the eigenvalue problem (4.1) has the static HF solutions ϱ^0. For systems showing spontaneous fission the static HF solution ϱ^0 may be classically stable but becomes unstable if quantum tunnelling is switched on. The imaginary-time mean-field equation (4.1) has then non-trivial periodic solutions of period T, which for $T \to \infty$ describe the time evolution of the system from the static HF state at $\tau = -\frac{T}{2}$ to the scission point at $\tau = 0$ where the system is bounced off. At $\tau = \frac{T}{2}$ it reaches again the static HF solution. A closer examination of the imaginary-time mean-field equation [18] shows that the bounce solution $\varrho(\tau)$ approaches the static HF solution ϱ^0 for $|\tau| \to \frac{T}{2} \to \infty$ in the direction of the lowest RPA mode ω_0 like $\exp(-\omega_0 \tau\ \ )$.

Prof. Negele has found this bounce solution numerically for a number of light nuclei [25]. These solutions give a dynamical picture of the fission process. The numerical solution of eqs. (4.1) by iteration does not show the difficulties appearing in the real time case, since they follow here from a minimum principle.

If one includes the leading order quantum correction to the "classical" bounce solution one is able to derive an expression for the life-time $1/\Gamma$ of the decaying many-body system, which in zero order (neglect of the contributions from the fluctuations) is given by [18]

$$\Gamma \sim \exp\Big[-\int_{-\frac{T}{2}}^{\frac{T}{2}} d\tau \sum_{\nu} n_{\nu} \langle f_{\nu}(-\tau) | \partial_{\tau} | f_{\nu}(\tau) \rangle \Big], \quad T \longrightarrow \infty \, . \tag{4.2}$$

In this way the life-time of spontaneously decaying many-body systems can be evaluated from the underlying two-body interaction.

5. Reactions between self-bound many-body systems

5.1. Preliminaries

The mean-field description of the scattering of composite many-body systems is a much more involved problem than those considered before but also the most interesting one. As it is well known the initial-value TDHF is capable of reproducing the cross features of the inclusive data of heavy-ion reactions but does not allow an exclusive description of reactions. The latter is connected with the difficulties of defining a S-matrix within the usual initial-value TDHF, which are mainly based on the following facts: The initial-value TDHF generates a single determinantal wave function which at later times is supposed to contain all the effects of the collision. This wave function evolves in a non-linear way and if it is expanded with respect to an orthogonal set of channel states, the modules of the expansion coefficients will not be asymptotically constant. This fact can also be expressed by saying that the TDHF wave function contains "spurious cross channel correlations" [33] which obviously are a direct consequence of the lack of the superposition principle, in contrast to the exact Schrödinger theory. Thus a "S-matrix" constructed by projecting the TDHF wave functions on channel states would not be asymptotically constant, which means that the results predicted by the TDHF theory for physical measurements would depend on the precise position of the measuring apparatus. The single determinantal description of the initial-value TDHF precludes therefore any exclusive descriptions of multi-channel reactions.

Toward a mean-field reaction theory in ref. [33] TDHF analogs of the "in"- and "out"-states of the Schrödinger theory were used to define a mean-field S-matrix theory. While in this way the spurious cross channel correlations were excluded the resulting S-matrix was not unambiguously defined and an additional time-averaging had to be introduced, which however is not entirely understood yet.

A more prosperous way to a mean-field theory of exclusive reactions is offered by the path integral approach, which overcomes the single determinantal description of the initial-value TDHF in that for each quantum process it defines uniquely a corresponding time-dependent mean-field, as we have seen before. In what follows let me demonstrate how an exclusive mean-field description of reaction processes can be brought about [26].

Let us consider a reaction of the type

$$a + b \longrightarrow A + B \tag{5.1a}$$

where a,b,A and B are self-bound systems of identical particles with particle numbers N_a, N_b, N_A and N_B, respectively. For simplicity we shall not consider here reactions leading to more than two final fragments. The extension of the approach given below to such reactions is, however, straightforward. Furthermore we shall restrict ourselves to non-relativistic processes so that the total number of particles N is conserved.

$$N_a + N_b = N = N_A + N_B \tag{5.1b}$$

The S-matrix for reactions between composite systems of identical particles is defined in the interaction picture by [27]

$$S_{fi} = \lim_{T\to\infty} C_i C_f \langle \Psi_f | \mathcal{U}_f(0,T) \mathcal{A}\, \mathcal{U}(T,-T) \mathcal{A}\, \mathcal{U}_i(-T,0) | \Psi_i \rangle \tag{5.2}$$

where U denotes the time-evolution operator of the total (reacting) system while $U_i = U_a U_b$ and $U_f = U_A U_B$ are the evolution operators of the independent fragments in the initial and final channel. A denotes the antisymmetrization operator of the N-particle system. The channel functions $\Psi_i = \Psi_a \Psi_b$ and $\Psi_f = \Psi_A \Psi_B$ are supposed to be properly antisymmetrized in the s.p. indices of the individual fragments. C_i and C_f are combinatorial factors ensuring that the states $\widetilde{\Psi}_i = C_i \mathcal{A}\, \Psi_i$ and $\widetilde{\Psi}_f = C_f \mathcal{A}\, \Psi_f$ are normalized to unity.

In the exact Schrödinger theory the channel functions Ψ_c, c = i,f are eigenfunctions of the channel Hamiltonians

$$H_i = H_a + H_b , \qquad H_f = H_A + H_B \tag{5.3}$$

Denoting the corresponding eigenvalues by E_c the expression for the S-matrix (5.2) can be simplified to

$$S_{fi} = \lim_{T\to\infty} e^{iE_fT} \langle \tilde{\Psi}_f | \mathcal{U}(T,-T) | \tilde{\Psi}_i \rangle e^{iE_iT} \tag{5.4}$$

If one tries to approximate the transition amplitude $\langle \tilde{\Psi}_f | U | \tilde{\Psi}_i \rangle$ in the time-dependent mean-field theory offered by the path integral approach presented above the resulting S-matrix would not be asymptotically constant since the non-linear time-evolution of the mean-field causes non-trivial oscillations of the mean-field amplitude, which cannot be compensated for by the phase factor exp[i E_fT + i E_iT] even if Ψ_c are exact eigenstates of H_c. Alhassid and Koonin have shown for a Fermi system interacting with an external field that asymptotically constant mean-field excitation amplitudes can be obtained within the path integral approach if the preparation stages are simultaneously treated with the reaction process. It turns out the same trick works also in the more involved problem of the collision of two composite systems [26]. In fact a mean-field reaction theory can be formulated in the following way: One starts from the interaction picture S-matrix (5.2) and represents each of the time-evolution operators in S by means of the path integral formula (2.3) as a superposition of s.p. evolution operators. Application of the SPA to all of these path integrals should define an appropriate mean-field approximation to the S-matrix.

If one really tries to perform this program one immediately faces the following two difficulties:

i) Application of the path integral formula (2.3) assumes that the evolution operator i.e. the Hamiltonian of the considered many-body system is given in its second quantization representation. But in second quantization the Hamiltonians of the individual fragments have exactly the same representation as the Hamiltonian of the total system, which makes the extraction of the channel interaction $H - H_c$ impossible.

ii) One needs an explicit representation of the many-body channel states Ψ_c, which in general are not known. One has therefore to find appropriate approximate many-body functions which should fulfil the basic requirements of the formal scattering theory to the channel eigenstates being eigenfunctions of the momentum and the channel Hamiltonian H_c. Both problems have been solved in ref. [26].

5.1.1. Second quantization representation of the interaction picture S-matrix

A suitable field operator representation of the interaction picture S-matrix can be derived by introducing different sets of field operators for the individual subsystems (fragments) a, b, A, B and the total system, which we denote, respectively, by (a, a^+), (b, b^+), (A, A^+), (B, B^+) and (Ψ, Ψ^+), and imposing appropriate commutation relations between them: All the different types of field operators, which we denote by the common letter ξ = a, b, A, B, Ψ, have the same vacuum $|o\rangle$ ($\xi(x)|o\rangle = o$) and satisfy the usual fermion anticommutation relations, e.g.

$$\{\Psi(x), a(x')\} = \{\Psi^+(x), a^+(x')\} = 0, \qquad \{\Psi(x), a^+(x')\} = \delta(x, x') \tag{5.5a}$$

except that the field operators referring to different fragments of the same channel anticommute, i.e.

$$\{a(x), b(x')\} = \{a^+(x), b^+(x')\} = \{a(x), b^+(x')\} = 0, \qquad \{A(x), B(x')\} = \{A^+(x), B^+(x')\} = \{A(x), B^+(x')\} = 0, \tag{5.5b}$$

which ensures the independence of the different fragments in the free channel propagation during the preparation stages.

With the so defined commutation relations we can write the free channel Hamiltonian H_c, where the channel interaction between the fragments is "switched off", simultaneously with the total Hamiltonian H, which includes the full interaction, in a field operator representation by setting $H_\xi = H(\xi, \xi^+)$, ξ = a, b, A, B and $H = H(\Psi^+, \Psi)$, where $H(\xi, \xi^+)$ is the Hamiltonian of the system ξ in the usual second quantization representation (2.2). For the interaction picture S-matrix (5.2) one gets in this field operator representation the following expression

$$S_{fi} = \lim_{T\to\infty} \langle \Psi_f | \mathcal{U}_f(0,T)\, \mathcal{U}(T,-T)\, \mathcal{U}_i(-T,0) | \Psi_i \rangle , \qquad (5.6)$$

where the time-evolution operators are expressed here by the field operators via the corresponding Hamiltonians. Note the antisymmetrization operators as well as the related normalization factors disappeared now, since the identity of the particles is already taken into account in the field operator representation.

The representation (5.6) is exact and holds independently of the chosen channel functions Ψ_c, which are only assumed to be antisymmetric with respect to the permutations of the particle indices of the same fragment but not with respect to permutations of particle indices referring to different fragments, so that they can be expressed by the field operators of the individual fragments.

Note, during the so-called preparation stages described by the channel evolution operators $U_{c=i,f}$ the fragments of a given channel move completely independent of each other ("free" channel motion). These stages are necessary in order to eliminate the changes in the wave function originated by the time-evolution of the independent fragments which are not the result of the scattering process but which would also evolve in the isolated fragments. Due to the anticommutation relations (5.5b) during these stages even the Pauli principle is "switched off" between the particles belonging to different fragments. This ensures that the effects of the exchange correlation between particles of different fragments during the scattering process itself (described by U) are taken into account by the S-matrix.

5.1.2. Projected channel eigenstates

A suitable representation of the channel eigenstates can be obtained by projecting arbitrary many-body wave functions (e.g. Slater determinants) of the independent fragments a(A), b(B) of the channel i(f) $|\phi_{i(f)}\rangle = |\phi_{a(A)}\rangle |\phi_{b(B)}\rangle$ onto states with good momentum and channel energy. Let P_c, R_c and p_c, r_c, respectively, denote the operators of the momenta and coordinates of the center of mass and the relative motion of the two fragments in the channel c. The projected channel functions with c.m. and relative momentum K_c and k_c, respectively, then read

$$|\Psi_{c\,\varepsilon_c K_c k_c}\rangle = \exp\left[i K_c R_c + i k_c r_c\right] | \chi_{c\,\varepsilon_c}\rangle \qquad (5.7a)$$

where

$$|\chi_{c\mathcal{E}_c}\rangle \sim \delta(H_c-\mathcal{E}_c)\,\delta(P_c)\,\delta(p_c)\,|\phi_c\rangle \tag{5.7b}$$

is the intrinsic function, which by construction is an eigenfunction of H_c and P_c, p_c with eigenvalues $\mathcal{E}_c$ and zero, respectively. For definiteness we have here supposed that $\mathcal{E}_c$ belongs to the continuous part of the spectrum. (Otherwise the projector $\delta(H_c - \mathcal{E}_c)$ has to be modified, see ref. [26]). The basis functions ϕ_c can be chosen, in principle, arbitrarily as long as we use the exact projector $\delta(H_c - \mathcal{E}_c)$. If we later approximate the δ-functions the concrete choice of the ϕ_c's becomes material. Intuitively we suggest to choose the ϕ_c's in such a way that the expectation values of P_c, p_c and H_c in these states coincide with the corresponding eigenvalues of the intrinsic functions $\chi_{c\mathcal{E}_c}$, and furthermore that they are localized at the center of mass.

5.1.3. The interaction picture S-matrix in the representation of the projected channel states

Using the Fourier integral representation of the δ-functions in (5.7b) one derives for the S-matrix elements between the projected channel states (5.7) the following (second quantization) representation

$$S_{fi} = \delta(E_i - E_f)\,\delta(K_i - K_f)\,\mathcal{S}_{fi} \tag{5.8}$$

where up to a normalization factor the reduced S-matrix $\mathcal{S}_{fi}$ reads (cf. ref. [26])

$$\mathcal{S}_{fi} = \int \frac{d\tau}{\sqrt{2\pi}}\, e^{-iE\tau} \int \frac{dq_i\, dq_f\, dQ}{(2\pi)^{9/2}} \lim_{T\to\infty} S^T_{fi}(q_f,q_i,Q,\tau) \tag{5.9}$$

Here $E = E_i = E_f$ and

$$S^T_{fi}(q_f,q_i,Q,\tau) = \langle \phi_{f k_f q_f Q}|\,U_f(\tfrac{\tau}{2},T)\,U(T,-T)\,U_i(-T,-\tfrac{\tau}{2})\,|\phi_{i k_i q_i -Q}\rangle \tag{5.10}$$

is the S-matrix between the wave packet like states

$$|\phi_{c k_c q_c Q}\rangle = e^{ik_c r_c}\, e^{-iq_c p_c - i\frac{Q}{2}P_c}\,|\phi_c\rangle . \tag{5.11}$$

Eqs. (5.8), (5.9) express the S-matrix elements between the projected channel eigenstates, as a superposition of the S-matrix elements (5.10) between the localized, wave packet like states $|\phi_c\, q_c\, k_c\, Q\rangle$ (5.11). For the latter ones a mean-field approximation is more appropriate since a mean-field of finite bound Fermi systems should be localized in space.

Note that the energy projection does not introduce any new difficulty but merely shifts the time-argument in the channel evolution operators.

5.2. Time-dependent mean-field theory for reaction processes

To find an appropriate mean-field approximation to the S-matrix we apply now the path integral formula (2.3) to each of the many-body time-evolution operators U_a, U_b, U_A, U_B and U appearing in eq. (5.6). Denoting the corresponding auxiliary fields (integration variables) by ρ_a, ρ_b, ρ_A, ρ_B and ρ we get for the reduced S-matrix between the wave packet like states (5.10)

$$\mathcal{S}^T_{fi}(q_f, q_i, Q, \tau) = (\mathrm{Det}\, V)^{\frac{5}{2}} \int \mathcal{D}\rho_A \mathcal{D}\rho_B \mathcal{D}\rho \mathcal{D}\rho_a \mathcal{D}\rho_b \exp\left[\tfrac{i}{2} \oint \rho V \rho\, dt\right] \mathcal{S}^T_\rho(q_f, q_i, Q, \tau) \tag{5.12}$$

where the single particle S-matrix element $\mathcal{S}^T_\rho$ is given by

$$\mathcal{S}^T_\rho(q_f, q_i, Q, \tau) = \langle \phi_{f k_f q_f Q} | \mathcal{U}_{\rho_f}(\tfrac{\tau}{2}, T)\, \mathcal{U}_\rho(T, -T)\, \mathcal{U}_{\rho_i}(-T, -\tfrac{\tau}{2}) | \phi_{i k_i q_i -Q} \rangle \tag{5.13}$$

where $U_{\rho_i} \equiv U_{\rho_a} U_{\rho_b}$, $U_{\rho_f} \equiv U_{\rho_A} U_{\rho_B}$ and the phase is defined by

$$\oint \rho V \rho\, dt = \int_T^{\frac{\tau}{2}} dt\, (\rho_A V \rho_A + \rho_B V \rho_B) + \int_{-T}^{T} dt\, \rho V \rho + \int_{-\frac{\tau}{2}}^{-T} dt\, (\rho_a V \rho_a + \rho_b V \rho_b) \tag{5.14}$$

The path integral representation expresses the many-body S-matrix as a sum of an infinite number of analogous single-particle "S-matrices" determined by the five density fields ρ, ρ_a, ρ_b, ρ_A and ρ_B. The weight of a particular set of density fields (ρ-trajectory) is again controlled by the two-body interaction and is given here by $\exp(\frac{i}{2}\oint \rho V \rho\, dt)$ (see eq. (5.14)). The s.p. S-matrix $\mathcal{S}^T_\rho$ (5.13) describes the following three evolution stages which can be illustrated by the loop shown in fig. 1(a): i) Initial preparation stage: The two subsystems (colliding fragments) a and b, which are initially localized at $\langle R_i \rangle = -\frac{Q}{2}$, $\langle r_i \rangle = q_i$ start at time $t = -\frac{\tau}{2}$ with average momenta $P_i = 0$, $p_i = k_i$ and evolve by the action of U_{ρ_i}

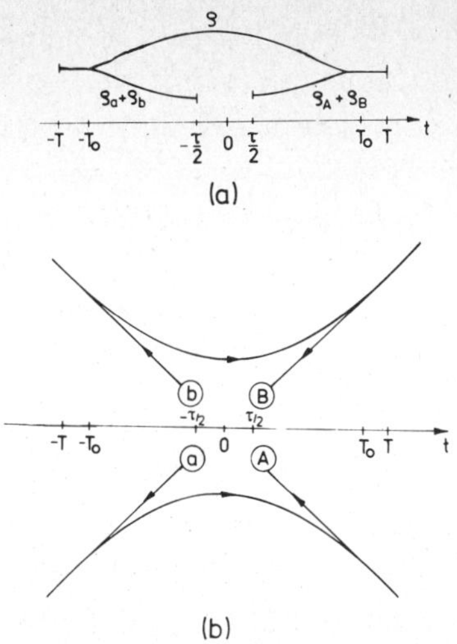

Fig. 1

(a) Graphical illustration of the mean-field solutions of eqs. (5.15).

(b) Mean-field trajectories of the centers of mass of the individual fragments for the case where the fragment velocities are constant during the preparation stages.

independently of each other backward in time, each system in its own "external" field $h[\varrho_a]$ and $h[\varrho_b]$, respectively, up to a time $t = -T < -\frac{\tau}{2}$ $(T > 0)$ (see fig. 1(b). ii) Reaction process: Due to the action of $U_\varrho(T,-T)$ the system evolves then forward in time where all particles move in the same s.p. field $h[\varrho(t)]$ and, in general, a rearrangement of the initially two separated density distributions a and b will take place. iii) Final preparation stage: At time T the system evolves due to the action of $U_{\varrho_f}(\frac{\tau}{2}, T)$ again backward in time up to $t = \frac{\tau}{2}$ where the "A- and B-particles" move again completely independently of each other in their "own" mean-fields $h[\varrho_A(t)]$ and $h[\varrho_B(t)]$, respectively. For convenience, we shall label the three evolution stages by an index c = i, "no index" and f, respectively.

Performing the functional integrals (5.12) in the SPA one obtains five coupled TDMF equations (cf. eq. (2.7)), which can be cast into the following form

$$\begin{aligned}
\varrho_a(x,x',t) &= \langle a^+(x')a(x)\rangle_i(t) \\
\varrho_b(x,x',t) &= \langle b^+(x')b(x)\rangle_i(t) \\
\varrho(x,x',t) &= \langle \psi^+(x')\psi(x)\rangle(t) \\
\varrho_A(x,x',t) &= \langle A^+(x')A(x)\rangle_f(t) \\
\varrho_B(x,x',t) &= \langle B^+(x')B(x)\rangle_f(t)
\end{aligned} \tag{5.15}$$

Here the bracket $\langle O\rangle_c$ is a shorthand notation for the mean-field value of an operator O

$$\langle O\rangle_c(t) = \frac{\langle\tilde{\phi}_c(t)|\,O\,|\phi_c(t)\rangle}{\langle\tilde{\phi}_c(t)|\phi_c(t)\rangle}\,, \tag{5.16}$$

where the wave functions $\phi_c(t)$, $\tilde{\phi}_c(t)$ are defined by a successive action of the s.p. time-evolution operators on the localized states (5.11). They can be obtained by solving the time-dependent Schrödinger equations

$$i\partial_t |\phi_c(t)\rangle = \hat{h}_c |\phi_c(t)\rangle \tag{5.17a}$$

$$-i\partial_t \langle\tilde{\phi}_c(t)| = \langle\tilde{\phi}_c(t)|\hat{h}_c \tag{5.17b}$$

along the loop shown in fig. 1(a) with the initial conditions

$$|\phi_i(-\tfrac{\tau}{2})\rangle = |\phi_{i k_i q_i - Q}\rangle \tag{5.18a}$$

$$\langle\tilde{\phi}_f(\tfrac{\tau}{2})| = \langle\phi_{f k_f q_f Q}| \tag{5.18b}$$

where

$$\hat{h}_i = \hat{h}_a[\rho_a] + \hat{h}_b[\rho_b] \tag{5.19a}$$

$$\hat{h} = \hat{h}_\psi[\rho] \tag{5.19b}$$

$$\hat{h}_f = \hat{h}_A[\rho_A] + \hat{h}_B[\rho_B] \tag{5.19c}$$

and the ρ's are given by the self-consistent conditions (5.15). The numerical solutions of the coupled eqs. (5.17), (5.15) can be performed by iteration, whereas in the first step one inserts for the ρ's in eq.(5.17a) and eq.(5.17b) (cf. eq.(5.19)) the values corresponding to the two related TDHF initial-value problems defined by eqs. (5.17a), (5.18a) and eqs. (5.17b), (5.18b), respectively.

From the structure of the temporally non local eqs.(5.15) immediately follows that the mean-field solutions satisfy the following relations

$$\begin{aligned} \rho_a(-t) + \rho_b(-t) &= \rho(-t) \\ \rho_A(t) + \rho_B(t) &= \rho(t) \end{aligned} \tag{5.20}$$

for t = T. Invoking further the principle of the attenuation of the correlation one can show [26] that the relations holds for all $t \leq T$ larger than some characteristic time T_o (see fig. 1(a)). As a consequence of these asymptotic properties of the mean-field solutions any evolution generated by U (T,-T) if T is increased beyond T_o is exactly canceled by either U_{ρ_i} or U_{ρ_f}. From this follows that the mean-fields $\rho(t)$ for $|t| < T_o$ are independent of T as long as $T > T_o$. The last property, together with eq. (5.20), establishes the asymptotical constancy (i.e. the T-independence for $T > T_0$) of the resulting mean-field S-matrix, which is given in the lowest order SPA by the integrand of eq. (5.12) taken at the stationary phase point ϱ (i.e. the mean-field solution of eq. (5.15))

$$S_{fi}^{T}(q_f,q_i,Q,\tau) = \exp\left[\frac{i}{2}\oint \varrho V \varrho \, dt\right] S_{\varrho}^{T}(q_f,q_i,Q,\tau) \tag{5.21}$$

So far we have found a TDMF approximation to the S-matrix between the localized states $S_{fi}^{T\to\infty}$ (5.10). In order to obtain the desired S-matrix between the projected channel eigenstates $\mathcal{S}_{fi}$ (5.9) we have to perform the integrations over the parameters arising from the projection: $\int d\tau \, dq_i q_f \, dQ$. Consistent with the evaluation of the functional integrals we perform these integrations in the SPA. The stationary phase conditions read:

$$\begin{gathered} E = E_{HF}[\varrho(t)], \quad K = \frac{1}{2}\left[\langle P_i\rangle_i\left(-\frac{\tau}{2}\right) + \langle P_f\rangle_f\left(\frac{\tau}{2}\right)\right] = 0, \\ k_i = \langle p_i\rangle_i\left(-\frac{\tau}{2}\right), \quad k_f = \langle p_f\rangle_f\left(\frac{\tau}{2}\right) \end{gathered} \tag{5.22}$$

where $E_{HF}[\varrho]$ is the Hartree-Fock energy functional, which is conserved for the mean-field solutions along the whole ϱ-loop. These equations, which must be solved simultaneously with the TDMF equation (5.15), determine the parameters τ, q_i, q_f and Q in such a way that: i) the Hartree-Fock energy of the resulting mean-field solutions coincides with the given channel energy; ii) the mean-field values of the momentum of the relative motions in the initial and final channel are equal to the given channel momenta k_i and k_f, respectively; iii) the average of the mean-field value of the total momentum at $t = \frac{\tau}{2}$ and $t = -\frac{\tau}{2}$ equals to the given total channel momentum $K = 0$. These results hold independently of the specific choice of the basis states ϕ_c. If the basis states of the individual fragments ϕ_ξ, from which we have projected the channel eigenstates, are Slater determinants the mean-field value of the total momentum is conserved during the reaction process but not necessarily during the preparation stages.

If the considered channel energy corresponds to the discrete part of the intrinsic spectrum the parameter τ does not regulate the mean-field energy but can be choosen freely. The choice of τ then merely influences the actual values of the parameters q_i, q_f and Q. The condition to the mean-field energy must then be fulfilled by a suitable choice of the basis functions ϕ_ξ.

In the SPA to the projection integrals the S-matrix between the projected channel states $\mathcal{S}_{fi}$ is approximated by that particular S-matrix element between the localized wave-packet like states for which the mean-field values of momentum and energy at the preparation times $t = \pm \frac{\tau}{2}$ coincide with their corresponding exact channel eigenvalues. This S-matrix element can be cast into the form

$$\mathcal{S}_{fi} = \exp\Big[i \int_{\infty}^{\frac{\tau}{2}} dt \, \frac{\langle \tilde{\phi}_f(t) | i\partial_t | \phi_f(t) \rangle}{\langle \tilde{\phi}_f(t) | \phi_f(t) \rangle}$$

$$+ i \int_{-\infty}^{\infty} dt \, \frac{\langle \tilde{\phi}(t) | i\partial_t | \phi(t) \rangle}{\langle \tilde{\phi}(t) | \phi(t) \rangle} + i \int_{-\frac{\tau}{2}}^{-\infty} dt \, \frac{\langle \tilde{\phi}_i(t) | i\partial_t | \phi_i(t) \rangle}{\langle \tilde{\phi}_i(t) | \phi_i(t) \rangle} \tag{5.23}$$

$$- i \log \langle \tilde{\phi}(t) | \phi(t) \rangle \Big],$$

which bears a strong resemblence to the S-matrix in the semiclassical potential scattering theory [29] except for the additional terms arising from the explicit treatment of the preparation stages. The last term in eq. (5.23) renders the S-matrix invariant with respect to time-dependent but space-independent gauge transformations (cf. ref. [26])

6. Applications

Since realistic calculations within the mean-field S-matrix theory presented above are rather expensive it is useful to study at the beginning model situations where the coupled temporally non-local mean-field equations (5.15) can be solved analytically.

6.1. One-dimensional system with attractive δ-interaction

In ref. [30] the TDMF-S-matrix theory was applied to the "scattering" of two self-bound one-dimensional N-particle systems interacting via an attractive δ-force

$$V(x,x') = -\tilde{g}\,\delta(x,x'). \tag{6.1.}$$

The particlesare supposed to have an internal symmetry group of dimension N, so that N fermions can be placed in the same spatial s.p. state. For a saturated system of n times N particles the TDMF equation reduces to the n-component non-linear (cubic) Schrödinger equation

$$i\hbar \frac{\partial}{\partial t}\psi_\ell(x,t) = \left[-\frac{\hbar^2}{2m}\frac{\partial^2}{\partial x^2} - \tilde{g}N\sum_{k=1}^{n}|\psi_k(x,t)|^2\right]\psi_\ell(x,t),\ \ell=1,\dots,n \tag{6.2}$$

which can be solved analytically as well as the corresponding static mean-field equation. The time-dependent equation (6.2) has two types of solutions: the trivially time-dependent Galilei translation of the static mean-field solution of the saturated N particle system (n = 1, called one-soliton solution) and the two-soliton solution (n = 2), which describes the elastic scattering of two saturated self-bound N-particle systems. This process was studied within the TDMF-S-matrix theory. As basis states of the fragments, ϕ_ξ, the static mean-field solutions have been chosen. For the elastic scattering the mean-field solutions of the individual fragments of the preparation stages can be constructed from the one-soliton solution, while the mean-field of the scattering process is given by the two-soliton solution. Self-consistency can be reached, i.e. the temporally non-locally self-consistent condition (5.15) can be fulfilled, by matching the two one-soliton solutions of the two fragments of the preparation stages with the two-soliton solution of the reaction process at $t = \pm T$ ($T \to \infty$). From the resulting TDMF S-matrix ($\mathcal{S}_{fi} = \exp(i\,2\,\delta_{(N)})$) one extracts the following expression for the phase shift

$$\delta_{(N)} = -N \arctan \frac{Nm\tilde{g}}{2\hbar k_0} \tag{6.3}$$

where $Nk_0 = k_i = k_f$ is the momentum of the relative motion. For N = 1 this is the exact quantum mechanical phase shift for the one-dimensional "scattering" of two particles with mass m interacting via the attractive δ-force (6.1).

The case N = 1 needs some comments since it implies a mean-field description of the one-particle systems during the free channel motion. In the usual TDH(F) theory derived from the variational principle the case N = 1 leads to the free particle motion since the (self-)interaction of a particle with its own mean-field is excluded there. In the mean-field Hamiltonian (2.5) derived in the path integral approach [8] this self-interaction is present and plays an important role in the description of two-particle scattering within the TDMF-S-matrix theory, by which the equation of motion of the free particle

$$i\hbar \frac{\partial}{\partial t}\psi(x,t) = -\frac{\hbar^2}{2m}\frac{\partial^2}{\partial x^2}\psi(x,t)$$

is transformed into the (one-component)non-linear Schrödinger equation (6.2). Thus the self-interaction converts the free (plane wave) motion of an individual particle to a stable non-decaying wave packet (one-soliton solution of eq. (6.2)). This localized state description of the independent particle motion during the preparation stages is necessary to compensate for the oscillations introduced into the S-matrix by the mean-field description of the real scattering process[x)].

For a weak two-body interaction $\frac{mN\tilde{g}}{2\hbar k_o} \ll 1$ the scattering phase reduces to

$$\delta_{(N)} = -N^2 \frac{m\tilde{g}}{2\hbar k_0} = N^2 \delta_{(1)} , \qquad (6.4)$$

which means that the phase shift for the scattering of two-weakly bound N-particle systems is the sum of the phase shifts of the N^2 individual scattering acts of each particle of the one weakly bound system with each particle of the other system.

This model study also nicely shows the physical meaning of the parameters q_i, q_f, Q and τ appearing in the S-matrix between the localized states: $\overset{-}{(+)}\frac{Q}{2}$ is here the mean-field value of the position of the center of mass of the colliding (scattered) fragments at $t = \overset{-}{(+)}\frac{\tau}{2}$ while $q_i(q_f)$ gives their relative distance at that time. In the present case where the exchange terms are neglected the center of mass rests at zero ($Q = 0$) during the whole time evolution.

The parameter τ gives the time the reaction process needs in excess to the free channel motion starting at a (given) relative distance q_i and terminating at q_f. In the present case we can freely choose either $q_i = q_f$ or τ. If we choose $\tau = 0$ (see fig. 2(a)) the preparation stages last as long as the reaction process (cf. ref. [26]). But the system travels then during the reaction process the distance $q_f - q_i$ in addition to the distance passed through during the free channel motion of the preparation stages. The time delay Δt of the reaction process is therefore given by $\Delta t = -(q_f - q_i)m/2k_o$. (Actually we have here a time advance ($\Delta t < 0$) due to the attractive interaction.) Alternatively we may put $q_i = q_f = 0$ (see fig. 2(b)). The corresponding τ coincides then with the time delay Δt.

---

x) Our model study shows also that for large N the mean-field energy with the self-interaction included, $E^N_{MF}[\varrho^o] = -mg^2 N^3/(24\hbar^2)$, better approximates the exact ground state energy $E_N = -mg^2 \cdot N(N^2-1)/(24\hbar^2)$ than the usual Hartree energy $E^N_H = -mg^2 \cdot N \cdot (N-1)^2/(24\hbar^2)$ where the self-interaction is excluded.

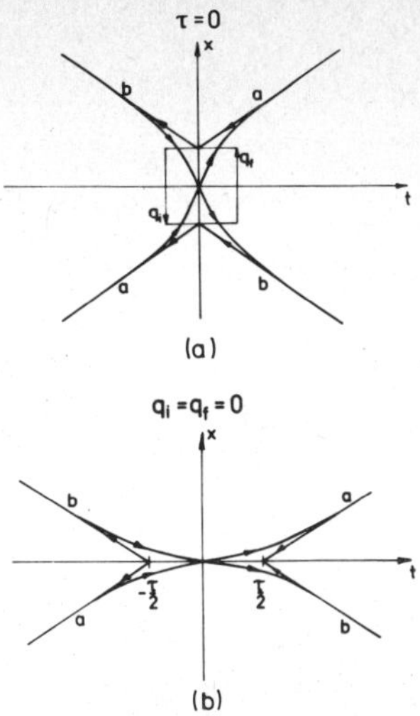

Fig. 2

Graphical illustration of the mean-field trajectories for the one-dimensional model for different choices of the localization parameters:
(a) $\tau = 0$, (b) $q_i = q_f = 0$

6.2. Low energy elastic heavy ion collision

As a second illustrative example of the TDMF S-matrix theory let me consider low energy elastic collisions [31]. I shall assume that the relative motion proceeds adiabatically slow so that the intrinsic excitations can be neglected. Under this assumption the TDMF equations can be solved analytically provided one neglects also the exchange amplitudes, which are of minor importance for sub-Coulomb barrier collisions. For this purpose one considers the mean-field values of the total momenta and the positions of the fragments as collective coordinates and transforms the TDMF equation into the moving system, which is defined by vanishing mean-field values of the momenta and coordinates. With the neglect of the intrinsic excitations the mean-field solutions in the intrinsic system are given by the corresponding static HF solutions. One obtains then a pure potential scattering picture describing the collision of the two corresponding static HF bags. The TDMF equations (5.15) reduce to the canonical equation of motion for the mean-field momentum p(t) and coordinate q(t) of the relative motion, whereas the scattering potential is given by the familiar folding potential. The phase shift extracted from the corresponding TDMF S-matrix coincides with the WKB result [31]

$$\delta_{WKB} = -\int_{-\infty}^{\infty} dt\, \dot{p}(t)\, q(t) \tag{6.5}$$

7. Conclusion

The quantum mean-field theory I have presented in my talk yields a unified description of low-energy nuclear dynamics. Whether this theory can give an overall und detailed understanding of the different nuclear phenomena remains to be proved by numerical calculations. If the answer should be negative because of practical limitations in the numerical calculations I would be rather pessimistic for future progress in the microscopic understanding of large amplitude phenomena in nuclear many-body systems since the mean-field approach is probably the simplest theory, which has any chance of incorporating the essential physics of the problem.

References

[1] J.W. Negele, Lectures at the NATO-NSF Summer School on Theoretical Methods in Medium Energy and Heavy-Ion Physics Madison, Wisconsin, June 1978

[2] D.J. Thouless, Nucl. Phys. 21 (1961) 225

[3] P. Bonche, S.E. Koonin and J.W. Negele, Phys. Rev. C13 (1976) 1226

[4] H. Kleinert, Phys. Lett. 69B (1977) 9; H. Reinhardt, Nucl. Phys. A298 (1978) 77

[5] H. Reinhardt, J. Phys. G 5 (1979) L91

[6] H. Reinhardt, Nucl. Phys. A331 (1979) 353; H. Kleinert and H. Reinhardt, Nucl. Phys. A332 (1979) 331

[7] H. Reinhardt, Nucl. Phys. A346 (1980) 1

[8] H. Reinhardt, Proceedings of International Summer School, "Critical Phenomena in Heavy-Ion Collisions", 25 August - 10 September 1980, Poiana Brashov, Romania, eds. A.A. Raduta and G. Stratan; H. Reinhardt, Fortschr. d. Physik 30 (1982) 127

[9] S. Levit, Phys. Rev. C21 (1980) 1594

[10] S. Levit, J. Negele and Z. Paltiel, Phys. Rev. C21 (1980) 1603

[11] J.W. Negele, Proceedings of International Summer School, "Critical Phenomena in Heavy-Ion Collisions", 25 August - 10 September 1980, Poiana Brashov, Romania, ed. A.A. Raduta and G. Stratan

[12] M. Kuratsuji and T. Suzuki, Phys. Lett. 92B (1980) 19

[13] J.-P. Blaizot and M. Orland, J. Phys. Lett. 41 (1980) L53

[14] Y. Ohnuki and T. Kashiwa, Progr. Theor. Phys. 60 (1978) 548

[15] J. Hubbard, Phys. Rev. Lett. 3 (1959) 77; R.L. Stratonovich, Dokl. Acad. Nauk SSSR 115 (1957) 1097

[16] P. Ring and P. Schuck, "The Many-Body Problem", Springer Verlag, New York, 1980

[17] R. Balian and E. Brezin, Nuovo Cimento 64B (1969) 37

[18] H. Reinhardt, Nucl. Phys. A367 (1981) 269

[19] H. Reinhardt, Nucl. Phys. A369 (1981) 109

[20] A.K. Kerman and S. Levit, Phys. Rev. C24 (1981) 1029

[21] K. Goeke, P.-G. Reinhard, H. Reinhardt, Nucl. Phys. A

[22] P. Arve and H. Reinhardt, Phys. Lett. 105B (1981) 249

[23] H. Reinhardt, Nucl. Phys. A381 (1982) 217

[24] H.J. Mang, Phys. Reports 18C (1975) 327

[25] S. Levit, J.W. Negele and Z. Paltiel, Phys. Rev. C22 (1980) 1979; J.W. Negele, this conference

[26] H. Reinhardt, NBI preprints (1981) No. 81-43 and No. 81-44 and Nucl. Phys, A, in press

[27] C.J. Joachim, "Quantum Collision Theory". North Holland, Publ. Comp. Amsterdam, 1975
[28] Y. Alhassid and S.E. Koonin, Phys. Rev. C23 (1981) 1590
[29] W.H. Miller, J. Chem. Phys. 53 (1970) 1949
[30] H. Reinhardt, submitted to Phys. Lett. B
[31] H. Reinhardt, submitted to Nucl. Phys. A
[32] K.-K. Kan, J.J. Griffin, P.C. Lichtner and M. Dworzecka, Nucl. Phys. A332 (1979) 109; K.-K. Kan, Phys. Rev. C24 (1981) 279
[33] J.J. Griffin, P.C. Lichtner and M. Dworzecka, Phys. Rev. C21 (1980) 1351; J.J. Griffin, M. Dworzecka, P.C. Lichtner and K.-K. Kan, Phys. Lett. 93B (1980) 235

CROSS-SECTIONS FROM TDHF CALCULATIONS

E. A. Remler†

Department of Physics

College of William and Mary

Williamsburg, Virginia 23185

I. INTRODUCTION

This paper connects TDHF to reaction theory. Its approach differs that of previous related work[1,2] and leads to cross section formula which are free of some of their difficulties.

Arbitrarily many equivalent formula can be constructed relating exact solutions of the multiparticle Schroedinger equation to cross-sections. They are generally equivalent only in terms of exact solutions. Formula to be used with approximate solutions should not critically depend on properties of the exact solution poorly represented in the approximate one.

For example, scattering amplitudes are defined directly via the asymptotic values of exact wave functions. But since all presently known approximations for multiparticle wave functions fail asymptotically, this formula cannot be directly calculated. The usually used formula, of form $T_{fi} = \int \psi_f V_f \psi_i^{(+)}$, avoids this problem because it is insensitive to the asymptotic values of an approximate $\psi_i^{(+)}$. Another, of form $S_{fi} = (\psi_f^{(-)}, \psi_i^{(+)})$, similarly does not critically probe asymptotic values of $\psi_i^{(+)}$ or $\psi_f^{(-)}$. The manifestation of the analogous deficiency in TDHF is the appearance of the "spurious cross channel correlations". Analogs of both of the above formula, the first for inclusive and the second for exclusive cross sections, are used to circumvent this.

Another TDHF deficiency is its inability to describe the time evolution of a pure as opposed to a mixed state. Even though, neglecting dissipation, a TDHF solution can formally be put in the form of a wave function and, even if its initial Hartree-Fock (HF) states be considered reasonable approximate pure state wave packets, it seems unlikely that the putative wave function will retain much meaning after some time development. Clearly, without a wave function an S matrix cannot even be defined. This problem is circumvented by using analogs of the aforementioned formula in which density operators replace wave functions and as a result, the cross section replaces T_{fi} or S_{fi} respectively.

The next section begins by deriving a very general exact formula relating two body final state cross sections to density operators. Information from dynamics for a reaction $A + B \to C + D$ appears only in the form of density overlap, $\mathrm{tr}\rho_{CD}\rho_{AB}$, where $\rho_{AB}(\rho_{CD})$ is the density of the incoming (outgoing) channel at time $T_{AB}(-T_{CD})$. After a little manipulation, a simple and physically transparent formula is obtained.

---

†Work supported by the National Science Foundation under Grant Phy-80-06210

Two potentially serious problems however are associated with this formulation. The first arises from the fact that the trace is independent of $T_{AB}-T_{CD}$ when exact densities are used but not after being approximated via TDHF.[8] A reasonable criterion is used to optimize its choice. The practical utility of this formula with respect to TDHF applications depends on its sensitivity to variations about the optimum choice of $T_{AB}-T_{CD}$. This remains to be investigated.[9]

Another problem arises from the fact that the formula requires a statistical averaging over two body final states. The theoretical uncertainty surrounding the very meaning of TDHF causes the relevant averaging functions to be similarly uncertain. Reasonable conjectures can be made and in addition, self consistency checks are available. Despite these problems the generality of the formula strongly suggests that if TDHF is at all applicable to such exclusive processes (an open question), it must be so in the form obtained here.

Section III treats inclusive cross sections for composite fragments - a measurement which should be more amenable to TDHF. The general approach has been described elsewhere[3] and has been applied to cascade[4] and hydrodynamic[5] models. Only points particular to TDHF applications are discussed here. Essentially none of the problems encountered in the previous section occur here.

Much of this paper can be generalized rather easily - Section II extended to more than two bodies, section III to more than one body. Both sections easily accommodate TDHF plus incoherent dissipation models. Certainly the basic formula used in section II and probably that used in section III can also serve in the context of mean field theory.

II. AVERAGED EXCLUSIVE CROSS-SECTIONS

Let $\phi = \underset{\sim}{r}, \underset{\sim}{k}$ denote a point in six dimensional phase space and $\rho_A(\phi_A)$ a density operator describing a mixture of eigenstates of nucleus A centered at ϕ_A. There are other parameters besides ϕ_A for ρ_A defining average values of internal energy, spin, etc., which are left implicit. This density is related to the density centered at the phase space origin by

$$\langle \underset{\sim}{k}''_A |\rho_A(\phi_A)| \underset{\sim}{k}'_A \rangle = \exp(-i(\underset{\sim}{k}''_A - \underset{\sim}{k}'_A)\cdot \underset{\sim}{r}_A) \quad \langle \underset{\sim}{k}''_A - \underset{\sim}{k}_A |\rho_A(0)| \underset{\sim}{k}'_A - \underset{\sim}{k}_A \rangle \quad . \tag{2.1}$$

The probability distribution function for A in state $\rho_A(\phi_A)$ is denoted by

$$g_A(\underset{\sim}{k}'_A - \underset{\sim}{k}_A) = \langle \underset{\sim}{k}'_A |\rho_A(\phi_A)| \underset{\sim}{k}'_A \rangle \tag{2.2}$$

and is assumed to be peaked at zero momentum and about the averages of its implicit arguments.

If $\rho_B(\phi_B)$ is similarly constructed for nucleus B and $\rho_{AB}(\phi_A,\phi_B)$ is the antisymmetrized direct product of ρ_A and ρ_B then as long as ϕ_A is sufficiently well separated from ϕ_B,

$$<\underset{\sim}{k}_A'' , \underset{\sim}{k}_B'' | \rho_{AB}(\phi_A,\phi_B) | \underset{\sim}{k}_A' , \underset{\sim}{k}_B' > \cong <\underset{\sim}{k}_A'' | \rho_A(\phi_A) | \underset{\sim}{k}_A' > <\underset{\sim}{k}_B'' | \rho_B(\phi_B) | \underset{\sim}{k}_B' > . \tag{2.3}$$

Let $\rho_{AB}(\phi_A,\phi_B;T)$ be the exact density evolved after time T from $\rho_{AB}(\phi_A,\phi_B)$. For large enough T it will consist of well separated packets of collision products. Let C and D be two possible collision products in a two-nuclei final state and $\rho_{CD}(\phi_C,\phi_D)$ a density for them constructed as above. Then $\mathrm{tr}[\rho_{CD}(\phi_C,\phi_D)\rho_{AB}(\phi_A,\phi_B;T)]$ will be non-negligible only at well separated ϕ_C and ϕ_D. Using Eqs.(2.1-2.3) we obtain ($h = 2\pi$),

$$\int d\underset{\sim}{r}_C \, d\underset{\sim}{r}_D \, h^{-6} \, \mathrm{tr}[\rho_{CD}(\phi_C,\phi_D)\rho_{AB}(\phi_A,\phi_B;T)] =$$

$$\int d\underset{\sim}{k}_C' \, d\underset{\sim}{k}_D' \, g_C(\underset{\sim}{k}_C - \underset{\sim}{k}_C') \, g_D(\underset{\sim}{k}_D - \underset{\sim}{k}_D') \quad <\underset{\sim}{k}_C' , \underset{\sim}{k}_D' | \rho_{AB}(\phi_A,\phi_B;T) | \underset{\sim}{k}_C' \; \underset{\sim}{k}_D' > . \tag{2.4}$$

If ϕ_A and ϕ_B are well separated pre-collision configurations then the matrix element in Eq. (2.4) is just the resultant final state momentum distribution in the C + D channel. Integrating over impact parameter $\underset{\sim}{b}_B$ thus gives the averaged cross-section for beam nucleus B on target A;

$$\int d\underset{\sim}{b}_B \, d\underset{\sim}{r}_C \, d\underset{\sim}{r}_D \, h^{-6} \mathrm{tr}[\rho_{CD}(\phi_C,\phi_D)\rho_{AB}(\phi_A,\phi_B;T)] =$$

$$\int \prod_I d\underset{\sim}{k}_I' \, g_I(\underset{\sim}{k}_I - \underset{\sim}{k}_I') \, \sigma(\underset{\sim}{k}_C' , \underset{\sim}{k}_D' ; \underset{\sim}{k}_A' , \underset{\sim}{k}_B') \overset{\Delta}{=} \bar{\sigma}(\underset{\sim}{k}_C, \underset{\sim}{k}_D ; \underset{\sim}{k}_A, \underset{\sim}{k}_B) , \tag{2.5}$$

where I = A,B,C,D, and σ is the momentum differential cross section for the reaction $A + B \to C + D$ and $\bar{\sigma}$ is the cross section averaged with g_I as weights. The averaging is implicitly over internal energy, nuclear spin, etc. The trace may be expressed more symmetrically by applying a time displacement to its argument;

$$\mathrm{tr}[\rho_{CD}(\phi_C,\phi_D)\rho_{AB}(\phi_A,\phi_B;T)] = \mathrm{tr}[\rho_{CD}(\phi_C,\phi_D;-T_{CD})\rho_{AB}(\phi_A,\phi_B;T_{AB})] \tag{2.6}$$

where in order for Eq. (2.5) to be valid we require $T_{AB} + T_{CD} = T >$ the time needed for A + B to collide starting from ϕ_A, ϕ_B and C + D to separate.

Our intent is to approximate this trace with the trace of corresponding density products formed from TDHF solutions. Thus let $\psi_{AB}^{HF}(\phi_A,\phi_B;t)$ be a TDHF solution evolved from the antisymmetrized product $A\psi_A^{HF}(\phi_A)\psi_B^{HF}(\phi_B) = \psi_{AB}^{HF}(\phi_A,\phi_B;0)$ of HF solutions centered at ϕ_A and ϕ_B. The corresponding density operator is $\rho_{AB}^{HF}(\phi_A,\phi_B;t) = \psi_{AB}^{HF}(\phi_A,\phi_B;t)\psi_{AB}^{HF}(\phi_A,\phi_B;t)^{\dagger}$. Defining a similar final state density ρ_{CD}^{HF}, we now assume that for appropriate values of the parameters $\phi_A, \ldots, \phi_D, T_{AB}$, T_{CD} to be discussed,

$$\mathrm{tr}\rho_{CD}\rho_{AB} \cong \mathrm{tr}\rho_{CD}^{HF}\rho_{AB}^{HF} . \tag{2.7}$$

Note that this does not require ψ_{AB}^{HF}, ψ_{CD}^{HF} to approximate exact time dependent wave functions of the system or ρ_{AB}^{HF}, ρ_{CD}^{HF}, exact time dependent densities, but only that the inner products defined in Eq. (2.7) be approximated. This seems to be the least one must ask of TDHF if it is to be applicable to two body final state cross sections. The approximate trace is given in terms of the determinant of the matrix O of overlaps of initial and final state orbitals, $\psi^{HF}_{AB,i}$ and $\psi^{HF}_{CD,j}$ respectively, as follows.

$$O_{ji} = (\psi^{HF}_{CD,j}(\phi_C,\phi_D;-T_{CD}),\ \psi^{HF}_{AB,i}(\phi_A,\phi_B;T_{AB})) \qquad (2.8)$$

$$\mathrm{tr}\rho^{HF}_{CD}\rho^{HF}_{AB} = |\mathrm{Det}\ O|^2 \quad . \qquad (2.9)$$

The cross section formula developed so far

$$\bar{\sigma} \cong \int d\underset{\sim}{b}_B\, d\underset{\sim}{r}_C\, d\underset{\sim}{r}_D\, h^{-6}\mathrm{tr}\rho^{HF}_{CD}\rho^{HF}_{AB} \qquad (2.10)$$

with $\bar{\sigma}$ given by Eq. (2.5) and the trace by Eqs. (2.8) and (2.9) is an almost necessary consequence of the (albeit perhaps overly optimistic) assumption that TDHF has any relevance to such averaged exclusive cross sections. Two major questions however remain. (1) What are g_A . . . g_D. (2) What to use for T_{AB} and T_{CD}. The answers to these, in contrast, are not so well determined.

There is virtually no theory available on which to base an answer to the first question. A "maximalist" opinion of the meaning of TDHF might be that if ψ^{HF}_I are e.g. HF ground state wave functions then the g_I average only over three momentum and total spin projection. Their form is then given by the calculable total momentum and spin distribution in a HF packet. A similar statement would hold for transitions to excited states.

But this is not necessarily true. The g_I might average also over many nuclear eigenstates of internal energy. Their shape in this case is unknown. One can hope and might reasonably expect that their internal energy distribution is obtainable by simple statistical arguments or that experiment is insensitive to it, or both. After discussion of the second question we shall see that some theoretical experiments may shed light on their shape.

We turn now to the division of T into T_{AB} and T_{CD}. It should be chosen to minimize the error in Eq. (2.7). For this question some simple intuitive arguments are available.

We are free to choose for simplicity $\underset{\sim}{r}_A = (\underset{\sim}{0}, Z_A)$, $\underset{\sim}{r}_B = (\underset{\sim}{b}_B, Z_B)$ so that $|Z_A - Z_B|$ is the minimum initial distance between target A and projectile B. Clearly this must be chosen to be sufficiently greater than the entrance channel interaction's range R_{AB}. Let the initial velocities be $\underset{\sim}{v}_A = \underset{\sim}{k}_A/M_A = (0,v_A)$ and $\underset{\sim}{v}_B = \underset{\sim}{k}_B/M_B = (\underset{\sim}{0}, -v_B)$ where v_A and v_B are their magnitudes. For a pre-collision configuration this requires $Z_A < 0$, $Z_B > 0$. If in addition we set $Z_A/v_A + Z_B/v_B = 0$, then non-interacting nuclei would achieve closest approach after a time $|Z_A/v_A| = Z_B/v_B$. About this time ρ^{HF}_{AB} will typically be most dense and begin to generate outgoing components. After this time any meaningfulness of ρ^{HF}_{AB} may be expected to deteriorate rapidly as ρ_{AB} begins to develop distinct final state channels which ρ^{HF}_{AB} cannot describe.

Consider the choice of this time for T_{AB}. Then $|\mathrm{Det}\ O|$ will be non-negligible only for post-collision configurations ϕ_C, ϕ_D which, projected backwards in time by T_{CD}, describe a coalescent system near that of A + B at T_{AB}. What ϕ_C, ϕ_D

do this? Consider first the case in which $\underset{\sim}{v}_C$ and $\underset{\sim}{v}_D$ are colinear and oppositely directed. Then their projected positions $Z_C = \underset{\sim}{r}_C \cdot \underset{\sim}{v}_C / v_C$, $Z_D = \underset{\sim}{r}_D \cdot \underset{\sim}{v}_C / v_C = -\underset{\sim}{r}_D \cdot \underset{\sim}{v}_D / v_D$ should satisfy $Z_C/v_C \cong |Z_D|/v_D \cong T_{CD}$. Large enough T_{CD} therefore guarantees a separation $|Z_C - Z_D|$ sufficiently larger than the final channel interaction range R_{CD}. We can always adjust T and $|Z_A - Z_B|$ so that $T_{AB} = T_{CD} = \frac{1}{2}T$.

Each HF packet translates itself with unchanging shape when isolated. Thus for large enough $|Z_C - Z_D|$

$$\rho^{HF}_{CD}(\underset{\sim}{r}_C, \underset{\sim}{k}_C, \underset{\sim}{r}_D, \underset{\sim}{k}_D; -\tfrac{1}{2}T + \tau) = \rho^{HF}_{CD}(\underset{\sim}{r}_C + v_C\tau, \underset{\sim}{k}_C, \underset{\sim}{r}_D + \underset{\sim}{v}_D\tau, \underset{\sim}{k}_D; -\tfrac{1}{2}T) \tag{2.11}$$

with a similar expression for ρ^{HF}_{AB}. This is the analog of the "intertwining relation" in ordinary scattering theory. This allows a spatial displacement to be converted to a time displacement as follows. Let $\underset{\sim}{b}_C$ and $\underset{\sim}{b}_D$ be impact parameter type two-vectors perpendicular to the respective $\underset{\sim}{v}_C$ and $\underset{\sim}{v}_D$ and let

$$\lambda = (\underset{\sim}{r}_C \cdot \underset{\sim}{v}_C + \underset{\sim}{r}_D \cdot \underset{\sim}{v}_D)/v_{CD}$$

$$\eta = (\underset{\sim}{r}_C \cdot \underset{\sim}{v}_D - \underset{\sim}{r}_D \cdot \underset{\sim}{v}_C)/v_{CD}$$

$$v_{CD} = (v_C^2 + v_D^2)^{1/2} \tag{2.12}$$

Considering $\underset{\sim}{r}_C$ and $\underset{\sim}{r}_D$ as functions of $\underset{\sim}{b}_C, \underset{\sim}{b}_D, \lambda, \eta$ the intertwining relation allows us to write Eq. (2.10) in the form

$$\bar{\sigma} = v_{CD} h^{-6} \int d\underset{\sim}{b}_B d\underset{\sim}{b}_C d\underset{\sim}{b}_D d\eta d\tau \times \mathrm{tr}\, \rho^{HF}_{CD}(\phi_C, \phi_D; -\tfrac{1}{2}T + \tau) \rho^{HF}_{AB}(\phi_A, \phi_B; \tfrac{1}{2}T) \tag{2.13}$$

The spatial integral for C and D is now over a five dimensional hyperplane $\lambda = \frac{1}{2}v_{CD}T$. The integral for B is over the two dimensional impact parameter hyperplane $Z_B = \frac{1}{2}v_B T$ with $\underset{\sim}{r}_A$ fixed at $(\underset{\sim}{0}, -\frac{1}{2}v_A T)$ so that the λ parameter for the pre-collision configuration is

$$(\underset{\sim}{r}_A \cdot \underset{\sim}{v}_A + \underset{\sim}{r}_B \cdot \underset{\sim}{v}_B)/v_{AB} = -\tfrac{1}{2}Tv_{AB} \quad . \tag{2.14}$$

The λ hyperplane for C and D was chosen so that freely travelling nuclei would reach it in a time $\frac{1}{2}T$ from their point of closest approach. Thus we expect the integrand in Eq. (2.13) to be maximum somewhere around $\tau = 0$ and $\eta = \underset{\sim}{b}_C = \underset{\sim}{b}_D = 0$. The nuclear separation at this point is $|Z_C - Z_D| = \frac{1}{2}T(v_C + v_D)$. Thus T must be chosen sufficiently greater than both $R_{AB}/\frac{1}{2}(v_A + v_B)$ and $R_{CD}/\frac{1}{2}(v_C + v_D)$ but is otherwise arbitrary. Having picked such a T, all the remaining hyperplane parameters Z_A, Z_B, λ are determined as indicated in terms of it.

Recall now that the choice of a fixed T_{AB} was somewhat arbitrary and has resulted in an expression, Eq. (2.13), which is asymmetric between AB and CD within the trace. That expression uses $\tau < 0$ values where ρ^{HF}_{CD} is in its rapidly deteriorating range of validity. The total error in the trace can be minimized when the error in ρ^{HF}_{AB} and ρ^{HF}_{CD} is equalized. This can be done by changing T_{AB} so that it is λ (or τ) dependent while keeping T fixed. Since to lowest order the error will vary linearly in τ, on the average we can expect to improve accuracy by replacing $T_{AB} \to \frac{1}{2}T - \frac{1}{2}\tau$, $T_{CD} \to -\frac{1}{2}T + \frac{1}{2}\tau$ in Eq. (2.13):

$$\bar{\sigma} = v_{CD}\ h^{-6} \int d\underset{\sim}{b}_B d\underset{\sim}{b}_C d\underset{\sim}{b}_D d\eta d\tau$$

$$\times \left|\mathrm{Det}(\psi^{HF}_{CD,j}(\phi_C,\phi_D;\ -\tfrac{1}{2}(T-\tau)),\ \psi^{HF}_{AB,i}(\phi_A,\phi_B;\ +\tfrac{1}{2}(T-\tau)))\right|^2 \quad . \qquad (2.15)$$

Note that the integrand of this 8 dimensional integral should be smoothly peaked about the zero of its arguments. No oscillations are generally expected. Therefore it should be approximable by numerical interpolation techniques using relatively few grid points.

We must finally address the question of the choice of $\underset{\sim}{v}_C$ and $\underset{\sim}{v}_D$. Recall they were arbitrarily taken to be opposite and collinear. In fact the integral determining $\bar{\sigma}$ in Eq. (2.5) can be done for arbitrary $\underset{\sim}{k}_C$ and $\underset{\sim}{k}_D$. Consider center of mass scattering and the "maximalist" interpretation of the meaning of the TDHF solutions. In this case a choice of $\underset{\sim}{k}_A \ . \ . \ . \ \underset{\sim}{k}_D$ which satisfies four momentum conservation will agree with that made in the preceding discussion. Assuming the transition matrix is relatively slowly varying we can write (neglecting spin factors)

$$\sigma(\underset{\sim}{k}_C,\underset{\sim}{k}_D;\underset{\sim}{k}_A,\underset{\sim}{k}_B) \stackrel{\Delta}{=} \Gamma(\underset{\sim}{k}_C\underset{\sim}{k}_D;\underset{\sim}{k}_A\underset{\sim}{k}_B)\delta^4(k_C+k_D-k_A-k_B) \qquad (2.16)$$

$$\bar{\sigma} \cong \Gamma\int\prod_i d\underset{\sim}{k}'_I g_I(\underset{\sim}{k}'_I-\underset{\sim}{k}_I)\delta^4(k_C+k_D-k_A-k_B) \qquad (2.17)$$

With the "maximalist" interpretation, g_I are known and the behavior of $\bar{\sigma}$ as $\underset{\sim}{k}_I$ vary near the 4 momentum shell can be computed by Eq. (2.17) and compared to the computed integral in Eq. (2.15). This is a self consistency check. Without the maximalist interpretation such a comparison provides information which may be used to empirically determine the form of the g_I. This point requires further investigation and may lead to some insight into the meaning of the TDHF approximation.

III. INCLUSIVE CROSS-SECTIONS

If, as before, a TDHF solution is written as $\psi^{HF}_{AB}(\phi_A\phi_B;t)$, then the number density of nucleons can be written as $(\psi^{HF}_{AB},\hat{\psi}^{\dagger}_{\underset{\sim}{k}}\hat{\psi}_{\underset{\sim}{k}}\psi^{HF}_{AB})$ where $\hat{\psi}_{\underset{\sim}{k}}$ denotes a momentum $\underset{\sim}{k}$ annihilation operator. Clearly, this expression does not distinguish asymptotically bound from free nucleons. To emphasize this we call this number density, primordial. Assuming ϕ_A,ϕ_B corresponds to a well separated initial state configuration, the limit of the primordial nucleon number density as $t \to \infty$ is denoted as $\tilde{\sigma}(\underset{\sim}{k};\phi_A\phi_B)$ and is called the primordial inclusive nucleon cross-section (for initial state ϕ_A,ϕ_B). The usual cross-section for asymptotically free nucleons is written without a tilde. The difference between these is the cross-section for bound ("virtual") nucleons;

$$\tilde{\sigma}(\underset{\sim}{k};\phi_A\phi_B) = \sigma(\underset{\sim}{k};\phi_A\phi_B) + \sigma^{(v)}(\underset{\sim}{k};\phi_A\phi_B) \quad . \qquad (3.1)$$

The expression for $\tilde{\sigma}$ can be generalized by writing it in terms of the density operator. Thus

$$\tilde{\sigma}(\underset{\sim}{k};\phi_A\phi_B) = \lim_{t\to\infty} \mathrm{tr}[\hat{\psi}^\dagger_{\underset{\sim}{k}}\hat{\psi}_{\underset{\sim}{k}}\rho_{AB}(\phi_A,\phi_B;t)] \quad . \tag{3.2}$$

When ρ_{AB} is exact, so is $\tilde{\sigma}$. Use of ρ^{HF}_{AB} gives an approximate $\tilde{\sigma}$. Equation (3.2) can also be used with generalizations of TDHF which include incoherent ("two body collision") terms where ρ^{HF}_{AB} but not ψ^{HF}_{AB} has meaning.

It will be useful to consider an alternate form of Eq. (3.2); leaving ϕ_A, ϕ_B, A, B implicit, we can write

$$\Delta\tilde{\sigma}(\underset{\sim}{k}) = \int_0^\infty \partial_t \mathrm{tr}[\hat{\psi}^\dagger_{\underset{\sim}{k}}\hat{\psi}_{\underset{\sim}{k}}\rho(t)]dt = \int_0^\infty \mathrm{tr}([i\hat{H},\hat{\psi}^\dagger_{\underset{\sim}{k}}\psi_{\underset{\sim}{k}}]\rho(t))dt = \int_0^\infty \mathrm{tr}([i\hat{V},\hat{\psi}^\dagger_{\underset{\sim}{k}}\hat{\psi}_{\underset{\sim}{k}}]\rho(t))dt \quad , \tag{3.3}$$

$$\Delta\tilde{\sigma}(\underset{\sim}{k}) \triangleq \tilde{\sigma}(\underset{\sim}{k}) - \mathrm{tr}[\hat{\psi}^\dagger_{\underset{\sim}{k}}\psi_{\underset{\sim}{k}}\rho(0)] \quad , \tag{3.4}$$

where $\hat{H}$ is the Hamiltonian and $\hat{V}$ is the potential

$$\hat{V} = \frac{1}{4}\int d\underset{\sim}{k}_1 d\underset{\sim}{k}_2 d\underset{\sim}{k}'_1 d\underset{\sim}{k}'_2 \hat{\psi}^\dagger_{\underset{\sim}{k}'_2}\hat{\psi}^\dagger_{\underset{\sim}{k}'_1}<\underset{\sim}{k}'_1\underset{\sim}{k}'_2|V|\underset{\sim}{k}_2\underset{\sim}{k}_1>\hat{\psi}_{\underset{\sim}{k}_1}\hat{\psi}_{\underset{\sim}{k}_2} \tag{3.5}$$

Consider the effect of approximating ρ by ρ^{HF} in Eq. (3.3). If ρ^{HF} is of thermal Hartree-Fock[6,7] form then

$$\mathrm{tr}(\hat{\psi}^\dagger_{\underset{\sim}{k}'_N}\cdots\hat{\psi}^\dagger_{\underset{\sim}{k}'_1}\hat{\psi}_{\underset{\sim}{k}_1}\cdots\hat{\psi}_{\underset{\sim}{k}_N}\rho^{HF}) \triangleq <\underset{\sim}{k}_N\cdots\underset{\sim}{k}_1|n^{(N)}|\underset{\sim}{k}'_1\cdots\underset{\sim}{k}'_N> = \mathrm{Det}<\underset{\sim}{k}_i|n|\underset{\sim}{k}'_j> \quad , \tag{3.6}$$

$$<\underset{\sim}{k}'|n|\underset{\sim}{k}> \triangleq \mathrm{tr}[\hat{\psi}^\dagger_{\underset{\sim}{k}}\hat{\psi}_{\underset{\sim}{k}'}\rho^{HF}] \tag{3.7}$$

This rule includes the case $\rho^{HF}_{AB} = \psi_{AB}\psi^\dagger_{AB}$. Under this assumption, Eq. (3.5) into Eq. (3.3) yields

$$\Delta\tilde{\sigma}(\underset{\sim}{k}) = \int_0^\infty <\underset{\sim}{k}|[n,iv]|\underset{\sim}{k}>dt \tag{3.8}$$

where v is a mean field potential

$$<\underset{\sim}{k}'|v|\underset{\sim}{k}> = \int d\underset{\sim}{p}d\underset{\sim}{p}'<\underset{\sim}{p}'\underset{\sim}{k}'|V|\underset{\sim}{k}\underset{\sim}{p}><p|n|p'> \tag{3.9}$$

This result suggests that V should be replaced by the effective (e.g. G-matrix) interaction whenever ρ is approximated by ρ^{HF}. This turns v into the TDHF potential and if n is the TDHF generated number density, Eq. (3.8) implies

$$\Delta\tilde{\sigma}(\underset{\sim}{k}) = \int_0^\infty <\underset{\sim}{k}|[-ih,n]\underset{\sim}{k}>dt = \int_0^\infty <\underset{\sim}{k}|\partial_t n|\underset{\sim}{k}>dt \tag{3.10}$$

where h is the TDHF Hamiltonian. This is exactly the expected result.

We are now prepared to generalize the preceding to composite fragment production. The generalization of Eq. (3.2) is

$$\tilde{\sigma}_C(\underset{\sim}{k};\phi_A\phi_B) = \lim_{t\to\infty} \mathrm{tr}(\hat{\psi}^\dagger_{C,\underset{\sim}{k}}\hat{\psi}_{C,\underset{\sim}{k}}\rho_{AB}(\phi_A,\phi_B;t)) \tag{3.11}$$

where $\hat{\psi}_{C,\underset{\sim}{k}}$ annihilates a composite C of momentum $\underset{\sim}{k}$;

$$\hat{\psi}_{C,\underset{\sim}{k}} \triangleq \int d\underset{\sim}{k}_1\cdots d\underset{\sim}{k}_N<C,\underset{\sim}{k}|\underset{\sim}{k}_1\cdots\underset{\sim}{k}_N>/N! \quad \hat{\psi}_{\underset{\sim}{k}_N}\cdots\hat{\psi}_{\underset{\sim}{k}_1} \tag{3.12}$$

Similarly, Eq. (3.1) suitably generalized still hold;[3] $\tilde{\sigma}_C$ includes contributions from virtual C's in larger composites which must be subtracted out to obtain the true inclusive cross-section σ_C.

As with Eq. (3.3), we re-express Eq. (3.11) as a difference,

$$\Delta\tilde{\sigma}_C(\underset{\sim}{k}) = \int_0^\infty \mathrm{tr}([i\hat{H},\hat{\psi}^\dagger_{C,\underset{\sim}{k}}\hat{\psi}_{C,\underset{\sim}{k}}]\rho(t))\, dt \quad . \tag{3.13}$$

After some algebra, the generalization of the last step in Eq. (3.3) is found to be

$$\Delta\tilde{\sigma}_C(\underset{\sim}{k}) = \int_0^\infty \mathrm{tr}(:[i\hat{V},\hat{\psi}^\dagger_{C,\underset{\sim}{k}}\hat{\psi}_{C,\underset{\sim}{k}}]:\rho(t))dt \tag{3.14}$$

where colons denote normal ordering. This form contains the information that $\hat{\psi}_{C,\underset{\sim}{k}}$ annihilates an exact eigenstate of $\hat{H}$.

As before we may now approximate ρ by ρ^{HF} and replace V by the effective interaction. Using Eq. (3.6) and some further algebra, we can write the result as

$$\Delta\tilde{\sigma}_C(\underset{\sim}{k}) = \int_0^\infty \langle C,\underset{\sim}{k}|[n^{(N)},i(v-V')]|\underset{\sim}{k},C\rangle/(N!)^2 dt \quad . \tag{3.15}$$

The notation is now as follows: The ket $|\underset{\sim}{k},C\rangle$ has components $\langle \underset{\sim}{k}_1 \ldots \underset{\sim}{k}_N|\underset{\sim}{k},C\rangle$ as defined by Eq. (3.12) and is antisymmetric in $\underset{\sim}{k}_1 \ldots \underset{\sim}{k}_N$. The matrix $n^{(N)}$ is defined by Eq. (3.6). The matrix v is now a sum of N(TDHF) potentials each defined as in Eq. (3.9), each operating on a separate momentum. The matrix V' is a sum over $\frac{1}{2}N(N-1)$ two body operators, one for each distinct pair of momenta $V' = \sum_{i<j} V'_{ij}$,

$$\langle \underset{\sim}{k}'_i \underset{\sim}{k}'_j|V'_{ij}|\underset{\sim}{k}_j \underset{\sim}{k}_i\rangle = \int d\underset{\sim}{q}_j d\underset{\sim}{q}'_j \langle k'_j|n|q'_j\rangle\langle \underset{\sim}{k}'_i \underset{\sim}{q}'_j|V_{ij}|\underset{\sim}{q}_j \underset{\sim}{k}_i\rangle\langle \underset{\sim}{q}_j|n|\underset{\sim}{k}_j\rangle \quad , \tag{3.16}$$

To obtain Eq. (3.15) the pure state condition $n^2 = n$ has been used. Thus n is a projection and $[n^{(N)},iV']$ can be considered as the time rate of change of $n^{(N)}$ due to interactions between fragment constituents. The fact that V' is subtracted in Eq. (3.15) shows that fragment production is determined by the potential felt by fragment constituents due to particles exterior to the fragment. This is exactly what one expects on the basis of multiparticle scattering theory.[3]

Since $h_0 + v$ generates the time rate of change in $n^{(N)}$,

$$\int_0^\infty \langle C,\underset{\sim}{k}|[n^{(N)},iv]|\underset{\sim}{k},C\rangle dt = \int_0^\infty \langle C,\underset{\sim}{k}|\partial_t n^{(N)}-[n^{(N)},ih_0]|\underset{\sim}{k},C\rangle\, dt \quad . \tag{3.17}$$

Assuming $\lim_{t\to\infty} \langle C,\underset{\sim}{k}|n^{(N)}|\underset{\sim}{k},C\rangle = 0$, we find

$$\tilde{\sigma}_C(\underset{\sim}{k}) = \int_0^\infty \langle C,\underset{\sim}{k}|[n^{(N)},-ih']|\underset{\sim}{k},C\rangle/(N!)^2 dt \tag{3.18}$$

where $h' = h_0 + V'$. For purposes of computation one might calculate a virtual change in $|\underset{\sim}{k},C\rangle$ over a time segment dt by

$$\delta'_t|\underset{\sim}{k},C\rangle = -ih'|\underset{\sim}{k},C\rangle dt \quad . \tag{3.19}$$

Letting t become a time segment index, Eq. (3.18) becomes

$$\tilde{\sigma}_C(\underset{\sim}{k}) = \sum_t \delta'_t \langle C,\underset{\sim}{k}|n^{(N)}|\underset{\sim}{k},C\rangle/(N!)^2 \tag{3.20}$$

Noting that $n^{(N)}$ is the sum of N! terms each of which contributes the same we get finally

$$\tilde{\sigma}_C(\underset{\sim}{k}) = \sum_t \delta'_t \sum_{(\alpha_1\ldots\alpha_N)} |\langle\alpha_N\ldots\alpha_1|\underset{\sim}{k},C\rangle|^2 \tag{3.21}$$

where we have used the fact that

$$n = \sum_{\alpha} |\alpha\rangle\langle\alpha| \tag{3.22}$$

each $|\alpha\rangle = \psi_{\alpha}(\phi_A,\phi_B;t)$, being an orbital at time interval t. The sum over $(\alpha_1 \ldots \alpha_N)$ extends over all sets of N (out of $N_A + N_B$) orbitals which do not differ by a permutation amongst themselves.

REFERENCES

1. James J. Griffin, Peter C. Lichtner and Maria Dworzecka, Phys. Rev. C21, 1351 (1980).
2. Y. Alhassid and S. E. Koonin, Phys. Rev. C23, 1590 (1981).
3. E. A. Remler, Ann. Phys. 136, 293 (1981).
4. M. Gyulassy, E. Remler, K. Frankel, Proceedings of the Workshop on Nuclear Dynamics, Granlibakken, (1982) LBL-14138.
5. E. A. Remler, Phys. Rev. C, in press.
6. A. Huber in Mathematical Methods in Solid State and Superfluid Theory (R. C. Clark and G. H. Derrick, eds.) Plenum Press, New York, 1968.
7. E. A. Remler, Nuclear Kinetic Theory, William and Mary Preprint, 1981.
8. If mean field theory is used here instead of TDHF this problem does not arise.
9. Note however that the trace does not oscillate as in the analogous case discussed in Reference 1.

PHYSICAL ASYMPTOTICITY IN NONLINEAR COLLISION THEORIES AND TIME AVERAGING OF GAUGE INVARIANT PERIODIC QUANTIZED SOLUTIONS†

James J. Griffin
Department of Physics and Astronomy, and Institute for Physical Sciences and Technology
University of Maryland
College Park, MD 20742 U.S.A.

Introduction

In recent years several methods of utilizing nonlinear mean field propagation in time to describe nuclear reactions have been studied. They include the Initial Value Time-Dependent Hartree-Fock (TDHF) method,[1,2] the S-Matrix Time-Dependent Hartree-Fock (TD-S-HF) method,[3,4] and the Functional Integral Stationary Phase (FISP) method.[5-7] In each, one-body fields which are functionals of the time-dependent solutions are used to describe approximately the effects of the two-body interactions of the exact system.

We here address the question when such theories have the qualitative property of physical asymptoticity (discussed more precisely below), without which their predictions can, as of matter of principle, be of little practical value. We show that physical asymptoticity can be assured only when the approximate S-matrix amplitudes are constant, and for nonlinear theories also only by a proper a priori choice of the asymptotic reaction channel states. Such a choice therefore becomes an essential element of the reaction theory itself.

In the single-determinantal TDHF case the need to select channel states leads to the question of which single-determinantal bound state solutions are the proper analogs of exact eigensolutions, an answer to which was provided by the Gauge Invariant Periodic Quantization (GIPQ) method.[8-10] But when the generalized GIPQ method is applied in the full space of the Schrödinger equation, it yields not just all of the exact eigenstates, but an infinity of additional, physically spurious states. Thus arises a fundamental many-to-one ambiguity in the correspondence between GIPQ solutions and exact eigensolutions.

Some results of a recent study[11] of time averaging as an interpretative procedure for extracting the physical content of the bound state GIPQ solutions are reported here. In the full Schrodinger space, direct time averaging annihilates every physically spurious GIPQ solution, and projects each of the others onto the corresponding exact eigensolution. Thus, time averaging emerges from this test as a plausible general criterion for rejecting physically spurious GIPQ solutions. If it suffices for that role, then it could provide a correspondence (e.g.) between the exact eigenstates and the physical GIPQ solutions.

As an incidental by-product, the present analysis suggests a new, properly asymptotic, Hartree-Fock Stationary Phase S-matrix amplitude[12] which utilizes the FISP method only within the collision interval; outside of the collision interval

---

†Work supported in part by U. S. Department of Energy.

properly selected asymptotic channel solutions propagate by the Dirac-TDHF dynamics.[13] This new amplitude therefore joins that of the TD-S-HF theory as the second S-matrix approximant to be obtained from a physically asymptotic nonlinear theory.

Physical Asymptoticity

Physical Asymptoticity is that property of a reaction theory which guarantees that its predictions for the internal properties of the emergent reactants will be independent of the precise (distant) location of the measuring apparatus. As we shall illustrate in the case of the Schrödinger theory, this property follows from the constancy of the S-matrix expansion coefficients ("coefficient asymptoticity") and from the existence of channel states in which packets translate through space without alteration of the internal properties of the fragment ("channel asymptoticity"). The latter property stems essentially from the translational invariance of the interactions in the Hamiltonian, which implies the complete dynamical separation of the motion of its center of mass from the internal motions of a composite fragment. Thus once such a fragment is isolated, its center of mass translation and its internal dynamical evolution proceed separately and independently. Hence the familiar freedom always to choose the overall "center of mass" coordinate system without loss of generality, but also the result for "scattering systems" that the internal motion of each isolated fragment will evolve without coupling to the uniform translational motion of its center of mass.

"Scattering systems" are those[14] for which the interactions are of sufficiently short range that the complete Hamiltonian reduces at large fragment separations to the sum of isolated Hamiltonians, one for each fragment; i.e., for which

$$H = H_c^0 + V_c \xrightarrow[|\vec{R}_{12}| \to \infty]{} H_c^0 = H_{c,1}^0 + H_{c,2}^0 \tag{1}$$

where $H_{c,i}^0$ is the sum of the center of mass (kinetic energy) Hamiltonian of fragment i in channel c and the internal Hamiltonian of that fragment, and $\vec{R}_{12}$ is the fragment separation. Since our discussion is not concerned with the nuances of marginally short range forces, we assume that V_c is zero outside of some finite range, so that for each channel the fragment separation space is divided sharply into an external region and a collision volume. Correspondingly, for our localized packets,† the time is separated into the early and late asymptotic intervals and a collision interval.

For such systems, the independence of the center of mass and internal motions applies not just to the system as a whole, but also to individual channel fragments

†Throughout this paper we discuss the time-dependent scattering theory, and assume that the center of mass motions of the composite reactants are described by localized packets.

in the external region where the condition (1) prevails. Thus (1) guarantees that each channel wave function can be written in the external region as a product of packets describing the motions of the centers of mass and internal wave functions for each of the products.

Physical Asymptoticity in the Exact Schrödinger Theory

In the time-dependent Schrödinger description of a scattering system, an S-matrix amplitude is given by the overlap,

$$S_{fi} = \langle \Psi_f^{(-)}(\vec{x},t) | \Psi_i^{(+)}(\vec{x},t) \rangle \tag{2}$$

between one exact solution, $\Psi_i^{(+)}$, initialized early as well-separated approaching packets, $\Phi_i(\vec{x},t)$, and another, $\Psi_f^{(-)}$, initialized late as emerging packets, $\Phi_f(\vec{x},t)$. Thus

$$\lim_{t \to \mp\infty} \Psi_c^{(\pm)}(\vec{x},t) \to \Phi_c(\vec{x},t). \tag{3}$$

The channel labels c must suffice to initialize a unique solution of the time-dependent Schrödinger equation, including the internal wave function of each fragment and the motion of its center of mass. Since S_{fi} in (2) is constant in time (by the Schrödinger equation itself), the coefficients in

$$\Psi_i^{(+)}(\vec{x},t) = \sum_f S_{fi} \Psi_f^{(-)}(\vec{x},t) \tag{4}$$

$$\xrightarrow[t \to +\infty]{} \sum S_{fi} \Phi_f(\vec{x},t) \tag{5}$$

are constant in time.

If the channel label f includes the specification of a set of eigenvalues, say $\{m_f\}$, of an exhaustive set of internal operators $\{M\}$, then the probability of measuring such a set is simply $P_{\{m_f\}} = |S_{fi}|^2$ and remains constant as the packets of Φ_f propagate outward in time.

But if instead of f, one had initially chosen some other set of channel labels, say g, which did not include an exhaustive set of internal eigenvalues (or perhaps not the particular exhaustive set of eigenvalues desired), then a time-independent unitary transformation, A, among the internal channel states will form the desired channel states, f, from the states, g, so that the new basis, obtained by the replacement,

$$X_g(\vec{x},t) = \sum_f A_{fg} \Phi_f(\vec{x},t), \tag{6}$$

is labelled directly by the measurables $\{m_f\}$ of f. Because of the linearity of H, the Φ_f are surely Schrödinger solutions if only the X_g were solutions. Because of the constancy of the unitary transformation U_{fg}, which depends upon the guaranteed asymptotic separation of the internal and translational motion, the final probability of measuring the eigenvalues $\{m_f\}$ is still constant in time,

$$P_{\{m_f\}} = |\sum_g A_{fg} S_{gi}|^2 \tag{7}$$

so that the probability of measuring the set $\{m_f\}$ is independent of time and therefore of the location of the measuring apparatus. Thus it follows that the Schrödinger system is physically asymptotic.

Asymptoticity is Ubiquitous in Schrödinger Theory

We note that the physical asymptoticity of the Schrödinger scattering systems requires (a) constant S matrix, (b) the asymptotic separation of the translational and internal motions of the separate fragments, and either (c) an a priori choice of channel states labelled by the eigenvalues of the desired exhaustive set of measurements or (d) a unitary transformation relabelling the channels by such a set.

But "coefficient asymptoticity", (a), is guaranteed by the definition (2) of the S matrix and the time-dependent Schrödinger equation. And "channel asymptoticity" is implied by properties (b) and (d), of which (b) follows from the translational invariance of the interactions and the asymptotic condition (1) for scattering systems, and (d) is always an available option, by virtue of the linearity of the theory. Thus physical asymptoticity is automatically a property of the Schrödinger collision theory of scattering systems. This may partially explain why it is sometimes either taken erroneously for granted or overlooked in the study of alternative collisions theories.

Nonlinear Approximate Theories May Not Be Physically Asymptotic

In contrast with the Schrödinger theory, approximate theories are likely to lack one or more of these conditions for physical asymptoticity unless explicitly constructed to include them. In particular since nonlinearity voids the superposition principle, option (d) above is not available in mean field theories. This implies that properly asymptotic channels, if they exist at all, must be selected a priori, and used exclusively to describe all motions in the asymptotic regions (1). Otherwise channel asymptoticity will be lacking. But independently, coefficient asymptoticity may fail if in the approximate analog of the expansion (5), the coefficients, comprising the approximate S-matrix elements of the theory, are not constant in time. We emphasize that physical asymptoticity, which speaks to measurements in the asymptotic region, requires both coefficient asymptoticity and channel asymptoticity.

Because TDHF Translations Separate from Internal Motions, Channel Asymptotic TDHF Solutions Exist

In the Dirac-TDHF dynamics a set of channel asymptotic states always exists, due to a remarkable parallel between the TDHF system and the Schrödinger system, as follows: under TDHF time evolution, the expectation value of every one-body operator which commutes with the exact Hamiltonian remains constant.[15] Since the total momentum, $\vec{P}$, is such a one-body operator, its expectation value remains constant under TDHF evolution. The result is that a whole continuum of TDHF solutions initialized at t_i by the functions,

$$\Phi_{\vec{K}}(\vec{x},t_i) = \Phi_0(\vec{x},t_i)\exp i\vec{K}\cdot\vec{R} \tag{8}$$

(where $\vec{R} = A^{-1}\sum \vec{x}_i$ is the center of mass coordinate and $\vec{x}$ denotes the whole set of A coordinates, $\vec{x}_i$), and parametrized continuously by the mean total momentum vector, $\vec{K}$, correspond to precisely the same time-dependent internal state. For if $\Phi_0(\vec{x},t)$ is a zero-momentum time-dependent constant-<H> Hartree-Fock solution,[16] then

$$\Phi_{\vec{K}}(\vec{x},t) = \Phi_0(\vec{x}_i - \hbar\vec{K}t/AM,t)\exp i(\vec{K}\cdot\vec{R} - \hbar^2K^2t/2AM) \tag{9}$$

is another TDHF solution whose internal evolution in the fragment's average center of mass frame is identical with that of Φ_0 in the laboratory frame.

Thus it follows that from each TDHF solution whose measurable internal physical properties remain constant in time, translating solutions whose measurable internal properties are guaranteed to remain constant in time may be constructed by (9). In this way, each stationary Hartree-Fock solution becomes a prospective channel asymptotic state for a physically asymptotic TDHF collision theory. And if periodic states are interpreted in terms of the progressive phase averages, discussed below, over these periods,[4,11] they, too, may serve as channel asymptotic states when made to translate as in (9) above.[17]

Initial Value TDHF Theory is Non-Asymptotic

Nevertheless, the Initial Value TDHF reaction theory, which has in recent years been the object of the extensive numerical computation,[2] is decidedly not a physically asymptotic theory. Initial Value TDHF assumes an initial state in which two well-separated Hartree-Fock ground states translate towards one another, and utilizes the Dirac-TDHF time evolution to carry the system into and through the collision and the re-separation of the density into spatially disconnected parts, which then translate apart.

Although the assumed initial state happens to be an acceptably asymptotic channel, so that its early time behavior incorporates correct channel asymptoticity, Initial Value TDHF does not lead to physically asymptotic late time solutions because the expansion, analogous to (5) above of its late-time solution onto any set of channel asymptotic states will in general exhibit coefficients with nontrivial time dependence, and have been found to do so in the many numerical calculations discussed in Ref. 2. Therefore, its predictions for the results of internal physical measurements change* as the wave packet propagates outward in time. Thus the predictions of initial value TDHF for physical measurements on the emergent fragments depend upon the precise location of the measuring apparatus, even when it is in the asymptotic region well outside the collision volume.

*excepting those[15] which correspond to one-body operators which commute with H.

TD-S-HF Theory is Asymptotic By Construction

Like Initial Value TDHF theory,[1,2] the TD-S-HF theory[3,4] also assumes that all time evolution proceeds according to the Dirac-TDHF equation.[13] But it builds the theory entirely upon solutions which, either at early or at late times, reduce to the channel asymptotic TDHF solutions discussed above. Since the overlap of such solutions (analogous to Eqs. (2) for S) is not time independent, TD-S-HF utilizes a time average over the collision interval to define constant coefficients for the expansion analogous to (5). It therefore achieves channel asymptoticity by the explicit choice of the solutions allowed, and coefficient asymptoticity by using time averaging to extract a constant S-matrix analog.

Although this procedure is intuitive and heuristic, it results in a proper physically asymptotic theory whose predictions for measurements of the internal properties of the emergent fragments, unlike those of physically non-asymptotic theories, are stable over the range of space and time outside the collision volume. One might also mention that it has also been constructed so as to reduce precisely to the correct exact theory whenever the system is exactly described by the Dirac-TDHF theory.

Functional Integral Stationary Phase Mean Field Theories Are Not Asymptotic

The FISP theory of many-body collisions is constructed by approximating the exact Schrödinger expression for the time-propagator by a stationary phase approximation to a functional integral.[5] It leads to a coupled pair of first-order time-dependent equations which somewhat resemble Dirac-TDHF equations, and opens a whole range of interesting questions and possibilities in the many-body reaction theory.

As regards asymptoticity, however, there seems no reason to believe that FISP theory is either channel asymptotic or coefficient asymptotic. In a later application,[6] the possibility of coefficient asymptoticity was kept open by utilizing a kind of approximate interaction representation, in which the calculated S matrix becomes independent of the asymptotic initial and final time limits of the calculation by virtue of the mutual cancellation of the time dependences of two approximate factors. However, this particular restructuring omits any consideration of channel asymptoticity and is therefore insufficient to assure physical asymptoticity.

Indeed, even if a selection criterion upon the initial and final states allowed in the FISP S-matrix theory were adopted, under which they were required to be channel asymptotic single determinants of the form (9), it seems unlikely that the FISP time evolution equations can allow them to remain so, since it offers no apparent mechanism for keeping the internal and translational motions uncoupled during the propagation outside the collision interval. On the other hand, the FISP approximation to the dynamics can be incorporated into a new, physically asymptotic description by reverting to Dirac-TDHF propagation in the asymptotic region, as we now discuss.

A Physically Asymptotic Hartree-Fock Stationary Phase (AHFSP) Theory

Just as the TD-S-HF theory was structured explicitly to provide physical asymptoticity together with Dirac-TDHF propagation in time, one can also construct a physically asymptotic theory in which the Functional Integral Stationary Phase approach provides the description of the dynamical evolution during the collision.[12] In this approach, the FISP method is utilized to define the amplitude that a state, $\Phi_i(\vec{x},T_1)$, at the beginning of the collision interval propagates into a state $\Phi_f(\vec{x},T_2)$ at the end of the collision interval, thereby defining a definite constant S-matrix approximent, and guaranteeing coefficient asymptoticity for the theory. Then channel asymptoticity is guaranteed by utilizing exclusively the channel asymptotic TDHF solutions discussed above, propagated by the Dirac-TDHF equation, outside of the collision interval.

The resulting Asymptotic Hartree-Fock Stationary Phase S-matrix element is given by the expression

$$S_{fi}^{AHFSP} = \langle \Phi_f^{(-)}(\vec{x},T_2)|U_{fi}^{FISP}(T_2,T_1)|\Phi_i^{(+)}(\vec{x},T_1)\rangle \tag{10}$$

and provides the coefficient of the channel asymptotic TDHF solution, $\Phi_f^{(-)}(\vec{x},t)$, in the expansion analogous to (5). Here $U_{fi}^{FISP}(T_2,T_1)$ symbolizes the Functional Integral Stationary Phase approximate transition amplitude[5] from Φ_i at T_1 to Φ_f at T_2.

It is noteworthy that the amplitude S_{fi}^{AHFSP} is determined entirely by the dynamical behavior within the collision interval, (T_1,T_2). This feature is agreeably commensurable with the fact that in the exact theory all of the essential physical behavior is determined during the collision interval while only trivial translation and unperturbed time evolution occurs outside that interval.

The determination of the S matrix entirely within the collision interval is a feature that S_{fi}^{AHFSP} shares with the corresponding TD-S-HF amplitude, given by the expression,

$$S_{fi}^{TDSHF} = (T_2-T_1)^{-1}\int_{T_1}^{T_2} dt'\langle \Phi_f^{(-)}(\vec{x},T_2)|U_f^{TDHF}(T_2,t')^{\dagger}\, U_i^{TDHF}(t',T_1)|\Phi_i^{(+)}(\vec{x},T_1)\rangle. \tag{11}$$

Here, e.g., the symbol, $U_i^{TDHF}(t',T_1)$, represents the Dirac-TDHF propagation of $\Phi_i^{(+)}(\vec{x},t)$ from time T_1 to time t'.

It is especially interesting to inquire about possible relationships between these two expressions, which constitute the only examples so far of S-matrix approximants from physically asymptotic nonlinear theories.

Time Averaged Periodic Solutions as Stationary Solutions

Early in the study of mean field reaction theories, the possibility of building asymptotic channel states on periodic TDHF solutions was advanced.[3] It was physically motivated by the fact that properly time averaged properties of such solutions might serve as the TDHF analogs of stationary exact eigenvalues, allowing thereby approximate periodic states to serve the role of exact stationary eigenstates.

Subsequently, it was realized that the casting of periodicity as a physically essential property implied that it ought to be gauge invariant, like all the other physical properties of the nonrelativistic theory. This condition lead to a discrete quantization of the periodic solutions, and to the Gauge Invariant Periodic Quantization (GIPQ) method.[8-10] Later it emerged that the FISP approximation to the poles of $(E-H)^{-1}$ yields precisely the same spectrum as the (GIPQ) method.[18,19] We turn now to a brief summary of some recent developments[11] in the GIPQ method.

Gauge Invariant Periodic Quantization of Exact Schrödinger Solutions

The Gauge Invariant Periodic Quantization procedure seeks time-dependent GIPQ solutions of the form

$$\Phi(\vec{x},t) = \Phi_{GP}(\vec{x},t) \exp(-i\langle H\rangle t/\hbar). \tag{11}$$

where Φ_{GP} is the time-periodic analog of an exact gauge invariant stationary eigensolution.[8-10] The method may be applied to the TDHF single-determinantal subspace, or to any time-dependent parametric subspace of the full space.[10] In particular, Φ may be defined in the full space of the exact Schrödinger eigensolutions. In that case, one would expect the set GIPQ solutions to reduce to the set of exact eigensolutions.

For an exact Schrödinger solution, of the form

$$\Psi(\vec{x},t) = \sum_n a_n \psi_n(x) \exp(-i E_n t/\hbar) \tag{12}$$

the Gauge Invariant Periodic Quantization condition requires the form (11):

$$\Psi(\vec{x},t) = \{\sum_n a_n \psi_n(x) \exp(-it[E_n-E])\}\{\exp(-iEt/\hbar)\} \tag{13}$$

where the first factor is periodic and where $\langle H\rangle = E = \sum_n |a_n|^2 E_n$. Periodicity requires that the frequency of every term must be an integral multiple, say k_n, of some fundamental frequency, Ω, so that

$$(E_n-E) = \hbar k_n \Omega \qquad (k_n \text{ an integer}) \tag{14}$$

for every n value for which $a_n \neq 0$.

Spurious GIPQ States Occur But Are Deleted by Progressive Phase Averaging

The exact eigensolutions $E = E_n$, $a_n = 1$, are evidently always included among the GIPQ solutions honoring (13). But in addition, an infinity of two-component solutions exists which are not exact eigensolutions, and which therefore are to be considered physically spurious. They pose the question how one is to select the physical GIPQ solutions from the spurious ones.

From the same physical viewpoint as that which suggested periodic functions in the first place[3,17] (namely, that under a time averaged interpretation periodic functions can describe constant physical characteristics), one naturally inquires about time averaging, and finds immediately that the time average of the Gauge Invariant factor in (13) is given by[11]

$$\overline{\Psi_{GP}(\vec{x},t)} = \lim_{\tau\to\infty} (2\tau)^{-1} \int_{t-\tau}^{t+\tau} dt' \sum a_n \psi_n(x) \exp(-it'[E_n - E]/\hbar) \tag{15a}$$

$$= \sum_n a_n \psi_n(x)\ \delta(E_n - E), \tag{15b}$$

equal to zero unless the energy equals one of the eigenergies, and equal to $a_n\psi_n(x)$ when $E = E_n$. Thus time averaging in the full space annihilates all the spurious GIPQ solutions, and projects the (unnormalized) exact spatial eigenfunctions from the non-spurious ones, providing thereby a valid criterion for selecting the physical GIPQ solutions out of the set of all GIPQ solutions.

This exercise in the full eigenfunction space therefore provides no deterrent to the prospect that in more restricted approximate subspaces, time averaging might also serve to eliminate spurious solutions and to provide a correspondence between the physical GIPQ solutions and the (perhaps only approximate) eigensolutions obtained by other methods (such as, e.g., stationary state diagonalization via RPA, Tamm-Dancoff, etc.).

It should be noted that for periodic solutions time averaging is equivalent to averaging over a real-valued "progressive phase" parameter which increases monotonically by 2π during each period. Indeed, this view of the averaging as a "progressive phase" rather than a time averaging is distinctly preferable when one addresses the collision problem, since the translating periodic solutions depend upon time both because of the genuine physical translation in space and time and because of the periodic internal oscillation of the solution, which is viewed as non-physical behavior to be removed by the averaging. Then the progressive phase prescription of the average neatly selects and properly executes only the average only over the periodic variation.

Summary

The property of physical asymptoticity, which guarantees that a reaction theory's predictions for physical measurements of internal fragment properties shall not depend upon the precise location of the measuring apparatus, has here been analyzed in some detail. Since the lack of such a property reduces most of the nontrivial physical content of an approximate many-body reaction theory to practical unobservability, this property warrants some attention.

We find that physical asymptoticity is guaranteed in the Schrödinger collision theory of scattering systems with translationally invariant interactions by the constancy of the S-matrix elements (which we label "coefficient asymptoticity") and by the decoupling of the internal and translational motions for well-separated fragments ("channel asymptoticity"). Both conditions are necessary for physical asymptoticity.

For nonlinear approximants to the Schrödinger theory, a specific a priori selection of asymptotic channel states seems necessary to achieve channel asymptoticity. Within the Dirac-TDHF dynamics such a set does exist, so that channel asymptotic single-determinantal propagation can be described by Dirac-TDHF time evolution. But it is not clear that such a set exists for the propagation of well-separated fragments under the FISP approximate time evolution.

A review of nonlinear mean field many-body reaction theories shows that Initial Value TDHF (by far the most computed theory) is non-asymptotic, as is the FISP theory (even in the coefficient asymptotic form of Alhassid and Koonin,[6] since that description is still channel non-asymptotic). The TD-S-HF theory is asymptotic by construction.

A new Asymptotic Hartree-Fock Stationary Phase (AHFSP) description[12] which utilizes Dirac-TDHF channel solutions and propagation outside the collision interval, and the FISP method to describe the dynamical evolution within the collision interval, is reported. Together with the TD-S-HF theory it constitutes the second example of a physically asymptotic nonlinear many-body reaction theory. Although the S-matrix approximants of these theories are evidently different, their careful comparison should prove enlightening.

The Gauge Invariant Periodic Quantized solutions of the exact Schrödinger problem are considered to test whether they include as they ought the exact eigenfunctions. They do, but include as well an infinity of spurious solutions. However, time averaging over the periodic motions (or better stated, "progressive phase averaging") annihilates all the spurious solutions, and projects the physical solutions onto the exact (but unnormalized) eigensolutions. These results offer no discouragement for the prospect that in more restrictive approximate subspaces progressive phase averaging of GIPQ solutions may reject physically spurious solutions and provide a correspondence between the remaining solutions and the (exact or approximate) eigensolutions obtained by stationary state methods.[11]

REFERENCES

1. P. Bonche, S. E. Koonin and J. W. Negele, Phys. Rev. C13 (1976) 1226.
2. K.T.R. Davies, K.R.S. Devi, et al., to be published in Heavy Ion Science, Vols. I & II - Nuclear Science, ed. by D. A. Bromley (Plenum Publ. Co., New York, 1982). This review includes an extensive list of references related to the Initital Value TDHF calculations to date.
3. James J. Griffin, P. C. Lichtner and M. Dworzecka, Phys. Rev. 21 (1980) 1351.
4. J. J. Griffin, M. Dworzecka, et al., Phys. Lett. 93B (1980) 235.
5. S. Levit, Phys. Rev. C21 (1980) 1594.
6. Y. Alhassid and S. E. Koonin, Phys. Rev. C23 (1981) 1590.
7. Y. Alhassid, B. Müller and S. E. Koonin, Phys. Rev. C23 (1981) 487.
8. K.-K. Kan, J. J. Griffin, et al., Nucl. Phys. A332 (1979) 109.
9. K.-K. Kan, Phys. Rev. C24 (1981) 789; C22 (1980) 2228; C21 (1980) 1098.
10. K.-K. Kan, Phys. Rev. C24 (1981) 279.
11. K.-K. Kan, J. J. Griffin, et al., U. of Md. Tech. Report #ORO 5126-44, to be published.
12. J. J. Griffin and M. Dworzecka, Proc. 2nd International Nuclear Dynamics Workshop, Granlibakken, CA, 1982 (Lawrence Berkeley Labs Conf. Report, 1982), and U. of Md. Tech. Report #ORO 5126-157.
13. P. A. M. Dirac, Proc. Cambridge Phil. Soc. 26 (1930) 376.
14. R. Haag, in Lectures in Theoretical Physics, Vol. III, p. 326, ed. W. E. Brittin (Interscience Publ. Inc., New York, 1961).
15. Cf. the appendix of Ref. 3 above for further discussion.
16. P. C. Lichtner, J. J. Griffin, et al., Phys. Lett. 88B (1979) 221; Phys. Rev. C20 (1979) 845.
17. See Section VI of Ref. 3.
18. H. Reinhardt, Nucl. Phys. A331 (1979) 353.
19. S. Levit, J. W. Negele and Z. Paltiel, Phys. Rev. C21 (1980) 1603.

SELF-CONSISTENT COLLECTIVE COORDINATE METHOD FOR THE MAXIMALLY DECOUPLED COLLECTIVE MOTION

T. Marumori, F. Sakata*, T. Une and Y. Hashimoto
Institute of Physics, University of Tsukuba, Ibaraki 305
*Institute for Nuclear Study, University of Tokyo, Tanashi, Tokyo, 188

1. INTRODUCTION

In recent years, the concept of a collective path or, more generally, of a collective subspace in the many-particle Hilbert space has been increasingly arousing interest in attempts of microscopic description of the large amplitude collective motion like vibrations of soft nuclei, fission, heavy-ion reactions, etc., and the problem of how to determine the collective subspace (rather than assuming it a priori) has been discussed by many authors[1)∿8)].

The main purpose of this talk is to develop a new quantum theory which is capable by itself of determining the maximally decoupled collective subspace as well as a maximally decoupled collective Hamiltonian, on the basis of a fundamental principle called the invariance principle of the Schrödinger equation[4),6),7),8)].

First, the basic ideas of the theory are explained by formulating the invariance principle within the framework of the time-dependent Hartree-Fock (TDHF) theory[7)]. In order to overcome the essential drawback of the TDHF theory that the theory itself does not give any prescription how to quantize the "classical" collective motion derived by the theory, we then formulate the invariance principle within the framework of the unitary-transformation method with auxiliary variables[9),10),11)]. It is demonstrated that the requirement of the invariance principle of the Schrödinger equation is completely equivalent to requiring existence of an invariant collective subspace of the Hamiltonian, and the unitary-transformation method with auxiliary variables associated with the invariance principle enables us to develop a full quantum theory of the maximally decoupled collective motion.

2. BASIC IDEAS
- SEMI-CLASSICAL THEORY WITHIN THE FRAMEWORK OF THE TDHF[7] -

An essential element in the study of the microscopic dynamics underlying the collective mode of motion is to find the "particle-collective coupling" which is just what organizes self-consistently the collective modes under consideration out of the particle excitations. A clue for such an investigation is provided by the cranking model[12] for the collective rotation. In the cranking model, one considers the particle motion in such a mean field with a deformed equilibrium that is uniformly rotating with frequency ω_{rot}. In the rotating (time-dependent) coordinate frame, the "particle-collective coupling", which organizes the collective rotation out of the particle excitations, manifests itself as the Coriolis coupling $-\vec{\omega}_{rot}\cdot\vec{J}$. ($\vec{J}$ is the total angular-momentum operator of particles, and we use $\hbar=1$ here and hereafter.) This coupling in the rotating frame gives rise to an increase in the energy of the particle motion, which is identified with the collective rotational energy,

$$\mathcal{H}_{rot}(\omega_{rot}) \equiv \langle\phi(\omega_{rot})|\hat{H}|\phi(\omega_{rot})\rangle_\beta - \langle\phi_0|\hat{H}|\phi_0\rangle_\beta \quad . \qquad (2.1)$$

Here $|\phi_0\rangle_\beta$ is the Hartree-Fock ground state with deformation β and $|\phi(\omega_{rot})\rangle_\beta$ is the Hartree-Fock state in the rotating frame satisfying

$$\delta\langle\phi(\omega_{rot})|\hat{H} - \vec{\omega}_{rot}\cdot\vec{J}|\phi(\omega_{rot})\rangle_\beta = 0 \quad , \qquad (2.2)$$

where $\hat{H}' \equiv \hat{H} - \vec{\omega}_{rot}\cdot\vec{J}$ is the time-displacement operator describing the time-evolution of the system in the rotating frame.

Now let us generalize this idea in order to describe the large-amplitude collective motion around the Hartree-Fock ground state $|\phi_0\rangle$ with a "spherical" equilibrium. An essential difference of this case from the above cranking model is in the fact that we do not know, in advance, the "particle-collective coupling" organizing the large amplitude collective vibration under consideration, out of the particle excitations. In order to find the "particle-collective coupling" in this case, we therefore introduce a generalized moving frame by a time-dependent unitary transformation with a finite number of pairs of parameters $\{\eta_r^*(t),\eta_r(t)\}$,

$$|\phi(\eta_r{}^*(t),\eta_r(t))\rangle = \hat{U}^{-1}(\eta_r{}^*(t),\eta_r(t))|\phi_0\rangle,$$
$$\hat{U}^{-1} = \exp\{i\hat{G}_0(\eta_r{}^*(t),\eta_r(t))\}, \tag{2.3}$$

where $|\phi(\eta_r^*(t),\eta_r(t))\rangle$ is the Hartree-Fock state in the moving frame. The set of parameters $\{\eta_r^*(t),\eta_r(t)\}$, which corresponds to a c-number version of collective bosons in the Heisenberg representation

$$\eta_r{}^*(t) \leftrightarrow b_r^+(t) \equiv \exp\{i\mathbb{H}_B t\} b_r^+ \exp\{-i\mathbb{H}_B t\}\ , \tag{2.4}$$

with a collective boson Hamiltonian $\mathbb{H}_B(b_r^+,b_r)$, specifies the time-dependent variations of the mean field associated with the collective motion described by a set of collective coordinates $\alpha_r(t)$ and their conjugates $\pi_r(t)$,

$$\alpha_r(t) = \frac{1}{\sqrt{2}}(\eta_r^*(t)+\eta_r(t)),\ \pi_r(t) = \frac{i}{\sqrt{2}}(\eta_r^*(t)-\eta_r(t)). \tag{2.5}$$

In order to simplify the presentation of the theory, hereafter we restrict ourselves to a single pair of parameters $(\eta^*(t),\eta(t))$.

Since the time-dependence of the parameters $(\eta^*(t),\ \eta(t))$ completely specifies the time-dependence of $|\phi(\eta^*,\eta)\rangle$ in Eq. (2.3), we have

$$i\frac{\partial}{\partial t}|\phi(\eta^*,\eta)\rangle = \{i\dot{\eta}\cdot\hat{O}_0^+(\eta^*,\eta)-i\dot{\eta}^*\hat{O}_0(\eta^*,\eta)\}|\phi(\eta^*,\eta)\rangle, \tag{2.6}$$

where the operator $\hat{O}_0{}^+(\eta^*,\ \eta)$ is the local infinitesimal generator with respect to η, and is defined by

$$\hat{O}_0{}^+(\eta^*,\eta) \equiv \{\frac{\partial}{\partial\eta}\hat{U}^{-1}(\eta^*,\eta)\}\hat{U}(\eta^*,\eta). \tag{2.7}$$

In order to determine the motion of the moving frame specified by the time-dependence of $(\eta^*(t),\ \eta(t))$ as well as the structure of the operators $(\hat{O}_0{}^+(\eta^*,\eta),\ \hat{O}_0(\eta^*,\eta))$, we employ the <u>invariance principle of the Schrödinger equation</u>[4),6),7),8)] which specifies the concept of the <u>maximally decoupled</u> collective path. The principle can be simply stated as follows: <u>The time-dependence of the parameters</u> $(\eta^*(t),\eta(t))$ <u>must be introduced so as to keep the Schrödinger equation invariant</u>. Within the framework of the Hartree-Fock theory, the principle is expressed as

$$\delta_0\langle\phi(\eta^*,\eta)|\{(i\frac{\partial}{\partial t}-\hat{H})|\phi(\eta^*,\eta)\rangle\} = 0,\quad \text{and}\quad \text{h.c.}, \tag{2.8}$$

with the boundary conditions $\hat{U}^{-1}(\eta^*=0,\ \eta=0) = 1$ and $\delta\langle\phi_0|\hat{H}|\phi_0\rangle = 0$ at $\eta^*=\eta=0$, where the variation $|\delta_0\phi(\eta^*,\eta)\rangle$ is defined by

$$|\delta_0\phi(\eta^*,\eta)\rangle \equiv \hat{U}^{-1}(\eta^*,\eta)|\delta\phi_0\rangle.$$

With the use of Eq. (2.6), Eq. (2.8) can be written as

$$\delta_0\langle\phi(\eta^*,\eta)|\hat{H}-i\dot{\eta}\hat{O}_0^+(\eta^*,\eta)+i\dot{\eta}^*\hat{O}_0(\eta^*,\eta)|\phi(\eta^*,\eta)\rangle = 0, \text{ i.e.,}$$

$$\delta\langle\phi_0|e^{-i\hat{G}_0}\hat{H}e^{i\hat{G}_0}-i\dot{\eta}e^{-i\hat{G}_0}\frac{\partial}{\partial\eta}e^{i\hat{G}_0}-i\dot{\eta}^*e^{-i\hat{G}_0}\frac{\partial}{\partial\eta^*}e^{i\hat{G}_0}|\phi_0\rangle = 0 . \tag{2.9}$$

Corresponding to Eq. (2.2), thus, the "particle-collective" coupling in the moving frame, which is highly nonlinear in general, manifests itself as the coupling $\hat{H}'-\hat{H} \equiv -i\dot{\eta}\hat{O}_0^+(\eta^*,\eta)+i\dot{\eta}^*\hat{O}_0(\eta^*,\eta)$. This coupling gives rise to an increase in the energy of the particle motion, which is identified with the energy of the collective motion

$$\mathcal{H}_0(\eta^*,\eta) \equiv \langle\phi(\eta^*,\eta)|\hat{H}|\phi(\eta^*,\eta)\rangle-\langle\phi_0|\hat{H}|\phi_0\rangle. \tag{2.10}$$

This corresponds to the c-number version of the collective boson Hamiltonian $\mathbb{H}_B(b^+,b)$ in Eq. (2.4); $\mathcal{H}_0(\eta^*,\eta) \leftrightarrow \mathbb{H}_B(b^+,b)$.

At this stage, we require the following <u>self-consistency condition</u> on the collective motion: The collective Hamiltonian identified by Eq. (2.10) must self-consistently determine the time-dependence of the moving frame (i.e., the time-dependence of $(\eta^*(t),\ \eta(t))$) by the canonical equation of motion

$$i\dot{\eta} = \frac{\partial}{\partial\eta^*}\mathcal{H}_0(\eta^*,\eta), \qquad -i\dot{\eta}^* = \frac{\partial}{\partial\eta}\mathcal{H}_0(\eta^*,\eta) . \tag{2.11}$$

It is easily shown[7)] that this self-consistency condition is fulfilled when we choose the parameters (η^*,η) so as to satisfy

$$\langle\phi(\eta^*,\eta)|[\hat{O}_0(\eta^*,\eta),\hat{O}_0^+(\eta^*,\eta)]|\phi(\eta^*,\eta)\rangle$$

$$= \langle\phi_0|[e^{-i\hat{G}_0}\frac{\partial}{\partial\eta}e^{i\hat{G}_0},e^{-i\hat{G}_0}\frac{\partial}{\partial\eta^*}e^{i\hat{G}_0}]|\phi_0\rangle = 1. \tag{2.12}$$

It can be also verified that such a choice of the parameters (η^*,η) is generally possible. (See Appendix C of Ref. (7)). It is now clear that the problem to solve a set of the basic equations (2.9), (2.10),

(2.11) and (2.12) self-consistently can be reduced to finding the hermitian operator $\hat{G}_0(\eta^*,\eta)$ satisfying these equations.

One of the simplest ways to determine $\hat{G}_0(\eta^*,\eta)$ is a method called an (η^*,η)-expansion[7]: Since $\hat{G}_0(\eta^*,\eta)$ is restricted to be a one-body operator, we can make the following expansion of $\hat{G}_0(\eta^*,\eta)$ with respect to (η^*,η);

$$\hat{G}_0(\eta^*,\eta) = \Sigma_\lambda\{g_\lambda(\eta^*,\eta)\hat{X}^+_\lambda + g^*_\lambda(\eta^*,\eta)\hat{X}_\lambda\},$$
$$g_\lambda(\eta^*,\eta) = \sum_{n\geq 1} g_\lambda(n),\quad g_\lambda(n) \equiv \sum_{(r+s=n)} {}_{r,s} g_\lambda(r,s)(\eta^*)^r(\eta)^s, \tag{2.13}$$

where we have used the complete set of the RPA (or the Tamm-Dancoff) eigenmodes $\{\hat{X}^+_\lambda,\hat{X}_\lambda\}$ instead of the particle-hole pairs $\{c^+_\mu c_i,\ c_i{}^+c_\mu\}$,

$$\langle\phi_0|[\hat{X}_\lambda,\hat{X}^+_{\lambda'}]|\phi_0\rangle = \delta_{\lambda\lambda'},\quad \langle\phi_0|[\hat{X}_\lambda,\hat{X}_{\lambda'}]|\phi_0\rangle = 0,$$
$$\langle\phi_0|[\hat{X}_\lambda,[\hat{H},\hat{X}^+_{\lambda'}]]|\phi_0\rangle = \omega_\lambda\delta_{\lambda\lambda'},\quad (\omega_\lambda > 0). \tag{2.14}$$

(We adhere to the convention of denoting occupied single-particle orbits by the indices $i,j,\cdots$, and unoccupied single-particle orbits by the indices $\mu,\nu,\cdots$.) Since the basic equations (2.9) with (2.11) and (2.12) with the (η^*,η)-expansion (2.13) are supposed to be valid for continuous ranges of η^* and η, we can equate the coefficients of each power of (η^*,η) in these equations to zero. Thus, by starting with the coefficients with the lowest power of (η^*,η) and by proceeding to the higher (η^*,η)-coefficients step by step, we can determine the unknown quantities $g_\lambda(r,s)$ of $\hat{G}_0(\eta^*,\eta)$ in Eq. (2.13) as well as the collective Hamiltonian $\mathcal{H}_0(\eta^*,\eta)$ self-consistently. The important task in this expansion method is to set up some specific conditions which characterize the collective motion under consideration in "small-amplitude" (i.e., small η) limit. In the case of the collective band whose lowest excited state in the small-amplitude limit is described by the lowest-energy eigenmode $\hat{X}^+_{\lambda_0}$ in Eq. (2.14), we may set up the condition so as to satisfy $g_\lambda(1,0) = g_\lambda(0,1)=0$ for $\lambda \neq \lambda_0$.

3. UNITARY-TRANSFORMATION METHOD WITH AUXILIARY BOSONS ASSOCIATED WITH THE INVARIANCE PRINCIPLE OF THE SCHROEDINGER EQUATION

We are now in a stage to formulate the above basic ideas in a form of full quantum theory. For this purpose, it is convenient to employ the unitary-transformation method with auxiliary variables[9),10),11)]. Let us suppose that the nuclear many-body system under consideration exhibits a class of collective excitation spectra and all the collective states can be expanded in terms of a set of orthonormal state vectors

$$\{|n\rangle,\quad n = 0,1,2,\cdots\},\tag{3.1}$$

which defines a collective subspace. The orthogonal complement of the collective subspace is then specified to be spanned by a set of orthonormal state vectors

$$\{|i\rangle\};\langle n|i\rangle = 0 \quad \text{for} \quad |n\rangle\in\{|n\rangle\}.\tag{3.2}$$

The collective subspace is characterized by idealized creation and annihilation operators $\hat{K}^+$ and $\hat{K}$, which may be complicated functions of fermion operators and are defined by

$$\hat{K}^+ \equiv \sum_n \sqrt{n+1}\,|n+1\rangle\langle n|,\quad \hat{K} \equiv \sum_n \sqrt{n+1}\,|n\rangle\langle n+1|\tag{3.3a}$$

$$[\hat{K},\ \hat{K}^+] = 1\cdot\hat{P}\ .\tag{3.3b}$$

Here $\hat{P}$ is the projection operator into the collective subspace; $P = \Sigma_n|n\rangle\langle n|$. We also denote, hereafter, the projection operator into the orthogonal complement of the collective subspace as $\hat{Q} = 1 - \hat{P} = \Sigma_i|i\rangle\langle i|$.

In the unitary-transformation method with auxiliary bosons, we introduce auxiliary bosons (b^+,b) satisfying

$$[b,b^+] = 1,\tag{3.4}$$

as a new degree of freedom independent of the fermion degrees of freedom, and the state vectors are extended into a product space $\{|\Psi\rangle\!\rangle\} \equiv \{|\Psi\rangle\}\otimes\{|\Phi)_B\}$ of the fermion-state space $\{|\Psi\rangle\}$ and the boson-state space $\{|\Phi)_B\}$. We then employ a unitary transformation satisfying the following properties;

$$\hat{\mathbb{V}}^{-1}\hat{K}^{+}\hat{\mathbb{V}} = b^{+}\cdot\hat{P}, \quad \hat{\mathbb{V}}^{-1}b^{+}\hat{P}\hat{\mathbb{V}} = \hat{K}^{+}, \tag{3.5a}$$

$$\hat{\mathbb{V}}\cdot\hat{Q} = \hat{Q}\cdot\hat{\mathbb{V}} = \hat{Q}, \tag{3.5b}$$

$$\hat{\mathbb{V}}^{-1}|n_1>\otimes|n_2)_B = |n_2>\otimes|n_1)_B, \tag{3.5c}$$

where $\{|n)_B\}$ is a set of orthonormal boson states spanning the boson-state space $\{|\Phi)_B\}$;

$$|n)_B = \frac{1}{\sqrt{n!}}(b^{+})^{n}|0)_B, \quad b|0)_B = 0. \tag{3.6}$$

A formal solution of $\hat{\mathbb{V}}$ is given by[8]

$$\hat{\mathbb{V}} = \exp(i\pi b^{+}b)\cdot\exp\{\tfrac{\pi}{2}(\hat{K}^{+}b-\hat{K}b^{+})\}\cdot\hat{P} + \hat{Q}. \tag{3.7}$$

Equation (3.5) implies that the unitary transformation $\hat{\mathbb{V}}$, which we call <u>the transformation into "collective representation"</u>, interchanges the collective subspace $\{|n>\}$ and the boson space $\{|n)_B\}$ while it keeps the orthogonal complement $\{|i>\}$ invariant.

Since we have not yet made any specification of the collective subspace so far, the Hamiltonian of the system is generally decomposed into three parts;

$$\begin{aligned}
&\hat{H} = \hat{H}_{coll} + \hat{H}_{intr} + \hat{H}_{coupl}\,,\\
&\hat{H}_{coll} \equiv \hat{P}\hat{H}\hat{P} - \varepsilon_{intr}\hat{P}, \quad \hat{H}_{intr} \equiv \hat{Q}\hat{H}\hat{Q} + \varepsilon_{intr}\hat{P}\,,\\
&\hat{H}_{coupl} \equiv \hat{Q}\hat{H}\hat{P} + \hat{P}\hat{H}\hat{Q}\,.
\end{aligned} \tag{3.8}$$

Thus, the Hamiltonian in the collective representation is given by

$$\hat{\mathbb{V}}^{-1}\hat{H}\hat{\mathbb{V}} = \mathbb{H}_B(b^{+}b)\cdot\hat{P} + \hat{H}_{intr} + \hat{\mathbb{V}}^{-1}\hat{H}_{coupl}\hat{\mathbb{V}}\,, \tag{3.9}$$

where the boson Hamiltonian $\mathbb{H}_B(b^{+},b)\hat{P}\equiv\hat{\mathbb{V}}^{-1}\hat{H}_{coll}\hat{\mathbb{V}}$ manifests itself as a collective Hamiltonian. However, it must be emphasized that, only if the coupling $\hat{\mathbb{V}}^{-1}\hat{H}_{coupl}\hat{\mathbb{V}}$ is "weak", the collective motion under consideration is to have physical reality.

We are now in a position to discuss the problem of how to specify the maximally decoupled collective subspace as well as the maximally decoupled collective (boson) Hamiltonian. The fundamental principle for this specification is <u>the invariance principle of the Schrödinger equation</u>. To formulate the principle in the present full quantum case, we employ the boson operators in the Heisenberg representation

$b^+(t) = \exp(i\mathbb{H}_B t)b^+\exp(-i\mathbb{H}_B t)$, satisfying the equation of motion, $idb^+(t)/dt = [b^+(t), \mathbb{H}_B(b^+,b)]$. This corresponds to Eq. (2.4). The collective boson Hamiltonian $\mathbb{H}_B(b^+,b)$ is not yet specified and will be self-consistently determined later. We then introduce a time-dependent unitary transformation

$$\hat{\mathbb{V}}(t) = e^{i\mathbb{H}_B t}\hat{\mathbb{V}}e^{-i\mathbb{H}_B t} \equiv \hat{\mathbb{V}}(b^+(t), b(t)), \tag{3.10}$$

where $\hat{\mathbb{V}}$ is the unitary transformation into the collective representation given by Eq. (3.7). The invariance principle is now formulated as follows: <u>The boson Hamiltonian</u> $\mathbb{H}_B(b^+,b)$, <u>which specifies the time-dependence of the boson operators</u> $(b^+(t), b(t))$, <u>must be determined in such a way that the Schrödinger equation after the time-dependent unitary transformation</u> $\hat{\mathbb{V}}(t)$ <u>remains invariant</u>, i.e.,

$$i\frac{\partial}{\partial t}|\tilde{\Psi}(t)\rangle\rangle_c = \hat{H}|\tilde{\Psi}(t)\rangle\rangle_c, \qquad |\tilde{\Psi}(t)\rangle\rangle_c = \hat{\mathbb{V}}^{-1}(t)|\Psi(t)\rangle\rangle, \tag{3.11}$$

where $|\Psi(t)\rangle\rangle$ satisfies the original Schrödinger equation

$$i\frac{\partial}{\partial t}|\Psi(t)\rangle\rangle = \hat{H}|\Psi(t)\rangle\rangle, \quad |\Psi(t)\rangle\rangle = |\Psi(t)\rangle \otimes |\Phi)_B. \tag{3.12}$$

Equation (3.11) can be expressed as

$$\begin{bmatrix} \hat{P}\hat{\mathbb{L}}\hat{P} & \hat{P}\hat{\mathbb{L}}\hat{Q} \\ \hat{Q}\hat{\mathbb{L}}\hat{P} & \hat{Q}\hat{\mathbb{L}}\hat{Q} \end{bmatrix} \begin{bmatrix} \hat{P}\cdot\exp\{-i\cdot\hat{H}_{coll}t\}|\Psi(t)\rangle\rangle_c \\ \hat{Q}\cdot\exp\{-i\cdot\mathbb{H}_B t\}|\Psi(t)\rangle\rangle_c \end{bmatrix} = 0, \quad |\Psi(t)\rangle\rangle_c \equiv \hat{\mathbb{V}}^{-1}|\Psi(t)\rangle\rangle, \tag{3.13}$$

where

$$\hat{\mathbb{L}} \equiv \hat{\mathbb{V}}^{-1}\{\hat{H}+\mathbb{H}_B(b^+,b)\}\hat{\mathbb{V}} - \{\hat{H}+\mathbb{H}_B(b^+,b)\}. \tag{3.14}$$

With the use of Eq. (3.5b), we have $\hat{Q}\hat{\mathbb{L}}\hat{Q}=0$ so that Eq. (3.13) can be decomposed into the <u>maximal-decoupling condition</u>

$$\hat{P}\hat{\mathbb{L}}\hat{Q} = \hat{Q}\hat{\mathbb{L}}\hat{P} = 0, \qquad \text{i.e.,} \tag{3.15a}$$

$$\hat{P}\hat{H}\hat{Q} = \hat{Q}\hat{H}\hat{P} = 0, \tag{3.15b}$$

and the definition of the maximal decoupled collective boson Hamiltonian

$$\hat{P}\hat{\mathbb{L}}\hat{P}=0, \qquad \text{i.e.,} \tag{3.16a}$$

$$\mathbb{H}_B(b^+,b)\hat{P}=\hat{P}\{\hat{\mathbb{V}}^{-1}\hat{H}\hat{\mathbb{V}}-\hat{H}+\hat{\mathbb{V}}^{-1}\mathbb{H}_B(b^+,b)\hat{\mathbb{V}}\}\hat{P}=\hat{P}\{\hat{\mathbb{V}}^{-1}\hat{H}\hat{\mathbb{V}}-\varepsilon_{intr}\}\hat{P}, \tag{3.16b}$$

where we have used the relation $\hat{\mathbb{V}}^{-1}\mathbb{H}_B(b^+,b)\hat{\mathbb{V}}\hat{P} = \hat{H}_{coll}$ together with the expression (3.8) of the Hamiltonian. Equation (3.15) clearly demonstrates that the collective subspace $\{|n>\}$ associated with the invariance principle must be an invariant subspace of the Hamiltonian.

The physical meaning of the invariance principle is now clear: The invariance principle of the Schrödinger equation is completely equivalent to requiring existence of the unitary transformation $\hat{\mathbb{V}}$ into the collective representation which leads us to the invariant collective subspace as well as the maximally decoupled collective boson Hamiltonian $\mathbb{H}_B(b^+,b)$. Thus, in the collective representation associated with the invariance principle, the transformed Hamiltonian $\hat{\mathbb{V}}^{-1}\hat{H}\hat{\mathbb{V}}$ is given by

$$\hat{\mathbb{V}}^{-1}\hat{H}\hat{\mathbb{V}}=\mathbb{H}_B(b^+,b)\hat{P}+\hat{H}_{intr}, \quad \hat{H}_{intr}=\hat{Q}\hat{H}\hat{Q}+\varepsilon_{intr}\hat{P}. \tag{3.17}$$

4. BASIC EQUATIONS IN THE THEORY

In order to compensate for the redundancy in the degrees of freedom due to the introduction of the auxiliary bosons, it is necessary to impose a subsidiary condition which physical state vectors $\{|\Psi_{phys}>>\}$ in the enlarged fermion-boson product space have to fulfill. The simplest one is

$$b|\Psi_{phys}>> = 0, \tag{4.1}$$

which simply implies that the physical state vectors are given with the boson vacuum by

$$\{|\Psi_{phys}>>\} = \{|\Psi>\}\otimes|0)_B, \quad b|0)_B = 0. \tag{4.2}$$

In the collective representation, we thus obtain

$$\{\hat{P}|\Psi_{phys}>>_c\} = \{|\Phi)_B\}\otimes|0>, \quad \hat{K}|0> = 0. \tag{4.3}$$

The invariance principle (3.15a) and (3.16a) connected with the state $|0>$ is now written as

$$\hat{Q}\cdot[\hat{\mathbb{V}}^{-1}\{\hat{H}+\mathbb{H}_B(b^+,b)\}\hat{\mathbb{V}} - \{\hat{H}+\mathbb{H}_B(b^+,b)\}]|0> = 0, \tag{4.4}$$

$$\hat{P}\cdot[\hat{\mathbb{V}}^{-1}\{\hat{H}+\mathbb{H}_B(b^+,b)\}\hat{V} - \{\hat{H}+\mathbb{H}_B(b^+,b)\}]|0> = 0, \tag{4.5}$$

$$\mathbb{H}_B(b^+,b) = \langle 0|\hat{\mathbb{V}}^{-1}\hat{H}\hat{\mathbb{V}}|0\rangle - \langle 0|\hat{H}|0\rangle, \tag{4.6}$$

where we have supposed $(0|\mathbb{H}_B(b^+,b)|0)_B=0$, i.e., $\langle 0|\hat{\mathbb{V}}^{-1}\mathbb{H}_B(b^+,b)\hat{\mathbb{V}}|0\rangle \equiv \langle 0|\hat{H}_{coll}(\hat{K}^+,\hat{K})|0\rangle = 0$. (This supposition is always fulfilled when the boson Hamiltonian has a form $\mathbb{H}_B = \sum_{r,s} h_{rs}(b^+)^r(b)^s$ with $h_{00} = 0$.)

In order to find the unitary transformation $\hat{\mathbb{V}}$ associated with the invariance principle without having to know explicit functional forms $(\hat{K}^+, \hat{K})$, it is important to express $(\hat{K}^+, \hat{K})$ in terms of "local infinitesimal generators" $(\hat{\mathbb{O}}^+, \hat{\mathbb{O}})$ of $\hat{\mathbb{V}}$ with respect to (b^+,b), which are defined by

$$\hat{\mathbb{O}}^+ = \hat{\mathbb{V}}^{-1}[\hat{\mathbb{V}},b^+], \quad \hat{\mathbb{O}} = \hat{\mathbb{V}}^{-1}[\hat{\mathbb{V}},b] \quad . \tag{4.7}$$

With the use of the fundamental property (3.5a) of $\hat{\mathbb{V}}$, we then can express $(\hat{K}^+, \hat{K})$ as

$$\hat{K}^+ = (b^+-\hat{\mathbb{O}}^+)\cdot\hat{P}, \quad \hat{K} = (b-\hat{\mathbb{O}})\cdot\hat{P}. \tag{4.8}$$

Summarizing the above discussion, we have a set of the basic equations to be solved for the determination of $\hat{\mathbb{V}}$ and $\mathbb{H}_B(b^+,b)$;
(I) <u>the self-consistency condition for the maximally decoupled collective boson Hamiltonian (4.6)</u>, (II) <u>the invariance principle of the Schrödinger equation</u> (4.4) and (4.5) and (III) <u>the subsidiary condition for the physical states</u> (4.3) <u>associated with</u> (3.3), i.e.,

$$\begin{aligned} \hat{K}|0\rangle &\equiv (b-\hat{\mathbb{O}})|0\rangle = 0, \\ \langle 0|[\hat{K},\hat{K}^+]|0\rangle &= 1. \end{aligned} \tag{4.9}$$

Here, it is rather convenient to reformulate the set of these basic equations by employing the following theorem concerning the coherent state:

[<u>theorem</u>] The normal-ordered form of any function $\mathbb{F}(b^+,b)$ is given by

$$\mathbb{F}(b^+\!\!: b) = \mathbb{N}\cdot\{F(\eta^*,\eta)\}, \quad F(\eta^*,\eta) \equiv (\eta|\mathbb{F}(b^+,b)|\eta)_B, \tag{4.10}$$

where $|\eta)_B$ is the coherent state with a complex c-number η ,

$$|\eta)_B \equiv \exp(b^+\eta - b\eta^*)|0)_B, \quad b|\eta)_B = \eta|\eta)_B, \tag{4.11}$$

and $\mathbb{N}$ is an operator which arranges all c-number η in $F(\eta^*,\eta)$ to the right and all η^* to the left and then replaces η^* and η by the boson operators b^+ and b respectively.

By making use of this theorem, the basic equations can be written as

$$\mathbb{H}_B(b^+:b)=\mathbb{N}\cdot\{\ \mathcal{H}_B(\eta^*,\eta)\}$$
$$\mathcal{H}_B(\eta^*,\eta)=<0|\{(\eta|\hat{\mathbb{V}}^{-1}\hat{H}\hat{\mathbb{V}}|\eta)_B\}\ |0>\ -<0|\hat{H}|0>, \quad \text{(I)}$$

$$[(\eta|\hat{\mathbb{V}}^{-1}\{\hat{H}+\mathbb{H}_B(b^+,b)\}\hat{\mathbb{V}}|\eta)_B - \{\hat{H}+\mathcal{H}_B(\eta^*,\eta)\}]|0> =0, \quad \text{(II)}$$

$$\hat{K}|0>\equiv\{\eta-\hat{O}(\eta^*,\eta)\}|0>=0\ ,$$
$$<0|[\hat{O}(\eta^*,\eta),\hat{O}^+(\eta^*,\eta)]|0> =1, \qquad \hat{O}(\eta^*,\eta)\equiv(\eta|\hat{\mathbb{O}}|\eta)_B\,. \quad \text{(III)}$$

From Eqs. (II) and (III), of course we obtain

$$[\hat{K},\{(\eta|\hat{\mathbb{V}}^{-1}(\hat{H}+\mathbb{H}_B)\hat{\mathbb{V}}|\eta)_B-(\hat{H}+\mathcal{H}_B)\}]|0> = 0. \quad (4.12)$$

So far we have not yet made any essential approximation in deriving the basic equations (I), (II) and (III). At this stage, we make the following <u>fundamental approximation: The state $|0>$ is approximated by a single slater determinant $|\phi>$</u>. Generally this single slater - determinant state <u>need not</u> satisfy the condition $\delta<\phi|\hat{H}|\phi>=0$, because the state $|0>$ generally <u>need not</u> an eigenstate of $\hat{H}$ and is merely a state satisfying

$$\hat{H}_{intr}|0> = \varepsilon_{intr}|0>\ . \quad (4.13)$$

(As is easily seen from Eq. (3.17), any state belonging to the collective subspace $\{|n>\}$ can satisfy Eq. (4.13).) In this approximation, the basic equations (I), (II) and (III) are reduced respectively to

$$\mathbb{H}_B(b^+:b) = \mathbb{N}\cdot\{\mathcal{H}_B(\eta^*,\eta)\},$$
$$\mathcal{H}_B(\eta^*,\eta) = <\phi|\{(\eta|\hat{\mathbb{V}}^{-1}\hat{H}\hat{\mathbb{V}}|\eta)_B\}|\phi>-<\phi|\hat{H}|\phi>, \quad \text{(I')}$$

$$\delta<\phi|\{(\eta|\hat{\mathbb{V}}^{-1}\hat{H}\hat{\mathbb{V}}|\eta)_B-\hat{H}+(\eta|\hat{\mathbb{V}}^{-1}\mathbb{H}_B\hat{\mathbb{V}}|\eta)_B\}|\phi> = 0, \quad \text{(II')}$$

$$\{\eta-\hat{O}(\eta^*,\eta)\}|\phi> = 0, \quad \hat{O}(\eta^*,\eta) \equiv (\eta|\hat{\mathbb{V}}^{-1}[\hat{\mathbb{V}},b]|\eta)_B,$$
$$<\phi|[\hat{O}(\eta^*,\eta),\hat{O}^+(\eta^*,\eta)]|\phi> = 1. \quad \text{(III')}$$

and Eq. (4.12) is written as

$$<\phi|[\hat{O}(\eta^*,\eta),\{(\eta|\hat{\mathbb{V}}^{-1}(\hat{H}+\mathbb{H}_B)\hat{\mathbb{V}}|\eta)_B-\hat{H}\}]|\phi> = 0. \qquad (4.14)$$

5. DETERMINATION OF THE MAXIMALLY DECOUPLED COLLECTIVE BOSON HAMILTONIAN

We are now in a position to discuss the problem of determining the maximally decoupled collective boson Hamiltonian $\mathbb{H}_B(b^+,b)$ by solving the set of the approximate basic equations (I'), (II') and (III'). Although the set of the basic equations (I), (II) and (III) has been derived by employing the fundamental property (3.5) of the unitary transformation $\hat{\mathbb{V}}$, where the idealized operators $(\hat{K}^+,\hat{K})$ have been imaged as if they were known, we must not employ the property in solving the set of the basic equations (I'), (II') and (III') to keep a consistency of the method.

Thus, the problem of solving the set of the basic equations can be reduced to finding a hermitian operator $\hat{\mathbb{G}}(b^+:b)$, which is defined through

$$\hat{\mathbb{V}} = \exp\{i\hat{\mathbb{G}}(b^+:b)\} \qquad (5.1)$$

and is a normal-ordered operator with respect to (b^+,b). With the use of a transformation

$$\begin{aligned} b^+ &= \eta^*+a^+, \quad b = \eta+a \ , \\ [a,a^+] &= 1, \quad a|\eta)_B = 0 \ , \end{aligned} \qquad (5.2)$$

let us make the Taylor expansion of $\hat{\mathbb{G}}(b^+:b)$ around (η^*,η)

$$\begin{aligned} \hat{\mathbb{G}}(b^+:b) &= \hat{G}(\eta^*,\eta)+\sum_{m\geq 1}\frac{1}{m!} : \left(a^+\frac{\partial}{\partial\eta^*} + a\frac{\partial}{\partial\eta}\right)^m : \hat{G}(\eta^*,\eta) \ , \\ \hat{G}(\eta^*,\eta) &\equiv (\eta|\hat{\mathbb{G}}(b^+:b)|\eta)_B \quad , \end{aligned} \qquad (5.3)$$

where the symbol : : means the normal product with respect to (a^+,a). With the use of Eq. (5.3), we then can express

$$(\eta|\hat{\mathbb{V}}^{-1}\hat{H}\hat{\mathbb{V}}|\eta)_B,\quad (\eta|\hat{\mathbb{V}}^{-1}\mathbb{H}_B\hat{\mathbb{V}}|\eta)_B \quad\text{and}\quad \hat{O}(\eta^*,\eta)\equiv(\eta|\hat{\mathbb{V}}^{-1}[\hat{\mathbb{V}},b]|\eta)_B$$

in the basic equations in terms of $\hat{G}(\eta^*,\eta)$ and its derivatives with respect to (η^*,η).

For example, in the case of the "semi-classical" approximation[8], <u>in which only the first order terms of the Taylor expansion of $\hat{\mathbb{G}}(b^+:b)$ is retained</u>:

$$\hat{\mathbb{G}}(b^+:b) \approx \hat{G}(\eta^*,\eta)+\left(a^+\frac{\partial}{\partial\eta^*} + a\frac{\partial}{\partial\eta}\right)\hat{G}(\eta^*,\eta), \tag{5.4}$$

we obtain

$$\begin{aligned}
(\eta|\hat{O}|\eta)_B &\approx -e^{-i\hat{G}}\frac{\partial}{\partial\eta^*}\, e^{i\hat{G}} \equiv \hat{O}^{(0)}(\eta^*,\eta),\\
(\eta|\hat{V}^{-1}\hat{H}\hat{V}|\eta)_B &\approx e^{-i\hat{G}}\hat{H}e^{i\hat{G}},\\
(\eta|\hat{V}^{-1}\mathbb{H}_B\hat{V}|\eta)_B &\approx \mathcal{H}_B(\eta^*,\eta) - \frac{\partial}{\partial\eta^*}\mathcal{H}_B(\eta^*,\eta)\cdot\hat{O}^{(0)+}(\eta^*,\eta)\\
&\quad - \frac{\partial}{\partial\eta}\mathcal{H}_B(\eta^*,\eta)\cdot\hat{O}^{(0)}(\eta^*,\eta)\ .
\end{aligned} \tag{5.5}$$

where $\hat{G}$ stands for $\hat{G}(\eta^*,\eta)$ and we have used the property of the coherent state; $(\eta|[\mathbb{H}_B,a^+]|\eta)_B = \partial\,\mathcal{H}_B(\eta^*,\eta)/\partial\eta$.

Thus, in the semi-classical approximation, the set of the basic equations (I'), (II') and (III') can be written in the forms;

$$\mathcal{H}_B(\eta^*,\eta) = \langle\phi|(e^{-i\hat{G}}\hat{H}e^{i\hat{G}}-\hat{H})|\phi\rangle \equiv \mathcal{H}_0(\eta^*,\eta), \tag{5.6}$$

$$\begin{aligned}
\delta\langle\phi|\{e^{-i\hat{G}}\hat{H}e^{i\hat{G}}-\hat{H}\} &- \frac{\partial}{\partial\eta^*}\mathcal{H}_B(\eta^*,\eta)\cdot\hat{O}^{(0)+}(\eta^*,\eta)\\
&- \frac{\partial}{\partial\eta}\mathcal{H}_B(\eta^*,\eta)\cdot\hat{O}^{(0)}(\eta^*,\eta)|\phi\rangle = 0,
\end{aligned} \tag{5.7}$$

$$\langle\phi|[\hat{O}^{(0)}(\eta^*,\eta),\ \hat{O}^{(0)+}(\eta^*,\eta)]|\phi\rangle = 1, \tag{5.8}$$

and Eq. (4.14) is reduced to

$$\langle\phi|[\hat{O}^{(0)}(\eta^*,\eta),\{e^{-i\hat{G}}\hat{H}e^{i\hat{G}}-\hat{H}\}]|\phi\rangle = \frac{\partial}{\partial\eta^*}\mathcal{H}_B(\eta^*,\eta) \tag{5.9}$$

The forms of these equations (5.6), (5.7) and (5.8) are completely the same as Eq. (2.10), Eq. (2.9) with (2.11) and Eq. (2.12), except that the single slater determinant $|\phi\rangle$ need not be the Hartree-Fock ground state $|\phi_0\rangle$ and $\hat{G}(\eta^*,\eta)$ has not yet specified to be a one-body operator.

In Section 2, we have discussed the (η^*,η)-expansion method to determine $\hat{G}_0(\eta^*,\eta)$ as well as $H_0(\eta^*,\eta)$. Provided that $\hat{G}(\eta^*,\eta)$ is assumed to be a one-body operator, we can also employ the (η^*,η)-expansion method in solving the set of basic equations (I') and (II') and (III'). The "quantum" effects are then obtained by taking account of the higher order terms of the Taylor expansion of $\hat{\mathbb{G}}(b^+:b)$ in Eq. (5.3).

6. CONCLUDING REMARKS

Finally, we would like to make two remarks.

(i) In this talk, we have discussed the set of basic equations (I'), (II') and (III'), which is obtained from the set of (I), (II) and (III) by assuming the state $|0\rangle$ to be a single slater determinant . When we make a further assumption specifying the collective subspace so as to satisfy

$$|\phi\rangle = |0\rangle, \quad |n\rangle = \frac{1}{\sqrt{n!}}(\hat{K}^+)^n|\phi\rangle \in \{|np\text{-}nh\rangle\}, \tag{6.1}$$

where $\{|np\text{-}nh\rangle\}$ is a subspace consisting of n-particle-n-hole excited states with respect to the single slater determinant $|\phi\rangle$, the set of basic equations can be greatly simplified. In this case, we can show[13] that the set of basic equations leads us to a Marumori-Yamamura-Tokunaga-type boson-mapping theory[14] <u>of the maximally decoupled collective subspace</u>.

(ii) In this talk, we have used the simplest subsidiary condition (4.1), i.e., $b|0)_B=0$. In the collective representation, the correspondent of the boson vacuum $|0)_B$ is the state $|0\rangle$ which satisfies Eq. (4.13). As has been discussed in Section 4, however, an "intrinsic" state $|\Psi_{intr}\rangle$ which fulfills

$$\hat{H}_{intr}|\Psi_{intr}\rangle = \varepsilon_{intr}|\Psi_{intr}\rangle \tag{6.2}$$

<u>need not</u> to be the state $|0\rangle$, because any state in the collective subspace $\{|n\rangle\}$ can fulfill Eq. (6.2). If the system under consideration undergoes a phase transition to a "deformed" phase, therefore, it is rather convenient to start with a subsidiary condition

$$(b-\beta)|\Psi_{phys}(\beta)\rangle\rangle = 0 \tag{6.3}$$

which implies that the physical state vectors are given with a coherent state with "deformation" β by

$$\{|\Psi_{phys}(\beta)\rangle\rangle\} = \{|\Psi\rangle\}\otimes|\beta)_B \quad ,$$
$$|\beta)_B = \exp(b^+\beta - b\beta^*)\ |0)_B \ . \tag{6.4}$$

This case will be discussed elsewhere.

References

1) F. Villars, Nucl. Phys. A285 (1977) 269.
E. Moya de Guerra and F. Villars, Nucl. Phys. A285 (1977) 297.
F. Villars, in Dynamic Structure of Nuclear States, ed. by D. J. Rowe (University of Toronto Press, 1972).
2) M. Baranger and M. Vếneroni, Ann. of Phys. A114 (1978) 123.
D. M. Brink, M. J. Giannoni and M. Vếneroni, Nucl. Phys. A258 (1976) 237.
3) D. J. Rowe and R. Bassermann, Can. J. Phys. 54 (1976) 1941.
4) T. Marumori, Proceedings of International Conference on Selected Topics in Nuclear Structure (June 1976, Dubna).
T. Marumori, Prog. Theor. Phys. 57 (1977) 112.
5) K. Goeke and P. G. Reinhard, Ann. of Phys. 112 (1978) 328.
P. G. Reinhard and K. Goeke, Phys. Letters 69B (1977) 17.
P. G. Reinhard and K. Goeke, Nucl. Phys. A312 (1978) 121.
K. Goeke and P. G. Reinhard, Ann. of Phys. 124 (1982) 289.
6) T. Marumori, A. Hayashi, F. Sakata and A. Kuriyama, Prog. Theor. Phys. 63 (1980) 1576.
7) T. Marumori, T. Maskawa, F. Sakata and A. Kuriyama, Prog. Theor. Phys. 64 (1980) 1294.
8) T. Marumori, F. Sakata, T. Une, Y. Hashimoto and T. Maskawa, Prog. Theor. Phys. 66 (1981) 1651.
9) T. Marumori, J. Yukawa and R. Tanaka, Prog. Theor. Phys. 13 (1955) 442.
T. Marumori, Prog. Theor. Phys. 14 (1955) 608.
T. Marumori and E. Yamada, Prog Theor. Phys. 14 (1955) 1557.
S. Hayakawa and T. Marumori, Prog. Theor. Phys. 18 (1957) 396.
10) S. Tomonaga, Prog. Theor. Phys. 13 (1955) 467.
T. Miyazima ans T. Tamura, Prog. Theor. Phys. 15 (1956) 255.
11) H. Lipkin, A. de Shalit and I. Talmi, Phys. Rev. 103 (1956) 1773.
F. Villars, Ann. Rev. Nucl. Sci. 7 (1957) 185.
12) D. R. Inglis, Phys. Rev 96 (1954) 1054.
13) T. Marumori, F. Sakata, T. Une and Y. Hashimoto, to be published.
14) T. Marumori, M. Yamamura and T. Tokunaga, Prog. Theor. Phys. 31 (1964) 1009.
T. Marumori, M. Yamamura, A. Tokunaga and K. Takada, Prog. Theor. Phys. 32 (1964) 726.
See also, T. Marumori, K. Takada and F. Sakata, Supplement of Prog. Theor. Phys. 71 (1982).

QUANTIZED ATDHF: THEORY AND REALISTIC APPLICATIONS TO HEAVY ION FUSION

F. Grümmer
Institut für Kernphysik
Kernforschungsanlage Jülich
D-5170 Jülich, West Germany

K. Goeke
Institut für Kernphysik
Kernforschungsanlage Jülich
D-5170 Jülich, West Germany

and

Physik-Department
Universität Bonn
D-5300 Bonn, West Germany

P.-G. Reinhard[+]
Institut für Kernphysik
Universität Mainz
D-6500 Mainz, West Germany

Abstract

The quantized ATDHF theory is reviewed and discussed in the context of the generator coordinate method. This allows for a derivation which does not require an a posteriori quantization process. The ATDHF equations are then solved numerically on a coordinate and momentum grid in fully three dimensional geometry. The theory is applied to various heavy ion systems, where potentials, mass parameters and quantum corrections are evaluated and compared to conventional results from constrained Hartree-Fock. Subbarrier fusion cross sections are calculated and compared with experiment.

1. Introduction

For a couple of years adiabatic microscopic theories have been discussed as a tool to describe in a parameterfree way quantum effects of low energy and subbarrier heavy ion collisions. Actually the interest was mostly focussed on two problems, namely the quantization and the numerical evaluation of the collective path. The first one, the quantization, is to our belief basically solved, as we have emphasized already some time ago. We take here the opportunity to recall that the quantized ATDHF theory can fully be derived in the framework of the generator coordinate method (GCM) and the Gaussian overlap approximation (GOA), i.e., by means of a theory which is already quantized from the beginning. The real new thing in the present contribution is the presentation of realistic calculations of the collective path, the potentials, mass parameters and quantum corrections (zero point energies) of colliding heavy ion systems and the evaluation of fusion cross sections. This will

---

[+] Fellow of the Heisenberg-Stiftung

actually demonstrate that quantized ATDHF together with the corresponding validity condition is indeed a solvable theory which allows, e.g., the use of present time well established density dependent forces and coordinate space techniques. The paper is constructed in the following way: Section 2 recalls the Gaussian overlap approximation in the GCM and review the derivation of quantized ATDHF by means of GCM. Section 3 explains the way how the optimal collective path is obtained numerically and shows some results regarding the masses and potentials of the systems $\alpha+\alpha\leftrightarrow{}^{8}Be$, ${}^{12}C+{}^{12}C\leftrightarrow{}^{24}Mg$, and ${}^{16}O+{}^{16}O\leftrightarrow{}^{32}S$. Here also the constrained HF method, yielding a different collective path, is discussed. Section 4 shows the application to subbarrier fusion cross sections; some conclusions and an outlook are given in section 5.

2. The Derivation of Quantized ATDHF from GCM

Quantized ATDHF is a consistent microscopic theory for large amplitude collective motion. It provides first an equation determining an optimal collective path, second a unique quantized collective Hamiltonian (for the motion within that collective basis), third a simple prescription for evaluating any desired expectation value and transition moment within the collective states, and fourth it allows to check a posteriori the validity of the collective description. The theory has been derived in three independent ways. We will give here a short account of the derivation from the generator coordinate method (GCM) which is the most satisfying for its conceptual clarity.

We start out from the concept of a collective path $\{|\Phi_q\rangle\}$, which is a set of collectively deformed Slater determinants (or BCS states), where q labels somehow the deformation. This set $\{|\Phi_q\rangle\}$ is supposed to be an expansion basis along which the actual collective motion does evolve. This means quantum mechanically that we consider the superposition

$$|\psi\rangle = \int dq \; |\Phi_q\rangle \; f(q) \qquad (2.1)$$

where the actual collective dynamics is described by f(q). As a means to optimize the collective superposition f(q) as well as the basis $\{|\Phi_q\rangle\}$ we are going to use the stationary variational principle

$$\langle\delta\psi|\hat{H}-E|\psi\rangle = 0 \qquad (2.2)$$

In this general form, however, the equations turn out to be impractical, since complicated kernels $\langle\Phi_q|H|\Phi_{q'}\rangle$ arise, and, even worse, they are not specific enough for a variation $\langle\delta\Phi_q|$ since the ansatz (2.1) still embraces the full Hilbert space if no further assumption for the basis $\{|\Phi_q\rangle\}$ is made.

An obvious requirement for a tractable $|\Phi_q\rangle$ is that it is analytical in q. Furthermore, the $\{|\Phi_q\rangle\}$ ought to represent a series of collective deformations, where a small displacement of many particles adds up coherently to a substantial reordering of the matter. Hence the overlaps $\langle\Phi_q|\hat{O}_p|\Phi_{q'}\rangle$ should be continuous and fall off rapidly with rising (q-q'). Therefore, it is reasonable to assume Gaussian overlap for the norm kernel

$$\langle\Phi_q|\Phi_{q'}\rangle = \exp\left(-\frac{\lambda(\frac{q+q'}{2})}{4}(q-q')^2\right) \tag{2.3}$$

where $\lambda = 2\langle\Phi_q|\overleftarrow{\partial}_q\overrightarrow{\partial}_q|\Phi_q\rangle$ and a Gaussian times a polynomial in (q-q') for the overlap $\langle\Phi_q|\hat{O}|\Phi_{q'}\rangle$ of any operator O different from unity.

Working out this Gaussian overlap approximation (GOA) we obtain for any operator $\hat{O}$ the expectation value[1)]

$$\langle\psi|\hat{O}|\psi\rangle_{GOA} = \int dq\, g^*(q)\,\{(1-\frac{1}{4\lambda}\partial_q^2)\langle\Phi_q|\hat{O}|\Phi_q\rangle$$

$$-:\frac{d}{dq},\langle\Phi_q|\overleftarrow{\partial}_q\bar{O}-\bar{O}\overrightarrow{\partial}_q|\Phi_q\rangle : + \frac{1}{2}:(\frac{d^2}{dq^2}+\frac{\lambda}{2}),\langle\Phi_q|\overleftarrow{\partial}_q^2\bar{O}-2\overleftarrow{\partial}_q\bar{O}\overrightarrow{\partial}_q+\hat{O}\overrightarrow{\partial}_q^2|\Phi_q\rangle:\}g(q) \tag{2.4}$$

where ∂_q acts only on an $|\Phi_q\rangle$ or $\langle\Phi_q|$ while $\frac{d}{dq}$ is acting on everything at the right. The :...: is the normal ordering, $:(\frac{d}{dq})^n+\lambda,A(q): = 2^{-n}\{\frac{d}{dq},\ldots,\{\frac{d}{dq},A\}\ldots\}+\lambda A$. The g(q) is the collective wave function, related to the superposition f(q) by ref. 1)

$$g(q) = \int dq'\,(\lambda/\pi)^{1/4}\exp\left(-\frac{\lambda}{2}(q-q')^2\right)f(q') \tag{2.5}$$

as follows from the GOA.

If g(q) is normalized as $\int dq\, g^*(q)g(q) = 1$, the $|\psi\rangle$ is normalized to unity, $\langle\psi|\psi\rangle = 1$. The $\bar{O}$ is a shorthand for $\bar{O} = \hat{O} - \langle\Phi_q|\hat{O}|\Phi_q\rangle$. The eq. (2.4) is the key formula for all further developments. It can easily be generalized to the case of a many parameter GCM.

The GCM-GOA now is a useful starting point for all further developments. In particular, it allows to apply the variational principle

$$\langle\delta\psi|H-E|\psi\rangle_{GOA} = 0 \tag{2.2b}$$

for both variations, $\langle\delta\Phi_q|$ and δg^*, leading eventually to unambiguous results and workable equations. To derive the quantized ATDHF we make some further assumptions about the collective basis. In order to avoid the well known shortcoming of GCM to yield for Slater determinants $|\Phi_q\rangle$ in case of pure translation not the proper collective mass resulting from Galilei invariance, we are using a dynamic basis $\{|\Phi_{qp}>\}$ where p labels the momentum conjugate to q (this is a generalization of the Thouless-Peierls double projection method). Such a dynamic basis, however, is hard to treat in general. We therefore consider only the adiabatic expansion of it

$$|\Phi_{qp}> = |\Phi_q> + i\, p\, Q(q)|\Phi_q> \tag{2.6}$$

In that small-p expansion the p-dynamics can be solved[1] explicitly by means of GOA and RPA techniques, leading to an RPA correlated basis $\{|\tilde{\Phi}_q>\}$ where the dynamic information up to order p^1 is built in by the redundancy property

$$\lambda\, \hat{Q}(q)|\tilde{\Phi}_q> = \partial_q|\tilde{\Phi}_q> \tag{2.7}$$

For the new basis we obtain[1] any expectation value as

$$<\psi|\hat{O}|\psi>_{GOA} = \int dq\, \tilde{g}^*(q)\{(1 - \frac{1}{4\lambda}\partial_q^2|<\tilde{\Phi}_q|\hat{O}|\tilde{\Phi}_q> + :\frac{d}{dq}<\tilde{\Phi}_q|[\bar{\hat{O}},\hat{Q}(q)]|\tilde{\Phi}_q>:$$
$$+ \frac{1}{2} : |-\frac{d^2}{dq^2} - \frac{\lambda}{2}|<\tilde{\Phi}_q|[\hat{Q}(q),[\hat{O},\hat{Q}(q)]]|\tilde{\Phi}_q>:\}\tilde{g}(q) \tag{2.8}$$

Now we apply the path variation $<\delta\tilde{\Phi}_q|$ for the case $\hat{O} = \hat{H}\text{-}E$ requiring in addition independency, which means that the set $\{|\tilde{\Phi}_q>\}$ should be a basis for a whole collective spectrum and therefore be independent of the actual collective state g(q). This yields[2] the ATDHF equations for the states $|\Phi_q>$, which are in this GCM context the single particle vacua associated to the correlated states $|\tilde{\Phi}(q)>$. The $\{|\Phi_q>\}$ set up the collective path and obey the well known ATDHF equations[3,4].

$$<\Phi_q|a^+a(\hat{H}-\partial_q V\hat{Q})|\Phi_q> = 0 \tag{2.9a}$$

$$<\Phi_q|a^+a([H,\hat{Q}] - \frac{1}{M}\overleftrightarrow{\partial}_q)|\Phi_q> = 0 \tag{2.9b}$$

where V is the collective potential

$$V(q) = <\Phi_q|\hat{H}|\Phi_q> \tag{2.9c}$$

M is the collective mass

$$M^{-1}(q) = <\Phi_q|[\hat{Q}(q),[\hat{H},\hat{Q}(q)]]|\Phi_q> \tag{2.9d}$$

and $\hat{Q}$ is normalized according to

$$<\Phi_q|\hat{Q}(q)\overrightarrow{\partial}_q + \overleftarrow{\partial}_q\hat{Q}(q)|\Phi_q> = -1 \tag{2.9e}$$

This is a coupled set of equations determining at once the path $|\Phi_q\rangle$, the $\hat{Q}(q)$, the mass M(q) and the potential V(q).

Using $\hat{O} = \hat{H}-E$, the variation of eq. (2.8) with respect to $\delta g^*(q)$ yields immediately the quantized collective Hamiltonian[1,2)]

$$(\hat{H}_{coll}-E)g(q) = 0 \tag{2.10a}$$

with

$$\hat{H}_{coll} = -\frac{1}{8}\left\{\frac{d^2}{dq^2}\frac{1}{M} + 2\frac{d}{dq}\frac{1}{M}\frac{d}{dq} + \frac{1}{M}\frac{d^2}{dq^2}\right\} + V(q) - Z(q) \tag{2.10b}$$

where Z(q) includes the quantum corrections (zero-point energies) contained in the states $|\Phi_q\rangle$,

$$Z(q) = \frac{\lambda}{4M} + \frac{1}{4\lambda}\partial_q^2 V + Z_{rot} + Z_{trans} \tag{2.10c}$$

Apparently in the present derivation the quantization is already naturally built in and no further a posteriori step is necessary.

In practice, we will consider always more than one collective coordinate, at least the two rotational angles ϑ,φ together with the axial deformation q. Therefore the Z(q) contains the corresponding rotational zero-point energy Z_{rot} and, because of spurious center-of-mass motion, also a corresponding Z_{trans}. Actually, in eq. (2.10) the correlated $|\tilde{\Phi}_q\rangle$ do not appear explicitly but only the Slater determinants $|\Phi_q\rangle$. The big difference to an ordinary GCM starting directly from $|\Phi_q\rangle$ rather than from $|\tilde{\Phi}_q\rangle$ or $|\Phi_{qp}\rangle$ lies in the expression for the mass parameter M in eqs. (2.10b), (2.10c). This mass is now the asymptotically correct mass being of Peierls-Thouless rather than of Peierls-Yoccoz form. It is the well known ATDHF mass obtained by GCM. The other corrections, originating from a GCM over $\tilde{\Phi}_q$ rather than Φ_q are negligible. For details of this procedure see ref. 1).

If we have determined the basis according to the ATDHF equations (2.9) and solved the collective Schrödinger equation (2.10), we can evaluate any desired expectation value according to eq. (2.8). In other words, eq. (2.8) provides us with a translation of any operator $\hat{O}$ into collective space

$$\begin{aligned}\hat{O}_{coll} &= (1 - \frac{1}{4\lambda}\partial_q^2)\langle\Phi_q|\hat{O}|\Phi_q\rangle + :\frac{d}{dq}\langle\Phi_q|[\bar{O},\hat{Q}(q)]|\Phi_q\rangle: \\ &- \frac{1}{2} : \left[\frac{d^2}{dq^2} + \frac{\lambda}{2}\right]\langle\Phi_q|[\hat{Q}(q),[\bar{O},\hat{Q}(q)]]|\Phi_q\rangle :\end{aligned} \tag{2.11a}$$

and any transition moment is easily evaluated as a simple q-integration

$$\langle\psi_f|\hat{O}|\psi_i\rangle_{GOA} = \int dq\, g_f^*(q)\, \hat{O}_{coll}\, g_i(q) \tag{2.11b}$$

Finally we can derive a criterion for the validity of the collective description thus obtained[5,6]. To this end one evaluates perturbationally the coupling of the space $\{|\Phi_q\rangle\}$ to the 1p-1h neighbourhood $\{|\Phi_q\rangle, a_p^+ a_h|\Phi_q\rangle\}$. This eventually leads to the so-called "validity condition"

$$\left(\frac{p^2}{2}\right)^2 \langle\Phi_q|\hat{X}_\perp^2|\Phi_q\rangle \ll 1 \tag{2.12a}$$

where X is the solution of

$$[H,\hat{X}]_{ph} = -i\, [[H,\hat{O}],Q]_{ph} - \frac{i}{M}(\partial_q Q)_{ph} \tag{2.12b}$$

and

$$\hat{X}_\perp = \hat{X} - \hat{Q}\, \langle\Phi_q|\overleftarrow{\partial}_q\hat{X} + \hat{X}\overrightarrow{\partial}_q|\Phi_q\rangle \tag{2.12c}$$

Thus altogether we have set up a consistent theory which yields eqs. (2.9) for an optimal collective basis, eqs. (2.10) for a quantized collective Hamiltonian, eqs. (2.11) for evaluating transition moments and expectation values, and finally eqs. (2.12) for an a posteriori check.

3. Numerical Solution of the ATDHF Equations

The first and indeed most difficult step consists in solving the ATDHF eqs. (2.9) in order to obtain the optimal collective path $\{|\Phi_q\rangle\}$. The only way, which has led to numerical results in realistic cases, has been suggested in ref. 7). It consists in transforming[7] eqs. (2.8) to one single differential equation for $\{|\Phi_q\rangle\}$

$$\frac{\partial}{\partial q}|\Phi_q\rangle = \bar{\varepsilon}(q)\, [H,H_{ph}]_{ph}|\Phi_q\rangle \tag{3.1}$$

with

$$\bar{\varepsilon}(q) = M(q)\left(\frac{\partial V}{\partial q}\right)^{-1} \tag{3.2}$$

Apparently eq. (3.1), being equivalent to the ATDHF eqs. (2.9), is an initial value problem: For any choice of an initial value $|\Phi_{q_0}\rangle$ one obtains a collective path $\{|\Phi_q^{(q_0)}\rangle\}$, which shows that the old formulations of ATDHF[3] were not unique. An illustration of several solutions is given in fig. 1, where some resulting trajectories of eq. (3.1) are evaluated in a two dimensional landscape model[8]. One realizes that all solutions converge towards the special solution of eq. (3.1) extending from the saddle point towards the HF minimum. As has been shown in ref. 9)

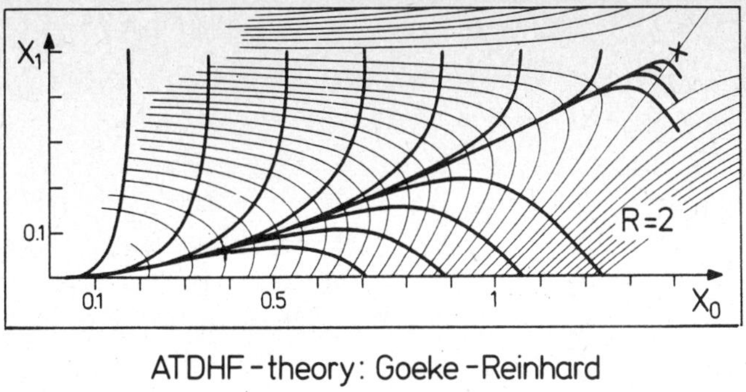

Figure 1

these solutions are, in differential geometrical terms, "fall lines with respect to the mass tensor" and the fall line from the saddle to the HF minimum is the fall line closest to a geodesic between these points. It is very important, and shows the relevance of the validity condition, that the fall line from saddle to HF can be shown[8] to fulfil best the validity condition. Hence it is the desired optimal collective path. It should be remarked that the ATDHF equations need the complement[7] of the validity condition[5,6] to yield this unique result. Thus for the numerical procedure one has first to find the saddle point and then one can solve eq. (3.1).

The eq. (3.1) is solved by means of the Euler-Step-Method. There one solves eq. (3.1) in successive steps n = 0,1,2,... with

$$|\Phi^{(n+1)}\rangle = \{1 - \varepsilon\, [H,H_{ph}]_{ph}\}\,|\Phi^{(n)}\rangle \tag{3.3}$$

Here the point n=0 is close to the saddle point and $\varepsilon = 10^{-4}\ \mathrm{MeV}^{-2}$ is a small stabilizing constant. In the present calculations eq. (3.3) is solved using the Bonche-Koonin-Negele force[10] completed by a direct Coulomb potential. The evaluation is done in the coordinate space using a three dimensional grid of dimensions 24x16x16 with a meshsize of 1 fm. For the evaluation of the kinetic energy the wave functions are Fourier-transformed and smoothed by restricting the Fourier components to $k \le k_{max}$ with $(\hbar k_{max})^2/2m \le 130$ MeV. This also simplifies the storage in the computer. The wave functions are furthermore simplified by assuming quartet symmetry (averaged Coulomb) and good parity of the single particle wave functions with respect to reflections at the x-y-, y-z- and x-z-planes.

Fig. 2 shows for the α-α system the energies $\langle\Phi_n|H|\Phi_n\rangle$ associated to various $|\Phi^{(n)}\rangle$ where different initial starting points were chosen. One sees immediately how the saddle point is identified and how the various trajectories converge towards the optimal path, i.e., the one from saddle point towards HF or towards fragmentation. The wave functions $|\Phi^{(n)}\rangle$ are labelled with the distance R of the ions extracted from the total quadrupole moment by means of

$$\frac{1}{2} A R^2 = \langle\Phi|r^2 Y_2|\Phi\rangle - Q_1 - Q_2$$

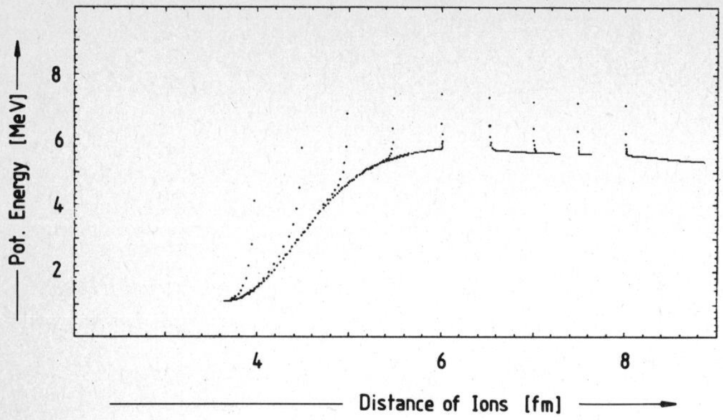

Figure 2

where A is the total nucleon number and Q_1 and Q_2 are the quadrupole moments of the initial fragments. This labelling is convenient and allows an immediate interpretation of the collective wave function in the asymptotic region of separate fragments. However in principle any other labelling could be used since the physical properties of the system as e.g. energies, transition moments, penetration coefficients, etc. are independent of the label.

The final results of the $\alpha+\alpha \leftrightarrow {}^8Be$ system are given in fig. 3, where the classical potential V(R), the quantum corrected V(R)-Z(R) (see eq. (2.10)), and the collective mass M(R) associated to the relative motion are given. One notices that the M(R) approaches the correct asymptotic value of μ = 2 m for large separation distances of the ions. The M(R) is by no means constant but shows a clear bump at R = 4.7 fm with a value of M(R) = 2.8 m. This bump indicates at which distance the internal rearrangements on the way from the separated system to the compound nucleus take place. The fig. 4 shows the zero-point energies for the various degrees of freedom and the total zero-point energy. Apparently the Z(R) reaches a value of about 2 MeV and is therefore not at all negligible, as it is often assumed in the literature.

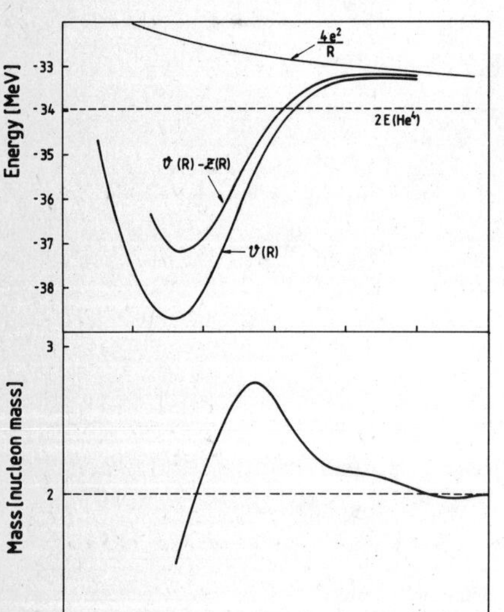

Figure 3

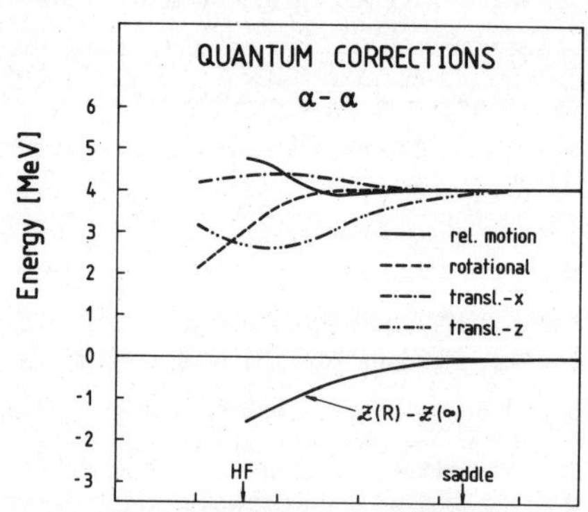

Figure 4

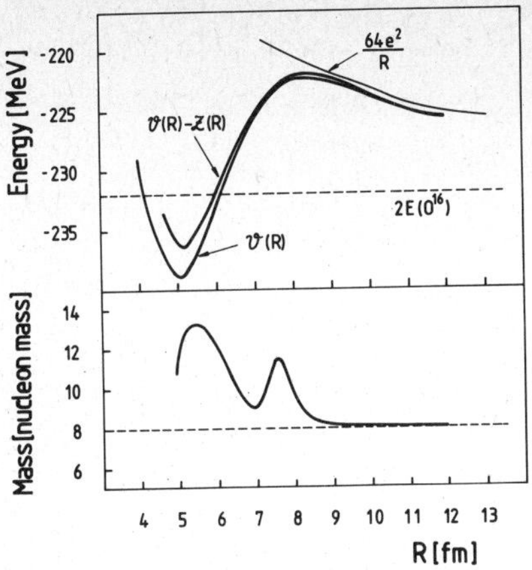

Figure 5

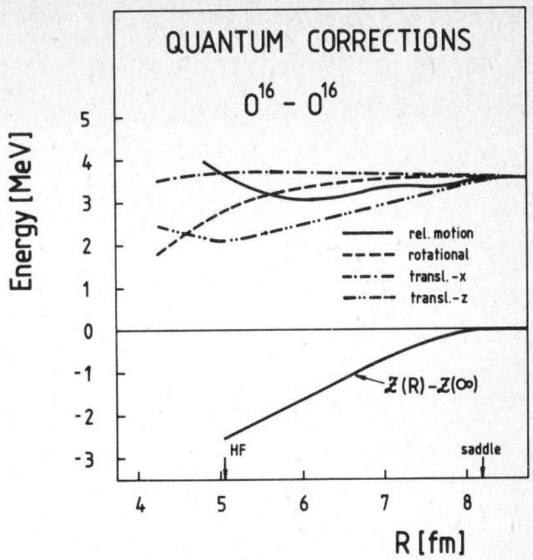

Figure 6

The collective potentials and the mass parameter for the $^{16}O+^{16}O \leftrightarrow {}^{32}S$ fusion reaction are given in fig. 5 and the zero-point energies of this process in fig. 6. Here the mass parameter shows two bumps, one close to the HF-point and one half-way to the saddle point. The fact that the potentials rise few distances smaller than the HF-point at R = 5 fm is due to the reflection symmetries used and due to the lack of pairing correlations. The potential V(R)-Z(R) should lie below the classical potential V(R), however the V(R) is shifted such that it exhibits the pure Coulomb tail as V(R)-Z(R) does. Similar curves for the $^{12}C+^{12}C \leftrightarrow {}^{24}Mg$ system are given in figs. 7 and 8. There one has to distinguish between an axial and a triaxial configuration depending on whether the two oblate ^{12}C nuclei are parallel to each

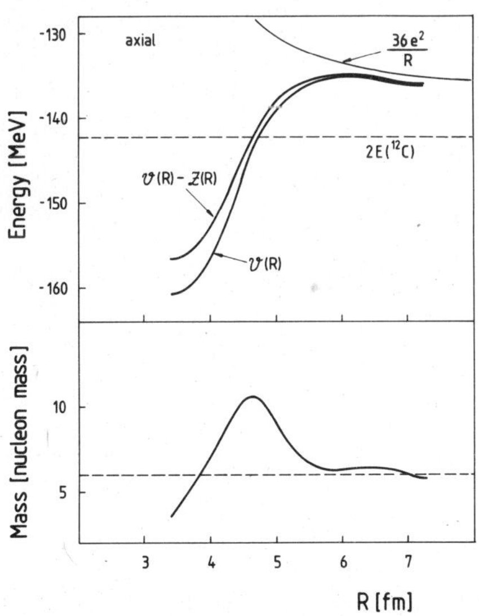

Figure 7

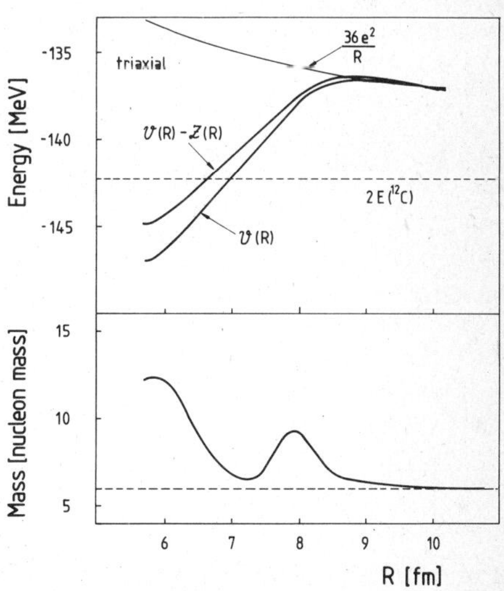

Figure 8

other or whether they are lying in a plane. Apparently the corresponding potentials and masses are quite different. The saddle point for the triaxial configuration, e.g., lies at a distance which is 3 fm larger than for the axial configuration. In addition the masses show different peaks.

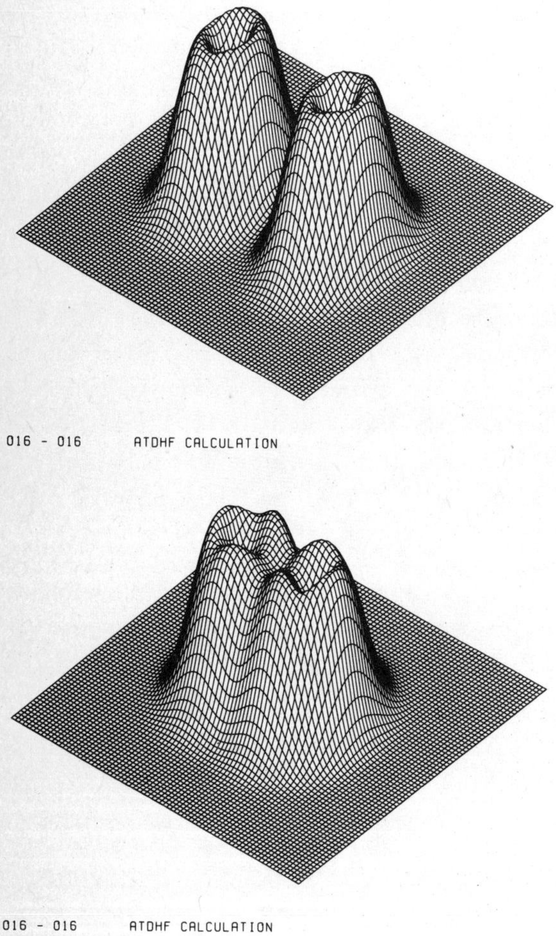

Figure 9

The fig. 9 shows density distributions for the $^{16}O+^{16}O \leftrightarrow {}^{32}S$ system at two relevant R-values close to the saddle point and close to the HF-point of the $^{16}O+^{16}O$ system, respectively. The central depression in the ^{16}O fragments is due to the BKN-interaction used which lacks the ls-term. One realizes that the neck degrees of freedom are fully incorporated. Fig. 10 shows a cut through the density distributions along the axis connecting the fragments, which is chosen to be the z-axis. One realizes even for an adiabatic reaction a small density pile up in the central region at R = 6.6 fm compared to the relaxed HF-state at R = 5.0 fm. Figs. 11 and 12 show contour plots of the density of the $^{12}C+^{12}C \leftrightarrow {}^{24}Mg$ system.

It is interesting to compare the collective path obtained by the present ATDHF method with the one obtained by conventional constrained Hartree-Fock techniques. The standard choice for the constraint in the present systems is the quadrupole operator. Flocard et al.[11,12] have done calculations for the $^{16}O+^{16}O \leftrightarrow {}^{32}S$, $^{12}C+^{12}C \leftrightarrow {}^{24}Mg$ and recently for the $^{40}Ca+^{40}Ca \leftrightarrow {}^{80}Zr$ system. Besides the method used the technicalities of these calculations were similar to the ones used here. Thus the results are directly comparable. Whereas the potentials are not much different the mass parameters turn

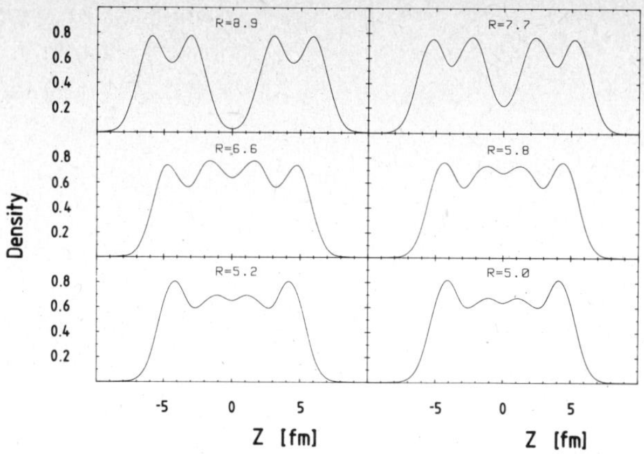

Figure 10

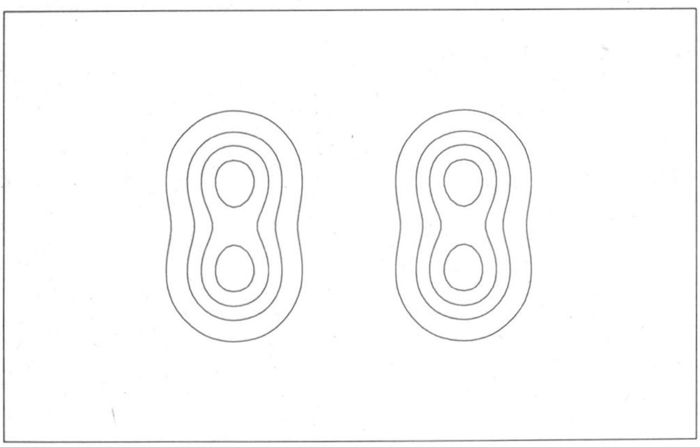

Figure 11

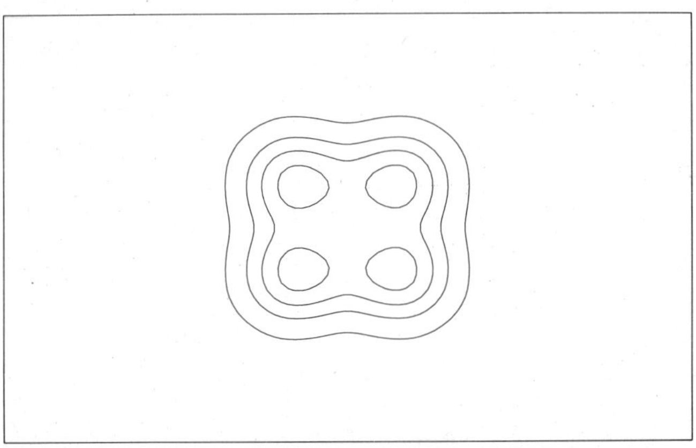

Figure 12

out to be very sensitive to the path used, CHF or ATDHF. This can be seen at fig. 13 which can be directly compared to Heenen's fig. 1 in these proceedings. Apparently the masses are totally different, showing strange peaks in the CHF case. This actually demonstrates that CHF is a rather poor approximation to ATDHF. Indeed the constraining operator $\hat{Q}$ of eq. (2.9a) is allowed to be nonlocal and R-dependent in contrast to the constraint r^2Y_{20} as used in CHF. That indeed $\hat{Q}$ is highly nonlocal has long time ago been demonstrated in ref. 13) for the monopole vibration of ^{16}O using the Skyrme-3 interaction. In fig. 14, not given in ref. 13), the quantity $\chi(r,r')/\dot{q}$ is plotted which is proportional to Q(r,r') in this example. The case $^{16}O+^{16}O \leftrightarrow ^{32}S$ is by no means an exception. Figs. 15 and 16 show similar comparisons for the $\alpha+\alpha \leftrightarrow ^{8}Be$ and $^{12}C+^{12}C \leftrightarrow ^{24}Mg$ systems.

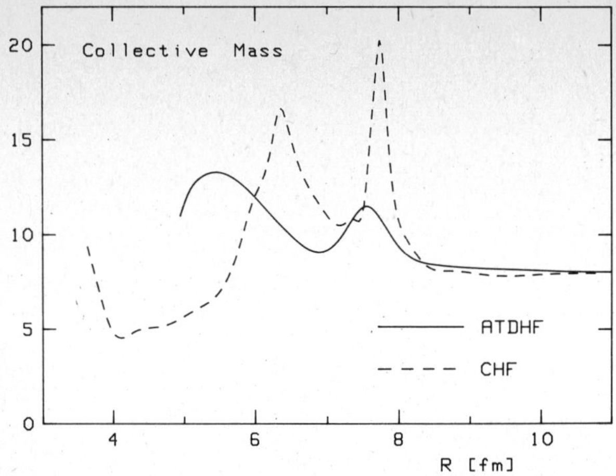

Figure 13

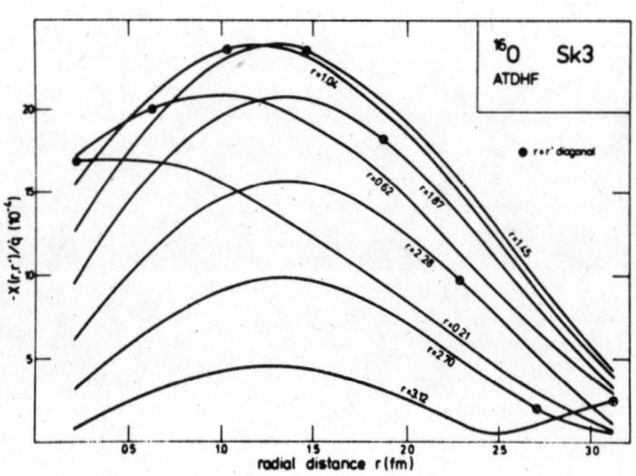

Figure 14

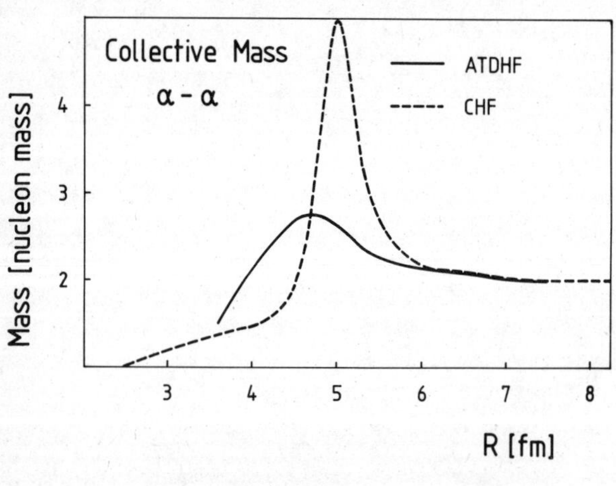

Figure 15

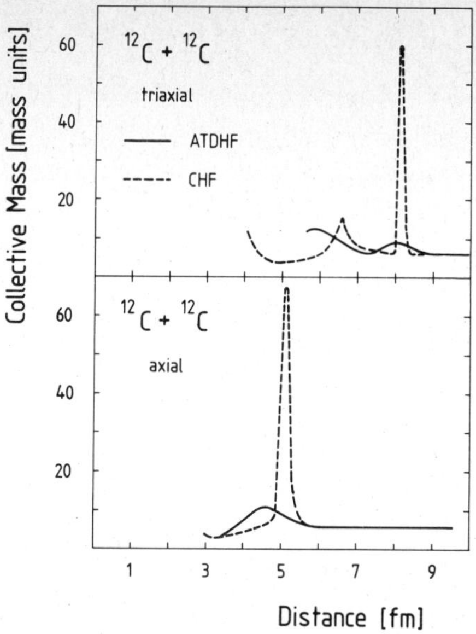

Figure 16

4. Application of Quantized ATDHF to Nuclear Fusion

Except the mass parameter M(R), being asymptotically correct, the classical potential V(R) and the quantum corrections Z(R), we also have calculated the rotational moment of inertia of the total system, $\Theta(R)$, in the GCM or Peierls-Yoccoz approximation. This allows one to complete the collective Hamiltonian of eq. (2.10) by a centrifugal term $L(L+1)/2\Theta(R)$ in order to evaluate also higher partial waves. If one inserts all these quantities into the eq. (2.10) one can easily evaluate transmission coefficients corresponding to the penetration through the barrier by using WKB-techniques:

$$T_L(E_{CM}) = \exp\{-2\, I_L(E_{CM})\} \tag{4.1}$$

with

$$I_L(E_{CM}) = \int_a^b \{ \frac{2M(R)}{\hbar^2} (V(R)-Z(R) + \frac{\hbar^2}{2\Theta(R)} L(L+1)-E_{CM})\}dR \tag{4.2}$$

where a and b are the classical turning points. From these quantities one can easily evaluate the fusion cross section for identical fragments:

$$\sigma_{fus}(E_{CM}) = \frac{2}{k^2} \sum_{L=even} (2L+1)T_L(E_{CM}) \tag{4.3}$$

For the $^{16}O+^{16}O \leftrightarrow ^{32}S$ fusion the cross section of quantized ATDHF in comparison with the experimental values is given in fig. 17. One realizes that one can evaluate σ_{fus} even at very low energies where the experimental counting rate is too low to allow for reliable measurements. This is perhaps relevant for the description of thermonuclear burning in stellar systems which, if one includes the Gamow-peak, requires cross sections at a few MeV. A more detailed comparison with experimental data can be done if one considers instead of the fusion cross section the astrophysical S-factor defined as

$$\sigma_{fus}(E_{CM}) = \frac{1}{E_{CM}} S(E_{CM}) \exp\{-\frac{2\pi Z_1 Z_2 e^2}{\hbar v}\}$$

with v being the relative velocity. The S-factor for quantized ATDHF is given in fig. 18 in comparison to the experimental data and the values extracted from CHF.

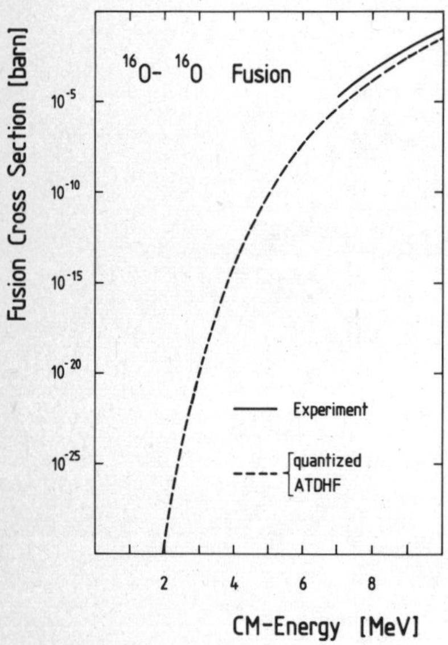

Figure 17

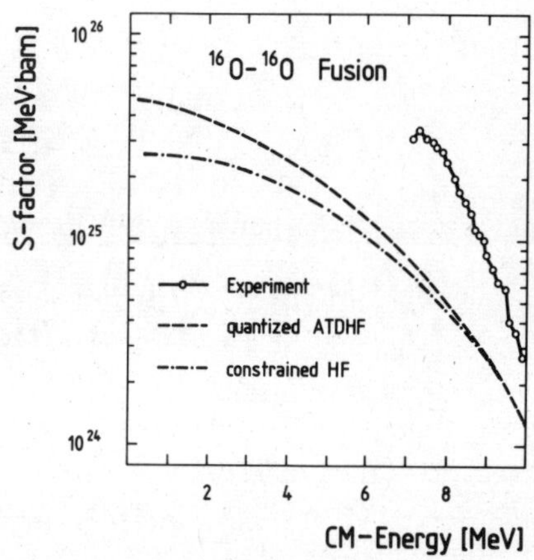

Figure 18

Although quantized ATDHF is closer to experiment than CHF still a factor 2-5 is missing and the trend versus E_{CM} is poorly reproduced. The reason for this is still to be found out. Such an investigation certainly has to incorporate a detailed study of the TDHF fusion cross sections slightly above the barrier. It is very probable that the lack of the ls-component of the nuclear interaction has a strong impact on the subbarrier results, perhaps more than on superbarrier properties.

5. Conclusion

The present article shows that quantized ATDHF is indeed a theory which besides conceptual merits allows also for clean numerical solutions.

Quantized ATDHF provides a variational way to extract the optimal collective path from which a fully quantized collective Hamiltonian with one dynamical collective coordinate can be extracted. The resulting collective wave functions allow one a clear interpretation in terms of matrix elements of observables. All quantities necessary can be and have been evaluated using techniques in a three dimensional coordinate and momentum grid (lattice) similar to techniques known from TDHF calculations. In contrast to TDHF the domain of applications of quantized ATDHF are typical quantum properties as e.g. subbarrier fusion cross sections, examples to which are given in this article, where the light ion systems $\alpha+\alpha \leftrightarrow {}^{8}Be$, ${}^{12}C+{}^{12}C \leftrightarrow {}^{24}Mg$ and ${}^{16}O+{}^{16}O \leftrightarrow {}^{32}S$ are considered. More detailed studies concerning a greater variety of nuclear systems, better interactions, less restrictive symmetries and the inclusion of pairing correlations are required and will be the task for the near future.

References

1) K. Goeke, P.-G. Reinhard, Ann. Phys. 124 (1980) 249.
2) P.-G. Reinhard, K. Goeke, Phys. Rev. C20 (1979) 1546.
3) F. Villars, Nucl. Phys. A285 (1977) 269.
4) K. Goeke, P.-G. Reinhard, Ann. Phys. 112 (1978) 328.
5) P.-G. Reinhard, K. Goeke, Phys. Lett. 69B (1977) 17.
6) P.-G. Reinhard, K. Goeke, Nucl. Phys. A312 (1978) 121.
7) P.-G. Reinhard, J.A. Maruhn, K. Goeke, Phys. Rev. Lett. 44 (1980) 1740.
8) K. Goeke, P.-G. Reinhard, D.J. Rowe, Nucl. Phys. A359 (1981) 408.
9) D.J. Rowe, A. Ryman, J. Math. Phys. 23 (1982) 732.
10) P. Bonche, S. Koonin, J. Negele, Phys. Rev. C13 (1976) 1226.
11) H. Flocard, P.H. Heenen, D. Vautherin, Nucl. Phys. A339 (1980) 336.
12) P.H. Heenen, H. Flocard, D. Vautherin, these proceedings.
13) K. Goeke, Phys. Rev. Lett. 38 (1977) 212.

A microscopic calculation of potentials and mass parameters for heavy-ion reactions

P.-H. HEENEN [+]

Physique Nucléaire Théorique CP 229, Campus de la Plaine U.L.B., Bd. du Triomphe , B 1050 Brussels, Belgium

H. FLOCARD and D. VAUTHERIN

Division de Physique Théorique [x], Institut de Physique Nucléaire, F - 91406 Orsay, Cedex , France.

Many methods are used to compute the potential energy of a heavy-ion system as a function of the interdistance R between the ions. However, the dependance on R of the mass parameter M associated with the kinetic energy is generally neglegted, M being assumed to be equal to the reduced mass. The adiabatic time-dependent Hartree-Fock method (1) - ATDHF - provides a framework to derive the potential and the mass parameter in a consistent way. The use of ATDHF for the study of a heavy-ion collision is greatly simplified by the fact that the interdistance between the ions is the most important collective variable. It is therefore sufficient to solve the ATDHF equations along a fixed path. This simplification decreases the amount of numerical work. It also permits to derive a potential energy directly comparable with the real part of the experimental optical potentials.

In this communication, we review the potentials and mass parameters calculated for the $^{16}O + ^{16}O$ and $^{40}Ca + ^{40}Ca$ systems (2) by ATDHF . We shall show that the fusion cross sections calculated with these potentials and mass parameters reproduce very well the experimental data.

The collective path of the collision process is generated by means of a set of constrained Hartree-Fock calculations :

$$\left(-\frac{\hbar^2}{2m}\Delta + U(r) - \lambda \hat{Q}(r)\right)\varphi_i(r,\lambda) = \varepsilon_i(\lambda)\,\varphi_i(r,\lambda)$$

where Q(r) is a quadrupole constraint related to the distance between the ions by :

$$\frac{1}{2} A R^2(\lambda) = \langle \Phi(\lambda) | \hat{Q} | \Phi(\lambda) \rangle$$

The Slater determinant $\Phi(\lambda)$ is composed of the individual wave functions $\varphi_i(r,\lambda)$

With the BKN force (3), the Hartree-Fock potential U(r) is given by :

$$U(r) = \frac{3}{4} t_0\, \rho(r) + \frac{3}{16} t_3\, \rho^2(r) + \int d\vec{r}'\, \rho(\vec{r}')\, \frac{a V_0 \exp -\frac{|\vec{r}-\vec{r}'|}{a} + \frac{e^2}{4}}{|\vec{r}-\vec{r}'|}$$

[+] Chercheur agréé FNRS

[x] Laboratoire associé au CNRS

Each orbitals being assumed to be occupied by two neutrons and two protons, the density ρ (r) is given by :

$$\rho(\lambda, r) = 4 \sum_i |\varphi_i(\lambda, r)|^2$$

The collective potential is the expectation value of the many-body hamiltonian with respect to $\Phi(\lambda)$

$$V(R) = \langle \Phi(\lambda) | H | \Phi(\lambda) \rangle$$

while the mass parameter is the polarizability with respect of the operator:

$$\hat{P}(\lambda) = i\hbar \left[\rho(\lambda), \frac{\partial \rho}{\partial \lambda} \right]$$

The calculation of this polarizability requires the solution of another set of Hartree-Fock equations with a second constraint :

$$\left(-\frac{\hbar^2}{2m} \Delta + U(r) - \lambda \hat{Q} - \mu \hat{P} \right) \varphi_i(r, \lambda, \mu) = \varepsilon_i \varphi_i$$

The mass parameter M(R) is then given by :

$$M(R) = \left(\frac{\partial}{\partial \mu} \langle \Phi(\lambda, \mu) | \hat{P} | \Phi(\lambda, \mu) \rangle \right)_{\mu=0}$$

where $\Phi(\lambda, \mu)$ is the Slater build from the $\varphi_i(r, \lambda, \mu)$
The solutions of the Hartree-Fock equation were performed in the coordinate space, assuming reflection symmetries of the individual wave functions with respect to the three planes x = 0, y = 0 and z = 0.

The potential and mass parameter are shown in fig. 1 forthe $^{16}O + ^{16}O$ system and in fig. 2 for the $^{40}Ca + ^{40}Ca$ system. The barrier radii are in good agreement with the radii of the optical potentials deduced from elastic scattering data (4), (5). However, the barrier radius of $^{40}Ca + ^{40}Ca$ (R_B = 10.4 fm) is larger than the values determined from various phenomenological potentials (6). This larger value is due to the very rapid formation of the neck between the ^{40}Ca nuclei, lowering the total energy of the system. The kink found in the potential energy is also due to this very rapid formation of the neck.

The mass parameters of both systems exhibit large peaks in the region where the ions are touching. In the case of the BKN interaction, the ATDHF mass parameter is equal to the cranking one :

$$M(R) = 2\hbar^2 \sum_{i=1}^{A} \sum_{j>A} \frac{|\langle \varphi_j(\lambda, r) | \frac{\partial}{\partial R} | \varphi_i(\lambda, r) \rangle|^2}{\varepsilon_j - \varepsilon_i}$$

One sees that M(R) has a peak either when an unoccupied level comes closed to an occupied one or when the individual wave-functions change rapidly. No crossings occur in the region where the mass increases. In the case of the $^{16}O + ^{16}O$ system, it is possible to relate each peak with the rearrangement of an individual wave function due to the Pauli principle. The most external peak is due to the individual wave functions with the largest elongation along the collision axis. The peaks are no more

Fig. 1

Potential and mass parameter obtained for the $^{16}O + ^{16}O$ system.

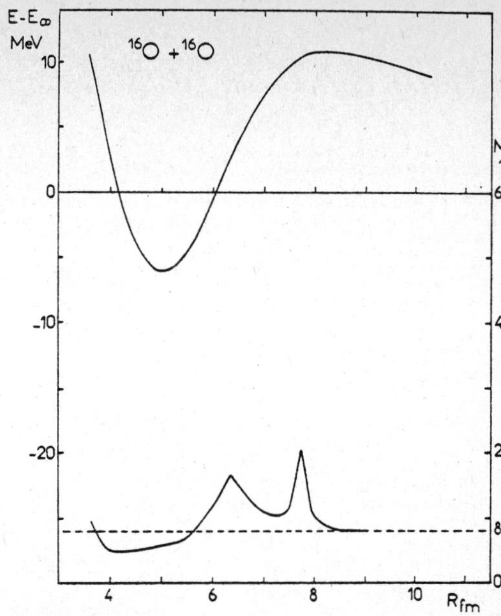

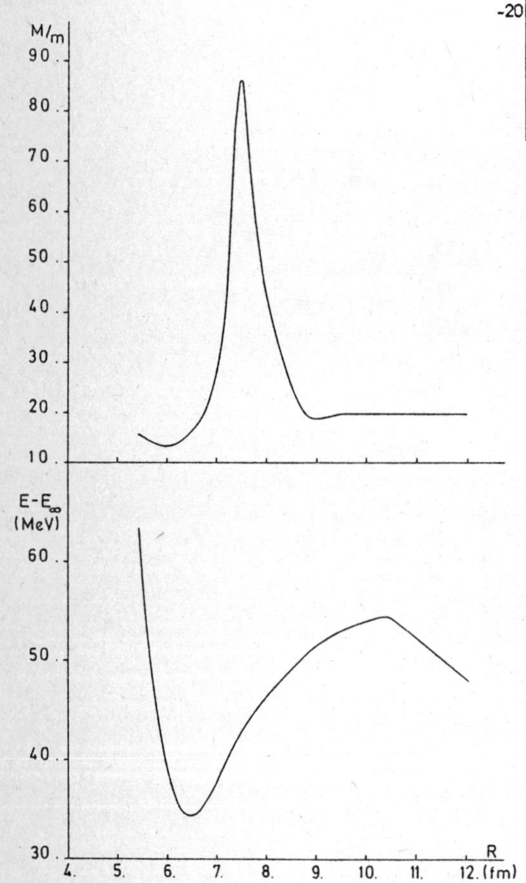

Fig. 2

Potential and mass parameter obtained for the $^{40}Ca + ^{40}Ca$ system.

resolved in the $^{40}Ca + ^{40}Ca$ system, the number of individual states being too large. To show the effect of the mass parameter on observable quantities, we have calculated the fusion cross section of $^{40}Ca + ^{40}Ca$ for which experimental data exist over a large energy range above the barrier (6). We have used a barrier penetration model proposed by Descouvemont et al. (7) in which the flux transmitted down to a critical radius R_c is assumed to be completely absorbed. To calculate the transmission coefficient down to R_c, the collective Bohr Hamiltonian :

$$H(R,\dot{R}) = \frac{1}{2} M(R)\, \dot{R}^2 + V(R)$$

must be requantized. We used the Pauli prescription, leading to the Schrödinger equation :

$$\left\{ - \left[\left(\frac{\hbar^2}{2M(R)}\right)^{\frac{1}{4}} \frac{\partial}{\partial R} \left(\frac{\hbar^2}{2M(R)}\right)^{\frac{1}{4}} \right]^2 + V(R) + \frac{\hbar^2}{2M_0} \frac{\ell(\ell+1)}{R^2} \right\} \Psi_\ell(R) = E\, \Psi_\ell(R)$$

where M_o is the reduced mass of the system.

The fusion cross sections calculated with the reduced mass and with the ATDHF mass are compared with experiment in fig. 3. It can be seen that the cross section

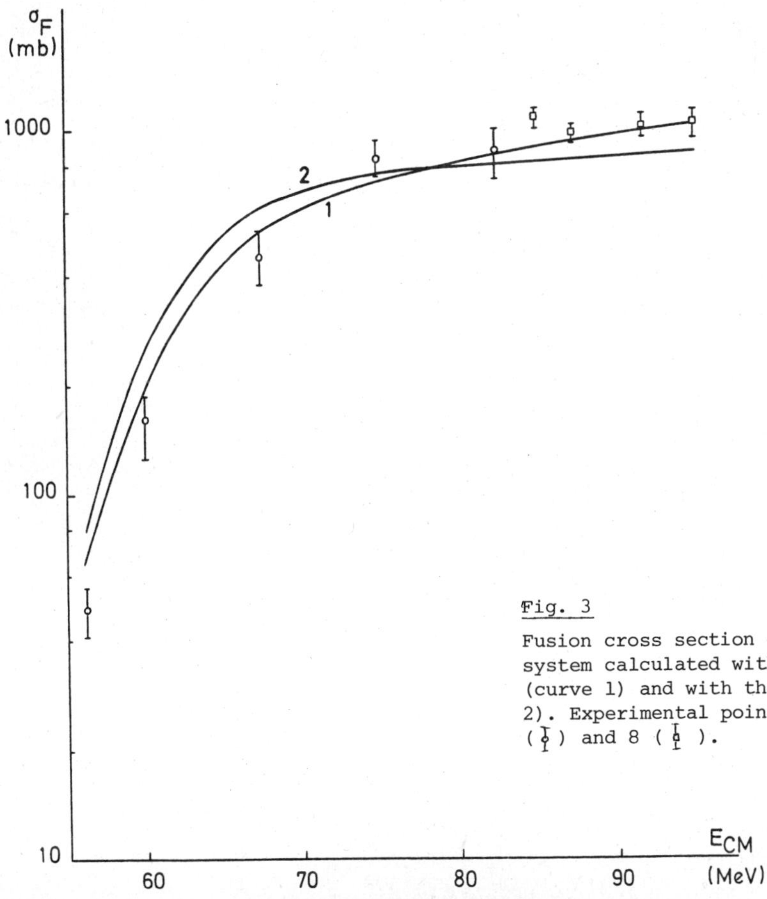

Fig. 3

Fusion cross section of the $^{40}Ca + ^{40}Ca$ system calculated with the ATDHF mass (curve 1) and with the reduced mass (curve 2). Experimental points are from ref. 6 (⌶) and 8 (⌶).

obtained with a constant mass is in reasonable agreement with experiment. However, it overestimates the cross section in the vicinity of the Coulomb barrier. The peak in the mass parameter decreases the probability of penetrating the barrier at low energy. It brings the fusion cross section calculated with the ATDHF mass in closer agreement with experiment.

This result indicates that the effect of the effective mass is not negligible for the fusion of two nuclei as heavy as ^{40}Ca. For collision more peripherical than fusion like elastic or inelastic scattering, the effect of the mass parameter is probably much weaker for ^{40}Ca + ^{40}Ca than for ^{16}O + ^{16}O. As can be seen from figs 1 and 2 , the peaks of the ^{16}O + ^{16}O mass are located at relatively larger interdistances than the peak of the ^{40}Ca + ^{40}Ca mass. It is therefore probable that the effect of M(R) is more weakened by absorption for heavy systems than for light systems.

References

1) M. BARANGER and M. VENERONI Ann. of Phys. 114 (1978) 123
2) H. FLOCARD, P.-H. HEENEN and D. VAUTHERIN, Nucl. Phys. A339 (1980) 336 and to be published
3) P. BONCHE, S.E. KOONIN and J.W. NEGELE, Phys. Rev. C13 (1976) 1226
4) A. GOBBI et al. Phys. Rev. C7 (1973) 30
5) M. RICHTER et al. Nucl. Phys. A278 (1977) 163
6) E. TOMASI et al. Nucl. Phys. A373 (1982) 341
7) P. DESCOUVEMONT, D. BAYE and P.-H. HEENEN to appear in Z. Phys.
8) H. DOUBRE et al. Phys. Lett. 73B (1978) 135.

THEORY OF LARGE AMPLITUDE COLLECTIVE VIBRATIONS: CLASSICAL ANALOG OF THE CONCEPT OF COLLECTIVE PATH.

J. da Providência and J.N. Urbano
Departamento de Física
Universidade de Coimbra
P 3000 Coimbra, Portugal

1. Introduction

Marumori's and Villars theories of large amplitude collective oscilations[1,2] are here examined within the framework of time dependent Hartree-Fock (TDHF) theory. It is not our purpose to provide new results. We believe, however, it is useful to translate to a language as simple and familiar as possible, the concepts involved in those theories. In particular, the invariance principle postulated by Marumori is here explained in elementary terms. In section 2, TDHF theory is briefly reviewed and applied to the description of large amplitude collective oscillations.
In section 3, an interpretation of the equations for the "collective path" and of the concept of collective variables is developed on the basis of Hamilton equations of Classical Mechanics.

2. TDHF Theory of large amplitude collective motion

Time-dependent Hartree-Fock equations, which govern the time evolution of Slater determinants $|\Phi(t)\rangle$, may be derived from the action principle

$$\delta \int_{t_1}^{t_2} L\, dt = 0 \tag{2.1}$$

with

$$L = i\langle\Phi|\dot{\Phi}\rangle - \langle\Phi|H|\Phi\rangle, \tag{2.2}$$

where H is the hamiltonian.
These equations are summarized by the variational equation

$$i(\langle\Phi|\delta S|\dot{\Phi}\rangle + \langle\dot{\Phi}|\delta S|\Phi\rangle) = \langle\Phi|\left[\delta S, H\right]|\Phi\rangle, \tag{2.3}$$

δS being an arbitrary one body operator.

If $|\Phi(t)\rangle$ is a TDHF state, solution of eq. (2.3) and we require that also the Slater determinant

$$|\Psi(t)\rangle = \exp\ (iS(t))\,|\Phi(t)\rangle$$

is a TDHF state, S(t) being an infinitesimal hermitian one body operator, the linear equation for the time evolution of S(t), which arises from such a requiremente, is written, in variational form,

$$i\langle\Phi|[\delta S,\dot{S}]|\Phi\rangle = \langle\Phi|[\delta S,[H,S]\,]|\Phi\rangle, \tag{2.4}$$

where δS is again an arbitrary one body operator. Although this equation is formally analogous to the RPA equations, there is an important difference. Indeed, since $|\Phi(t)\rangle$ is a function of time, the equation satisfied by $S(t)$ is a linear differential equation with time dependent coefficients.

We observe in passing that it is possible to rewrite eq. (2.3) in a more familiar form if we introduce an hermitian one-body operator $K(t)$ such that

$$|\Phi(t+\delta t)\rangle = \exp\,(-i\delta t K(t))\,|\Phi(t)\rangle. \tag{2.5}$$

We have, therefore,

$$\partial_t|\Phi(t)\rangle = -iK(t)|\Phi(t)\rangle\ . \tag{2.6}$$

We obtain, finally

$$\langle\Phi(t)|[\delta S,[K(t)-H]\,]|\Phi(t)\rangle = 0\ . \tag{2.7}$$

We discuss now large amplitude collective motion. We denote by Σ a set of normalized Slater determinants such that their time evolution as TDHF states takes place inside Σ. If $|\Phi(0)\rangle$ belongs to Σ then $|\Phi(t)\rangle$ remains in Σ for all times t. We assume, for simplicity, that the elements of the invariant set Σ are labeled by two real parameters α,β.

The time evolution of the parameters α,β is determined by the following effective Lagrangean

$$\mathcal{L}(\alpha,\beta,\dot{\alpha},\dot{\beta}) = i[\,\langle\Phi(\alpha,\beta)|\partial_\alpha\Phi(\alpha,\beta)\rangle\,\dot{\alpha} + \langle\Phi(\alpha,\beta)|\partial_\beta\Phi(\alpha,\beta)\rangle\dot{\beta}] - \langle\Phi(\alpha,\beta)|H|\Phi(\alpha,\beta)\rangle\ . \tag{2.8}$$

The parametrization α,β of Σ is chosen such that

$$i\langle\Phi(\alpha,\beta)|\partial_\alpha\Phi(\alpha,\beta)\rangle = 0, \tag{2.9}$$

$$i\langle\Phi(\alpha,\beta)|\partial_\beta\Phi(\alpha,\beta)\rangle = \alpha. \tag{2.10}$$

This choice is permissible according to Darboux theorem, and implies the following relation

$$i[\langle\partial_\alpha\Phi|\partial_\beta\Phi\rangle - \langle\partial_\beta\Phi|\partial_\alpha\Phi\rangle] = 1. \tag{2.11}$$

Then we have

$$\mathcal{L} = \alpha\dot{\beta} - \mathcal{H}(\alpha,\beta), \tag{2.12}$$

$$\dot{\alpha} = -\frac{\partial\mathcal{H}}{\partial\beta}\ ,\ \dot{\beta} = \frac{\partial\mathcal{H}}{\partial\alpha} \tag{2.13}$$

with

$$\mathcal{H}(\alpha,\beta) = \langle\Phi(\alpha,\beta)|H|\Phi(\alpha,\beta)\rangle\ . \tag{2.14}$$

Since we have

$$\partial_t|\Phi(\alpha,\beta)> = |\partial_\alpha\Phi(\alpha,\beta)>\dot{\alpha} + |\partial_\beta\Phi(\alpha,\beta)>\dot{\beta}$$

$$= -|\partial_\alpha\Phi(\alpha,\beta)>\partial_\beta\mathcal{H} + |\partial_\beta\Phi(\alpha,\beta)>\partial_\alpha\mathcal{H}, \qquad (2.15)$$

the condition that $|\Phi(\alpha,\beta)>$ is a TDHF state becomes

$$-(<\Phi(\alpha,\beta)|\delta S|\partial_\alpha\Phi(\alpha,\beta)> + <\partial_\alpha\Phi(\alpha,\beta)|\delta S|\Phi(\alpha,\beta)>)\partial_\beta\mathcal{H}$$

$$+ (<\Phi(\alpha,\beta)|\delta S|\partial_\beta\Phi(\alpha,\beta)> + <\partial_\beta\Phi(\alpha,\beta)|\delta S|\Phi(\alpha,\beta)>)\partial_\alpha\mathcal{H}$$

$$+ i<\Phi(\alpha,\beta)|[\delta S,H]|\Phi(\alpha,\beta)> = 0 . \qquad (2.16)$$

This is equivalent to the equation which has been proposed by Marumori[1)] for the collective path, and reduces to a simpler form if we introduce hermitian one body operators $A(\alpha,\beta)$, $B(\alpha,\beta)$, such that

$$|\Phi(\alpha+\delta\alpha,\beta)> = \exp(-i\delta\alpha B(\alpha,\beta))|\Phi(\alpha,\beta)>, \qquad (2.17)$$

$$|\Phi(\alpha,\beta+\delta\beta)> = \exp(i\delta\beta A(\alpha,\beta))|\Phi(\alpha,\beta)> . \qquad (2.18)$$

We have, therefore

$$\partial_\alpha|\Phi(\alpha,\beta)> = -iB(\alpha,\beta)|\Phi(\alpha,\beta)> , \qquad (2.19)$$

$$\partial_\beta|\Phi(\alpha,\beta)> = iA(\alpha,\beta)|\Phi(\alpha,\beta)> , \qquad (2.20)$$

so that eq. (2.16) reduces to

$$<\Phi(\alpha,\beta)|[\delta S, (B\partial_\beta\mathcal{H} + A\partial_\alpha\mathcal{H} - H)]|\Phi(\alpha,\beta)> = 0 . \qquad (2.21)$$

This is another version of the equation of the path. With the help of the operators A, B, eq. (2.11) may also be given a more meaningful form,

$$-i<\Phi|[B,A]|\Phi> = 1 . \qquad (2.22)$$

3. Classical analog of the concept of collective degrees of freedom

Let us consider the classical Hamiltonian

$$H = \frac{1}{2}\sum_{i,j=1}^{n} p_i (M^{-1})_{ij}\, p_j + V(q_1,\ldots,q_n), \qquad (3.1)$$

where the matrix M_{ij} may be some function of the coordinates $q_1,\ldots,q_n$. We denote by Σ a surface such that if $(p_1(0),\ldots,p_n(0), q_1(0),\ldots,q_n(0))$ belongs to Σ, then $(p_1(t),\ldots,p_n(t), q_1(t),\ldots,q_n(t))$ will remain on Σ for all times t. Here, the functions $p_i(t)$, $q_i(t)$, $i = 1,\ldots,n$, describe a possible motion of the system under investigation, i.e, they are a solution of Hamilton equations

$$\dot{p}_i = -\frac{\partial H}{\partial q_i}\,,\quad \dot{q}_i = \frac{\partial H}{\partial p_i}\,. \tag{3.2}$$

There are many such surfaces, but it may happen that only a few of them are physically interesting. Let the equations of the surface be

$$p_i = f_i(\alpha,\beta),\ q_i = g_i(\alpha,\beta),\ i = 1,\ldots,n. \tag{3.3}$$

The parameters α,β may, under some circumstances, be regarded as collective variables or generator coordinates.

It is well known that Hamilton equations are derived from an action principle related to the following Lagrangean

$$L = \sum_{i=1}^{n} p_i\,\dot{q}_i - H. \tag{3.4}$$

An effective Lagrangean $\mathcal{L}$, describing the evolution in time of the parameter α,β is obtained when we combine equations (3.3) and (3.4),

$$\mathcal{L} = \Big(\sum_{i=1}^{n} f_i \frac{\partial g_i}{\partial \alpha}\Big)\,\dot{\alpha} + \Big(\sum_{i=1}^{n} f_i \frac{\partial g_i}{\partial \beta}\Big)\,\dot{\beta} - \mathcal{H}(\alpha,\beta) \tag{3.5}$$

where $\mathcal{H}(\alpha,\beta)$ is given by

$$\mathcal{H}(\alpha,\beta) = \frac{1}{2}\sum_{i,j} f_i(\alpha,\beta)\,(M^{-1})_{ij}\, f_j(\alpha,\beta) + V(g_1(\alpha,\beta),\ldots,g_n(\alpha,\beta)). \tag{3.6}$$

Also in M_{ij} the replacement $q_i = g_i\ (\alpha,\beta)$ is assumed. According to Darboux theorem, the parametrization α,β of the surface Σ may be chosen in such a way that

$$\sum_{i=1}^{n} f_i \frac{\partial g_i}{\partial \alpha} = 0 \tag{3.7}$$

$$\sum_{i=1}^{n} f_i \frac{\partial g_i}{\partial \beta} = \alpha. \tag{3.8}$$

If this choice is made, the effective Lagrangean becomes

$$\mathcal{L} = \alpha\,\dot{\beta} - \mathcal{H}(\alpha,\beta) \tag{3.9}$$

so that α is the effective momentum, β is the effective coordinate and $\mathcal{H}(\alpha,\beta)$ is the effective Hamiltonian corresponding to the particular collective motion under consideration. The time evolution of the parameters α,β is described by Hamilton equations

$$\dot{\alpha} = -\frac{\partial \mathcal{H}}{\partial \beta}\,,\quad \dot{\beta} = \frac{\partial \mathcal{H}}{\partial \alpha}\,. \tag{3.10}$$

These equations are of interest even if the surface Σ is not exactly, but only approximately, invariant. From eqs. (3.7) and (3.8) the following normalization condition (analogous to eqs. (2.11) or (2.22)) is obtained

$$\sum_{i=1}^{n} \left(\frac{\partial f_i}{\partial \alpha} \frac{\partial g_i}{\partial \beta} - \frac{\partial f_i}{\partial \beta} \frac{\partial g_i}{\partial \alpha}\right) = 1 \ . \tag{3.11}$$

We present now the classical equivalent to Marumori's equations of the collective path.

Imposing that the exact equations of motion are satisfied by the parametrization α, β, we find

$$\frac{\partial f_i}{\partial \alpha} \dot{\alpha} + \frac{\partial f_i}{\partial \beta} \dot{\beta} = - \left(\frac{\partial H}{\partial q_i}\right)_{(f,g)} \ ,$$

$$\frac{\partial g_i}{\partial \alpha} \dot{\alpha} + \frac{\partial g_i}{\partial \beta} \dot{\beta} = \left(\frac{\partial H}{\partial p_i}\right)_{(f,g)} \ , \tag{3.12}$$

where $i = 1, \ldots, n$, and the subscript (f,g) means that the replacement $p_i = f_i$ and $q_i = g_i$ has been made. A convenient set of equations for the surface Σ is now obtained if we take into account eqs. (3.10)

$$\left(\frac{\partial f_i}{\partial \alpha} \frac{\partial g_i}{\partial \beta} - \frac{\partial f_i}{\partial \beta} \frac{\partial g_i}{\partial \alpha}\right) \frac{\partial \mathcal{H}}{\partial \alpha} = \frac{\partial H}{\partial q_i} \frac{\partial g_i}{\partial \alpha} + \frac{\partial H}{\partial p_i} \frac{\partial f_i}{\partial \alpha} \ ,$$

$$\left(\frac{\partial f_i}{\partial \alpha} \frac{\partial g_i}{\partial \beta} - \frac{\partial f_i}{\partial \beta} \frac{\partial g_i}{\partial \alpha}\right) \frac{\partial \mathcal{H}}{\partial \beta} = \frac{\partial H}{\partial q_i} \frac{\partial g_i}{\partial \beta} + \frac{\partial H}{\partial q_i} \frac{\partial g_i}{\partial \beta} \ . \tag{3.13}$$

Analogous sets of equations have been given in ref. 3). These equations for an invariant surface Σ are equivalent to Marumori's equations of the path. This may be seen as follows. Introduce generators

$$A = A\ (p_1, \ldots, p_n, q_1, \ldots, q_n)\ , \quad B = B\ (p_1, \ldots, p_n, q_1, \ldots, q_n)$$

such that

$$\frac{\partial h}{\partial \alpha} = \{B,h\}, \quad \frac{\partial h}{\partial \beta} = - \{A,h\},$$

where $h = h\ (p_1, \ldots, p_2, q_1, \ldots, q_n)$.

Then eqs. (3.12) may be written

$$\{B,p_i\} \frac{\partial \mathcal{H}}{\partial \beta} + \{A,p_i\} \frac{\partial \mathcal{H}}{\partial \alpha} = \{H,p_i\}$$

$$\{B,q_i\} \frac{\partial \mathcal{H}}{\partial \beta} + \{A,q_i\} \frac{\partial \mathcal{H}}{\partial \alpha} = \{H,q_i\} \ . \tag{3.14}$$

These equations are clearly of the same form as eqs. (2.21).

We assume that H is invariant under time reversal. The invariant surface Σ will reflect the time reversal simmetry of the Hamiltonian, so that

$$f_i(\alpha,\beta) = - f_i(-\alpha,\beta)$$

$$g_i(\alpha,\beta) = g_i(-\alpha,\beta) \; .$$

We look now for a power series solution of eqs. (3.13) by expanding $f_i(\alpha,\beta)$ and $g_i(\alpha,\beta)$ in powers of α. Retaining only the zeroth and first order terms we arrive at equations which are equivalent to Villars[2)] equations. We write, therefore,

$$\begin{aligned} f_i(\alpha,\beta) &= \alpha\, f_i^{(1)}(\beta) + \mathcal{O}(\alpha^3) \\ g_i(\alpha,\beta) &= g_i^{(0)}(\beta) + \mathcal{O}(\alpha^2) \end{aligned} \tag{3.15}$$

From eq. (3.11) we find

$$\sum_{i=1}^{n} f_i^{(1)}(\beta)\, \frac{dg_i^{(0)}(\beta)}{d\beta} = 1 \; , \tag{3.16}$$

while the Hamiltonian $\mathcal{H}(\alpha,\beta)$ becomes replaced by

$$\mathcal{H}(\alpha,\beta) = \frac{\alpha^2}{2\mathcal{M}} + \mathcal{V}(\beta) \tag{3.17}$$

where

$$\frac{1}{2\mathcal{M}} = \sum_{i,j} f_i^{(1)}(\beta)\, f_j^{(1)}(\beta)\, (M^{-1})_{ij} \; , \tag{3.18}$$

$$\mathcal{V}(\beta) = V\,(g_1^{(0)}(\beta),\ldots,g_n^{(0)}(\beta)) . \tag{3.19}$$

In this approximation, eqs. (3.13) are replaced by

$$\frac{dg_i^{(0)}}{d\beta}\,\frac{1}{\mathcal{M}} = \sum_j (M^{-1})_{ij}\, f_j^{(1)}$$

$$f_i^{(1)} \frac{\partial \mathcal{V}}{\partial \beta} = \frac{\partial V}{\partial q_i} \; .$$

One easily verifies that these equations are analogous to Villars equations for the collective path[2,3].

We end this section with a reference to the stability of a particular solution $p_i(t)$, $q_i(t)$, $i = 1,\ldots,n$, of eqs. (2.2). Let $\delta p_i(t)$, $\delta q_i(t)$ be infinitesimal quantities such that also $p_i(t) + \delta p_i(t)$, $q_i(t) + \delta q_i(t)$, is a solution of eqs. (3.2). It follows that $\delta p_i(t)$, $\delta q_i(t)$ satisfy the following equations

$$\delta\dot{p}_i = - \sum_j \frac{\partial^2 H}{\partial q_i \partial p_j}\, \delta p_j - \sum_j \frac{\partial^2 H}{\partial q_i \partial q_j}\, \delta q_j$$

$$\dot{\delta q_i} = \sum_j \frac{\partial^2 H}{\partial p_i \partial p_j} \delta p_j + \sum_j \frac{\partial^2 H}{\partial q_i q_j} \delta q_j .$$

The trajectory $p_i(t)$, $q_i(t)$ is stable if all possible quantities $\delta p_i(t)$, $\delta q_i(t)$ remain infinitesimal for all times.

It seems natural to require that the collective path is a surface Σ which not only is invariant bot also stable.

Aknowledgments The present research has been supported by Instituto Nacional de Investigação Cientifica (Portugal).

References

1) T. Marumori et. al., Prog. Theor. Phys. 64 (1980) 1294

2) F. Villars, Nucl. Phys. A 285 (1977) 269

3) L.P. Brito and C.A. de Sousa, J. Phys. A 14 (1981) 2239

OPTIMAL COLLECTIVE PATHS*

E.J.V. de Passos
Instituto de Fīsica da Universidade de São Paulo
CP 20516 São Paulo, S.P., Brasil

I. INTRODUCTION

In this work we consider microscopic theories appropriate for describing large amplitude nuclear collective motion such as fission or oscillations of transitional nuclei. Since an exact solution of the nuclear many-body problem is impossible, the approach is to isolate a subspace of the many-body Hilbert space maximally decoupled such that the time evolution of the states within the subspace is equivalent to the motion of a few collective degrees of freedom. In the ideal case this subspace, which we call the collective subspace, is an invariant subspace of the many-body Hilbert space.

One theory which can be used to construct the collective subspace is the generator coordinate method (GCM)[1,9-11]. In this method we consider the many-particle states which can be written as a linear superposition of the generator states $|\psi(\underset{\sim}{\alpha})\rangle$,

(1) $$|f\rangle = \int d\alpha \; f(\underset{\sim}{\alpha}) \; |\psi(\underset{\sim}{\alpha})\rangle$$

The generator states are chosen ā priori usually based on educated guesses as to the nature of the collective motion in consideration. Energy levels and wave functions are determined by the variational principle which leads to the GHW equation for $f(\underset{\sim}{\alpha})$

(2) $$\int d\underset{\sim}{\alpha}' \; (\langle\underset{\sim}{\alpha}|\hat{H}|\underset{\sim}{\alpha}'\rangle - E\langle\underset{\sim}{\alpha}|\underset{\sim}{\alpha}'\rangle) \; f(\underset{\sim}{\alpha}') = 0$$

The states which can be written as in eq. (1) span a subspace of the many-body Hilbert space which we identify with the collective subspace. The collective hamiltonian is defined by the restriction of the many-body hamiltonian to the collective subspace and the collective operators as any complete set of commuting observables where these observables are defined in the collective subspace[1,9-11].

If the generator states $|\psi(\underset{\sim}{\alpha})\rangle$ span the manifold of Slater determinants, the many-particle states given by the "ansatz" eq. (1) span the many-body Hilbert space. However the aim of the microscopic theories of large amplitude nuclear collective motion is to select an optimal submanifold of the manifold of Slater determinants on the bases of dynamical considerations as stated by Rowe and Bassermann[1-3], Villars[4], Goeke and Reinhard [5] and by Villars and Moya de Guerra[6] (see also references 7 and 8). This optimal submanifold of Slater deteminants is called the optimal collective path (O.C.P.). To construct the collective subspace we use the determinants of the O.C.P. as generator states in the "ansatz" of GHW, eq. (1). Even though it is possible to use the GCM and the variational principle to find not only the weight function but the generator states $|\psi(\underset{\sim}{\alpha})>$[2] the theories considered here uses a time-dependent approach based on the TDHF equations. They also restrict the determinants to the manifold of time-even Slater determinants (TED) since they are interested in a static optimal collective path (except references 1-3). For easier of presentation, we will suppose that it is sufficient to consider only one collective degree of freedom, the generalization to more than one being in principle straightforward. In this case the O.C.P. is a curve $|\psi_0(q)>$ in the manifold of TED. In the following development, based on reference 12, we will discuss a clear and concise way of deriving the equations which determine the O.C.P. of R-B, V-G-R and V-M studying the decoupling properties of the TDHF equations in the neighbourhood of a given Slater determinant $|\psi_0>$. This approach was used in reference 1 to derive the Rowe-Bassermann O.C.P. and one of the advantages of this approach is that it exhibits the physical content of the differences of the three theories mentioned above. We also argue that the aim of the theories of R-B, V-G-R and V-M is to construct the O.C.P. such that the generator of infinitesimal displacements along the path at each point $|\psi_0(q)>$ on the path and the force which acts on this determinant be always in the direction of a local normal mode and we investigate the conditions under which this is actually achieved.

II. TDHF EQUATIONS

The Thouless theorem says that any Slater determinant non-orthogonal to a given Slater determinant $|\psi_0>$ can be parametrised as

$$|\psi(\underset{\sim}{\alpha},\underset{\sim}{\Pi})> = e^{i\,\hat{S}(\underset{\sim}{\alpha},\underset{\sim}{\Pi})}\,|\psi_0> \tag{3}$$

where

$$\hat{S}(\alpha,\Pi) = \sum_j(\Pi_j\hat{q}_j - \alpha_j\hat{p}_j) \tag{4}$$

The one-body operators $\hat{q}_j$ and $\hat{p}_j$ have only particle-hole matrix elements with respect to $|\psi_o\rangle$ and they satisfy canonical commutation relations when the commutators act on $|\psi_o\rangle$[13],

$$\text{(5)}\qquad \begin{aligned} \left[\hat{q}_j,\hat{q}_k\right]|\psi_o\rangle &= \left[\hat{p}_j,\hat{p}_k\right]|\psi_o\rangle = 0 \\ \left[\hat{q}_j,\hat{p}_k\right]|\psi_o\rangle &= i\,\delta_{jk}\,|\psi_o\rangle \end{aligned}$$

Given any one-body operator its particle-hole component with respect to $|\psi_o\rangle$ can be written as

$$\text{(6)}\qquad \hat{F}^A = \sum_j \langle\psi_o|\left[\hat{F},-i\hat{p}_j\right]|\psi_o\rangle\hat{q}_j + \sum_j \langle\psi_o|\left[\hat{F},i\hat{q}_j\right]|\psi_o\rangle\hat{p}_j$$

The equation which gives the time-evolution of the Slater determinants is the TDHF equation, and it can be derived by the TDVP[4,13]

$$\text{(7)}\qquad \langle\delta\psi(t)|\ \hat{H} - i\vec{\partial}_t\ |\psi(t)\rangle = 0$$

If $|\psi_o\rangle$ is a time-even Slater determinant we can always choose the $\hat{q}_k$ and $\hat{p}_k$ as time-even and time-odd operators, respectively. Using this fact and the parametrization given by eqs. (3) and (4) it can be shown that the TDHF equations reduce, in the neighbourhood of $|\psi_o\rangle$, to

$$\text{(8)}\qquad \dot{\alpha}_k = \sum_j B_{kj}\,\Pi_j \qquad\qquad -\dot{\Pi}_k = -F_k + \sum_j V_{kj}\,\alpha_j$$

where

$$\text{(9)}\qquad \begin{aligned} F_k &= -\langle\psi_o|\left[\hat{H},-i\,\hat{p}_k\right]|\psi_o\rangle \\ B_{kj} &= B_{jk} = \langle\psi_o|\left[\hat{q}_k,\left[\hat{H},\hat{q}_j\right]\right]|\psi_o\rangle \\ V_{kj} &= V_{jk} = \langle\psi_o|\left[\hat{p}_k,\left[\hat{H},\hat{p}_j\right]\right]|\psi_o\rangle \end{aligned}$$

The eqs. (8) are analogous to classical canonical equations of motion whose hamiltonian is the expectation value of the many-body hamiltonian in the state $|\psi(\underset{\sim}{\alpha},\underset{\sim}{\Pi})\rangle$ expanded to second order in $\underset{\sim}{\alpha}$ and $\underset{\sim}{\Pi}$,

$$\text{(10)}\qquad \hat{H}(\underset{\sim}{\alpha},\underset{\sim}{\Pi}) = \langle\psi_o|\hat{H}|\psi_o\rangle - \sum_k F_k\alpha_k + \frac{1}{2}\sum_{k,j}\left(B_{kj}\Pi_k\Pi_j + V_{kj}\alpha_k\alpha_j\right)$$

Thus we can interpret the parameters $\underset{\sim}{\alpha}$ and $\underset{\sim}{\Pi}$ as local canonical variables at the point $|\psi_o\rangle$. These local canonical variables define the phase-space at $|\psi_o\rangle$. Furthermore to each local canonical variable α_k , Π_k we associate the pair of local canonical operators $\hat{q}_k$ and $\hat{p}_k$. The eqs.(8) show that F_k is the component in the k direction of the force which acts on $|\psi_o\rangle$ and that B_{kj} and V_{kj} are the components of the inverse inertia tensor and of the elastic tensor. Also they show that in general $|\psi_o\rangle$ is not in equlibrium and has null velocity.

If $|\psi_o\rangle$ is any determinant the TDHF trajectory which goes through it does not have any special decoupling property, when this trajectory is examined in the neighbourhood of $|\psi_o\rangle$. On the other hand an O.C.P. should be distinguished by the decoupling properties of the collective and non-collective degrees of freedom. In the next section we are going to show how we can derive the equations which determine the optimal collective paths of R-B, V-G-R and V-M by studying the de-coupling properties of the TDHF trajectory, when this trajectory is examined in the neighbourhood of a determinant on the path.

III. OPTIMAL COLLECTIVE PATHS FOR LARGE AMPLITUDE COLLECTIVE MOTION

III.A. O.C.P. of Rowe and Bassermann

To derive the O.C.P. of R-B we introduce the local normal modes which are defined as local canonical operators which makes the quadratic term of the hamiltonian (10) diagonal. As in our case a de-terminant along the O.C.P. is time-even, there is no coupling between coordinate and momentum and the quadratic term of the hamiltonian (10) can be diagonalized by a linear point transformation

$$\hat{Q}_i^L = \sum_j \hat{q}_j (a^{-1})_{ij} \qquad \hat{P}_i^L = \sum_j \hat{p}_j \, a_{ji} \tag{11}$$

This point transformation is such that the normal modes satisfy the equations

$$\langle\psi_o| \left[\hat{P}_i^L, \left[\hat{H}, \hat{P}_j^L\right]\right] |\psi_o\rangle = C_i \, \delta_{ij} \qquad \langle\psi_o| \left[\hat{Q}_i^L, \left[\hat{H}, \hat{Q}_j^L\right]\right] |\psi_o\rangle = B_i \, \delta_{ij} \tag{12}$$

We suppose that among the local normal modes there is one which near the minimum (solution of the HF eq.) is the lowest frequency mode which transforms into the only one unstable mode far from equilibrium. At each point this mode will be denoted by $\hat{Q}_o^L$, $\hat{P}_o^L$. The determinants of the O.C.P. of Rowe-Bassermann are selected by imposing that the force which acts on these determinants is in the direction of the local normal mode $\hat{Q}_o^L$, $\hat{P}_o^L$. This condition leads to

(13) $$\langle\psi_o| \left[\hat{H}, -i\,\hat{P}_j^L\right] |\psi_o\rangle = 0 \quad , \quad j \neq 0$$

Parametrizing the curve described by the determinants which have the above property by a label q, where q is the expectation value of a measured operator $\hat{D}$

(14) $$q = \langle\psi_o(q)|\hat{D}|\psi_o(q)\rangle \quad ,$$

eq. (13) can be written as a constrained equation where the operators $\hat{Q}_k(q)$ and $\hat{P}_k(q)$ are the local normal modes[1,13]

$$\langle\delta\psi_o(q)|\hat{H} + F_o^L(q)\hat{Q}_o^L(q)|\psi_o(q)\rangle = 0 \qquad \text{(I)}$$

(15) $$\langle\delta\psi_o(q)|\left[\hat{H},\hat{P}_k^L(q)\right] - i\,C_k(q)\hat{Q}_k^L(q)|\psi_o(q)\rangle = 0 \qquad \text{(II)}$$

$$\langle\delta\psi_o(q)|\left[\hat{H},\hat{Q}_k^L(q)\right] + i\,B_k(q)\hat{P}_k^L(q)|\psi_o(q)\rangle = 0 \qquad \text{(III)}$$

If, for some q, $F_o^L(q)$ is equal to zero, $|\psi_o(q)\rangle$ is in equilibrium and the eqs. (15) reduce to the R.P.A. Furthermore in equations (15)(II) and (15)(III) we can replace $\hat{H}$ by $H_{const}(q)$,

$$\hat{H}_{const}(q) = \hat{H} + F_o^L(q)\hat{Q}_o^L(q)$$

This shows that the R-B equations are a generalization of the R.P.A. to determinants outside an equilibrium point where the force of constraint which we should apply to the system in order for it be in equilibrium is along a local normal mode.

III.B. O.C.P. of Villars, Goeke and Reinhard

Identifying a determinant along the O.C.P. of V-G-R by a label q which is the expectation value of a measured operator $\hat{D}$,

$$q = \langle\psi_o(q)|\hat{D}|\psi_o(q)\rangle$$

the particle-hole component of the generator of infinitesimal displacements along the path at the point q is defined by the equation

(16) $$(|\psi_0(q+\Delta q) - |\psi_0(q)>)_{ph} = -i\Delta q\, \hat{P}_0^V(q)\,|\psi_0(q)>$$

and an operator canonically conjugate to $\hat{P}_0^V(q)$ by

(17) $$\left[\hat{Q}_0^V(q),\hat{P}_0^V(q)\right]|\psi_0(q)> = i|\psi_0(q)>$$

The determinants $|\psi_0(q)>$ on the O.C.P. of V-G-R satisfy the properties that, given a complete set of local canonical operators which includes $\hat{Q}_0^V(q)$, $\hat{P}_0^V(q)$:

i) The force which acts on $|\psi_0(q)>$ is in the direction of the degree of freedom associated to $\hat{Q}_0^V(q)$, $\hat{P}_0^V(q)$

ii) The inertia tensor is diagonal and the elastic tensor is diagonal for i, j $\neq 0$.

These conditions are expressed by the equations

$$<\delta\psi_0(q)|\hat{H} - \frac{dV^{VGR}(q)}{dq}\,\hat{Q}_0^V(q)|\psi_0(q)> = 0 \qquad \text{(I)}$$

(18) $$<\delta\psi_0(q)|\left[\hat{H},\hat{Q}_0^V(q)\right] + iB_0(q)\,\hat{P}_0^V(q)|\psi_0(q)> = 0 \qquad \text{(II)}$$

$$(|\psi_0(q+\Delta q)> - |\psi_0(q)>)_{ph} = -i\,\Delta q\,\hat{P}_0^V(q)|\psi_0(q)> \qquad \text{(III)}$$

with

$$V^{VGR}(q) = <\psi_0(q)|\hat{H}|\psi_0(q)>$$

Eq. (18)(I) is an equation of constraint and the constraint is determined by equation (18)(II). The two properties i) and ii) allow us the following interpretation: If the system is placed at rest somewhere on the O.C.P. of V-G-R it starts to move first along the path[13].

III.C. The O.C.P. of Villars and Moya de Guerra

In this case we impose that the particle-hole component of the generator of infinitesimal displacements along the O.C.P. of V-M at a point $|\psi_0(q)>$ on the path is in the direction of a local normal mode. This condition gives

(19) $$(|\psi_0(q+\Delta q)> - |\psi_0(q)>)_{ph} = -i\,\Delta q\,\hat{P}_0^L(q)|\psi_0(q)>$$

As in the two previous cases we can write the equation for the O.C.P. as a constrained equation where the $\hat{Q}_j^L(q)$ and $\hat{P}_j^L(q)$ are the local normal modes

(20)

$$<\delta\psi_o(q)|\hat{H} + \sum_k F_k^L(q)\ \hat{Q}_k^L(q)|\psi_o(q)> = 0 \qquad \text{(I)}$$

$$<\delta\psi_o(q)|\left[\hat{H},\hat{P}_k^L(q)\right] - i C_k(q)\hat{Q}_k^L(q)|\psi_o(q)> = 0 \qquad \text{(II)}$$

$$<\delta\psi_o(q)|\left[\hat{H},\hat{Q}_k^L(q)\right] + i B_k(q)\hat{P}_k^L(q)|\psi_o(q)> = 0 \qquad \text{(III)}$$

$$(|\psi_o(q+\Delta q)> - |\psi_o(q)>)_{ph} = - i\Delta q\ \hat{P}_o^L(q)|\psi_o(q)> \qquad \text{(IV)}$$

IV. PROPERTIES OF THE OPTIMAL COLLECTIVE PATHS

Given a path in the space of Slater determinants we can find at each point on this path the force, the local normal modes and the generator of infinitesimal displacements along the path. If $|\psi_o>$ is a stationary state there is no force acting on this determinant and we can always impose that the direction of the path at this point coincides with a local normal mode. When we are outside a stationary state, we can consider this determinant as instantaneously in equilibrium, as in the D'Alembert principle. As before we find the local normal modes at this point and impose that the direction of the path coincides with one of these normal modes. However the system placed at this point starts to move in the direction of the path only if the force which acts on this determinant is also along this normal mode. The O.C.P. considered here do not have this property. Using the developments of section III it can be easily seen that at each point on the O.C.P. of R-B the force is in the direction of a normal mode but the generator of infinitesimal displacements is not in this direction. At each point in the path of V-G-R the force and the generator of infinitesimal displacements are in the direction of a locally canonical degree of freedom but this degree of freedom is not a normal mode. And on the path of V-M, at each point, the generator of infinitesimal displacements is in the direction of a local normal mode but the force is not in this direction. This discussion suggests that the ultimate aim of the theories of O.C.P. considered here is to derive a path with the property that at each point the direction of the path is along a locally decoupled degree of freedom and that when the system is placed on the path it starts to move first in the direction of the path. When the above condition is satisfied the three O.C.P. coincide. To further clarify the physical meaning of the above condition it can be easily shown[2,14] that the path of R-B coincides at any point with a local valley of the local potential energy surface

$$V(\underset{\sim}{\alpha}) = - \sum_k F_k \alpha_k + \frac{1}{2} \sum_{k,j} V_{kj} \alpha_k \alpha_j \qquad (21)$$

in a space whose metric is the inverse inertia tensor at this point. The direction of the path of V-G-R coincides at any point with the direction of the local gradient line of the local energy surface eq.(21) in a space whose metric is the inertia tensor at this point[14]. Finally the direction of the path of V-M coincides at any point with the direction of a local normal mode. When the paths are identical it coincides at any point with a local valley line and the direction of the path at this point is along a local gradient line.

ACKNOWLEDGEMENTS

This report is based on a paper done in collaboration with F.F. de Souza Cruz.

* Work supported in part by CNPq and FAPESP (Brasil).

REFERENCES

1) D.J. Rowe and R. Bassermann - Can. J. Phys. 54 (1976), 1941
2) G. Holzwarth and T. Yukawa - Nucl. Phys. A219 (1974), 125
3) T. Marumori - Prog. Theor. Phys. 57 (1977), 112
4) F. Villars - Nucl. Phys. A285 (1977), 369
5) K. Goeke and P.G. Reinhard - Ann. of Physics 112 (1978), 328
6) E. Moya de Guerra and F. Villars - Nucl. Phys. A285 (1977), 297
7) M. Baranger and M. Veneroni - Ann. of Phys. 114 (1978), 123
8) T. Marumori, T. Maskawa, F. Sakata and A. Kuriyama - Progr. Theor. Phys. 64 (1980), 1294
9) E.J.V. de Passos and A.F.R. de Toledo Piza - Phys. Rev. C21 (1980), 425
10) E.J.V. de Passos and F.F. de Souza Cruz - Phys. Rev. C24 (1981), 690
11) B. Giraud and B. Grammaticos - Nucl. Phys. A255 (1975), 141
12) E.J.V. de Passos and F.F. de Souza Cruz - in preparation
13) P.G. Reinhard and K. Goeke - Nucl. Phys. A312 (1978), 121
14) K. Goeke, P.G. Reinhard and D.J. Rowe - Nucl. Phys. A359 (1981), 408.

ATDHF THEORY WITH A CONSISTENCY CONDITION

A.K. Mukherjee and M.K. Pal
Saha Institute of Nuclear Physics
Calcutta, India

The adiabatic time-dependent Hartree-Fock theory for the microscopic description of large amplitude collective motion was formulated in two seemingly different ways by (i) Villars[1] and by (ii) Baranger and Vênêroni[2]. The formal identity of these approaches has been demonstrated in ref. 3) and it has been shown that the solution of the zeroth and first order equations satisfying a consistency condition derived from the second-order equation follows the bottom of the potential energy valley.

In this paper we have shown the identity of the two approaches in a different way and in the process the ATDHF equations have been reduced to computationally tractable form. We have then discussed how the solution of the ATDHF equations along with the consistency requirement can be achieved in practice.

The decomposition of the single-particle density operator $\rho(t)$ in the forms

$$\rho(t) = e^{-i\chi(t)} \rho_0(t)\, e^{i\chi(t)} \tag{1}$$

where $\rho_0(t)$ and $\chi(t)$ are time-even hermitian operators, enables one to write the TDHF equation

$$i\dot{\rho} = [h, \rho] \tag{2}$$

as two coupled equations for the time-even and time-odd components ρ_+ and ρ_- of ρ. In the adiabatic approximation $\chi(t)$ is "small" and a series expansion in powers of χ enables one to pick up terms up to any desired order. Retaining terms up to the second order, the two coupled equations reduce to

time-even: $$i\dot{\rho}_1 = [h_0, \rho_0] + [h_0, \rho_2] + [h_1, \rho_1] + [h_2, \rho_0] \tag{3}$$

time-odd: $$i\dot{\rho}_0 = [h_0, \rho_1] + [h_1, \rho_0] \tag{4}$$

where

$$\rho_1 = [\chi, \rho_0] \tag{5}$$

$$\rho_2 = -\frac{1}{2}[\chi, [\chi, \rho_0]] \tag{6}$$

$$h_0 = t + \mathrm{Tr}\,\tilde{v}\,\rho_0 \tag{7}$$

$$h_1 = \mathrm{Tr}\,\tilde{v}\,\rho_1 \;;\; h_2 = \mathrm{Tr}\,\tilde{v}\,\rho_2 \tag{8}$$

As in ref. 3) substituting $\rho_0(t) = \rho_0(q)$ and $\chi = p\hat{q}$, where p is the small adiabaticity parameter and q the collective coordinate eq. (3) reduces to:

$$p\left[\left\{[h_0, \hat{q}] + \bar{h}_1 + i\frac{\hat{p}}{m}\right\}, \rho_0\right] = 0 \tag{9}$$

Here $\hat{q}$ and $\hat{p}$ are p-h type operators and the latter, in particular, is defined through

$$\frac{\partial \rho_0}{\partial q} = -i[\hat{P}, \rho_0] \tag{10}$$

The quantity $\bar{h}_1$ is given by

$$\bar{h}_1 = \frac{1}{p} h_1 = \mathrm{Tr}\,\tilde{v}\,[\hat{q}, \rho_0] \tag{11}$$

Similarly eq. (2) reduces to

$$[h_0 - \lambda\hat{q}, \rho_0] + p^2\left[\left\{-\frac{1}{2}\left([[h_0, \hat{q}], \hat{q}] + 2[\bar{h}_1, \hat{q}] + \bar{h}_2\right) + m^{-1}(q)\frac{\partial \hat{q}}{\partial q} - \frac{1}{2}\frac{\partial m^{-1}}{\partial q}\hat{q}\right\}, \rho_0\right] = 0 \tag{12}$$

where

$$\bar{h}_2 = -\frac{1}{2p^2} h_2 = [\hat{q}, [\hat{q}, \rho_0]] \tag{13}$$

In deriving eqs. (9) and (12) we have made use of the Hamilton equations

$$\dot{q} = \frac{\partial \mathcal{H}}{\partial p} = \frac{p}{m} \quad ; \quad \dot{p} = -\frac{\partial \mathcal{H}}{\partial q} = -\frac{\partial V}{\partial q} - \frac{p^2}{2}\frac{\partial m^{-1}}{\partial q}$$

corresponding to the collective Hamiltonian

$$\mathcal{H}(p, q) = \frac{p^2}{2m(q)} + V(q) \tag{14}$$

where

$$m^{-1}(q) = \mathrm{Tr}\, h_0\, [[\hat{q}, \rho_0], \hat{q}] + \mathrm{Tr}\,\mathrm{Tr}\,[\hat{q}, \rho_0]\,\tilde{v}\,[\hat{q}, \rho_0] \tag{15}$$

$$V(q) = \mathrm{Tr}\, t\, \rho_0 + \frac{1}{2}\mathrm{Tr}\,\mathrm{Tr}\,\rho_0\,\tilde{v}\,\rho_0 \tag{16}$$

We thus conclude as in ref. 3) that while the time-odd eq. (9) of Baranger-Vénéroni (BV) is of first order in the adiabaticity parameter p, the time-even eq. (12) mixes the zeroth- and second-order terms in p. In the spirit of the perturbation approach we write them as separate equations

$$p^0 \rightarrow \qquad [h_0 - \lambda\hat{q}, \rho_0] = 0 \tag{DI}$$

$$p^1 \rightarrow \qquad [\{[h_0, \hat{q}] + \bar{h}_1 + i\frac{\hat{p}}{m}\}, \rho_0] = 0 \tag{DII}$$

$$p^2 \rightarrow \qquad [\{-\tfrac{1}{2}([[h_0, \hat{q}], \hat{q}] + 2[\bar{h}_1, \hat{q}] + \bar{h}_2) + m^{-1}(q)\frac{\partial\hat{q}}{\partial q} - \frac{1}{2}\frac{\partial m^{-1}}{\partial q}\hat{q}\}, \rho_0] = 0 \tag{DIII}$$

Notice that in the basis of the eigenstates of ρ_0 only the particle-hole (p-h) matrix elements of the eqs. (DI), (DII) and (DIII) are relevant, since the p-p and h-h matrix elements of an operator of the form $[A, \rho_0]$ identically vanish.

In order to derive the consistency requirement we differentiate eq. (DI) with respect to q,

$$[\frac{\partial h_0}{\partial q} - \frac{\partial\lambda}{\partial q}\hat{q} - \lambda\frac{\partial\hat{q}}{\partial q}, \rho_0] + [h_0 - \lambda\hat{q}, \frac{\partial\rho_0}{\partial q}] = 0$$

which after some algebraic manipulations gives

$$[m^{-1}(q)\frac{\partial\hat{q}}{\partial q}, \rho_0] = \frac{1}{\lambda m}[\{\mathrm{Tr}\,\tilde{v}[-i\hat{p}, \rho_0] + [h_0, -i\hat{p}] - \frac{\partial\lambda}{\partial q}\hat{q}\}, \rho_0] - m^{-1}[[\hat{q}, -i\hat{p}], \rho_0] \tag{17}$$

Substituting eq. (17) in eq. (DIII) we have

$$[\{-\frac{\lambda^2}{2}([[h_0, \hat{q}], \hat{q}] + 2[\bar{h}_1, \hat{q}] + \bar{h}_2) + \frac{\lambda}{m}([h_0, -i\hat{p}] + \mathrm{Tr}\,\tilde{v}[-i\hat{p}, \rho_0]) - \lambda\,\omega_0(q)\hat{q}\}, \rho_0] = 0 \tag{DIII'}$$

where

$$\omega_0(q) = \frac{1}{2\lambda}\frac{d}{dq}(\lambda^2/m) \tag{18}$$

In ref. 3) we had the ATDHF equations of Villars

$$\langle\delta\Phi(q)|\,H - \lambda Q\,|\Phi(q)\rangle = 0 \tag{SI}$$

$$\langle \delta\Phi(q) | [H, iQ] - P/m(q) | \Phi(q) \rangle = 0 \qquad \text{(SII)}$$

along with the consistency requirement

$$\langle \delta\Phi(q) | \tfrac{1}{2}\lambda^2 [[H, iQ], iQ] + \frac{\lambda}{m} [H, -iP] - \lambda \omega_0(q) Q | \Phi(q) \rangle = 0 \qquad \text{(SIII')}$$

The similarity of the eqs. (DI), (DII) and (DIII') with eqs. (SI), (SII) and (SIII'), where $|\delta\Phi(q)\rangle$ are in general 1p-1h type states with respect to $|\Phi(q)\rangle$ as vacuum, can be easily seen from the following relations

$$\langle \Phi_{ph} | H | \Phi \rangle = \langle p | h_0 | h \rangle$$

$$\langle \Phi_{ph} | [H, Q] | \Phi \rangle = \langle \Phi_{ph} | [\sum_i t_i, \sum_j \hat{q}_j] + [\sum_{i<j} v_{ij}, \sum_k \hat{q}_k] | \Phi \rangle$$

$$= \langle \Phi_{ph} | \sum_i [t_i, \hat{q}_i] + \sum_{i<j} [v_{ij}, \hat{q}_i + \hat{q}_j] | \Phi \rangle$$

since the operators corresponding to different particles commute. Therefore,

$$\langle \Phi_{ph} | [H, Q] | \Phi \rangle$$

$$= \langle p | [t, \hat{q}] | h \rangle + \sum_{h'} [\langle ph' | [v_{12}, \hat{q}_1 + \hat{q}_2] | hh' \rangle - \langle ph' | [v_{12}, \hat{q}_1 + \hat{q}_2] | h'h \rangle]$$

where, since $\sum_i [t_i, \hat{q}_i]$ and $\sum_{i<j} [V_{ij}, \hat{q}_i + \hat{q}_j]$ are respectively one-body and two-body type symmetric operators, we have used the results

$$\langle \Phi_{ph} | \sum_i O_i | \Phi \rangle = \langle p | O | h \rangle$$

$$\langle \Phi_{ph} | \sum_{i<j} O_{ij} | \Phi \rangle = \sum_{h'} (ph' | O_{12} | hh')$$

Therefore

$$\langle \Phi_{ph} | [H, Q] | \Phi \rangle = \sum_{\alpha\beta} [(p\alpha | [v_{12}, q_1] | h\beta) \times$$

$$\times \langle \beta | \rho | \alpha \rangle + (p\alpha | [v_{12}, q_2] | h\beta) \langle \beta | \rho | \alpha \rangle] + \langle p | [t, q] | h \rangle$$

$$= \langle p | [t, \hat{q}] + Tr_2 [\tilde{v}_{12}, q_1] \rho_2 + Tr_2 [\tilde{v}_{12}, q_2] \rho_2 | h \rangle$$

$$= \langle p | [h, \hat{q}] + \bar{h}_1 | h \rangle$$

where we have used the definition of the trace

$$\langle a| Tr_2 \tilde{V}_{12} O_2 |b\rangle = \sum_{\alpha\beta} (a\alpha|V|b\beta) \langle\beta|O|\alpha\rangle$$

and the cyclic property of trace, which leads to

$$Tr\,[A,B]C = Tr\,(ABC - BAC) = Tr\,(BCA - CBA) = Tr\,[B,C]A$$

The round-bracket matrix element of V here stands for the usual "direct" minus "exchange" matrix elements.

In a similar manner it can be easily seen that

$$\langle\Phi_{ph}|[H,-iP]|\Phi\rangle = \langle p|[h_0,-i\hat{p}] + Tr\,\tilde{V}[-i\hat{p},\rho_0]|h\rangle$$

$$\langle\Phi_{ph}|[[H,iQ],iQ]|\Phi\rangle = -\langle p|[[h_0,\hat{q}],\hat{q}] + 2[\bar{h}_1,\hat{q}] + \bar{h}_2|h\rangle$$

Since the eqs. (DI), (DII) and (DIII) are one-body equations they can be easily reduced to equations for the single particle states $|\Phi_\alpha(q)\rangle$. Since $\hat{q}$ and $\hat{p}$ are p-h type operators eq. (DI) gives

$$\lambda\hat{q} = (1-\rho_0)h_0\rho_0 + \rho_0 h_0(1-\rho_0) \tag{19}$$

and eq. (DII) gives

$$-i\hat{p} = c(q)\{(1-\rho_0)h_0(1-2\rho_0)h_0\rho_0 - \rho_0 h_0(1-2\rho_0)h_0(1-\rho_0) + (1-\rho_0)\bar{h}_1'\rho_0 + \rho_0\bar{h}_1'(1-\rho_0)\} \tag{20}$$

where $c(q) = m/\lambda$ and we have substituted $\hat{q}$ from eq. (19). Recognizing that $-i\hat{p}|\Phi_\alpha(q)\rangle = \frac{\partial}{\partial q}|\Phi_\alpha(q)\rangle$, we have from eq. (20)

$$\frac{\partial}{\partial q}|\phi_h(q)\rangle = c(q)\{(1-\rho_0)h_0(1-2\rho_0)h_0 + (1-\rho_0)\bar{h}_1'\}|\phi_h(q)\rangle \tag{21a}$$

$$\frac{\partial}{\partial q}|\phi_p(q)\rangle = c(q)\{-\rho_0 h_0(1-2\rho_0)h_0 + \rho_0\bar{h}_1'\}|\phi_p(q)\rangle \tag{21b}$$

where $\bar{h}_1' = \frac{1}{\lambda}\bar{h}_1 = Tr\tilde{V}[h_0,\rho_0]$. In eq. (21a) since no particle states appear, in order to obtain the occupied states which are sufficient to evaluate the relevant quantities of the ATDHF theory, one needs not consider infinite number of particle states.

In ref. 3) it has been shown that solutions of eqs. (21) represent lines of force normal to the equipotential surfaces revealing the non-uniqueness of their solutions. It has been shown in ref. 3) that the particular solution of eq. (21) that satisfies the consistency condition (DIII') follows the bottom of the potential energy valley.

Substituting $\hat{q}$ and $\hat{p}$ from eqs. (19) and (20), eq. (DIII') reduces to

$$(1-\rho_0)\{h_0(1-\rho_0)h_0(1-2\rho_0)h_0 - h_0(1-2\rho_0)h_0\rho_0h_0 + h_0\bar{h}_1' - \\ - \bar{h}_1'h_0 - \omega_0(q)h_0 + \mathrm{Tr}\,\tilde{v}\hat{r}\}\,|\phi_h(q)\rangle = 0 \tag{22a}$$

$$\rho_0\{h_0(1-2\rho_0)h_0(1-\rho_0)h_0 - h_0\rho_0h_0(1-2\rho_0)h_0 + h_0\bar{h}_1' - \\ - \bar{h}_1'h_0 - \omega_0(q)h_0 + \mathrm{Tr}\,\tilde{v}\hat{r}\}\,|\phi_p(q)\rangle = 0 \tag{22b}$$

where the one-body operator $\hat{r}$ is given by

$$\hat{r} = (1-\rho_0)h_0(1-2\rho_0)h_0\rho_0 + \rho_0h_0(1-2\rho_0)h_0(1-\rho_0) - \\ - \rho_0h_0(1-\rho_0)h_0\rho_0 + (1-\rho_0)h_0\rho_0h_0(1-\rho_0) + (1-\rho_0)\bar{h}_1'\rho_0 - \\ - \rho_0\bar{h}_1'(1-\rho_0) \tag{23}$$

In order to solve the eqs. (21) the prescription

$$|\phi_\alpha(q+\delta q)\rangle = \left(1 - i\delta q\,\hat{p}(q)\right)|\phi_\alpha(q)\rangle \tag{24}$$

does not preserve the scalar product of the states. One has to use some unitary evolution operator

$$|\phi_\alpha(q+\delta q)\rangle = e^{-i\delta q\,\hat{p}(q)}\,|\phi_\alpha(q)\rangle \tag{25}$$

or some approximation[6)] to it. Given an initial set of states corresponding to an initial deformation

$$q_0 = \langle\Phi(q_0)|\,\hat{D}\,|\Phi(q_0)\rangle$$

one first calculates λ/m from the normalization condition

$$\frac{\partial q}{\partial q} = 1 = \langle\Phi(q)|\,[\hat{D}, -iP]\,|\Phi(q)\rangle$$

$$\text{or, } c^{-1}(q) = \frac{\lambda}{m} = \langle\Phi(q)|\,[\hat{D}, -ic^{-1}(q)P]\,|\Phi(q)\rangle$$

and then using eq. (25) the set of states corresponding to deformation $q_0+\delta q$ can be calculated and so on. There is another normalization condition

$$\langle \Phi(q) | [Q, -iP] | \Phi(q) \rangle = 1$$

which gives λ^2/m and thus at each step we obtain λ and m separately.

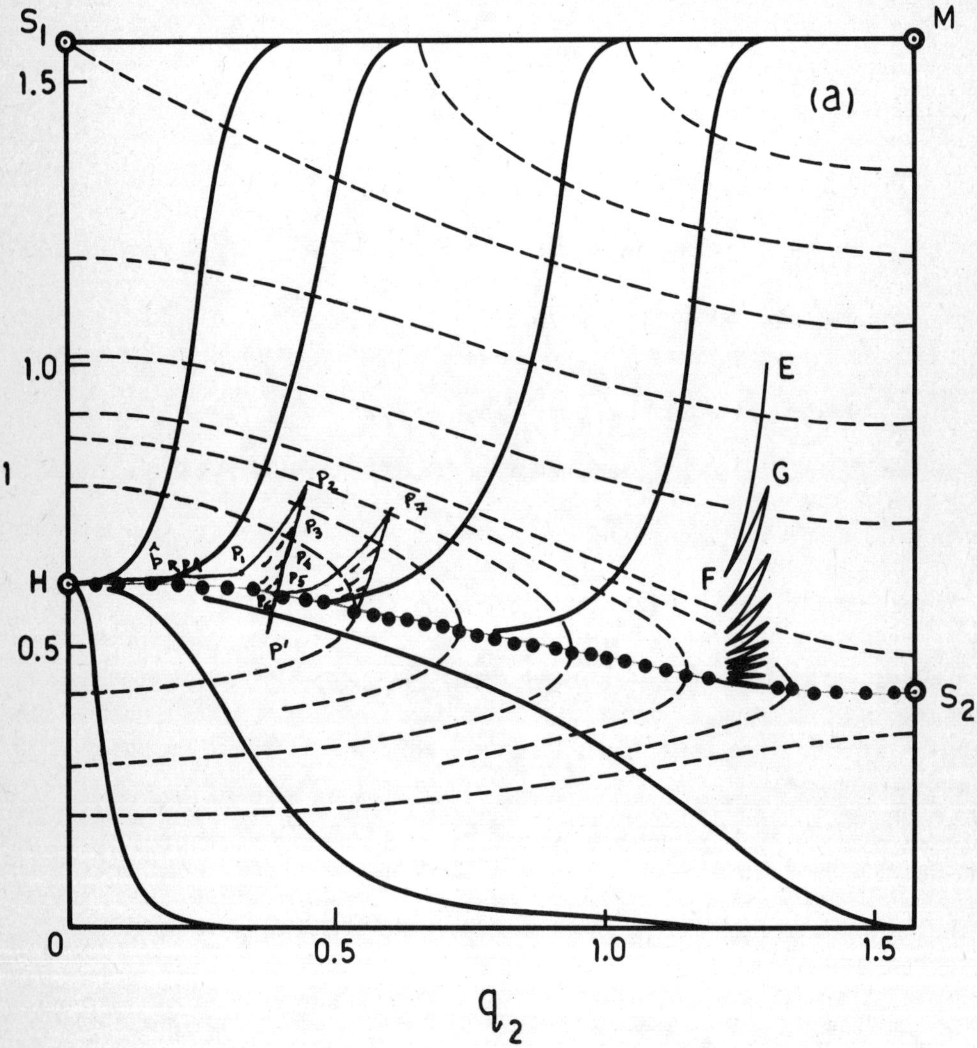

Figure: Schematic plot of ATDHF trajectory of the two-dimensional Lipkin model. Crossed line represents the valley path. The procedure to trace out the valley path is also shown (see the text). H is the static HF minimum and S_1 and S_2 are the saddle points of the potential.

Along with eq. (25) one has to satisfy the consistency conditions (22). The following procedure schematically explained in the figure can be adopted to achieve the consistency. With the displacement operator $\hat{p}_{RPA}$ of the lowest frequency RPA mode at the HF minimum we displace the static HF state by an amount δq and reach a point P_1 which is not a singular point of the eqs. (21). With P_1 as the initial condition we solve the eqs. (21) along which V(q) increases and after a few steps reach the point P_2 which in general does not satisfy eqs. (22). At P_2 we give a kick to the system with the momentum operator $\hat{p}$ corresponding to the point P_2 and at each step of this kick we solve the ATDHF equations for two or three steps and check the consistency conditions (22). Thus along the kick P_2P' we will reach a point P_6 on the valley path where the consistency condition is satisfied. With P_6 as the initial condition we solve eqs. (21) in a direction along which V(q) diminishes and trace out a portion of the valley path. Repeat the process with the point P_6 playing the role of P_1 one obtains the collective path along with the classical Hamiltonian defined in eq. (14).

The coordinate representation of the ATDHF equations with a generalized Skyrme force is in progress. However in order to apply ATDHF theory to nuclear fission some relevant features have to be incorporated. The collective coordinate q represents the fission degree of freedom. Though the vibration of the nucleus along the fission path is trivially included in the theory the rotational motion of the nucleus along its fission path has to be incorporated. This can be done by using cranked TDHF equation to derive the ATDHF equations and including a centrifugal term in the classical Hamiltonian (14).

Discussions with D.M. Brink, K. Goeke and a correspondence with P.-G. Reinhard are gratefully acknowledged.

References

1) F. Villars, Nucl. Phys. A285 (1977) 269.
2) M. Baranger and M. Vénéroni, Ann. Phys. 114 (1978) 123.
3) A.K. Mukherjee and M.K. Pal, Nucl. Phys. A373 (1982) 289.
4) A.K. Mukherjee and M.K. Pal, Phys. Lett. 100B (1981) 457.
5) K. Goeke et al., Phys. Rev. Lett. 44 (1980) 1740.
6) J.W. Negele, Lectures presented at the NATO-NSF Summer School on Theoretical Methods in Medium Energy and Heavy Ion Physics, Madison, Wisconsin, June 1978.
7) S. Krewald et al., Nucl. Phys. A281 (1977) 166.

QUANTIZED ATDHF

E. Moya de Guerra
Departamento de Física Teórica
Universidad Autónoma de Madrid
Canto Blanco, Madrid-34, Spain

Abstract

A new quantization procedure of ATDHF in the small amplitude limit is presented. Results concerning ground state energy and density are compared with those obtained in the G.C.M. and by time averaging of the time-dependent Hartree-Fock density. This quantization procedure is applied to the small amplitude oscillations perpendicular to the "fission path" in order to see to what extent coupling between large and small amplitude modes may result in damping-like terms.

I. Selfconsistent Microscopic Description of Zero Point Motion Effects

It should be by now well understood that the ATDHF approach is not merely intended to provide an approximate solution of time-dependent Hartree-Fock theory, but rather, to provide a self-consistent microscopic description of quantum collective nuclear modes. In dealing with collective nuclear phenomena one is generally interested in reducing the many-body Schrödinger equation to a few-body one, by deriving a collective Hamiltonian which depends on a few collective coordinates and is quadratic in the momenta. ATDHF approaches this problem in two steps.

In a first stage the TDHF equation is used to derive a classical Hamiltonian, $\mathcal{H}(p_i,q_i)$, by parametrizing the time-dependence of the Slater determinants $\phi(t)$ in terms of the collective coordinates, $q_i(t)$, which are supposed to be relevant to the problem at hand, and their canonical conjugate momenta, $p_i(t)$. Since the classical Hamiltonian is expected to be quadratic in the velocities, an expansion of the TDHF equation to first order in the velocities is sufficient to derive a set of self-consistency conditions for the static determinants $\phi(q_i)$ and constraining operators O_i at every point $\{q_i\}$; and hence the potential energy and inertial tensor governing the classical dynamics. To summarize, these conditions are |1|

$$\text{I} \quad \langle \delta\phi(q_i) | H - \sum_i \left(\frac{\partial V}{\partial q_i} Q_i\right) | \phi(q_i) \rangle = 0$$

$$\text{II} \quad \langle \delta\phi(q_i) | \{H, iQ_i\}_- - \sum_j \left(\frac{P_j}{M_{ij}}\right) | \phi(q_i) \rangle = 0$$

$$\text{III} \quad \langle \phi(q_i) | \{Q_i, -iP_j\}_- | \phi(q_i) \rangle = \delta_{ij}$$

with V, M_{ij} and Q_i, P_i as defined in ref. |1|.

In a second stage one must proceed to a quantal interpretation of this dynamics. While the first part of this program is now rather well established |1-3|, the second one remains open, and as a whole the theory has not been used in its full power. In these lectures I will focus on the problem of quantization. I shall first restrict the discussion to the case of small amplitude oscillations around the equilibrium shape in rigid nuclei. This, apart from being an interesting problem in itself, has the advantage that it allows us to compare easily with other theories, and in particular, with the generator coordinate method (GCM) which is in principle more reliable. Then I shall apply the method developed here to the problem of fission.

As can be easily shown |1| at the stable equilibrium point, $\{q_i = q_i^M\}$, condition I is just the static H-F equation for the unconstrained H-F ground state $\phi_o = \phi(q_i = q_i^M)$. In addition, its first derivative together with conditions II, III give the RPA equations for the normal modes of the system

$$\begin{aligned} &\langle \delta\phi_o | \{H, A_i^+\}_- - \omega_i A_i^+ | \phi_o \rangle = 0 \\ &\langle \delta\phi_o | \{H, A_i\}_- + \omega_i A_i | \phi_o \rangle = 0 \qquad (1) \\ &\langle \phi_o | \{A_i, A_j^+\}_- | \phi_o \rangle = \delta_{ij} , \end{aligned}$$

with

$$A_i^+ = \left(\frac{M_i K_i}{4}\right)^{1/4} Q_i - i\left(\frac{1}{4M_i K_i}\right)^{1/4} P_i , \quad A_i = (A_i^+)^+ \qquad (2)$$

$$\begin{aligned} &\omega_i = \sqrt{\frac{K_i}{M_i}} , \quad M_i^{-1} = \langle \phi_o | \{\{H, iQ_i\}_-, iQ_i\}_- | \phi_o \rangle \\ &K_i = \langle \phi_o | \{\{H, -iP_i\}_-, -iP_i\}_- | \phi_o \rangle . \end{aligned} \qquad (3)$$

The energy of the system to second order in the coordinates and momenta is given by

$$\mathcal{H}(p_i, q_i) = \langle \phi(t) | H | \phi(t) \rangle = \langle \phi_o | H | \phi_o \rangle + \sum_i \left(\frac{p_i^2}{2M_i} + \frac{1}{2} q_i^2 K_i\right) \qquad (4)$$

corresponding to the time-dependent Slater determinants

$$\phi(t) = \exp i \sum_i (p_i Q_i - q_i P_i)|\phi_o\rangle = (1+i\sum_i (p_i Q_i - q_i P_i)+\ldots)|\phi_o\rangle \quad (5)$$

where the one-body operators Q_i, P_i are given by the transformation (2) in terms of the solutions of the RPA equations:

$$A_i^+ = \sum_{ph} (X^i_{ph} a^+_p a_h - Y^i_{ph} a^+_h a_p). \quad (6)$$

To simplify the notation the assumption is made that $q_i^M = 0$ for all i, otherwise q_i in eqs. (4), (5) should be replaced by $q_i - q_i^M$. Henceforth we can proceed in two ways: 1) by considering the time evolution of the classical variables q_i, p_i governed by Hamilton's equations. This is the semiclassical approximation in which the collective variables are supposed to evolve in time according to classical mechanics, with inertial and restoring force parameters determined microscopically in a self-consistent way. 2) By quantizing the collective variables.

1) Semiclassical approximation

Let us consider the time dependent one-body density associated to the Slater determinant in eq.(5). It can be obtained as the expectation value of the one-body density operator in $\phi(t)$ (eq.(5)), where $q_i(t)$, $p_i(t)$ are the classical solutions of the equations of motion $\dot{q}_i = \frac{\partial \mathcal{H}}{\partial p_i}$, $\dot{p}_i = -\frac{\partial \mathcal{H}}{\partial q_i}$, i.e.,

$$q_i(t) = q_i^o \cos(\omega_i t), \quad p_i(t) = -M_i \omega_i q_i^o \,\mathrm{sen}(\omega_i t) \quad (7)$$

In particular for q_i^o corresponding to the zero-point amplitudes, $\overline{q_i^2} = \frac{1}{2} q_i^{o2} = \frac{\omega_i}{2K_i}$, for all i. The (local) time-dependent Hartree-Fock density is given in coordinate space by

$$\rho(\vec{r},t) = \langle\phi(t)|\hat{\rho}(\vec{r})|\phi(t)\rangle = \rho(\vec{r}) + \rho^1(\vec{r},t) \quad (8)$$

$\rho^1(\vec{r},t)$ in eq. (8) is the oscillating part,

$$\rho^1(\vec{r},t) = \frac{1}{\sqrt{2}} \sum_i \{e^{-i\omega_i t} \sum_{ph} (X^i_{ph}\, \varphi^*_h(\vec{r})\, \varphi_p(\vec{r}) + Y^i_{ph}\, \varphi^*_p\, \varphi_h) + c.c.\} \quad (9)$$

and states simply [4] that the density oscillates with the normal frequencies ω_i of the system. The static density $\rho(\vec{r})$ around which those oscillations take place is the time average of $\rho(\vec{r},t)$ in eq. (8) and to lowest order it is just the static H-F density ($\rho_{H-F}(\vec{r})$). However, keeping to quadratic terms in q_i^o, as in the classical Hamiltonian, one instead obtains

$$\rho(\vec{r}) = \rho_{H-F}(\vec{r}) + \frac{1}{2} \sum_i \sum_{\substack{ph\\p'h'}} \rho^{ph}_{p'h'}(\vec{r}) \, (X^{i*}_{ph} X^{i}_{p'h'} + Y^{i}_{ph} Y^{i*}_{p'h'}) \tag{10}$$

with

$$\rho^{ph}_{p'h'}(\vec{r}) = \delta_{hh'} \, \varphi^*_p(\vec{r}) \, \varphi_{p'}(\vec{r}) - \delta_{p'p} \, \varphi^*_{h'}(\vec{r}) \, \varphi_h(\vec{r}) \tag{11}$$

This corresponds to an energy

$$\mathcal{H}(p_i,q_i) = E_o + \frac{1}{2} \sum_i \omega_i \tag{12}$$

greater than the H-F energy, $E_o = \langle\phi_o|H|\phi_o\rangle$, which suggests that $\rho(\vec{r})$ in eq. (10) does not correspond to the ground state density but rather to an average over real excitations.

2) Quantization

The problem of quantization has already been discussed by Goeke and Reinhard [3]. Here we shall address this problem from a different point of view. Let us discuss first for simplicity the case of just one collective coordinate (q); suppose we have solved the set of self-consistency conditions I-III and therefore we know the optimal structure of the determinantal wave functions $\phi(p,q)$ and classical collective Hamiltonian $\mathcal{H}(p,q)$. The quantization of $\mathcal{H}(p,q)$ would provide us with a collective basis $f_n(q)$ with corresponding eigenvalues E_n. Then the eigenstates ψ_n of the nuclear Hamiltonian H, for that particular mode, could be constructed in a generator coordinate method fashion from $f_n(q)$ and $\phi(p,q)$ [5]. However, one is not really interested in what the exact form of these eigenstates is, but rather in observable quantities as the matrix elements of given operators θ between those states. Such matrix elements would be given by [1,5]

$$\langle\psi_n|\theta|\psi_m\rangle = \int dq \; f^*_n(q) \; \Theta(p,q) \; f_m(q) \tag{13}$$

with $\Theta(p,q) \equiv \Theta(-i \frac{\partial}{\partial q}, q)$ being the effective collective operator equivalent to θ. Then the whole problem of quantization is actually reduced to finding those effective collective operators from $\phi(p,q)$ as

$$\Theta(p,q) = \langle\phi(p,q)|\boldsymbol{\theta}|\phi(p,q)\rangle \tag{14}$$

The handling of eq. (14) requires some care for $\Theta(p,q)$ must represent an operator in collective space, not a function of time!

Concentrating now on the case of small amplitude oscillations it is clear what we should do: change the classical variables $p_i(t)$, $q_i(t)$ in eq. (5) into the operators

$$q_i = \sqrt{\frac{1}{2M_i\omega_i}}\,(c_i+c_i^+)\ , \qquad p_i = \sqrt{\frac{M_i\omega_i}{2}}\, i\,(c_i^+-c_i) \tag{15}$$

where $c_i^+(c_i)$ are creation (annihilation) operators of one quantum of the mode i (satisfying boson conmutation relations $\{c_i,c_j^+\}_- = \delta_{ij}$) and act only on the collective basis:

$$|\{n_i\}> = \prod_i \frac{(c_i^+)^{n_i}}{\sqrt{n_i!}}\,|0> \tag{16}$$

Then the effective collective operators $\mathcal{O}(p_i,q_i) \equiv \hat{\mathcal{O}}(c_i^+,c_i)$ are given by

$$\hat{\mathcal{O}}(c_i^+,c_i) = <\phi_o|\theta + \{\theta,iT\}_- + \tfrac{1}{2}\{\{\theta,iT\}_-,\ iT\}_-|\phi_o> \tag{17a}$$

where the two-fold operator $T = \sum_i (p_iQ_i - q_iP_i)$ can be written using the transformations (2) and (15) as

$$T = -i\sum_i (A_i^+c_i - A_ic_i^+) \tag{17b}$$

with the operators A_i^+, A_i defined on the s.p. H-F basis as in eq. (6). The matrix elements of the operator θ between "eigenstates" $\psi_{\{n_i\}}\psi_{\{m_i\}}$ can then be calculated as (see eq. (13))

$$<\psi_{\{n_i\}}|\theta|\psi_{\{m_i\}}> = <\{n_i\}|\hat{\mathcal{O}}(c_i^+,c_i)|\{m_i\}> \tag{18}$$

from the explicit expression of $\hat{\mathcal{O}}(c_i^+,c_i)$ that can be found in ref.|6|. As is shown in that reference the matrix elements in eq. (18) are identical to those obtained with the generator coordinate method. It has been pointed out|5| that after quantization ATDHF provides a description of excited states and energy levels analogous to that provided by the GCM, but that the latter method, being fully quantum mechanical, should be used to identify the zero point energy correction not included in the former. In what follows we shall show that with the quantization procedure presented here the effective collective operators in eq. (17) already contain the zero point corrections prescribed by the generator coordinate method. Let us first recall what the GCM results are for the correlated ground state energy ($E_{g.s.}$) and density ($\rho_{g.s.}$). To second order in the small amplitudes one gets in the usual quasi-boson approximation|7|

$$E_{g.s.} = E_o - \sum_i \omega_i \left(\sum_{ph}|Y^i_{ph}|^2\right) \tag{19}$$

$$\rho_{g.s.}(\vec{r}) = \rho_{HF}(\vec{r}) + \sum_i \sum_{\substack{ph\\p'h'}} Y^i_{ph}\, Y^{i*}_{p'h'}\, \rho^{ph}_{p'h'}(\vec{r}) \tag{20}$$

with $\rho^{ph}_{p'h'}(\vec{r})$ as defined in eq. (11). This correction to the H-F density due to RPA correlations, is equivalent to the sum of the graphs

for each vibrational mode in Nuclear Field Theory [8]. It is easily seen that these results are identical to the ones that follow from ATDHF using our quantization procedure. Using equations (17), together with the self-consistent conditions, the effective collective Hamiltonian and density operator in collective space are found to be:

$$\hat{\mathcal{H}}(c_i^+,c_i) = [E_0 - \sum_i \omega_i (\sum_{ph} |Y^i_{ph}|^2)] + \sum_i \omega_i \, c_i^+ c_i \tag{21}$$

$$\hat{\rho}_{(\vec{r})}(c_i^+,c_i) = [\rho_{HF}(\vec{r}) + \sum_i \sum_{\substack{ph \\ p'h'}} Y^i_{ph} \, Y^{i*}_{p'h'} \, \rho^{ph}_{p'h'}(\vec{r})]$$

$$+ \sum_i \{\sum_{ph} (X^i_{ph} \, \varphi^*_h(\vec{r}) \, \varphi_p(\vec{r}) + Y^i_{ph} \, \varphi^*_p(\vec{r}) \, \varphi_h(\vec{r}))\, c_i + h.c.\}$$

$$+ \sum_{ij} \{\sum_{\substack{ph \\ p'h'}} (X^{i*}_{ph} \, X^j_{p'h'} + Y^{i*}_{p'h'} \, Y^j_{ph}) \, \rho^{ph}_{p'h'}(\vec{r}) \, c_i^+ c_j\} + \ldots \tag{22}$$

The expectation values of these operators in the ground state $\psi_0 = \psi_{\{n_i=0\}}$ are then the terms within square brackets in eqs. (21) and (22), and are identical to eqs. (19) and (20), respectively.

It is now clear what the limitations of the semiclassical approximation discussed before are. Comparing eqs. (10) and (20) one sees that the semiclassical approximation is only valid (as it should be) for highly collective modes for which ω_i is much smaller than the average one particle - hole excitation energies, for only then $|X^i_{ph}| \simeq |Y^i_{ph}|$.

To summarize, we have shown that after a proper quantization ATDHF provides a description of vibrational modes that goes beyond the semiclassical description furnished by TDHF. This description is analogous in all respects to that provided by the GCM, with the advantage that it is simpler to handle and that having a parallel semiclassical interpretation is more appealing to intuition. This aspect may be very valuable in addressing more complex collective nuclear phenomena.

II. The Fission Path

The adiabatic limit of time-dependent Hartree-Fock has been mainly used to provide a basis for the description of large amplitude slow collective modes, under the assumption that they may be treated as completely decoupled (see refs. |1a| - |3|). In order to be able to describe phenomena such as friction or damping, the coupling between large amplitude modes and intrinsic excitations must be taken into account. In ref. |1b| a set of selfconsistency conditions was derived from conditions I-III in part I to determine as well the coupling between the "fission" mode and the intrinsic excitations (see eqs. IA - VIA in ref. |1b|). The simplifying assumption was made that a single collective coordinate is sufficient to describe the deformation of the nucleus in the fission (or fusion) mode, and the intrinsic excitations were described in the form of RPA modes at any given deformation. Calling q_1 the fission mode coordinate (defined as the line of minimal potential energy), q_K (with $K \geq 2$) the vibrational coordinates, and p_1, p_K their canonical conjugate momenta, the classical Hamiltonian was found to be |1b|:

$$\mathcal{H}(p_i,q_i) = \mathcal{H}_a(p_1,q_1) + \frac{1}{2} \sum_K \left(\frac{p_K^2}{M_K^o} + K_K\, q_K^2\right) + \frac{1}{2} p_1^2 \sum_K \frac{q_K}{M_{11}^{(K)}} +$$

$$+ p_1 \sum_{K,K'} \frac{q_K p_{K'}}{M_{K'1}^{(K)}} \tag{1}$$

where $\mathcal{H}_a(p_1,q_1) = \dfrac{p_1^2}{2M_1^o(q_1)} + V(q_1)$ represents the pure adiabatic motion in the fission direction and is identical to that in ref. |1a|. All the mass and restoring force parameters in eq. (1) are functions of q_1 and are to be determined together with the Slater determinants $\phi(q_1)$ and the 1 p-h operators Q_i^o , P_i^o in a selfconsistent way as prescribed in ref. |1b|. It is seen from eq. (1) that to lowest order in the small amplitudes and velocities the motion in the fission direction and the bound motion in the perpendicular surfaces, q_1 = cte, are decoupled except for the dependence of the vibrational frequencies

$\omega_K = \sqrt{\dfrac{K_K}{M_K^o}}$, on q_1. The explicit coupling comes from the dependence of the kinetic energy on the vibrational amplitudes. The Hamiltonian (1) corresponds to the time-dependent Slater determinant |1b|

$$\phi(t) = (1 - i \sum_{K\geq 2} q_K\, P_K^o + i \sum_{i\geq 1} p_i\, \tilde{Q}_i + \ldots)\, |\phi(q_i)\rangle \tag{2}$$

with $\tilde{Q}_i = Q_i^o + Q_1^o \sum_{K\geq 2} q_K \gamma_K^i$, where the q_1-dependent quantities γ_K^i are given by $\gamma_K^i = <\{Q_i^o, i \frac{\partial P_K^o}{\partial q_1}\}>$ and are crucial in defining the explicit coupling (the notation <A> is used for $<\phi(q_1)|A|\phi(q_1)>$).

The quantization procedure discussed above can now be applied to find the effective Hamiltonian at every point p_1, q_1 in the collective path ($\hat{\mathcal{H}}(c_K^+, c_K; p_1 q_1)$ by quantizing the variables of the bound motion in eq. (2). Using the tranformation (15) in part I for the normal modes $K(K\geq 2)$, we find

$$\hat{\mathcal{H}}(c_K^+, c_K; p_1, q_1) = \mathcal{H}'(p_1, q_1) + \sum_K \omega_K c_K^+ c_K + \frac{1}{2} \sum_K F_K(p_1, q_1)(c_K^+ + c_K) +$$

$$+ \frac{1}{2} \sum_{K,K'} G_{K,K'}(p_1, q_1)\, i(c_{K'}^+ c_K^+ - c_{K'} c_K + c_{K'}^+ c_K - c_K^+ c_{K'}) \quad (3)$$

where $\mathcal{H}'(p_1, q_1)$ can be interpreted as the classical Hamiltonian for the "frozen" motion in the fission direction, i.e., for the path along which the vibrational modes remain in their ground state. It is given by:

$$\mathcal{H}'(p_1, q_1) = V(q_1) + \frac{1}{2} \sum_K (\omega_K - \frac{<(P_K^o)^2>}{M_K^o} - K_K <(Q_K^o)^2> - K_{1K} <Q_K^o Q_1^o>) +$$

$$+ \frac{p_1^2}{2M_1^o} - \frac{p_1}{M_1^o} (\sum_K <\frac{\partial P_K^o}{\partial q_1} Q_K^o>) \quad (4)$$

and only differs from $\mathcal{H}_a(p_1, q_1)$ in the last term (which is linear in the collective velocity) and in the renormalization of the potential energy due to the zero point motion of the vibrational modes. Note that since $<\frac{\partial P_K^o}{\partial q_1} Q_K^o>$ must be imaginary, $\mathcal{H}'(p_1, q_1)$ is time-reversal invariant and therefore, although the equation of motion $\dot{p}_1 = -\frac{\partial \mathcal{H}'}{\partial q_1}$ contains a term linear in p_1, this term is imaginary and has, in principle, nothing to do with the usual viscosity term introduced by means of the Rayleigh dissipation function [9]. Analogous remarks stand for the whole effective Hamiltonian in eq. (3), where the coefficients in the coupling terms are given by:

$$F_K(p_1, q_1) = \frac{1}{\sqrt{2M_K^o \omega_K}} (\frac{p_1^2}{M_{11}^{(K)}} - \sum_{\ell\geq 2} \gamma_K^\ell <\frac{P_1^o P_\ell^o}{M_1^o} + K_\ell Q_\ell^o Q_1^o + K_{1\ell}(Q_1^o)^2>) \quad (5)$$

$$G_{K,K'}(p_1,q_1) = \frac{p_1}{M^{(K)}_{K'1}} \sqrt{\frac{M^{o}_{K'}\,\omega_{K'}}{M^{o}_{K}\,\omega_{K}}} \qquad (6)$$

showing its manifest time-reversal invariance. A complete quantum mechanical description of the collective (large amplitude) motion coupled to the intrinsic excitations can be finally obtained by quantizing the fission mode variables. However this is only necessary to describe purely quantum mechanical phenomena such as tunneling [10] and one would like to attach to a semiclassical description of the large amplitude mode in the classically accesible regions. For $M^{(K)}_{K'1} = 0$ the Hamiltonian (3) has the usual form [11] of that for a system consisting of a subsystem coupled to a thermal reservoir and the method of ensemble averages can be used to derive time-irreversible equations from the time-reversal ones. This Hamiltonian could then be taken as the starting point for models of damped collective motion based on linear response theory [12]. The fact that in our approach the inertia of the fission mode and the frequencies of the bound motion (as well as the coupling) are determined in a selfconsistent way at every point in the path would also allow to elucidate the regions at which the intrinsic excitations can be treated globally as a heat bath, as well as the modes that must be excluded from that treatment (and considered as collective) in a given region [13]. Another model with which one could establish connection is that of Broglia, Dasso and Winther for heavy ion collisions [14], where the relative motion is described in terms of classical equations and the nuclear degrees of freedom are described in terms of low-lying surface vibrations and damped giant resonances.

References

1a F. Villars, Nucl. Phys. A285 (1977) 269; 1b E. Moya de Guerra and F. Villars, Ibid. pg. 297.

2 M. Baranger and M. Veneroni, Ann. Phys. (N.Y.) 114 (1978) 123.

3 K. Goeke and P.G. Reinhard, Ann. Phys. (N.Y.) 112 (1978) 328.

4 D.J. Thouless, in "The Quantum Theory of Many-Body Systems" (Academic Press, 1961).

5 F. Villars, in "Nuclear Self-Consistent Fields" (North-Holland, 1975) pg. 3; M. Baranger, Ibid., pg. 19.

6 E. Moya de Guerra, Phys. Rev. Lett, to be published.

7 G. Ripka and J.P. Blaizot, in "Course de Physique Nucléaire Théorique", Vol. 1, CEA-N-2019 (1978); D.M. Brink and W. Weiguny, Phys. Lett. 26B (1968) 497; E. Moya de Guerra, An. Fís., to be published.

8 P.F. Bortignon, R. Broglia, D.R. Bes and R. Liotta, Phys. Rep. 30 (1977) 305; A. Winther, private communication.

9 K.T.R. Davies, A.J. Sierk, J.R. Nix, Phys. Rev. C13 (1976) 2385.

10 H. Hofmann, Nucl. Phys. A224 (1974) 116.

11 R.F. Fox, Phys. Rep. 48 (1978) 179.

12 H. Hofmann and P.J. Siemens, Nucl. Phys. A257 (1976) 165; Proc. of the IVth. Balaton Conf. on Nucl. Phys., Keszthely (Hungary 1979).

13 H.A. Weidenmuller, Progr. in Part. and Nucl. Phys. 3 (1980) 49.

14 R.A. Broglia, C.H. Dasso and A. Winther, Phys. Lett. 53B (1974) 301; 61B (1976) 113.

GENERALIZED DENSITY DYNAMICS

D. J. Rowe
Department of Physics, University of Toronto
Toronto, Ontario, M5S 1A7, Canada

THE HARTREE-FOCK DENSITY DYNAMICS

The HF approximation, in one form or another underlies most of many-body quantum mechanics. One reason for this is that the only many-body problem that can be solved, generally speaking, is the independent-particle problem. Another is that HF theory is a theory of Slater determinants, which represent particle-hole vacuum states and are therefore particularly simple to work with. In this talk, I shall consider HF as a dynamical theory of the one-body density.

A one-body density ρ contains all the one-body properties of the state ψ to which it corresponds. Thus if we define its components

$$\rho_{\beta\alpha} = \langle\psi|\ a^+_\alpha a_\beta\ |\psi\rangle, \tag{1}$$

then the expectation of any one-body operator X is given by

$$\langle\psi|\ X\ |\psi\rangle = \rho(X) = \sum_{\alpha\beta} X_{\alpha\beta}\,\rho_{\beta\alpha}\ . \tag{2}$$

Since most physical observables of interest (with the exception of the Hamiltonian) are one-body operators, the one-body density is a particularly useful entity.

Given a Slater determinant ψ, with density ρ, HF theory gives the dynamical equation

$$\frac{d}{dt}\langle\psi|X|\psi\rangle = i\langle\psi|[H_0,X]|\psi\rangle\ , \tag{3}$$

where $H_0=H_0(\rho)$ is the HF density dependent Hamiltonian. This equation has two very important features. Firstly, since $[H_0,X]$ is a one-body operator, for all one-body X, it follows that eq. (3) can be written

$$\dot\rho(X) = i\rho([H_0,\ X]), \tag{4}$$

which demonstrates that HF theory is a closed one-body density dynamics; i.e. one needs to know only the one-body density ρ to evaluate the r.h.s.

of eq. (4). A closed dynamical equation for the one-body density of the type (3) is sometimes called a kinetic equation.[1] Secondly, the HF Hamiltonian satisfies the equation

$$\langle\psi|[H_0, X]|\psi\rangle = \langle\psi|[H, X]|\psi\rangle , \qquad (5)$$

where H is the unconstrained many-body Hamiltonian. As a consequence, it follows that $\dot{\rho}(X)$ takes precisely the same value at each ρ in the constrained HF dynamics as it would in the full unconstrained quantum mechanics.

To put the HF density dynamics into perspective, it is profitable to consider the following three questions: (i) What is special about the set of one-body operators and hence about the one-body densities? (ii) Are there other subsets of observables of interest with corresponding densities? (iii) Can we construct for them a closed density dynamics?

One soon recognizes that what is special about the one-body operators is that they span a closed lie algebra; i.e. the commutator of two one-body operators is again a one-body operator. But the generators of any lie group of transformations of the many-particle Hilbert space span a closed lie-algebra. In particular, the generators of collective motions, e.g. su(3), sp(3,R), ..., are collective observables of considerable interest. We therefore consider how to define the densities and construct a density dynamics for any such Lie algebra of observables. Fuller details can be found in ref. 2, 3.

GENERALIZED DENSITIES

Let G be a lie group of unitary transformations of the many-particle Hilbert space $\mathbb{H}$. We denote the action of a group element $g \in G$ on a state $\psi \in \mathbb{H}$ simply by $g\psi$.

Let $\mathfrak{g}$ denote the lie algebra of G. The structure of $\mathfrak{g}$ is given in terms of a basis $\{E_\alpha\}$ by the commutation relations

$$[E_\alpha, E_\beta] = C^{\gamma}_{\alpha\beta} E_\gamma , \qquad (6)$$

where $\{C^{\gamma}_{\alpha\beta}\}$ are the structure constants. An arbitrary element $X \in \mathfrak{g}$ is a vector with components (X^α), defined in terms of the basis by the expansion

$$X = X^\alpha E_\alpha . \qquad (7)$$

To any state $\psi \in \mathbb{H}$ there now corresponds a density ρ defined by

$$\rho(X) = -i \langle\psi|\ X\ |\psi\rangle \qquad (8)$$

for all $X \in \mathbf{g}$. Note that the factor (-i), not present in eq. (2), is included to make $\rho(X)$ real and is necessitated by the fact that I am using the mathematicians convention of choosing X to be anti-Hermitian rather than Hermitian. If one prefers, one can replace X by $Y = -iX$ and regard the Hermitian operator Y as the physical observable.

Inserting the expansion (7) into (8) gives

$$\rho(X) = X^{\alpha}\ \rho_{\alpha}\ , \qquad (9)$$

where $\rho_{\alpha} = \rho(E_{\alpha})$, and reveals that ρ is a covector with components (ρ_{α}). Evidently ρ is a vector in $\mathbf{g}^{*}$, the dual space of the lie algebra $\mathbf{g}$.

Note that the definition (8) can be extended to a mixed (e.g. thermal) state in the standard way by

$$\rho(X) = -i \sum_{\lambda} C_{\lambda}\ \langle\psi_{\lambda}|\ X\ |\psi_{\lambda}\rangle\ , \qquad (10)$$

where (C_{λ}) are the (positive) weighting factors for the incoherent components of the mixed state.

GROUP ORBITS AND COADJOINT ORBITS

The set of Slater determinants is an orbit of the group of one-body unitary transformations. An orbit of a group G is a set of states

$$M = \{\psi(g) = g\psi\ ;\ g \in G\}, \qquad (11)$$

where ψ is any given fixed state in $\mathbb{H}$. Clearly G has many possible distinct orbits in $\mathbb{H}$ depending on the choice of representative state ψ. To each such orbit there corresponds a coadjoint orbit of densities.

$$N = \{\rho_{g}\ ;\ g \in G\}, \qquad (12)$$

where ρ_{g} is the density corresponding to $\psi(g)$. It follows from the definition (8) that

$$\rho_{g}(X) = \rho(g^{-1}\ X\ g). \qquad (13)$$

For densities of mixed states, one observes that eq. (13) still applies.

DENSITY DYNAMICS

We seek a dynamical equation that can be integrated to give paths $\rho(t)$ in g^*. Ideally, we would like paths that correspond precisely to solutions $\psi(t)$ of the full time-dependent Schrodinger equation.

The Schrodinger equation gives

$$\frac{d}{dt} \langle\psi| X |\psi\rangle = i\langle\psi|[H, X]|\psi\rangle, \tag{14}$$

where H is the Hamiltonian. Now if iH were an element of the Lie algebra, we could immediately set

$$\dot{\rho}(X) = \rho([iH, X]) \tag{15}$$

to obtain a density dynamics in accord with eq. (14). However, this is a trivial case which parallels the HF equation for a one-body Hamiltonian.

The interesting situation is when neither iH nor [iH, X] are in the lie algebra. The r.h.s. of eq. (15) is then not defined because [iH, X] is outside the domain of the density.

In HF theory the problem is solved by finding a one-body HF Hamiltonian H_0 that satisfies the equation

$$\langle\psi|[H_0, X]|\psi\rangle = \langle\psi|[H, X]|\psi\rangle, \tag{16}$$

for all one-body X, whenever ψ is a Slater determinant. We therefore enquire if there might similarly exist some $iH_0 \in g$ at each ψ on a G-orbit M that satisfies eq. (16) for all $X \in g$. For, if so, we can set

$$\dot{\rho}(X) = \rho([iH_0, X]), \tag{17}$$

in parallel with the HF eq. (4), and again obtain a density dynamics in accord with eq. (14). The answer is that there does exist such a Hamiltonian $H_0(\rho)$ if and only if the energy

$$\mathcal{H} = \langle\psi|H|\psi\rangle \tag{18}$$

is a single-valued function $\mathcal{H}(\rho)$ of the density for all $\psi \in M$.[3]

Note that the map $M \to N;\ \psi \to \rho$ is, in general, many-to-one in which case there are many energies to each ρ. In the special case that M and N are isomorphic, one can already construct a dynamics on M as was done in ref. 4. However, if the map $M \to N$ is many-to-one, it can be shown[3] that the dynamics on M is not well-defined (due to the fact that M is not a phase space). A very special situation arises when the energy is nevertheless a single-valued function of the density due to a symmetry leading to an invariance of the Hamiltonian. This is a most interesting situation considered by Dirac[5] in other language. However, it is straightforward and so we shall not focus specifically on it here. Instead we consider the general situation where $H(\rho)$ is not a priori well-defined by eq. (18). Eq. (18) is then replaced by

$$\mathcal{H}(\rho) = \langle\langle\psi|H|\psi\rangle\rangle \ , \tag{19}$$

by which we mean an average over all states $\psi \in M$ having the same density ρ.

We can now define $H_o(\rho)$ by

$$\langle\langle\psi|[H_o, X]|\psi\rangle\rangle = \langle\langle\psi|[H, X]|\psi\rangle\rangle \ , \tag{20}$$

which is easily shown to have a solution[3]. Thus we can again use eq. (17) to define a density dynamics.

The implication of this result is that the average dynamics for a set of states of density ρ is well defined whereas it was not defined for the individual members of the set. This is perhaps not surprising since, to distinguish members of the set, we need additional densities beyond those in $\mathfrak{g}^*$. An average dynamics is nevertheless useful because, in many practical situations, one may expect the time-derivatives of densities of states of the same density to differ only slightly. One might also contemplate treating first the average density dynamics with the intention of subsequently developing a theory for the fluctuations about the average.

Finally, we show that the above density dynamics is a Hamiltonian dynamics. Let $(\bar{E}_\nu)$ be a maximal subset of linearly independent vectors in $\mathfrak{g}$ having the property that

$$\rho([X,\bar{E}_\nu]) \neq 0 \qquad \text{for some } X \in \mathfrak{g}. \tag{21}$$

Substituting $g=\exp(x^\nu\bar{E}_\nu)$ in the expression (13) for ρ_g and differentiating gives

$$\frac{\partial\rho(\underline{X})}{\partial x^\nu} = \rho([X, \bar{E}_\nu]) \tag{22}$$

revealing that the vectors $(\bar{E}_\nu)$ are a basis of generators of infinitesimal displacement on N. It also follows from eq. (19) that

$$\frac{\partial H}{\partial x^\nu} = \langle\langle\psi|[H, \bar{E}_\nu]|\psi\rangle\rangle \tag{23}$$

Hence, by eq. (20), the equation of motion (17) gives

$$\dot{\rho}_\nu = \frac{\partial H}{\partial x^\nu} \tag{24}$$

or, in coordinate independent language,

$$\dot{\rho} = dH. \tag{25}$$

Now, for any smooth function $F(\rho)$ of the density,

$$\dot{F} = \frac{\partial F}{\partial \rho_\nu}\dot{\rho}_\nu = \frac{\partial F}{\partial \rho_\nu}\frac{\partial H}{\partial x^\nu}. \tag{26}$$

To evaluate $\partial F/\partial\rho_\nu$, observe that

$$\frac{\partial F}{\partial x^\nu} = \frac{\partial F}{\partial \rho_\mu}\sigma_{\mu\nu}, \tag{27}$$

where

$$\sigma_{\mu\nu} = \frac{\partial \rho_\mu}{\partial x^\nu} = \rho([\bar{E}_\mu, \bar{E}_\nu]). \tag{28}$$

It follows that

$$\frac{\partial F}{\partial \rho_\mu} = \frac{\partial F}{\partial x^\nu}\sigma^{\nu\mu}, \tag{29}$$

where $(\sigma^{\mu\nu})$ is the inverse of the matrix $(\sigma_{\mu\nu})$, which by eq. (21) must exist. Thus eq. (26) becomes a Hamilton's equation

$$\dot{F} = \{H, F\} \tag{30}$$

with Poisson bracket

$$\{H,F\} = \sigma^{\nu\mu} \frac{\partial H}{\partial x^\mu} \frac{\partial F}{\partial x^\nu} \quad . \tag{31}$$

APPLICATION

If one simply wanted the ground state of a one-body Hamiltonian H_0, it would make sense to seek it in the space of independent-particle states (Slater determinants). However, for the time-dependent problem, one may also be interested in the time evolution $\psi(t)$ of an initial state ψ that is not a Slater determinant. Since

$$\psi(t) = e^{-iH_0 t} \psi \tag{32}$$

and $\exp(-iH_0 t)$ is a one-body unitary transformation, it follows that the path $\psi(t)$ lies on an orbit of the one-body unitary group. This suggests that it might be useful, even when the Hamiltonian H is not one-body, to consider an approximate dynamics constrained to an orbit of the one-body unitary group. For example, it might be appropriate for describing the scattering of two heavy-ions like $^{12}C + ^{12}C$ with initial state given by the product of two shell-model ground states. This seems like a natural way of lifting the sometimes unrealistic TDHF restriction to Slater determinants.

However, as we have shown elsewhere[4], the dynamics on an orbit of wave functions is only defined when the orbit is a phase space which is an exceptional case. It is in fact the situation when the orbit of wave functions is isomorphic to the corresponding coadjoint orbit of densities[3]. On the other hand, coadjoint orbits are always phase spaces[3] and, as we have shown here, always support a well-defined Hamiltonian density dynamics.

Densities have another advantage over wave functions in that they can represent mixed states as well as pure states. Mixed states might correspond to thermal distributions or simply to one's incomplete knowledge of the system. One possibility is to let a density correspond to a whole shell-model space of states. Consider, for example, a shell model space corresponding to a partition of the single-particle Hilbert space into subspaces of occupied (hole) states, valence states and unoccupied (particle) states. The conventional procedure for partitioning the space is initiated by constructing an energy ordered basis of single-particle eigenstates of some physically reasonable shell-model Hamiltonian. But what is the optimal partition?

To answer this question, consider the one-body densities, corresponding to a particular shell-model partition, as a point on a manifold.

Since any other partition is obtained by some unitary transformation of the single-particle basis, it follows that the manifold in question is an orbit of the one-body unitary group. The optimal shell-model space can therefore be defined as the stationary points on this manifold. This gives the equation

$$\langle\langle\psi|\,[H,\, X]\,|\psi\rangle\rangle = 0 \tag{33}$$

for all one-body X, where the average is over all shell-model states. Eq. (33) obviously has a close resemblance to the HF equation and can likewise be solved by self-consistent field methods. A solution is obtained when the following off-diagonal single-particle matrix elements all vanish:

$$\begin{aligned}
\varepsilon_{hp} &= T_{hp} + \sum_{h'} V_{hh'ph'} + \frac{n}{N}\sum_{v} V_{hvpv} \\
\varepsilon_{hv} &= T_{hv} + \sum_{h'} V_{hh'vh'} + \frac{n}{N-1}\sum_{v'} V_{hv'vv'} \\
\varepsilon_{vp} &= T_{vp} + \sum_{h'} V_{vh'ph'} + \frac{n-1}{N-1}\sum_{v'} V_{vv'pv'}
\end{aligned} \tag{34}$$

where h,h' label hole states, v,v' valence states, and p,p' label particle states. n is the number of particles in the valence space and N is its dimensionality. T is the kinetic energy and V is the two body interaction.

One notes that in a mean field, rather than the above variational, theory the factors in front of the 3rd term would all be n/N. The differences can be interpreted as blocking effects.

REFERENCES

1. E.A. Remler, to be published.
2. G. Rosensteel and D.J. Rowe, Phys. Rev. A24 (1981) 673.
3. D.J. Rowe, M. Vassanji and D.J. Rowe, to be published.
4. D.J. Rowe and R. Basserman, Can. J. Phys. 54 (1976) 1941; A.K. Kerman and S.E. Koonin, Ann. Phys. (N.Y.) 100 (1976) 332; D.J. Rowe, A. Ryman and G. Rosensteel, Phys. Rev. A22 (1980) 2362; P. Kramer and M. Saraceno, Lecture Notes in Physics 140 (Springer-Verlag, 1981)
5. P.A.M. Dirac, Can. J. Math. 2 (1950) 129.

On the Information Content of the One-Body Density

H. Feldmeier*
Max-Planck-Institut für Kernphysik, Heidelberg, F R Germany

1. Introduction

Before two heavy ions collide they are assumed to be in their ground state and the relative motion can be regarded as a wave packet. Thus the whole system is prepared in a pure state. After the collision the system is highly excited and the measured data can be described best by a mixed state (statistical operator). During the reaction the system is closed and there is no external heat bath which could absorb information and give rise to a mixed state. It is rather the incomplete measurement which implies a loss of information and hence entropy production.

In this work we assume that the observer can measure the expectation values of all one-body operators. This is certainly much more information than can be achieved in heavy ion collisions. It is however just the information which equations of motion for the one-body density matrices are dealing with. We shall first construct the statistical operator of maximum entropy under the constraint of a given one-body density. From this we deduce the entropy and the two-body density matrix. We distinguish between the "grand canonical" ensemble and the one where the total number of particles is known. The two-body density of maximum entropy is applied to a model case and to heavy ions in thermal equilibrium.

2. The concept of maximum entropy

A formal way to reduce the information is the concept of maximum entropy [Ka67]. Assume the information I about the systems to be contained in a set of operators B_α and their expectation values.

$$I = \left\{ B_\alpha , \langle B_\alpha \rangle ; \alpha = 1, \ldots, N \right\} \tag{1}$$

Then the statistical operator R given in eq. 2 is the one of maximum entropy among all others which yield a given set of expectation values $\langle B_\alpha \rangle$.

$$R = e^{-\sum_{\alpha=1}^{N} \lambda_\alpha B_\alpha} \tag{2}$$

$$\langle B_\alpha \rangle = Tr\,(R B_\alpha) = \langle \Psi | B_\alpha | \Psi \rangle \tag{3}$$

The Lagrange parameters $\lambda_\alpha(t)$ have to be determined such that eq. 3 is fulfilled for $\alpha = 1, \ldots, N$. It is furthermore assumed that the expectation values are the same as for the exact state $|\Psi\rangle$ (eq. 3). The entropy S belonging to R is given by

$$S = -Tr\,(R \ln R) = \sum_{\alpha=1}^{N} \lambda_\alpha \langle B_\alpha \rangle \tag{4}$$

*Heisenberg Fellow.

The statistical operator R can also be regarded as the result of a projection of the pure state $|\Psi\rangle\langle\Psi|$ [Ro66]

$$|\Psi\rangle\langle\Psi| \xrightarrow{\;\mathcal{P}\;} R \tag{5}$$

One can of course calculate any other expectation value $\langle x\rangle = \mathrm{Tr}(Rx)$ with this statistical operator. If it turns out that $\langle x\rangle \approx \langle\Psi|\, x\, |\Psi\rangle$ it is said that $\{x,\langle x\rangle\}$ does not contain new information which is not already in the set I. If $\langle x\rangle$ differs significantly from the exact result (or measurement) $\{x,\langle x\rangle\}$ is informative and has to be included in I.

We want to follow these ideas by assuming the one-body information to be essential and construct at each time t R(t), determine the entropy S(t), and the two-body density [Fe81].

For a given single-particle basis

$$|\widetilde{\alpha,1}\rangle := \tilde{a}^{+}_{\alpha}\,|0\rangle \tag{6}$$

the one-body density matrix is defined as

$$\tilde{\rho}^{(1)}_{\alpha\beta}(t) = \langle\Psi(t)|\,\tilde{a}^{+}_{\beta}\,\tilde{a}_{\alpha}\,|\Psi(t)\rangle \tag{7}$$

If we suppose the system to have a definite particle number A, the information set is

$$I = \left\{ P\,\tilde{a}^{+}_{\beta}\,\tilde{a}_{\alpha}\,P\,,\ \tilde{\rho}^{(1)}_{\alpha\beta}(t)\ ;\ \alpha,\beta = 1,\dots,N \right\}, \tag{8}$$

where P projects on the subspace with particle number A.

$$P = \sum_{s} |\widetilde{s,A}\rangle\langle\widetilde{s,A}| \tag{9}$$

$|\widetilde{s,A}\rangle$ are Slater determinants created with A operators $\tilde{a}^{+}_{\alpha}$. The statistical operator of maximum entropy is herewith given by

$$R(t) = \sum_{ss'} |\widetilde{s,A}\rangle\langle\widetilde{s,A}|\, e^{-\sum\limits_{\alpha\beta}\lambda_{\alpha\beta}(t)\,\tilde{a}^{+}_{\beta}\,\tilde{a}_{\alpha}}\, |\widetilde{s',A}\rangle\langle\widetilde{s',A}| \tag{10}$$

It is more convenient to transform $\tilde{\rho}^{(1)}_{\alpha\beta}(t)$ into its eigenbasis

$$\rho^{(1)} = \sum_{\alpha\beta} |\widetilde{\alpha,1}\rangle\,\tilde{\rho}^{(1)}_{\alpha\beta}(t)\,\langle\widetilde{\beta,1}| = \sum_{\alpha} |\alpha,1\rangle_t\,\rho^{(1)}_{\alpha\alpha}(t)\,{}_t\langle\alpha,1| \tag{11}$$

and to represent R(t) with Slater determinants $|s,A\rangle_t$ built out of the single-particle states $|\alpha,1\rangle_t$ which depend on time however. In the following we omit the time argument again. With

$$a^{+}_{\alpha}\,|0\rangle = |\alpha,1\rangle \tag{12}$$

and $$B_\alpha = a^+_\alpha a_\alpha \tag{13}$$

we get $$R = \sum_s |s,A\rangle \, e^{-\sum_\alpha \lambda_\alpha b^{(s)}_\alpha} \langle s,A| . \tag{14}$$

$b^{(s)}_\alpha$ are the occupation numbers in $\langle s,A|$:

$$B_\alpha |s,A\rangle = b^{(s)}_\alpha |s,A\rangle \tag{15}$$

If we drop the condition of sharp particle number we consider the information I_{GC} equivalent to the grand canonical ensemble:

$$I_{GC} = \left\{ \tilde{a}^+_\alpha \tilde{a}_\beta \, , \, \tilde{\rho}^{(1)}_{\alpha\beta} \, ; \, \alpha,\beta = 1,\ldots,N \right\} \tag{16}$$

The statistical operator of maximum entropy becomes [Wi63]

$$R_{GC} = \frac{1}{Z_{GC}} e^{-\sum_{\alpha\beta} \chi_{\alpha\beta} \tilde{a}^+_\beta \tilde{a}_\alpha} = \frac{1}{Z_{GC}} e^{-\sum_\alpha \lambda_\alpha a^+_\alpha a_\alpha} \tag{17}$$

Especially the partition sum Z_{GC} assumes a simple form

$$Z_{GC} = \sum_{s,A} e^{-\sum_\alpha \lambda_\alpha b^{(s)}_\alpha} = \prod_\alpha \left(1 + e^{-\lambda_\alpha}\right) \tag{18}$$

Also the relation between the Lagrange parameter and the mean occupation numbers can be given analytically

$$\lambda_\alpha = \ln \frac{1 - \rho^{(1)}_{\alpha\alpha}}{\rho^{(1)}_{\alpha\alpha}} . \tag{19}$$

Using eq. 4 for the entropy we get

$$S_{GC} = - \sum_\alpha \left\{ \rho^{(1)}_{\alpha\alpha} \ln \rho^{(1)}_{\alpha\alpha} + \left(1-\rho^{(1)}_{\alpha\alpha}\right) \ln\left(1-\rho^{(1)}_{\alpha\alpha}\right) \right\} \tag{20}$$

Eq. 20 is exact and one does not need Stirling's formula to derive it, as done in [LF75].

3. Maximum entropy two-body density matrix

Utilizing the statistical operator given in eq. 14 or eq. 17 we can calculate the two-body density matrix as

$$\rho^{(2)}_{\alpha\beta\gamma\delta} = \mathrm{Tr}\left(R a^+_\delta a^+_\gamma a_\alpha a_\beta \right) \tag{21}$$

Choosing the representation in which $\rho^{(1)}$ is diagonal we get for the sharp particle number case

$$\rho^{(2)}_{\alpha\beta\gamma\delta} = \left(\delta_{\alpha\gamma} \delta_{\beta\delta} - \delta_{\alpha\delta} \delta_{\beta\gamma} \right) \sum_s b^{(s)}_\alpha b^{(s)}_\beta \, e^{-\sum_\varepsilon \lambda_\varepsilon b^{(s)}_\varepsilon} \tag{22}$$

The sum over s has to be performed for all possible Slater determinants with A particles. If we are dealing with N single-particle levels the sum has $\binom{N}{A}$ terms. For large A and N the summation becomes rather expensive. Therefore we propose in the following an approximation to eq. 22 relating $\rho^{(2)}$ directly to $\rho^{(1)}$. It turned out to work very well in cases where the sum (22) could be evaluated. The partition sum is given by

$$Z = \sum_s e^{-\sum_\alpha \lambda_\alpha b_\alpha^{(s)}} \tag{23}$$

The one-body density can be regained from Z by

$$\rho_{\alpha\alpha}^{(1)} = -\frac{1}{Z}\frac{\partial Z}{\partial \lambda_\alpha} \tag{24}$$

Similarly the two-body density is given by

$$\rho_{\alpha\beta\alpha\beta}^{(2)} = \frac{1}{Z}\frac{\partial^2 Z}{\partial\lambda_\alpha \partial\lambda_\beta}(1-\delta_{\alpha\beta}) = \left\{\rho_{\alpha\alpha}^{(1)}\rho_{\beta\beta}^{(1)} - \frac{\partial}{\partial\lambda_\alpha}\rho_{\beta\beta}^{(1)}\right\}(1-\delta_{\alpha\beta}) \tag{25}$$

The approximation consists in replacing $(\partial/\partial\lambda_\alpha)\rho_{\beta\beta}^{(1)}$ which is equal to $(\partial/\partial_{\lambda\beta})\rho_{\alpha\alpha}^{(1)}$ by a product of yet unknown numbers $\mu_\alpha\mu_\beta$

$$\rho_{\alpha\beta\alpha\beta}^{(2)} \approx \left\{\rho_{\alpha\alpha}^{(1)}\rho_{\beta\beta}^{(1)} - \mu_\alpha\mu_\beta\right\}(1-\delta_{\alpha\beta}) \tag{26}$$

The μ_α guarantee the two-body density to reduce properly to the given one-body density

$$\sum_\beta \rho_{\alpha\beta\alpha\beta}^{(2)} = (A-1)\,\rho_{\alpha\alpha}^{(1)} \tag{27}$$

Using eq. 27 we get a system of coupled quadratic equations

$$\mu_\alpha\left(\sum_\beta \mu_\beta - \mu_\alpha\right) = \rho_{\alpha\alpha}^{(1)}\left(1-\rho_{\alpha\alpha}^{(1)}\right). \tag{28}$$

A formal solution obeying physical conditions on $\rho^{(2)}$ is given by

$$\mu_\alpha = \frac{\gamma}{2}\left(1-\sqrt{1-\frac{4\rho_{\alpha\alpha}^{(1)}(1-\rho_{\alpha\alpha}^{(1)})}{\gamma^2}}\right) \quad \text{with} \quad \gamma = \sum_\beta \mu_\beta \tag{29}$$

γ can be obtained by starting an iteration with $\gamma^2 = \max[4\rho_{\alpha\alpha}^{(1)}(1-\rho_{\alpha\alpha}^{(1)})]$ calculating μ_α/γ via eq. 29, then determining a new γ^2 by

$$\gamma^2 = \frac{\sum_\alpha \rho_{\alpha\alpha}^{(1)}\left(1-\rho_{\alpha\alpha}^{(1)}\right)}{1-\sum_\alpha(\mu_\alpha/\gamma)^2} \tag{30}$$

and so on. Eq. 30 follows by summing eq. 28 over α. The μ_α are different from zero only around the Fermi edge, where $\rho_{\alpha\alpha}^{(1)}$ is in between 0 and 1. All μ_α are zero in the case of a single Slater determinant. Thus approximation (26) is exact in this case. It is also exact for the case where all $\rho_{\alpha\alpha}^{(1)}$ are equal.

In the case where only the one-body density is known and hence only the mean value <A> of the particle number (grand canonical) we get the two-body density $\Gamma^{(2)}$ using eq. 25 for Z_{GC} given in eq. 18.

$$\Gamma^{(2)}_{\alpha\beta\alpha\beta} = \rho^{(1)}_{\alpha\alpha}\,\rho^{(1)}_{\beta\beta}\,(1-\delta_{\alpha\beta}) \tag{31}$$

or in a general representation

$$\tilde{\Gamma}^{(2)}_{\alpha\beta\gamma\delta} = \tilde{\rho}^{(1)}_{\alpha\gamma}\,\tilde{\rho}^{(1)}_{\beta\delta} - \tilde{\rho}^{(1)}_{\alpha\delta}\,\tilde{\rho}^{(1)}_{\beta\gamma} \tag{32}$$

However, the grand canonical two-body density does not reduce properly to the given one-body density.

$$\sum_{\beta} \Gamma^{(2)}_{\alpha\beta\alpha\beta} = (\langle A\rangle - \rho^{(1)}_{\alpha\alpha})\,\rho^{(1)}_{\alpha\alpha} \tag{33}$$

The lack of information about the mass number results in a fluctuation of this quantity with a width

$$\sigma_A^2 = \langle (\sum_\alpha a^+_\alpha a_\alpha)^2\rangle - \langle \sum_\alpha a^+_\alpha a_\alpha\rangle^2 = \sum_\alpha \rho^{(1)}_{\alpha\alpha}(1-\rho^{(1)}_{\alpha\alpha}) \tag{34}$$

This fluctuation makes the product ansatz (32) worthless for calculating widths of extensive one-body quantities like momentum or mass of a subsystem. The variances will always contain a part which is due to the fluctuating total mass number. The effect is worst if the subsystem contains about half of the matter of the total system and is negligible if the subsystem is small compared to the rest. (See also eq. 42.)

4. Application in a model system

Within the frame of a simplified model [Fe80,BF82] we are going to study the loss of information as seen by an observer who is performing an incomplete measurement. The model describes in one spatial dimension the dynamical evolution of two interpenetrating Fermi gases. The fermions are enclosed in a box and interact via a two-body interaction of finite range. The time-dependent many-body wave function $|\Psi(t)\rangle$ is obtained by solving the Schrödinger equation numerically. A detailed description of this model is given in a contribution by P. Buck and H. Feldmeier [FB82] in these proceedings.

We use the exact wave function $|\Psi(t)\rangle$ and construct the statistical operator of maximum entropy at each time as described in chapter 3. The initial state consists in 4 particles in the left and 6 particles in the right half of the box. At t = 0 the wall separating them is removed instantaneously. We suppose that the observer measures the one-body density matrix. The results are summarized in fig. 1. The entropy S(t) is calculated from eq. 4 and shown in the upper part. At t = 0 the entropy does not vanish since the observer cannot recognize the two-body correlations which are already in the initial state. On the average the entropy increases with time and tends towards the entropy S_K of the canonical distribution for which the

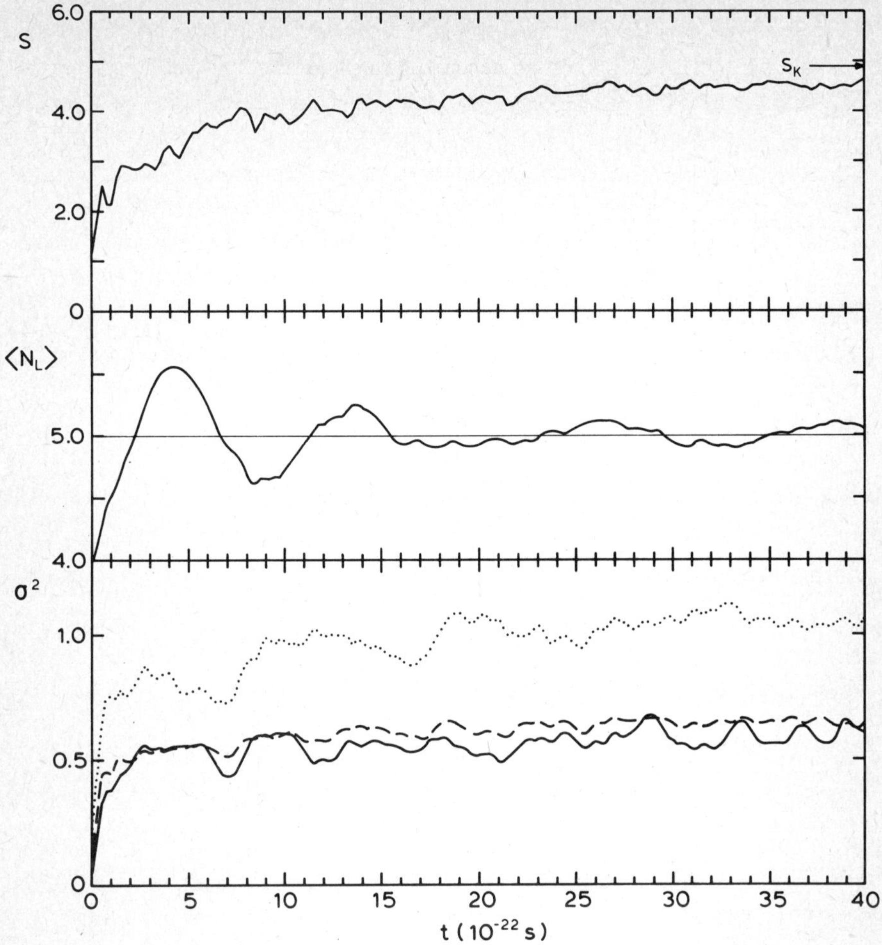

Fig. 1. Dissipative behavior of a model system with a total of 10 fermions (described in more detail in [FB82]). At time t = 0 a wall between 4 particles in the left and 6 particles in the right hand box is removed, giving rise to the time evolution of a wave function. S is the entropy as seen by an observer who measures the one-body density only. $\langle N_L \rangle$ is the mean particle number in the left box and σ^2 is the variance of N_L calculated in three different ways: —— σ^2 exact; ---- σ^2_{ME} deduced from the one-body information; $\cdots\cdot$ σ^2_{GC} calculated with the product ansatz.

information about the system consists only in knowing the Hamiltonian H and its expectation value or the temperature $1/\beta$.

$$R_K = \frac{1}{Z} e^{-\beta H} \quad ; \quad S_K = \beta \langle H \rangle + \ln Z \tag{35}$$

Eqs. 35 are special cases of eqs. 2 and 4 with $I_K = \{H,1;\ \langle H\rangle, \langle 1\rangle\}$. The unit operator 1 takes care of the normalization. Although physical intuition and Boltzmann's H theorem are in accord with this behavior we think that it is not trivial. The equation of motion is the Schrödinger equation, the state of the system is at any time a wave function, the loss of information is therefore not built into the theory but rather results from an incomplete measurement. There are altogether just 10 particles involved. Our observer who has access only to one-body information would classify the system as dissipative.

The equilibration of the number of particles $\langle N_L \rangle$ on the left side goes along with the saturation of the entropy. The lower part of fig. 1 shows the variances σ^2 of N_L calculated in three different ways. σ^2 denotes the exact variance

$$\sigma^2 = \langle \Psi | N_L^2 | \Psi \rangle - \langle \Psi | N_L | \Psi \rangle^2 \tag{36}$$

σ^2_{ME} is calculated using the statistical operator given in eq. 14 or equivalently the maximum entropy two-body density of eq. 22.

$$\sigma^2_{ME} = \mathrm{Tr}(R N_L^2) - (\mathrm{Tr}(R N_L))^2 \tag{37}$$

σ^2_{GC} utilizes the product ansatz (eq. 32) or equivalently the "grand canonical" statistical operator R_{GC}.

σ^2_{ME} and σ^2 are close to each other. This allows for our case the conclusion that the one-body information is already enough to calculate the two-body quantity $\langle N_L^2 \rangle$. σ^2_{GC} fails completely due to the superposed fluctuation of the total mass number as mentioned previously in chapter 3 (cf. eq. 34). The variance calculated with the approximated $\rho^{(2)}$ (eq. 26) differs so little from σ^2_{ME} that deviations would hardly show up on the graph.

5. Application to heavy ions

Very often in heavy ion physics the two excited nuclei are regarded as a gas of nucleons moving in a mean potential [AS76,GM76,HS76,Ra79] as illustrated in the left part of fig. 2. We are going to take this picture seriously and calculate the variance of the mass distribution. We assume that after the collision when the two nuclei are separated again they have reached thermal and chemical equilibrium. Therefore the mean occupation numbers ρ_α are given in both nuclei by the same Fermi distribution:

$$\rho_\alpha = \left(1 + e^{\frac{\varepsilon_\alpha - \nu}{T}}\right)^{-1} \tag{38}$$

Using the approximation (26) for the maximum entropy two-body density matrix the variance σ^2 is given by

$$\sigma^2 = \sum_{\alpha \in I_L} \left\{ \rho_\alpha (1-\rho_\alpha) - \mu_\alpha \left(\sum_{\beta \in I_L} \mu_\beta - \mu_\alpha \right) \right\} \tag{39}$$

With

$$\gamma_L = \sum_{\alpha \in I_L} \mu_\alpha \quad ; \quad \gamma_R = \sum_{\alpha \in I_R} \mu_\alpha \tag{40}$$

and use of eq. 28 σ^2 can be expressed as

$$\sigma^2 = \gamma_R \gamma_L \tag{41}$$

$I_L(I_R)$ denotes the set of states in the left (right) hand nucleus. The further assumptions of level densities $g(\varepsilon_F)$ which are at the Fermi edge proportional to the mass number of the nucleus and low temperature lead to

$$\sigma^2 = \frac{A_L A_R}{A} \frac{\gamma^2}{A} \approx \frac{A_L A_R}{A} g(\varepsilon_F)\, T \tag{42}$$

A_L and A_R are the mass numbers of the two nuclei. Analyzing eq. 30 it turns out that γ^2 can take values between 0 and A. Thus the first part of eq. 42 implies that the variance cannot exceed the reduced mass of the system (classical limit). The same has been shown to hold for a single Slater determinant [DD79]. It is also valid for a Fermi gas with temperature. Taking all the kinetic energy loss of the heavy ion

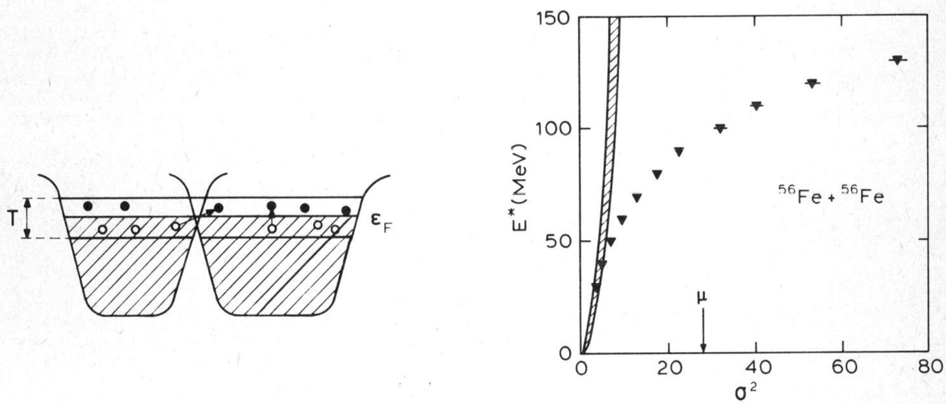

Fig. 2. Schematic of interacting nuclei (taken from [AS76]) and its implication to mass variances measured in heavy ion collisions. The hatched area results from eq. 42 using reasonable level densities $g(\varepsilon_F)$ at the Fermi edge. Triangles are measured data, taken from [BG79].

collision as heat we get the result shown in the right part of fig. 2. The experimental data contradict the assumed physical picture. Since time-dependent Hartree-Fock theories augmented by collision terms are very similar to the picture used here we do not expect them to describe mass fluctuations in heavy ion collisions. (See also [FB82], these proceedings.)

References

AS76 S. Ayik, B. Schürmann, W. Nörenberg; Z.Physik A277, 299 (1976).
BF82 P. Buck, H. Feldmeier; to be published.
BG79 H. Breuer et al.; Phys.Rev.Lett. 43, 191 (1979).
DD79 C.H. Dasso, T. Dossing, H.C. Pauli; Z.Physik A289, 395 (1979).
FB82 H. Feldmeier, P. Buck; these proceedings.
Fe80 H. Feldmeier; Report IKDA80/17, unpublished.
Fe81 H. Feldmeier; GSI annual report 1980, p. 90.
GM76 D. Glas, U. Mosel; Nucl.Phys. A264, 268 (1976).
HS76 H. Hofmann, J.P. Siemens; Nucl.Phys. A257, 165 (1976).
Ka67 A. Katz; Principles of statistical mechanics -The information theory approach; Freeman, San Francisco, 1967.
LF75 L.D. Landau, E.M. Lifschitz; Statistische Physik, Akademie-Verlag, Berlin 1975, p. 163.
Ra79 J. Randrup; Nucl.Phys. A327, 490 (1979).
Ro66 B. Robertson; Phys.Rev. 144, 151 (1966).
Wi63 E.H. Wichmann; J.Math.Physics 4, 884 (1963).

TDHF-LIKE EQUATIONS IN FIELD THEORY AND CONDENSED MATTER SYSTEMS

David K. Campbell
Center for Nonlinear Studies and
Theoretical Division
Los Alamos National Laboratory
Los Alamos, New Mexico 87545

"Es war getan, fast eh' gedacht."

Goethe

ABSTRACT

We discuss qualitatively a number of examples of TDHF-like equations that have arisen in model field theories and in (more physically relevant) quasi-one-dimensional condensed matter systems. These examples illustrate several phenomena seen in the (technically much more complicated) applications to nuclear physics and offer a new and perhaps invigorating perspective on TDHF.

I. INTRODUCTION

In this article, which is intended to provide a diversion from the focus of these proceedings on applications of TDHF to nuclear physics, I shall offer a very brief glimpse of some recent results (mostly not my own) on TDHF-like equations in field theory and condensed matter physics. Considerations of length have compelled me to select only three main topics -- the Gross-Neveu model,[1] a coupled electron-phonon model for the linear polymer polyacetylene $[(CH)_x]$,[2] and the ϕ^4 field theory coupled to fermions[3] -- and to treat even these limited topics sketchily. Hopefully, those intrigued by the results will pursue these subjects using the references in the incomplete but representative bibliography.

To anticipate here a question asked to me explicitly during the conference by Hubert Flocard, I should note that "the Fock is a fake", since none of the explicit calculations I describe includes exchange terms. In some cases -- for example for bound states in the Gross-Neveu model -- one can show that this absence of exchange terms is physically correct,[4] but in others -- scattering of the nonlinear excitations, for instance -- exchange terms should in general be included.

II. THE GROSS-NEVEU MODEL

The Gross-Neveu model,[1] a relativistic field theory in one space dimension and time, has proven to be one of the most productive theoretical laboratories of recent years. The Lagrangian density of the theory can be written in the (suggestive) form (see references 1 and 4 for details of notation)

$$(x,t) = \sum_{\alpha} \bar{\psi}^{(\alpha)}(x,t)\ (i\gamma^{\mu}\ \partial_{\mu} - g\sigma(x,t))\ \psi^{(\alpha)}(x,t) - \frac{1}{2}\,\sigma^{2}(x,t) \tag{1}$$

Here $\mu = (0,1)$ for time and space, respectively, $\psi^{+} = (u^{+}d^{+})$ is a two component Dirac field,[5] and $\alpha = 1, 2, \cdots N$ is an internal symmetry index corresponding to different "types" of fermions. As Hagen Kleinert mentioned in his contribution to the conference, the large N limit offers some simplifications and has been studied extensively.[1,6] The formal field equations which follow by varying the action corresponding to (1) are

$$\left\{ i\gamma_{o}\,\frac{\partial}{\partial t} + i\ \gamma_{1}\,\frac{\partial}{\partial x} - g\sigma(x,t)\right\} \psi(x,t) = 0 \tag{2.a}$$

and

$$g\sigma(x,t) = -g^{2}\ "\bar{\psi}\psi"\ . \tag{2.b}$$

Assuming that the formal expression "$\bar{\psi}\psi$" does not hide too many pathologies, these equations are clearly of the form of (a relativistic generalization of) the familiar TDHF equations, with a "self-consistent" mean field (σ) determined by the fermion field (ψ). In fact, Dashen, Hasslacher, and Neveu[4] have shown, using the stationary phase approximation to a functional integral (cf. John Negele's contribution to these proceedings), that equations (2) are the semi-classical approximation to the quantum field theory and that the solutions to (2) do give the bound states (i.e., the particle spectrum) of the quantum theory provided that (1) $\sigma(x,t)$ is a periodic function of time and (2) $\psi(x,t)$ is a set of wavefunctions, anti-periodic in time. The formal expression "$\bar{\psi}\psi$" becomes the sum over all "occupied" wave functions. If, for definiteness, we seek solutions for which $\sigma(x,t)$ is static (a special case of periodic), the occupied states become just the negative energy "Dirac sea" plus, for calculation of the mass spectrum, a single positive energy state (ψ_{o}) with occupation number $1 \leq n_{o} < N$. Thus, Eq. (2.b) becomes

$$Z(\Lambda)\ g\sigma = -g^{2}"\bar{\psi}\psi" = -g^{2} \sum_{occ.,\alpha} \bar{\psi}^{(\alpha)}\psi^{(\alpha)}$$

$$= -g^{2}\ n_{o}\ \bar{\psi}_{o}\psi_{o} - g^{2} \sum_{\varepsilon<0,\alpha} \bar{\psi}^{(\alpha)}\psi^{(\alpha)} \tag{2.b$'$}$$

Here $Z(\Lambda)$ is the (infinite as $\Lambda \to \infty$) renormalization required to make (2.b$'$) sensible when the momentum cut off $\Lambda \to \infty$ (in which limit there are an infinite number of states in the sum over occupied levels). I mention this technical point primarily to indicate a subtlety that besets any attempt to extend TDHF formalisms to include the "negative energy sea" of a relativistic theory. The generalization of (2.b′) for time periodic σ has the same structure,[4] except that the "energy" is replaced by the "Floquet index". For both the static and time-periodic cases, soliton-like solutions to Eqs. (2) are known and have been used to calculate the particle spectrum of the theory. Interestingly, one class of soliton solutions[4] ("bags" and "breathers") leads to a sequence of particles (the analogs of "nuclei") with 1, 2, ··· 3 (<N) fundamental constituents ("nucleons") as well as particles corresponding to the σ field ("mesons") and particles made of "nucleons", and "mesons" (the analogs of the "Δ" and other "isobars"). Amusingly, a second class of soliton solutions[4] (the "kinks") leads to one-dimensional analogs of Lee-Wick "density isomers" or "abnormal nuclei",[7] which are very massive in the large N limit. These results on bound states (at least for the "normal nuclei") are directly analogous to TDHF calculations of large amplitude collective oscillations[8] of a range of different nuclei. For application to the scattering of two "nuclei", which has not been studied in the Gross-Neveu model, the boundary conditions on and/or the form of the Eqs. (2) would be expected to change, along the lines suggested in Hugo Reinhardt's contribution to these proceedings.

One reason that scattering in the semi-classical approximation has not been investigated is that there exists an exact, analytic expression for the S-matrix of the full quantum field theory.[9,10] Thus one has the ideal case in which, as stressed here by John Negele, non-trivial tests of the semi-classical approximations can be made. For the bound states one finds that the spectrum of the full quantum theory is the same as that calculated semi-classically,[4] provided one replaces N by N-1.[9,10] Thus, as anticipated, for $N \to \infty$ the semi-classical limit is valid.

The analytic accessibility of the exact quantum S-matrix is a consequence of the complete integrability[9-12] of the Gross-Neveu model. This is a very special feature, and thus it is useful to note that apart from this exact analytic comparison, one can use recent (and more generally applicable) Monte Carlo calculations of the full quantum theory as tests of the TDHF-like approximations; this work is currently in progress.[13]

Finally, from the TDHF perspective a perhaps unfamiliar aspect of Eqs. (2) is the relativistic kinematics. It is thus interesting to observe that, for weakly bound "nuclei" -- here this is equivalent to the requirement $n_0/N \ll 1$ -- there is a well-defined non-relativistic reduction of Eqs. (2).[14] Introducing $V(x) \equiv g\sigma(x)-m$ and keeping only the u component of the wave function ψ_0 -- the d component is $O(1/m)$ in this limit -- one finds

$$i\frac{\partial u}{\partial t} - \frac{\partial^2 u}{\partial x^2} - V(x,t)\,u \cong 0 \qquad (3a)$$

and

$$V(x,t) \cong -g^2 n_o |u|^2 , \tag{3b}$$

which combine to yield precisely the famous nonlinear Schrödinger equation, familiar as the TDHF approximation to the one-dimensional δ-function potential system.

III. THE ELECTRON-PHONON MODEL OF TRANS-POLYACETYLENE ($(CH)_x$)

It is hopeless, given length constraints, for me to provide the full appropriate background for the following discussion of the coupled electron-phonon model[2] of trans-polyacetylene ($(CH)_x$); interested readers are referred to the literature.[15] To make the mathematical symbols clear, however, I must note that trans-$(CH)_x$ is a (quasi-) one dimensional (infinite, for theorists) polymeric chain consisting of (CH) units connected, in the ground state configuration, by alternating double bonds, as is shown in the chemical diagrams of Figs. 1b and 1c.

The kinematics of the microscopic electron-phonon model of $(CH)_x$ can be seen from Fig. 1a. Two of the three bonds available for each (CH) unit are treated as given and non-dynamic, and the problem of modelling the system is reduced to describing the coupled motion of the backbone "lattice" of (CH) units -- u_n will indicate the displacement of the nth (CH) unit from its equilibrium position -- and the one remaining electron per site (c_n^+(c_n) will indicate the operator creating (annihilating) an electron at the nth site). Actually c_n^+ and c_n should also have a spin

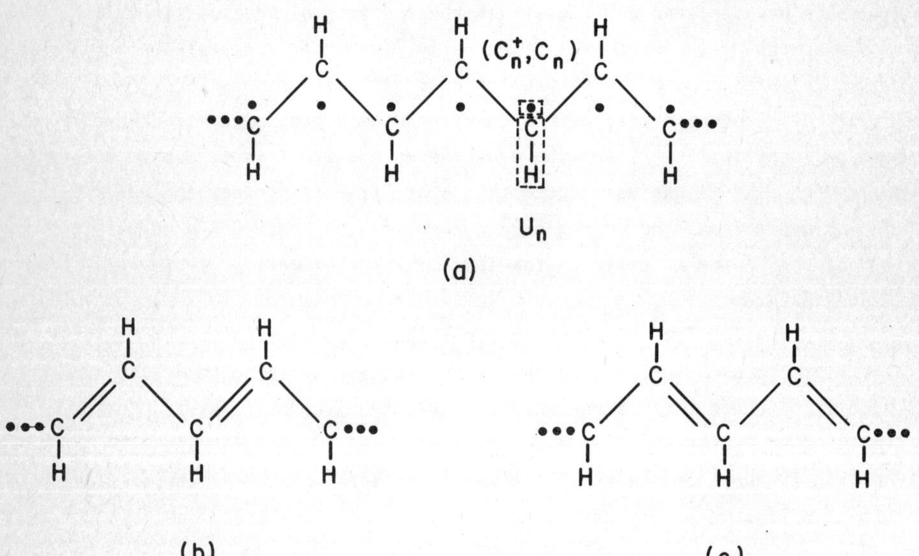

Fig. 1. (a) A sketch indicating the relation of dynamical variables (u_n, and (c_n,c_n)) to the chemistry of trans-$(CH)_x$; (b) and (c) short segments of the chemical structures corresponding to the two degenerate ground states of trans-$(CH)_x$.

label, s, corresponding to the two spin states of the electron. For simplicity, I will suppress this here.

The lattice Hamiltonian[2] describing $(CH)_x$ has the form

$$H = \frac{M}{2} \Sigma \left(\frac{\partial u_n}{\partial t}\right)^2 + \frac{K}{2} \Sigma (u_{n-1} - u_n)^2$$

$$- \Sigma\, t_o\, (c^+_{n-1}\, c_n + c^+_n\, c_{n-1}) \tag{4}$$

$$- \Sigma\, \alpha(u_{n-1} - u_n)(c^+_{n-1}\, c_n + c^+_n\, c_{n-1}) ,$$

Here the four terms represent respectively (1) the kinetic energy of the lattice, (2) the bond strain energy caused by lengthening or shortening distances between adjacent units, (3) the constant "hopping" of an electron from one site to the next and (4) the coupling of the lattice and electron motions. For trans- $(CH)_x$ the Hamiltonian in Eq. (4) must be supplemented by the requirement that precisely one half of the electron states - starting from the lowest state - are filled. From the equations of motion corresponding to (4) one can show analytically that the ground state is two-fold degenerate, as in Figs. 1b and 1c. To go much beyond this requires numerical studies,[2,16] and rather than pursue those results, I shall consider the continuum limit[17-20] -- the lattice spacing $a \to 0$ -- of H, in which analytic progress can be made. This has the dual advantages that the TDHF-like nature of the polyacetylene equations is obvious and further that an amusing connection[21,22] between polyacetylene and the Gross-Neveu model becomes evident. In the continuum limit, the Hamiltonian in (4) becomes[17-20]

$$H_c = \int dx \left\{ \frac{M}{2}\left(\frac{\partial \Delta(x,t)}{\partial t}\right)^2 + \frac{1}{2}\frac{\omega_Q^2}{g^2}\Delta^2(x,t) \right. \tag{5}$$

$$\left. + \psi(x,t)^+ \left[-iv_F\, \sigma_3 \frac{\partial}{\partial x} + \sigma_1\, \Delta(x,t)\right]\psi(x,t) \right\}$$

where $\Delta(x,t)$ is the (gap) parameter representing the lattice motion, σ_i is the $i^{\underline{th}}$ Pauli matrix, and $\psi^+ = (u^+ v^+)$ is a two component field[5] describing the electrons. The linear, Dirac-like kinetic energy for the electron is a result of expanding the lattice dispersion relation around the half-filled band and keeping only the leading term as $a \to 0$. Clearly, the general structure of this Hamiltonian is qualitatively like the energy functionals considered in TDHF. Specifically, the fermions (ψ) interact only via a mean field (Δ). Important differences from ordinary TDHF are (1) the requirement of filling the lowest one-half electron states -- the "negative energy sea" in the pseudo-relativistic analog or the "valence band" in solid state terminology -- and (2) the presence in the Hamiltonian of time-dependence ($\equiv \dot{\Delta}^2$) for the mean field. Sadly, the presence of the $\dot{\Delta}^2$ term has prevented analytic solutions of the equations of motion corresponding to H_c except in the case of static mean fields, i.e., $\dot{\Delta} = 0$. Of course numerical studies can be carried out either on the semi-classical, mean field equations (by equation of motion methods[16] like TDHF) or on the full quantum theory (by Monte Carlo methods)[23] and comparison of the approximate and exact solutions is of great current interest. Here, I will focus on static Δ, in which case the equations of motion become[17-20]

$$\varepsilon_n \psi_n = -iv_F \, \sigma_3 \frac{\partial}{\partial x} \psi_n + \Delta \, \sigma_1 \, \psi_n \tag{6.a}$$

and

$$\Delta = \sum_{\varepsilon_n < 0}{}^{\prime} \; \psi_n^+ \sigma_1 \psi_n \, , \tag{6.b}$$

where in both equations the spin index ($s = \pm 1/2$) on the fermion wave function ψ_n has been suppressed. Once more the TDHF-like -- here again, really only Hartree-like -- nature of the equations is evident. The summation in (6.b) runs, for the ground state, over all states in the valence band -- which, because of the linear kinetic energy in (6.a) must be cut off by hand at an energy corresponding to the full band width -- plus, for excitations, any additional occupied states. The structure of Eqs. (6) is clearly very similar to that of the static Gross-Neveu model Eqs. [(2)] -- with $N = 2$ to correspond to the two spin degrees of freedom -- and in fact one can show that, at this static, mean field level, the equations are identical.[21,22] The solutions corresponding to nonlinear excitations must therefore also be the same. Indeed, the "kinks" of the Gross-Neveu model are in $(CH)_x$ the "kink" solitons with the celebrated exotic spin-charge relations.[2,18,24] The "bag" solutions of the Gross-Neveu model become "polaron" solutions -- since $N = 2$ there is only one (electron) polaron state ($n_o = 1$) -- which, although more conventional than kinks, are also more generic in the sense that they are expected to occur in a much wider class of quasi-one-dimensional polymers.[21,22] Even the analog of the non-relativistic reduction of the Gross-Neveu model is of interest; here, this corresponds to reducing[14] the polaron of Eqs. (6) to the more conventional polaron,[25] satisfying the nonlinear Schrödinger equation, familiar in solid state physics.

Since we have stressed the similar structure of the electron-phonon model of $(CH)_x$ and the $N = 2$ Gross-Neveu model, it is crucial to note the manifest difference: namely, the $\dot{\Delta}^2$ term in (5), which has no counterpart in (1). This term leads to important differences in the time-dependent and fully quantum versions of these models.

Finally, in $(CH)_x$ there is an important open question directly analogous to the questions of "residual interactions" and "collision terms" in applications of TDHF to nuclear physics. Specifically, the Hamiltonians in (4) and (5) include only electron-phonon coupling and involve no direct electron-electron coupling. At least the Coulomb part of this coupling -- which would, loosely speaking, be of the form $(\psi^+\psi)^2$ -- is expected to be important, and it is an unsolved problem to include these terms correctly and non-perturbatively. Monte Carlo methods can, of course be used, but some clever, at least partially analytic, extended mean field theory is clearly desirable.

IV. $\lambda\phi^4$ FIELD THEORY WITH FERMIONS

Our last example of TDHF-like equations in field theory is the one space dimensional $\lambda\phi^4$ theory coupled to fermions.[3] The Lagrangian density is

$$(x,t) = \frac{1}{2}\left(\frac{\partial\phi}{\partial t}\right)^2 - \frac{1}{2}\left(\frac{\partial\phi}{\partial x}\right)^2 - \frac{\lambda}{4}(\phi^2-1)^2 + \sum_{\alpha=1}^{N} \bar{\psi}^{(\alpha)}(i\gamma\partial - g\phi)\psi^{(\alpha)} \tag{7}$$

The fermion term of the Lagrangian, including the linear coupling to the boson field ϕ, is clearly identical to that of the Gross-Neveu model. The purely bosonic part of has been used to describe the displacive limit of structural phase transitions in uniaxial ferroelectrics,[26] as a mean field theory of phase transitions involving one component order parameters ("Landau-Ginzburg theory"), and as a <u>phenomenological</u> -- as opposed to the previously discussed <u>microscopic</u> -- model of nonlinear excitations in polyacetylene.[27] The full Lagrangian has been studied, in the semi-classical approximation, as a "toy" model of nuclear physics in one dimension.[3,28] More specifically, the static field equations following from (7),

$$-\frac{\partial^2\phi}{\partial x^2} - \lambda\phi + \lambda\phi^3 = -g\text{"}\bar{\psi}\psi\text{"} \tag{8.a}$$

and

$$(i\gamma_1\,\partial_x - g\phi)\psi_i = \gamma_o\,\varepsilon_i\,\psi_i\,, \tag{8.b}$$

have been solved analytically[3] using the same methods applied to the Gross-Neveu model. Again, both sequences of "normal nuclei" and "kinks" ("abnormal nuclei") are found as bound states. Of interest here from the TDHF perspective are the nonlinear interactions ($\lambda\phi^3$) and propagation effects ($\partial^2\phi/\partial x^2$) in the "mean" field. Once again the full semi-classical approximation requires the inclusion of the infinite (for $\Lambda \to \infty$) negative energy sea. As sketched in the discussion of the non-relativistic reduction of the Gross-Neveu model, for weakly bound "nuclei", one can motivate a further approximation, in which only the single positive energy bound state (called ψ_0 previously) is kept in sum on the right-hand side of (8.a). Within this approximation -- which for definiteness we shall call the "Hartree approximation" -- two investigations closely parallelling familiar TDHF studies have been carried out. First, the fission decay of a large "nucleus" into two smaller "nuclei" has been calculated[29] using collective coordinate techniques for determining a "most probable" path in field space; this scheme is closely related to the adiabatic TDHF approach discussed in many contributions to these proceedings. Second, within the Hartree approximation, the time-dependent versions of Eqs. (8) have been used (without further justification) to study the scattering of two initially widely separated "nuclei".[30] The numerical results have shown the analog of the well-known "transparency"[8] of nuclear scattering in the TDHF approximation and, not surprisingly in view of our previous remarks, reduce to the analytically known results for the nonlinear Schrödinger equation for low relative velocities.

I will close with two comments on features found in the purely bosonic part of the Lagrangian in (8). Even in the absence of fermions, "kink" solitary wave solutions exist for the nonlinear equations for the field $\phi(x,t)$. The scattering of these kink solutions (actually, kink-antikink scattering) has been the subject of extensive study,[31-38] particularly because of an interesting sequence of resonances at low relative collision velocities. These resonances, for which a semi-phenomenological theory has recently been developed,[39] are the direct analogs of the fusion resonances seen at low collision velocities in one of the pioneering TDHF studies[8] of nuclear collisions. Finally, the use of the ϕ^4 theory (<u>without</u> fermions) as a <u>phenomenological</u> model -- with the field ϕ representing a bond length distortion -- of <u>trans</u>-$(CH)_x$[27] is an (indirect) example of the effective action approach suggested by Hagen Kleinert in these proceedings. In essence, the $(\phi^2-1)^2$ potential provides an approximate effective action that takes account of the effects of the electrons in the filled valence band in polyacetylene.[22]

I have presented a very brief "discussion" -- the term "index" may be more appropriate -- of a number of areas in field theory in which equations and/or results reminiscent of TDHF have arisen. Hopefully, the possibility of applications to a broader range of problems will stimulate further development of the concepts and methods of this interesting approach to quantum many body problems.

REFERENCES AND FOOTNOTES

1. D. J. Gross and A. Neveu, Phys. Rev. D 10, 3235 (1974).

2. W. P. Su, J. R. Schrieffer, and A. J. Heeger, Phys. Rev. Lett 42, 1698 (1979); Phys. Rev. B22, 2099 (1980).

3. D. K. Campbell and Y.-T. Liao, Phys. Rev. D14, 2093 (1976).

4. R. F. Dashen, B. Hasslacher, and A. Neveu, Phys. Rev. D12 2443 (1975).

5. In one space dimension the Dirac γ matrices can be represented by the Pauli matrices, σ_i. Note that, following the conventions in the literature, we use different representations for the γ matrices in the Gross-Neveu model and in the continuum electron-phonon model of polyacetylene. Hence the components of the wave functions in the two cases are linear combinations of each other.

6. J. F. Schonfeld, Nucl. Phys. B95, 148 (1975); R. G. Root, Phys. Rev. D11, 831 (1975).

7. T. D. Lee and G. C. Wick, Phys. Rev. D9, 2291 (1974).

8. P. Bonche, S. Koonin, and J. W. Negele, Phys. Rev. C13, 1226 (1976).

9. A. B. Zamolodchikov and Al. B. Zamolodchikov, Phys. Lett. 72B, 481 (1978); Ann. Phys. 20, 91 (1979).

10. R. Shankar and E. Witten, Nucl. Phys. B141, 349 (1978); E. Witten, Nucl. Phys. B142, 285 (1978).

11. V. E. Zakharov and A. V. Mikhailov, Commun. Math. Phys. 74, 21 (1980).

12. A. Neveu and N. Papanicolaou, Commun. Math. Phys. 58, 31 (1978).

13. J. Hirsch, D. Scalapino, and R. Sugar, in progress.

14. A. R. Bishop, D. K. Campbell, and K. Fesser, Phys. Rev. B, to be published.

15. For recent review see articles in Proc. Int. Conf. on Low-Dimensional Synthetic Metals (Helsingor, Denmark, August 1980) in Chemica Scripta 17 (1981); Physics in One Dimension, eds. J. Bernasconi and T. Schneider (Springer Verlag, 1981); and A. J. Heeger and A. G. MacDiarmid, p. 353-391 in The Physics and Chemistry of Low Dimensional Solids, ed. L. Alcâcer (Reidel, 1980).

16. W. P. Su and J. R. Schrieffer, Proc. Nat. Acad. Sci. 77, 5526 (1980).

17. S. Brazovskii, JETP Lett. 28 606 (1978) (ZhETF Pisma 28, 656 (1978)). JETP (Sov. Phys.) 51 342 (1980) (ZhETF (USSR) 78, 677 (1980)).

18. H. Takayama, Y. R. Lin-Liu, and K. Maki, Phys. Rev. B21 2388 (1980).

19. J. A. Krumhansl, B. Horovitz, and A. J. Heeger, Solid State Comm. 34, 945 (1980).

20. B. Horovitz, Solid State Comm. 34, 61 (1980); Phys. Rev. Lett. 46 742 (1981); Phys. Rev. B22, 1101 (1980).

21. S. Brazovskii and N. Kirova, JETP Lett. 33, 4 (1981) (ZhETF Pisma 33, 6 (1981).

22. D. K. Campbell and A. R. Bishop, Phys. Rev. B 24, 4859 (1981); Nucl. Phys. B200 (FS4), 297 (1982).

23. J. E. Hirsch and E. Fradkin, to be published (preprint NSF-ITP-82-34).

24. R. Jackiw and J. R. Schrieffer, Nucl. Phys. B190 (FS3), 253 (1982).

25. T. Holstein, Ann. Phys. 8, 325 (1959); Mol. Cryst. Liq. Cryst. 77, 235 (1981); T. Holstein and L. Turkevich, Phys. Reports (to be published).

26. R. A. Cowley, Adv. in Phys. 29, 1 (1980); A. D. Bruce, Adv. in Phys. 29, 111 (1980); A. Bruce and R. Cowley, Adv. in Physics 29, 218 (1980).

27. M. J. Rice, Phys. Lett. 71A, 152 (1979); M. J. Rice and J. Timonen, Phys. Lett. 73A, 368 (1979); E. J. Mele and M. J. Rice, Chemica Scripta 17, 21 (1981).

28. D. K. Campbell, p. 674 in Nuclear Physics with Heavy Ions and Mesons, Les Houches Session XXX, 1977, R. Balian, et al., eds. (North Holland, 1978).

29. D. K. Campbell, Ann. Phys. 129, 249 (1980).

30. D. K. Campbell, J. W. Negele, and M. Soldate, unpublished.

31. A. E. Kudryavtsev, JETP Lett. 22, 82 (1975).

32. B. S. Getmanov, JETP Lett. 24, 291 (1976).

33. V. G. Makhankov, Phys. Rep. 35C, 1 (1978).

34. M. J. Ablowitz, M. D. Kruskal, and J. F. Ladik, SIAM Journ. App. Math. 36, 478 (1979).

35. T. Sugiyama, Prog. Theor. Phys. 61, 1550 (1979).

36. R. Klein, W. Hasenfratz, N. Theodorakopoulos, and W. Wünderlich, Ferroelectrics 26, 721 (1980).

37. M. Moshir, Nuc. Phys. B185, 318 (1981).

38. C. A. Wingate, SIAM Journ. App. Math., in press.

39. D. K. Campbell, J. Schonfeld, C. A. Wingate, in preparation.

NUCLEAR COLLECTIVE MOTIONS IN SEMICLASSICAL TDHF

M. Di Toro
Istituto di Fisica Università di Catania, INFN Sez. di Catania
57 Corso Italia - 95129 Catania, Italy

Abstract

Using the Wigner transform of the one-body density matrix we extract a collective hamiltonian from the TDHF energy functional for slow nuclear collective motions. For a rotating nucleus and for small vibrations we find solutions of the Vlasov equation which allow an explicit evaluation of currents, inertial parameters and eigenfrequencies. The extension to generalized scaling modes is analysed to show that the TDHF-Fluid-Dynamical chain of equations can be exactly truncated. Quantum effects, one-body damping mechanisms and the elasticity limit are finally discussed.

1. Introduction

The fluid-dynamics approach seems to be a powerful tool to understand collective motions in nuclei in terms of macroscopic quantities as local density, current and pressure tensor. The microscopic foundation of nuclear fluid dynamics is the mean field theory (TDHF): his validity lies on the fact that, in low energy excitations of nuclei, the long mean free path regime prevents local equilibration; instead, dynamycal distortions of the local momentum distribution would have a crucial role. The starting point is the TDHF equation for the Wigner transform $f(\underline{r},\underline{p},t)$ of the one body density matrix, which is the quantum-mechanical analog of the classical distribution function[1])

$$\dot{f}(\underline{r},\underline{p},t)= \frac{2}{\hbar} \sin \left(\frac{\hbar}{2} \Lambda_{12}\right) h^{(1)}(\underline{r},\underline{p},t)\ f^{(2)}(\underline{r},\underline{p},t) \tag{1.1}$$

where $h=\frac{p^2}{2m} + w(\underline{r},\underline{p},t)$ is the Wigner transform of the self-consistent HF hamiltonian, $\Lambda_{12}=\underline{\nabla}_r^{(1)}\cdot\underline{\nabla}_p^{(2)}-\underline{\nabla}_p^{(1)}\underline{\nabla}_r^{(2)}$ and the differential operator is intended to act on the corresponding function: $\underline{\nabla}^{(i)}\rightarrow g^{(i)}$.

From eq. (1.1) one can generate in the usual way the chain of fluid-dynamical equations which express the conservation laws of mass, momentum, energy and so on.

Since

$$\frac{2}{\hbar} \sin \left(\frac{\hbar}{2} \Lambda_{12}\right) h^{(1)} f^{(2)}= \{h,f\}- \frac{h^2}{24}\Lambda_{12}^3\, h^{(1)} f^{(2)}+ \;\ldots\ldots \tag{1.2}$$

neglecting the second and higher order terms in $\hbar$ one obtains the semiclassical limit of TDHF, the well known Vlasov equation.

Moreover we remark that , due to the structure of the Λ_{12} operator, we do not have explicit quantum terms up to the second $\underline{p}$-moment dynamical equation derived from eq. (1.1), allowing a Galilei invariant non locality for the HF potential. This means that if we are able to truncate the chain at that order we get a completely classical set of equations, quantum effects being in the initial conditions and in the truncation procedure. In particular in the extreme case of a harmonic oscillator single particle potential the quantum terms are exactly dropped out and the TDHF dynamics becomes wholly classical.

This paper is devoted to a search for analytical solutions of the fluid dynamical equations for a wide class of nuclear collective motions.

2. Slow collective motions: adiabatic expansion

Taking in mind the remarks of the previous section we start our analysis from the Vlasov equation $\dot{f}=\{h,f\}$. In the case of small collective velocities

$$\omega\cdot A \ll v_F \tag{2.1}$$

where ω,A are respectively the collective frequencies and amplitudes and v_F is the Fermi velocity, we can follow an adiabatic expansion which leads in a natural way to a collective Hamilton dynamics [2,3]

$$f(\underline{r},\underline{p})=f_\circ(\underline{r},\underline{p})+\{f_\circ,\chi\}+\tfrac{1}{2}\{\chi,\{\chi,f_\circ\}\}+\ ..=f_\circ+f_1+f_2+\ldots.. \tag{2.2}$$

with $f_\circ,\chi$ $\underline{p}$-even functions in the phase space; $f_\circ$ being a time-dependent equilibrium distribution and χ proportional to the collective velocity field. Correspondingly the transform of the HF potential can be expanded

$$h(\underline{r},\underline{p})=\frac{p^2}{2m}+w(\underline{r},\underline{p})=h_\circ(\underline{r},\underline{p})+w_1(\underline{r},\underline{p})+w_2(\underline{r},\underline{p})+\ldots \tag{2.3}$$

The quantities with odd (even) labels are odd (even) with respect to the $\underline{p}\to-\underline{p}$ transformation.

The Vlasov equation becomes a set of coupled equations in the even and odd parts of the distribution function

$$\dot{f}_\circ=\{h_\circ,f_1\}+\{w_1,f_\circ\} \tag{2.4.1}$$

$$\dot{f}_1=\{h_\circ,f_\circ\}+\{h_\circ,f_2\}+\{w_2,f_\circ\}+\{w_1,f_\circ\} \tag{2.4.2}$$

A main result of this analysis is that the HF total energy functional of the system can be split in two parts [4]

$$E_{TOT}=U(f_\circ)+K(\chi,f_\circ)+O(\chi^4) \tag{2.5}$$

with

$$U(f_\circ)=\int_{\underline{r}}\int_{\underline{p}}\left(\frac{p^2}{2m}+\frac{1}{2}w_\circ\right)f_\circ$$

$$K(\chi,f_\circ)=\int_{\underline{r}}\int_{\underline{p}}\left[h_\circ f_2+\tfrac{1}{2}w_1 f_1\right]=\tfrac{1}{2}\int_{\underline{r}}\int_{\underline{p}}\chi\frac{\partial f_\circ}{\partial t}$$

where U, function of the "coordinate" $f_\circ$ only, and K, quadratic in χ, are interpreted respectively as collective potential and kinetic energy. The form (2.5) for the semiclassical collective hamiltonian is the starting point to study potential and inertial parameters for collective modes[4]. It is also possible to support the interpretation of eq. (2.5) through a Hamilton formulation of the dynamical equations (2.4) with $f_\circ,\chi$ playing a role of collective canonical coordinates[5]. A natural consequence of a Hamilton form is that the resulting collective motion is undamped: consistently with the adiabatic assumption we do not have one-body dissipation terms in the collective dynamics. The variational approach has also a practical use: for a wide class of collective motions it represents a valid tool to determine the field $\chi(\underline{r},\underline{p})$ without solving the adiabatic aquations (2.4)[5].

This is also the main aim of the next section.

3. Generalized scaling modes (GSM)

In this class of collective motions all the nucleons are looked to move coherently in a common velocity field. The time-even part of the density matrix is given in terms of the static solution

$$\rho_\circ(t)=\exp(i\ K/\hbar)\rho_{static}\ \exp(-i\ K/\hbar) \qquad (3.1)$$

where K is a time-odd hermitian operator of the form

$$K=\frac{1}{2}\left[\underline{p}\cdot\underline{s}(\underline{r},t)+\underline{s}(\underline{r},t)\cdot\underline{p}\right] \tag{3.2}$$

with $\underline{s}(\underline{r},t)$ a vector field which characterizes the collective motion.

Eq.(3.1) supplies the external information we need on the collective path in order to avoid to solve the second adiabatic equation (2.4.2) . The time derivative gives

$$i\hbar\dot{\rho}_o = \left[\rho_o, \dot{K}\right]$$

which in the Wigner transformed space, in the semiclassical limit, leads to

$$\frac{\partial f_o}{\partial t} = \{\underline{p}\cdot\underline{v}(\underline{r}), f_o\} \tag{3.3}$$

where we use the collective velocity field $\underline{v}(\underline{r},t)= -\dot{\underline{s}}(\underline{r},t)$.
From some properties of the Poisson brackets the collective kinetic energy assumes the form

$$K(\chi,f_o)=\frac{1}{2}\, m\int d\underline{r}\ \underline{v}(\underline{r})\cdot\underline{j}(\underline{r}) \tag{3.4}$$

where

$$\underline{j}(\underline{r})=\int_{\underline{p}}\frac{\underline{p}}{m}\, f(\underline{r},\underline{p}) = \int\frac{\underline{p}}{m}\{f_o,\chi\}$$

is the response current inside the system and in general is not parallel to $\underline{v}(\underline{r})$.

A second cosequence of eq. (3.3) is the relation

$$\{\chi,h_o\}=\underline{p}\cdot\underline{v}(\underline{r}) \tag{3.5}$$

obtained just comparing with the first adiabatic equation(2.4.1), in the case of local HF potential, $w_1(\underline{r},\underline{p})=0$. Eq. [3.5] can be used to find explicitly the χ field without solving the adiabatic equations.

Several cases of physical interest have the property (3.1). When

$\underline{s} = \alpha\, \underline{r} \times \underline{n}$ we describe a rotation through an angle α about the direction $\underline{n}$. In this case $\underline{v}(\underline{r}) = \underline{v}_{rigid}(\underline{r}) = \underline{\omega} \times \underline{r}$, with $\underline{\omega} = \dot{\alpha}\, \underline{n}$, angular velocity.

The collective mode obtained just scaling the coordinates corresponds to a displacement field $\underline{s} = \alpha\, \underline{\nabla}\, \phi$ where $\alpha(t)$ is the scaling parameter and $\phi(\underline{r})$ is a real function which depends on the studied collective motion. We have an irrotational velocity field $\underline{v}_{scaling} = -\dot{\alpha}\, \underline{\nabla}\, \phi(\underline{r})$ and the eq.(3.5) can be solved to give a scaling field $\chi(\underline{r},\underline{p}) = - m\, \dot{\alpha}\, \phi\, (\underline{r})$ and a collective kinetic energy(4)

$$K = \tfrac{1}{2}\, m \int_{\underline{r}} \rho_{\circ}\, \underline{v}^2_{scaling} \tag{3.6}$$

with inertial parameter

$$M_{scaling}(\alpha) = m \int_{\underline{r}} \rho_{\circ}\, (\underline{r},\alpha)\, (\underline{\nabla}\phi)^2 \tag{3.7}$$

4. Small amplitudes: semiclassical RPA

Scaling modes are meaningful for small amplitude oscillations. In this case the collective hamiltonian can be written in a harmonic form

$$E_{TOT} = \tfrac{1}{2}\, M\, \dot{\alpha}^2 + \tfrac{1}{2}\, C\alpha^2 \tag{4.1}$$

in terms of the scaling parameter $\alpha(t)$, and the frequencies are simply given by $\omega^2 = C/M$. The restoring parameter comes from second order variations of the collective potential

$$\tfrac{1}{2}\, C\, \alpha^2 = \delta^2\, U(f_{\circ}) = \int_{\underline{r}} \int_{\underline{p}} h_{st}\, \delta^2\, f_{\circ} + \tfrac{1}{2} \int_{\underline{r}} \int_{\underline{p}} \delta w_{\circ}\, \delta\, f_{\circ} \tag{4.2}$$

where, from eq. (3.1),

$$f_\circ = f_{st} + \{f_{st}, \xi\} + \frac{1}{2}\{\xi,\{\xi, f_{st}\}\} + \ldots = f_{st} + \delta f_\circ + \delta^2 f_\circ + \ldots.$$

with $\xi = \alpha \underline{p} \cdot \underline{\nabla}\phi$, and in correspondance the HF potential is $w_\circ = w_{st} + \delta w_\circ + \delta^2 w_\circ + \ldots\ldots$

For isoscalar modes and a Skyrme -Hartree-Fock potential

$$w_{HF} = \frac{3}{4} t_\circ \rho + \frac{3}{16} t_3 \rho^2 + \beta(\rho p^2 - 2m\underline{j}\cdot\underline{p} + 2m\tau) + \gamma\nabla^2\rho + V_{coul} \tag{4.3}$$

with $\beta = \dfrac{3t_1+5t_2}{16}$, $\gamma = \dfrac{5t_2-9t_1}{32}$, we get with some algebra a stiffness

$$C = \int_{\underline{r}} \left\{ \frac{K_s}{m^*}\left[2(\nabla_i u_i)^2 - u_i\nabla^2 u_i\right] - \frac{3}{4} t_\circ \rho_{st} u_i \nabla_i \left[\rho_{st}\nabla_k u_k\right] - \frac{3}{8} t_3 \rho_{st} u_i \nabla_i \left[\rho^2_{st}\nabla_k u_k\right] + \right.$$

$$\left. + \beta K_s \rho_{st}\left[7(\nabla_k u_k)^2 - 3u_i\nabla_i\nabla_k u_k\right] - \gamma\nabla_i(u_i\rho_{st})\left[u_i\nabla_i\nabla^2\rho_{st} - \nabla^2(\nabla_i(u_i\rho_{st}))\right] + \text{coulomb}\right\} \tag{4.4}$$

where $\underline{u} = \underline{\nabla}\phi$ and $k_s = \frac{1}{3}\int_{\underline{p}} P^2 f_{st}$ is proportional to the kinetic energy density in the static solution.

The form eq.(4.4) becomes very simple for low moltipoles[6]. For monopole ($\phi = r^2$) and quadrupole ($\phi = 2z^2 - x^2 - y^2$) modes we get frequencies exactly of the same form as in the quantum sum rule approach of Bohigas, Lane and Martorell[7]: quantum effects are not present for low multipoles in the scaling mode. This is a quite general result as we"ll show in the next section.

5. Small oscillations in generalized scaling: elasticity equation

In this section we release any semiclassical assumption. The fluid TDHF chain can be exactly truncated for small amplitude GSM.

In the Wigner space, at the first order in the $\underline{s}$ field, the $\underline{p}$-even part of the distribution function from eq.(3.1) satisfies the equation

$$\delta f_\circ = \frac{2}{\hbar} \sin \frac{\hbar}{2} \Lambda_{12} \quad f^{(1)}_{st} (\underline{p}.\underline{s})^{(2)} \tag{5.1}$$

which gives a transition density of the form $\delta\rho = \int_{\underline{p}} \delta f_\circ = \nabla_i(\rho_{st} s_i)$

Through the continuity equation we have in correspondance a current

$$\underline{j} = \rho_{st}\, \underline{v}\ , \quad \text{with } \underline{v} = - \dot{\underline{s}}$$

The first $\underline{p}$-moment of the Wigner-TDHF eq.(1.1), linearized for small oscillations, is a Euler-like equation of the form

$$m \frac{\partial}{\partial t} \underline{j} + \underline{\nabla}\, \delta \overleftrightarrow{\tau} + \delta\rho \underline{\nabla} w_{st} + \rho_{st} \underline{\nabla} \delta w = 0 \qquad (5.2)$$

where $\delta\tau_{ij}$ is the transition kinetic energy tensor and the HF potential $w = w_{st} + \delta w$ for simplicity has been assumed to be local.

Because the tensor $\delta\tau_{ij}$ can be written, from (5.1),

$$\delta\tau_{ij}(\underline{r},t) = \int_{\underline{p}} \frac{p_i p_j}{m}\, \delta f_\circ(\underline{r},\underline{p},t) =$$

$$\nabla_k (s_k \tau^{st}_{ij}) + \tau^{st}_{kj} \nabla_i s_k + \tau^{st}_{ik} \nabla_j s_k - \frac{\hbar^2}{4m} \nabla_k (\rho_{st} \nabla_i \nabla_j s_k) \qquad (5.3)$$

only in terms of the $\underline{s}$-field and the static kinetic energy tensor, we see that the fluid-dynamical chain is exactly truncated at the first $\underline{p}$-moment eq.(5.2) which now becomes a Schrodinger-like equation for the vector field $\underline{s}(\underline{r},t)$. With appropriate boundary conditions we can find solutions and eigenvalues which give respectively transition densities, flow patterns and frequencies of the various multipole oscillations.

The only quantum effect comes from the last term of the transition kinetic energy eq.(5.3): all the solutions with $\nabla_i \nabla_j s_k = 0$ give rise to a completely classical response. This is the case of the simple Tassie scaling model, discussed in the previous section, for isoscalar monopole and quadrupole resonances.

Remaining in the scaling picture $\underline{s}(\underline{r}) = \underline{\nabla}\phi(\underline{r})$, the eq.(5.2) supplies a general way to find the scalar field $\phi(\underline{r})$ for any multipole. A particular case was studied with a variational approach by Serr et al.[8] Moreover, if we specialize our HF potential in order to get the semiclassical energy density functional used by Holzwarth et al.[9], we easily get from eq. (5.2) the same fluid dynamical equations. This discussion shows that our result, derived in a direct way from a linearized TDHF theory is very general: 1) we can reach a fluid dynamical description of giant resonances with clear microscopic foundations; 2) we can use realistic interactions such as Skyrme-type forces.

Finally we would like to remark that assuming an isotropic static momentum distribution $\tau_{ij}^{st} = \frac{p_F^2}{5m} \rho_{st} \delta_{ij}$ and a uniform model density, eq.(5.2) takes the form of a Lamé equation for the displacement field in an elastic medium [10])

$$m\rho \frac{\partial^2}{\partial t^2} s_i = \lambda \nabla_i \underline{\nabla} \cdot \underline{s} + \mu \nabla_j (\nabla_i s_j + \nabla_j s_i) \qquad (5.4)$$

with shear modulus $\mu = \frac{p_F^2}{5m} \rho_{st}$ and longitudinal modulus $\lambda = \frac{\partial w_{st}}{\partial \rho} \rho_{st}^2 + \frac{p_F^2}{5m} \rho_{st}$. This means that there is no damping and the widths cannot be described in generalized scaling approach. Actually it can be shown that the heat flow tensor (third $\underline{p}$-moment of the distribution function) is exactly zero for GS modes: the coherence of the particle motion prevents the developing of the Landau damping, which is the only dissipation present in TDHF.

The work described here has been performed in part an a very pleasant collaboration with Prof. D.M.Brink of the Department of Theoretical Physics, Oxford University and with Drs.G.Giansiracusa, U.Lombardo, G.Russo and the graduate students A.Bonasera, M.Di Nardo of the University of Catania.

References

1) P.Ring and P.Schuck "The Nuclear Many Body Problem" Springer Verlag 1980 Chap. 13.3

2) M.Baranger and M.Vénéroni, Ann. Phys. 114(1978)123

3) M.J.Giannoni and P.Quentin, Phys. Rev. C21(1980)2060

4) D.M.Brink and M.Di Toro, Nucl. Phys. A372(1981)151

5) A.Bonasera, M.Di Toro, G.Giansiracusa and U.Lombardo "Variational derivation of a semiclassical theory of nuclear collective motions" Nuovo Cimento A (1982) in press

6) M. Di Nardo, Catania Thesis 1982, unpublished

7) O.Bohigas, A.M. Lane and J.Martorell, Phys. Rep. 51 268(1979)

8) F.E.Serr, G.F. Bertsch and J.Borysowicz, Phys. Lett. 92B 241(1980)

9) G.Holzwarth and G.Eckart, Nucl. Phys. A325(1979)1
G.Eckart, G.Holzwarth and J.P. da Providencia Nucl. Phys.A364(1981)1

10) G.F.Bertsch "Proceedings of the Nuclear Physics School-Erice 1979" Pergamon Press p. 483 Ed. D.Wilkinson.

BEYOND TDHF : SCHRÖDINGER TIME-EVOLUTION IN A TRUNCATED SUBSPACE

H. Schultheis and R. Schultheis
Institut für Theoretische Physik
Universität Tübingen
D-7400 Tübingen, West Germany

A.B. Volkov
Department of Physics
McMaster University
Hamilton, Ontario, Canada L8S 4M1

Abstract: We derive non-determinantal approximations to the Schrödinger time-evolution in a necessarily truncated subspace, that allow for a flux of probability between the subspace and the excluded space. A second-order time-differential equation in the subspace and a time-dependent optical potential are studied in a number of examples.

1. INTRODUCTION

Going beyond the Time-Dependent Hartree Fock approximation will generally require a big step backward in the dimensionality of the space within which the time evolution can be treated. While the time-evolution of a TDHF determinant is generally not confined to a predetermined subspace of the total Hilbert space, non-determinantal approximations to the time-dependent Schrödinger solution will necessarily be restricted to a finite basis of small enough dimension to make the computation feasible. The truncation of the Hilbert space of the system to a small finite subspace leads to the problem of the coupling between the selected subspace and the excluded space, and how to account for the corresponding flux of probability between the spaces.

In the present paper we study non-unitary approximation methods that allow for a flux of probability but are completely defined and solvable within the subspace, such that the computation requires no information about the neglected space. The approximations are related[1,2] to a series expansion of the true subspace projection of the Schrödinger time-evolution operator, augmented in some cases by an "optical" time-dependent imaginary potential.

2. TRUNCATION OF THE SCHRÖDINGER PROBLEM

In stationary quantum mechanics many approximation methods consist in truncating the Hilbert space of the system to a limited subspace, and then solving the truncated problem in the subspace rather than dealing with the original Hamiltonian in the full space. A similar truncation procedure, when applied to the time-dependent Schrödinger equation, however, leads to unphysical results because the solution of the truncated problem is unitary with respect to the truncated subspace, and does not allow for any flux of probability between the selected subspace and the rest of the Hilbert space. Therefore the solution of the ordinary time-dependent Schrödinger equation in a truncated subspace artificially conserves all the initial probability (and mass, charge,energy etc.) within the subspace for all times, and thus differs qualitatively from the non-unitary subspace part of the true Schrödinger time-evolution. In general this property is the more undesirable the higher the truncation because the subspace probability will remain constant irregardless of how small the selected subspace is. For instance, the extreme truncation of the time-dependent Schrödinger-equation to dimension 1 would always lead to no time evolution at all (apart from a phase) whereas the true Schrödinger solution will in general immediately generate flux into states other than the initial state.

3. SECOND-ORDER SUBSPACE TIME-EVOLUTION EQUATION

We consider a Hilbert space that is devided into two orthogonal subspaces with corresponding projection operators P and Q,

$$P^2 = P,\ Q^2 = Q,\ P + Q = 1,\ PQ = 0 = QP \tag{1}$$

where P denotes the selected subspace and Q the excluded space.

In order to obtain an approximation to the time evolution in the subspace we rewrite the Schrödinger equation

$$i \hbar \frac{\partial}{\partial t} \psi(t) = H \psi(t) \tag{2}$$

as a second-order equation

$$(i \hbar \frac{\partial}{\partial t})^2 \psi(t) = H^2 \psi(t), \tag{3}$$

project onto the P-space and insert $P^2 + Q^2 = 1$,

$$P\, H^2\psi = P\, H^2(P^2 + Q^2)\psi = (P\, H^2 P)P\psi + (P\, H^2 Q)Q\psi \,. \tag{4}$$

This yields for the P-space

$$-\hbar^2 \frac{\partial^2}{\partial t^2}\, \psi_P(t) = (H^2)_{PP}\, \psi_P + (H^2)_{PQ}\, \psi_Q \tag{5}$$

where the notation in analogy to Feshbach's optical model[3)] has been used,

$$(H^2)_{PQ} = P\, H^2 Q, \quad \psi_P = P\psi \quad \text{etc.} \tag{6}$$

Equation (5) describes the *exact* time evolution in the P-space for a given initial state and an initial time derivative as specified by eq. (2). The approximation consists in neglecting the $\psi_Q(t)$ term in eq. (5). In practical cases the P-space will always be selected such that it contains the initial state, $\psi_Q(0) = 0$. Therefore the $\psi_Q(t)$ term will vanish during the first time step and it should be reasonable to neglect it for sufficiently small intervals of time. This yields the second-order subspace time-evolution equation

$$-\hbar^2 \frac{\partial^2}{\partial t^2}\, \psi_P(t) = (H^2)_{PP}\, \psi_P(t) \tag{7a}$$

with the initial conditions

$$\psi(t{=}0) = \psi_P(t{=}0), \quad \psi_Q(t{=}0) = 0 \tag{7b}$$

and

$$i\,\hbar \frac{\partial}{\partial t}\, \psi_P(t) = H_{PP}\, \psi_P(t) \qquad (t{=}0). \tag{7c}$$

In general H_{PP} and $(H^2)_{PP}$ do not commute, and the approximate subspace time-evolution (7) is *non-unitary* as desired. In particular the exact flux of probability between the P- and Q-space is approximated rather accurately in many cases. This will be demonstrated below in a number of examples.

We note that a second (or higher order) equation is essential for a non-unitary solution. The ordinary time-dependent Schrödinger equation (2), when truncated in the same way as eq. (5) leads to the truncated Schrödinger equation

$$i\,\hbar \frac{\partial}{\partial t}\, \psi_P(t) = H_{PP}\, \psi_P(t) \tag{8}$$

with the undesirable unitary solutions.

4. SOLUTION OF THE SUBSPACE TIME-EVOLUTION EQUATION

Although the second-order subspace time-evolution equation (7) allows for a flux of probability out of the subspace (and back), it is completely defined within the subspace. In particular, the matrix elements of $(H^2)_{PP}$ can be computed in any P-space basis without information about the neglected Q-space. Equation (7) can easily be solved in the P-space representation that makes $(H^2)_{PP}$ diagonal. Thus the solution of eq. (7) requires one matrix diagonalization for all times, a numerical effort that does not exceed the one necessary for solving the (inappropriate) truncated Schrödinger equation (8).

For short time intervals an expansion to second order in time with the first and second derivative specified by eq. (7) is superior to diagonalizing $(H^2)_{PP}$,

$$\psi_P(t) = [1 - i\, H_{PP}\, t/\hbar - (H^2)_{PP}\, t^2/2\hbar^2]\psi_P\,(t=0) + \mathcal{O}\,(t^3). \qquad (9)$$

Such an expansion for short time steps may be important, e.g. for describing the first contact in a heavy-ion collision, or for an explicitely time-dependent Hamiltonian (e.g. in a semiclassical collision study with the center-of-mass motion given along some path).

5. EXAMPLES

5.1 Lipkin model

As a first example we consider the Lipkin model[4-7], a popular test case for time-dependent approximations to the exact Schrödinger solution. The model describes an exactly solvable many-body system consisting of N fermions on two N-fold degenerate levels evolving in time under the Hamiltonian

$$H = \frac{\varepsilon}{2} \sum_{p\sigma} \sigma a^{+}_{p\sigma}\, a_{p\sigma} + \frac{V}{2} \sum_{pp'\sigma} a^{+}_{p\sigma}\, a^{+}_{p'\sigma}\, a_{p'-\sigma}\, a_{p-\sigma} \qquad (10)$$

with the creation and anihilation operators of the single particle states $|p,\sigma\rangle$ where $p = 1,\ldots,N$ labels the degenerate states of both levels $\sigma = \pm 1$.

In the example of fig. 1 the system consists of 8 particles. The full Hilbert space of the system has dimension 9 and the subspace is truncated to dimension 5 ($|JM\rangle = |J-J\rangle,\ldots,|J0\rangle$ in the standard basis).

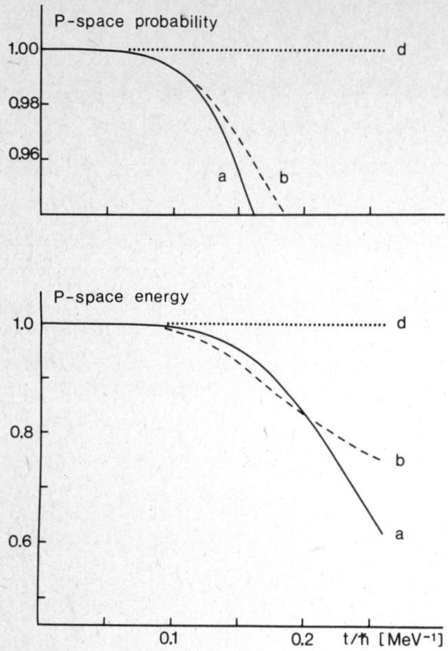

Fig. 1: Eight-particle Lipkin model (dimension 9) truncated to dimension 5.

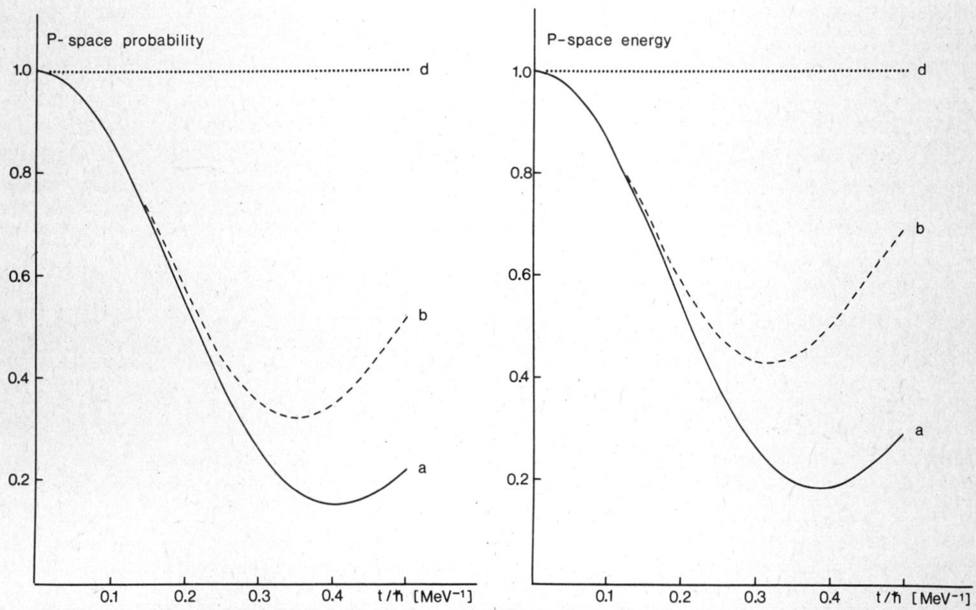

Fig. 2: Gaussian 100 x 100 Hamiltonian truncated to 10 x 10.

The initial state has M=-J and the parameter set is $\varepsilon=1$, V=1. The figure shows the loss of probability and the loss of energy out of the subspace as a function of time for both the exact Schrödinger time evolution (a) and the subspace approximation (b) of eq. (7). The approximation (b) remains rather accurate over a time interval during which a sizable flux of probability and energy out of the subspace occurs. This is much superior to the truncated Schrödinger calculation (d) of eq. (8) which results in no loss of probability and energy for all times. Other examples with different parameter sets and different initial conditions lead to qualitatively similar results.

5.2 Gaussian matrix parametrization

We have studied[2)] the behaviour of the subspace time evolution approximations for a number of schematic Hamiltonian matrices with exponential diagonal level density and a Gaussian fall-off for the off-diagonal matrix elements with or without random signs. This parametrization comprises constant and diagonal Hamiltonians, C-numbers and intermediate cases.

In the example plotted in fig. 2a the exact solution for a 100 x 100 Hamiltonian matrix is compared with the P-space approximation (7) for the first 10 x 10 components for an initial state peaked at the second and third basis state in the P-space. Here the probability in the P-space, as described by the non-linear approximation (b), is remarkably close to the exact result (a), even though a substantial flux of probability out of the P-space occurs. For instance, after a loss of 30% of the initial P-space probability the non-linear subspace approximation (7) remains within 1% of the exact result. This is clearly superior to the truncated result (d) which (arbitrarily) conserves the P-space probability. Although the non-linear approximation becomes less accurate for longer time intervals it still reproduces the qualitative behaviour of the exact solution. In particular a flux back from the neglected space into the P-space occurs almost simultaneously in the exact (a) and approximate (b) result.

The flux of probability between P- and Q-space is associated with a similar flux of energy, that is shown in fig. 2b. Here again the non-linear subspace approximation (b) is much superior to the truncated result (d). It is rather accurate over a sizable interval of time (e.g. 3% at an energy loss of 30%). Other cases are discussed in ref.2.

5.3 Exactly solvable Hamiltonians

For any Hamiltonian with the property

$$(H^2)_{PQ} = 0 = (H^2)_{QP} \tag{11}$$

the subspace approximation (7) leads to an *exact* solution in the P-space. In order to demonstrate that this class of Hamiltonian matrices comprises cases with substantial flux of probability between the spaces we consider the particularly simple example

$$(H_{PP})_{mn} = -\sqrt{\exp\ (-\alpha n)}\ \ \delta_{mn} = -(H_{QQ})_{mn}, \tag{12}$$

$$(H_{PQ})_{mn} = \gamma\ \delta_{mn} = (H_{QP})_{mn}. \tag{13}$$

Here the P-space probability can be given analytically as a function of time. The result for one set of parameters (α,γ) and different initial states β is given in figure 3. It shows that the probability tends to oscillate in time between the P- and Q-space. The Q-space probability can be quite large (even though the spaces are disconnected with respect to the second order equations and $\psi_Q(0) = 0$). Since in the case we have considered, the second-order approximation is *exact*, it is very much better than the usual truncation procedure (which would arbitrarily conserve the subspace probability and energy for all times).

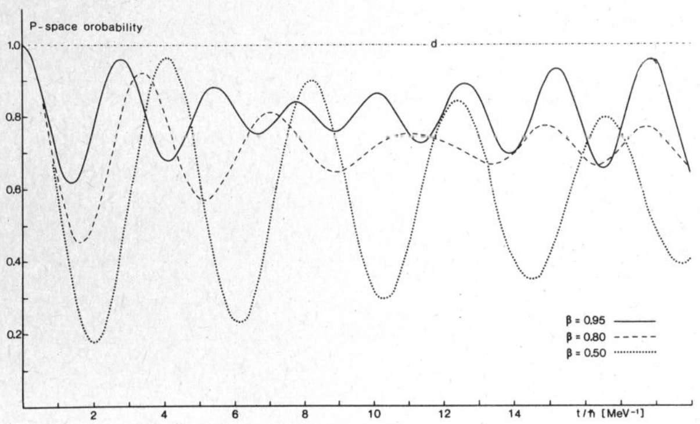

Figure 3: Coincidence of exact and approximate solution in the case $(H^2)_{PQ} = 0$, for different initial distributions β, and a 50% truncation to dimension 11. Curve d is the solution of the Schrödinger equation in the subspace.

6. PROJECTION OF THE TIME-EVOLUTION OPERATOR

An explanation why the approximation (7) is in many cases rather close to the exact time evolution in the subspace can be given in terms of the time-evolution operator

$$U(t) = \exp(-i\,H\,t/\hbar) = \sum_{n=0}^{\infty} (-i\,t/\hbar)^n/n!H^n. \tag{14}$$

The application of the projection and insertion of $P^2 + Q^2$, similar to eq. (4), yields

$$\psi_P(t) = \sum_{n=0}^{\infty} (-i\,t/\hbar)^n/n!\;(H^n)_{PP}\;\psi_P(0) \tag{15}$$

for an initial state $\psi(0) = \psi_P(0)$ in the P-space. This is the exact P-space time evolution for all times. Up to second order in time the expansion (15) is identical to the approximation (9) of sect. 3. Thus this approximation can be viewed as a first step away from the unitary solution of the truncated Schrödinger equation (8) towards the exact non-unitary solution (15) for the first time step. It turns out that, in many cases, the inclusion of the H^2 term alone already leads to a rather accurate approximation of the flux of probability into the Q-space.

7. TIME DEPENDENT "OPTICAL" POTENTIAL

The success and wide-spread use of the optical model to correct for the excluded channels in stationary scattering theory has motivated us to study if a similar, but time dependent, "optical" potential[7)] can be used to account for the flux of probability out of the truncated subspace in a time-dependent calculation.

The problem consists in adding to the truncated Hamiltonian an appropriate imaginary time-dependent matrix i W(t) such that the "optical" solution $|\phi\rangle$ of the modified subspace Schrödinger equation

$$i\,\hbar\,\frac{\partial}{\partial t}\,|\phi\,(t)\rangle = [H_{PP} + i\,W(t)]\;|\phi\rangle \tag{16}$$

fits the subspace part of the exact solution $|\psi(t)\rangle$ as closely as possible

$$|\phi(t)\rangle \approx P\;|\psi(t)\rangle\;. \tag{17}$$

Figure 4 shows an example of such an optical potential fit for a schematic heavy-ion model. Here the Hamiltonian

$$H = H^{(1)}_{sp} + H^{(2)}_{sp} + V(1,2) \tag{18}$$

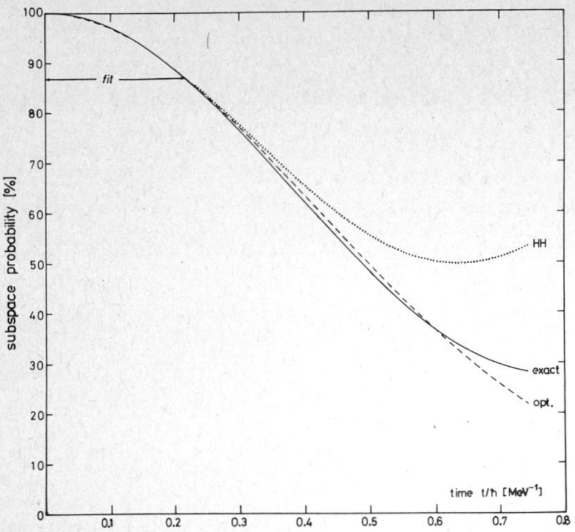

Figure 4: Optical potential fit for a schematic time-dependent heavy-ion collision model.

consists of single-particle (sp) terms for each ion and a two-body interaction term. The single-particle energies are taken to have a level-density form

$$H_{sp}^{(1)} \; |i\rangle = V_1 \exp\,(-i/a_1)\,|i\rangle \tag{18a}$$

and

$$H_{sp}^{(2)} \; |j\rangle = V_2 \exp\,(-j/a_2)\,|j\rangle \quad , \tag{18b}$$

and the interaction matrix is defined in the product basis $|i\rangle\,|j\rangle$ as

$$\langle ij|V|kl\rangle = V_o \exp\,\{-(|ij-kl|\;/a_o^2)\} \tag{19}$$

with the parameters V_o, V_1, V_2 and a_o, a_1, a_2. The interaction is chosen such that the matrix elements should be large between two ion states $|i\rangle|j\rangle$ and $|k\rangle|l\rangle$ that are either particularly simple (e.g. low excited) or both very complicated (e.g. highly excited), and small for combinations of simple and complicated states.

We have studied a number of different functional forms for the optical potential iW that may lead to a satisfactory description of the time-evolution of the system. It turns out that a rather simple parametrization is already sufficient in many cases: The matrix iW can be chosen diagonal in the subspace "energy" representation

$$H_{PP}|n\rangle = \varepsilon_n|n\rangle \qquad (|n\rangle \in P \text{ space}) \tag{20}$$

and linear in time, i.e.

$$\langle n'|iW|n\rangle = -i\hbar\,(\alpha_n + \beta_n\, t)\,\delta_{n'n} \quad , \qquad (21)$$

where α_n and β_n are the fit parameters of the optical potential.

In the example plotted for a statistical initial distribution the full space of dimension 36 (i.e., 6 levels in each ion) is truncated to dimension 16 (i.e. the first 4 levels in each ion). During the time interval of fig. 4 the probability in the subspace decreases from initially 100% to about 30% before a flux of probability back into the selected subspace sets in. The optical approximation is fitted for the initial time interval $t \leq 1.4 \cdot 10^{-22}$ sec during which about 13% of the subspace probability is lost. Although the further loss of probability is quite large, the optical approximation remains close to the exact solution until the flux of probability back into the selected subspace starts to dominate. The latter effect is, of course, outside the scope of an optical model. Thus the extrapolation achieved by the optical model remains rather accurate over three times the period of time of the initial fit, and for more than five times the loss of probability within the original fit interval.

For simplicity the initial optical fit in fig. 4 has been made to the *exact* solution, which, of course, will not be known in practical cases. Therefore the optical model approach requires an initial time-evolution approximation which the optical potential can then be fitted to. The figure shows for comparison the second-order subspace approximation discussed in the previous sections (here labelled HH). It almost coincides with the exact solution within the fit interval of the optical potential but starts to deviate much earlier than the optical solution. It is therefore evident that the subspace approximation when combined with the optical model fit converges over a much larger interval of time.

8. SUMMARY AND CONCLUSIONS

The examples of the present paper show that the second-order subspace approximation, eq. (7), avoids the undesirable unitarity of the time-dependent Schrödinger solution within a truncated subspace that leads to an unphysical conservation of probability, mass charge and energy etc. within the selected subspace. The second-order time-differential equations (7) are completely defined and solvable within any given subspace, and the numerical effort required is comparable to that of simply truncating the Schrödinger equation to the subspace. When expanded in powers of the Hamiltonian H (or time t) the second-order approximation coincides with the exact solution up to second order in H (and up to any order for certain classes of Hamiltonians). This may explain why the approximation is in many cases rather close to the exact result. In any case studied it turned out to be superior to the result of only truncating the time-dependent Schrödinger equation to the subspace (which approximates the exact time evolution only to first order in time).

We have also demonstrated that the approximation can be further improved by an appropriate time-dependent optical potential. According to our results, the optical model solution describes the flux of probability out of the selected subspace rather accurately but fails, as one would expect, when the flux back from the excluded space becomes important.

This work was supported by the Deutsche Forschungsgemeinschaft and the Natural Sciences and Engineering Research Council of Canada. Two of us (H. S. and R. S.) would also like to thank the Theory Group of the McMaster University for their hospitality and support.

REFERENCES

1. H. Schultheis, R. Schultheis and A.B. Volkov, Phys.Lett. 89B (1980) 165

2. H. Schultheis, R. Schultheis and A.B. Volkov, Ann.Phys. (N.Y.) to be published

3. H. Feshbach, Topics in the Theory of Nuclear Reactions, in "Documents of Modern Physics" (Gordon and Breach, New York, 1973) p. 171

4. H.J. Lipkin, Proc. of the Rutherford Jubilee Intern. Conf., Manchester, 1961 (Heywood, London, 1961) p. 275

5. A. Glick, H.J. Lipkin and N. Meshkov, Proc. of the Rutherford Jubilee Intern. Conf., Manchester, 1961 (Heywood, London, 1961) p. 299

6. H.J. Lipkin, N. Meshkov and A.J. Glick, Nucl. Phys. 62 (1965) 188

7. A.B. Volkov, Nucl.Phys. 43 (1963) 1

8. N. Austern, Ann.Phys. (N.Y.) 45 (1967) 113

LIST OF PARTICIPANTS

ARDE, P., Niels Bohr Institutet, Copenhagen, Denmark

AYIK, S., TU München, Garching, West Germany

BALAZS, N.L., SUNY at Stony Brook, USA

BOUTEN, M., Limburgs Universitair Centrum, Diepenbeck, Belgium

BRINGER, A., KFA Jülich, West Germany

BRINK, D.M., University of Oxford, England

BUCK, P., TH Darmstadt, West Germany

CAMPBELL, D.K., Los Alamos National Lab, USA

CUSSON, R.Y., Duke University, Durham, USA

DOBACZEWSKI, J., University of Warsaw, Poland

FABER, M., TU Wien, Austria

FELDMEIER, H., TH Darmstadt, West Germany

FLOCARD, H., CEN Orsay, France

GIRAUD, B., CEN Saclay, Gif-sur-Yvette, France

GOEKE, K., KFA Jülich, West Germany

GRIFFIN, J.J., University of Maryland, College Park, USA

GROSSMANN, M., Universität Münster, West Germany

GRÜMMER, F., KFA Jülich, West Germany

HAYASHI, A., TU München, Garching, West Germany

HEENEN, P.-H., Université Libre de Bruxelles, Belgium

IRVINE, J.M., University of Manchester, England

KLEINERT, H., FU Berlin, West Germany

KNOLL, J., GSI Darmstadt, West Germany

KÖHLER, S., University of Arizona, Tucson, USA

KOLB, D., Gesamthochschule Kassel, West Germany

KOONIN, S.E., Caltech, Pasadena, USA

KRIEGER, S.J., Lawrence Livermore Lab, USA

MARUMORI, T., University of Tsukuba, Ibaraki, Japan

MOYA DE GUERRA, E., Universidad Autônoma de Madrid, Spain

MUKHERJEE, A.K., Saha Institute of Nuclear Physics, Calcutta, India

NEERGARD, K., Universität Gießen, West Germany

NEGELE, J.W., MIT, Cambridge, USA

NEMES, M.C., CEN Orsay, France

NEMETH, J., Eötvös Universität, Budapest, Hungary

NIKLAS, M., Universität Münster, West Germany

NILSSON, B.S., Niels Bohr Institutet, Copenhagen, Denmark

NÖRENBERG, W., GSI Darmstadt, West Germany

PASSOS, de, E.J.V., Universidade de Sao Paulo, Brazil

PAULI, H.C., MPI für Kernphysik, Heidelberg, West Germany

PLOSZAJCZAK, M., Institute of Nuclear Physics, Krakow, Poland
PRAKASH, M., SUNY at Stony Brook, USA
PROVIDENCIA, da, J., Universidade de Coimbra, Portugal
PROVOOST, D., Limburgs Universitair Centrum, Diepenbeck, Belgium
REINHARD, P.-G., Universität Mainz, West Germany
REINHARDT, H., ZfK Rossendorf, GDR
REMLER, E.A., College of William and Mary, Williamsburg, USA
RICHERT, J., Centre de Recherches Nucléaires, Strasbourg, France
RING, P., TU München, Garching, West Germany
ROSINA, M., E. Kardelj University, Ljubljana, Yugoslavia
ROWE, D.J., University of Toronto, Canada
SAMI, T., Centre de Recherches Nucléaires, Strasbourg, France
SCHUCK, P., Institut Laue-Langevin, Grenoble, France
SCHÜTTE, G., Universität Heidelberg, West Germany
SCHULTHEIS, H., Universität Tübingen, West Germany
SCHULTHEIS, R., Universität Tübingen, West Germany
SOUZA CRUZ, de, F.F., Universidade de Sao Paulo, Brazil
SPETH, J., KFA Jülich, West Germany
STANCU, Fl., Université de Liège, Belgium
STINGL, M., Universität Münster, West Germany
SUZUKI, Toru, Niels Bohr Institutet, Copenhagen, Denmark
TOLEDO PIZA, de, A.F.R., Universidade de Sao Paulo, Brazil
TORO, di, M., Universitè di Catania, Italy
TRIPATHY, R.K., Institute of Physics, Bhubaneswar, India
URBANO, J.N., Universidade de Coimbra, Portugal
VENERONI, M., CEN Orsay, France
WANG, Shun-jin, GSI Darmstadt, West Germany
WEIDENMÜLLER, H.A., MPI für Kernphysik, Heidelberg, West Germany
WEIGEL, M., TU München, Garching, West Germany
WEIGUNY, A., Universität Münster, West Germany
WEISS, M.S., Lawrence Livermore Lab, USA
WOLSCHIN, G., MPI für Kernphysik, Heidelberg, West Germany
WONG, C.Y., Oak Ridge National Lab, USA
YADAV, H.L., GSI Darmstadt, West Germany

AUTHOR INDEX

R. Bass

Nuclear Reactions with Heavy Ions

1980. 176 figures, 31 tables. VIII, 410 pages
(Texts and Monographs in Physics)
ISBN 3-540-09611-6

The last decade has witnessed an astounding increase in heavy ion research. This book presents - from an experimentalist's point of view - a critical and coherent outline of the results of large scale heavy ion research in the area of low energy nuclear reactions in the 5–10 MeV per nucleon range. Using phenomenological models, the author explains these experimental results, achieving a good balance between a critically selected review and a textbook. This makes it attractive for the advanced student and the specialist alike.

P. Ring, P. Schuck

The Nuclear Many-Body-Problem

1980. 171 figures. XVII, 716 pages
(Texts and Monographs in Physics)
ISBN 3-540-09820-8

This book, while covering a fair amount of physical observations, stresses the methodology and technical aspects of the different theories presently used in the description of the nucleus. The authors present the more modern theories such as Boson expansions, generator coordinates, time-dependent Hartree-Fock method, and semiclassical models which so far have found only limited mention in textbooks. The book also covers subjects like the liquid drop and the shell model, both presented in a updated version in, for example, rotations and random phase approximation. The full presentation of mathematical details, illustrated by observational data, will help the student fully understand the present view on the nuclear many-body problem.

Zeitschrift für Physik A Atoms and Nuclei

EPS Europhysics Journal

Editor in Chief: H. A. Weidenmüller, Heidelberg

Editorial Board: P. Armbruster, E. Bodenstedt, H.-J. Gerber, A. Goldhaber, I. V. Hertel, M. Lefort, B. Povh, G. zu Putlitz, J. Specht

Zeitschrift für Physik appears in three parts:
- A: Atoms and Nuclei
- B: Condensed Matter
- C: Particles and Fields

Each part my be ordered separately.

Coordinating Editor for Zeitschrift für Physik, Sections A, B and C is O. Haxel, Heidelberg

Atomic Physics:
Properties of atoms and molecules · Spectra of inner and outer shells · μ-mesic, pionic and other exotic atoms · Hyperfine interactions · Atomic and molecular collisions · Atomic studies with heavy-ion collisions · Theory

Nuclear Physics:
Properties of nuclei · Nuclear structure and reactions · Heavy-ion reactions · Fission · Hadron-nucleus interactions · Theory

Special Features: Rapid publication (3–4 months for original articles, 6 weeks for short notes); no page charge; back volumes available, also in microform.

Language: More than 95% English.

Articles: Original reports and short notes.

Subscription information or sample copy upon request.

Springer-Verlag Berlin Heidelberg New York

Lecture Notes in Physics

Vol. 136: The Role of Coherent Structures in Modelling Turbulence and Mixing. Proceedings 1980. Edited by J. Jimenez. XIII, 393 pages. 1981.

Vol. 137: From Collective States to Quarks in Nuclei. Edited by H. Arenhövel and A. M. Saruis. VII, 414 pages. 1981.

Vol. 138: The Many-Body Problem. Proceedings 1980. Edited by R. Guardiola and J. Ros. V, 374 pages. 1981.

Vol. 139: H. D. Doebner, Differential Geometric Methods in Mathematical Physics. Proceedings 1981. VII, 329 pages. 1981.

Vol. 140: P. Kramer, M. Saraceno, Geometry of the Time-Dependent Variational Principle in Quantum Mechanics. IV, 98 pages. 1981.

Vol. 141: Seventh International Conference on Numerical Methods in Fluid Dynamics. Proceedings. Edited by W. C. Reynolds and R. W. MacCormack. VIII, 485 pages. 1981.

Vol. 142: Recent Progress in Many-Body Theories. Proceedings. Edited by J. G. Zabolitzky, M. de Llano, M. Fortes and J. W. Clark. VIII, 479 pages. 1981.

Vol. 143: Present Status and Aims of Quantum Electrodynamics. Proceedings, 1980. Edited by G. Gräff, E. Klempt and G. Werth. VI, 302 pages. 1981.

Vol. 144: Topics in Nuclear Physics I. A Comprehensive Review of Recent Developments. Edited by T.T.S. Kuo and S.S.M. Wong. XX, 567 pages. 1981.

Vol. 145: Topics in Nuclear Physics II. A Comprehensive Review of Recent Developments. Proceedings 1980/81. Edited by T. T. S. Kuo and S. S. M. Wong. VIII, 571-1.082 pages. 1981.

Vol. 146: B. J. West, On the Simpler Aspects of Nonlinear Fluctuating. Deep Gravity Waves. VI, 341 pages. 1981.

Vol. 147: J. Messer, Temperature Dependent Thomas-Fermi Theory. IX, 131 pages. 1981.

Vol. 148: Advances in Fluid Mechanics. Proceedings, 1980. Edited by E. Krause. VII, 361 pages. 1981.

Vol. 149: Disordered Systems and Localization. Proceedings, 1981. Edited by C. Castellani, C. Castro, and L. Peliti. XII, 308 pages. 1981.

Vol. 150: N. Straumann, Allgemeine Relativitätstheorie und relativistische Astrophysik. VII, 418 Seiten. 1981.

Vol. 151: Integrable Quantum Field Theory. Proceedings, 1981. Edited by J. Hietarinta and C. Montonen. V, 251 pages. 1982.

Vol. 152: Physics of Narrow Gap Semiconductors. Proceedings, 1981. Edited by E. Gornik, H. Heinrich and L. Palmetshofer. XIII, 485 pages. 1982.

Vol. 153: Mathematical Problems in Theoretical Physics. Proceedings, 1981. Edited by R. Schrader, R. Seiler, and D.A. Uhlenbrock. XII, 429 pages. 1982.

Vol. 154: Macroscopic Properties of Disordered Media. Proceedings, 1981. Edited by R. Burridge, S. Childress, and G. Papanicolaou. VII, 307 pages. 1982.

Vol. 155: Quantum Optics. Proceedings, 1981. Edited by C.A. Engelbrecht. VIII, 329 pages. 1982.

Vol. 156: Resonances in Heavy Ion Reactions. Proceedings, 1981. Edited by K.A. Eberhard. XII, 448 pages. 1982.

Vol. 157: P. Niyogi, Integral Equation Method in Transonic Flow. XI, 189 pages. 1982.

Vol. 158: Dynamics of Nuclear Fission and Related Collective Phenomena. Proceedings, 1981. Edited by P. David, T. Mayer-Kuckuk, and A. van der Woude. X, 462 pages. 1982.

Vol. 159: E. Seiler, Gauge Theories as a Problem of Constructive Quantum Field Theory and Statistical Mechanics. V, 192 pages. 1982.

Vol. 160: Unified Theories of Elementary Particles. Critical Assessment and Prospects. Proceedings, 1981. Edited by P. Breitenlohner and H.P. Dürr. VI, 217 pages. 1982.

Vol. 161: Interacting Bosons in Nuclei. Proceedings, 1981. Edited by J.S. Dehesa, J.M.G. Gomez, and J. Ros. V, 209 pages. 1982.

Vol. 162: Relativistic Action at a Distance: Classical and Quantum Aspects. Proceedings, 1981. Edited by J. Llosa. X, 263 pages. 1982.

Vol. 163: J. S. Darrozes, C. Francois, Mécanique des Fluides Incompressibles. XIX, 459 pages. 1982.

Vol. 164: Stability of Thermodynamic Systems. Proceedings, 1981. Edited by J. Casas-Vázquez and G. Lebon. VII, 321 pages. 1982.

Vol. 165: N. Mukunda, H. van Dam, L.C. Biedenharn, Relativistic Models of Extended Hadrons Obeying a Mass-Spin Trajectory Constraint. Edited by A. Böhm and J.D. Dollard. VI, 163 pages. 1982.

Vol. 166: Computer Simulation of Solids. Edited by C.R.A. Catlow and W.C. Mackrodt. XII, 320 pages. 1982.

Vol. 167: G. Fieck, Symmetry of Polycentric Systems. VI, 137 pages, 1982.

Vol. 168: Heavy-Ion Collisions. Proceedings, 1982. Edited by G. Madurga and M. Lozano. VI, 429 pages. 1982.

Vol. 169: K. Sundermeyer, Constrained Dynamics. IV, 318 pages. 1982.

Vol. 171: Time-Dependent Hartree-Fock and Beyond. Proceedings, 1982. Edited by K. Goeke and P.-G. Reinhard. VIII, 426 pages. 1982.

QC
793.3
C56T55
1982
c.1

DATE DUE

MAR 3 0 '84			

NASA
Ames Research Center Library
Moffett Field, CA 94035

ATL-12